Advances in Intelligent Systems and Computing

Volume 507

Series editor

Janusz Kacprzyk, Polish Academy of Sciences, Warsaw, Poland
e-mail: kacprzyk@ibspan.waw.pl

About this Series

The series "Advances in Intelligent Systems and Computing" contains publications on theory, applications, and design methods of Intelligent Systems and Intelligent Computing. Virtually all disciplines such as engineering, natural sciences, computer and information science, ICT, economics, business, e-commerce, environment, healthcare, life science are covered. The list of topics spans all the areas of modern intelligent systems and computing.

The publications within "Advances in Intelligent Systems and Computing" are primarily textbooks and proceedings of important conferences, symposia and congresses. They cover significant recent developments in the field, both of a foundational and applicable character. An important characteristic feature of the series is the short publication time and world-wide distribution. This permits a rapid and broad dissemination of research results.

Advisory Board

More information about this series at http://www.springer.com/series/11156

Suresh Chandra Satapathy
V. Kamakshi Prasad · B. Padmaja Rani
Siba K. Udgata · K. Srujan Raju
Editors

Proceedings of the First International Conference on Computational Intelligence and Informatics

ICCII 2016

 Springer

Editors

Suresh Chandra Satapathy
Department of Computer Science
  and Engineering
ANITS
Visakhapatnam, Andhra Pradesh
India

V. Kamakshi Prasad
Computer Science and Engineering
  Department
JNTUH College of Engineering Hyderabad
  (Autonomous)
Hyderabad, Telangana
India

B. Padmaja Rani
Computer Science and Engineering
  Department
JNTUH College of Engineering Hyderabad
  (Autonomous)
Hyderabad, Telangana
India

Siba K. Udgata
SCIS
University of Hyderabad
Hyderabad
India

K. Srujan Raju
CMR Technical Campus
Hyderabad
India

ISSN 2194-5357   ISSN 2194-5365   (electronic)
Advances in Intelligent Systems and Computing
ISBN 978-981-10-2470-2   ISBN 978-981-10-2471-9   (eBook)
DOI 10.1007/978-981-10-2471-9

Library of Congress Control Number: 2016951657

This Springer imprint is published by Springer Nature
The registered company is Springer Nature Singapore Pte Ltd.
The registered company address is: 152 Beach Road, #22-06/08 Gateway East, Singapore 189721, Singapore

Preface

The First International Conference on Computational Intelligence and Informatics (ICCII 2016) was held during May 28–30, 2016 at Hyderabad, hosted by the Department of Computer Science and Engineering, JNTUHCE, Hyderabad in association with Division-V (Education and Research) CSI. The proceedings of this conference contain 69 papers which are included in this volume. It proved to be a great platform for researchers from across the world to report, deliberate, and review the latest progress in the cutting-edge research pertaining to computational intelligence and its applications to various engineering fields. The response to ICCII 2016 was overwhelming. It received a good number of submissions from different areas relating to computational intelligence and its applications in main tracks; and after a rigorous peer-review process with the help of our program committee members and external reviewers, we finally accepted quality papers with an acceptance ratio of 0.25.

This conference is organized in the honor of Prof. A. Vinaya Babu, who is retiring from services, for his services rendered in the JNTUH University in general and the Department of Computer Science and Engineering, JNTUHCE, Hyderabad in particular.

Dr. L. Pratap Reddy, JNTUHCE, Hyderabad conducted a special session on "Information Theoretic Approach for Security". A special session on "Introduction to nature inspired optimization and machine learning" was conducted by Dr. S.C. Satapathy, ANITS, Vizag. Dr. V. Ravi, IDRBT delivered a talk on "Data Analytics" and Dr. P. Sateesh Kumar, IIT, Roorkee conducted a special session on "Cloud-based IoT for Agriculture". Dr. Swagatam Das, ISI, Kolkata conducted a special session on "Multi-objective Optimization and Unsupervised Learning: Some Issues and Challenges". These sessions were very informative and beneficial to the authors and delegates of the conference.

We take this opportunity to thank all keynote speakers and special session chairs for their excellent support to make ICCII 2016 a grand success. The quality of a refereed volume depends mainly on the expertise and dedication of the reviewers. We are indebted to the program committee members and external reviewers who not only produced excellent reviews but also did them in short time frame. We would also like to thank CSI Hyderabad, for coming forward to support us to organize this mega convention.

We express our heartfelt thanks to Chief Patrons, Smt. Shailaja Ramaiyer, IAS, Vice Chancellor, JNTUH, Prof. N.V. Ramana Rao, Rector, JNTUH, Prof. N. Yadaiah, Registrar, JNTUH, Prof. A. Govardhan, Principal, JNTUHCEH, faculty and administrative staff for their continuous support during the course of the convention.

We would also like to thank the authors and participants of this convention, who have considered the convention above all hardships. Finally, we would like to thank all the volunteers who spent tireless efforts in meeting the deadlines and arranging every detail to make sure that the convention runs smoothly. All the efforts are worth and would please us all, if the readers of this proceedings and participants of this convention found the papers and event inspiring and enjoyable. Our sincere thanks to the press, print, and electronic media for their excellent coverage of this convention.

Visakhapatnam, India   Suresh Chandra Satapathy
Hyderabad, India   V. Kamakshi Prasad
May 2016   B. Padmaja Rani
                                                                                      Siba K. Udgata
                                                                                      K. Srujan Raju

Contents

About the Editors

Dr. Suresh Chandra Satapathy is currently working as Professor and Head, Department of Computer Science and Engineering, Anil Neerukonda Institute of Technology and Sciences (ANITS), Visakhapatnam, Andhra Pradesh, India. He obtained his Ph.D. in Computer Science Engineering from JNTUH, Hyderabad and Master's degree in Computer Science and Engineering from National Institute of Technology (NIT), Rourkela, Odisha. He has more than 27 years of teaching and research experience. His research interests include machine learning, data mining, swarm intelligence studies, and their applications to engineering. He has more than 98 publications to his credit in various reputed international journals and conference proceedings. He has edited many volumes of Springer AISC and LNCS in past and he is also the editorial board member in reputed international journals. He is a senior member of IEEE and Life Member of Computer Society of India. Currently, he is the National Chairman of Division-V (Education and Research) of Computer Society of India.

Dr. V. Kamakshi Prasad is Professor and Head of Computer Science and Engineering Department at JNTUH College of Engineering Hyderabad. He completed his Ph.D. in speech recognition from IIT Madras, India. He did his M.Tech. in Andhra University and B.Tech. in K.L. College of Engineering. He has completed over 12 years in JNTU on various positions. He has 21 years of teaching and 11 years of research experience. Dr. Prasad has been teaching subjects such as speech processing, pattern recognition, computer networks, digital image processing, artificial neural, artificial intelligence and expert systems, computer graphics, object-oriented analysis and design through UML, and soft computing. He has supervised 12 Ph.D. and 2 MS students. His research areas include speech recognition and processing, image processing, neural networks, data mining, and ad hoc networks. He has authored two books published by Lambert Academic Publishing and over 50 papers in national and international journals.

Dr. B. Padmaja Rani is Professor of Computer Science and Engineering Department at JNTUH College of Engineering, Hyderabad. Her area of interest includes information retrieval embedded systems. She has published more than 25 papers in reputed journals and conferences in the areas of agile modeling, Web services and mining. She was the former Head of Department of CSE, JNTUH. She is a professional member of CSI.

Dr. Siba K. Udgata is Professor of School of Computer and Information Sciences, University of Hyderabad, India. He is presently heading the Centre for Modeling, Simulation and Design (CMSD), a high performance computing facility at University of Hyderabad. He has got his Master's followed by Ph.D. in Computer Science (mobile computing and wireless communication). His main research interests include wireless communication, mobile computing, wireless sensor networks, and intelligent algorithms. He was a United Nations Fellow and worked in the United Nations University/International Institute for Software Technology (UNU/IIST), Macau as a research fellow in the year 2001. Dr. Udgata is working as a principal investigator in many Government of India funded research projects mainly for development of wireless sensor network applications and application of swarm intelligence techniques. He has published extensively in refereed international journals and conferences in India as well as abroad. He was also in the editorial board of many Springer LNCS/LNAI and Springer AISC Proceedings.

Dr. K. Srujan Raju is the Professor and Head, Department of CSE, CMR Technical Campus, Hyderabad, India. Professor Raju earned his Ph.D. in the field of network security and his current research includes computer networks, information security, data mining, image processing, intrusion detection, and cognitive radio networks. He has published several papers in refereed international conferences and peer-reviewed journals and also he was on the editorial board of CSI 2014 Springer AISC series; 337 and 338 volumes. In addition to this, he has served as reviewer for many indexed journals. Professor Raju is also awarded with Significant Contributor and Active Member Awards by Computer Society of India (CSI) and currently he is the Honorary Secretary of CSI Hyderabad chapter.

GUI-Based Automated Data Analysis System Using Matlab®

Pritam Gayen, Pramod Kumar Jha and Praveen Tandon

Abstract Centre for Advanced Systems (CAS) is a newly sanctioned DRDO establishment engaged in the production of aerospace vehicle of national importance. Aerospace vehicles are fire and forget type and need rigorous testing and validation at ground before subjecting it to critical flight trial. Aerospace vehicles are very complex in nature and consist of various systems, subsystems, and assemblies that need to be tested both in standalone mode and in an integrated manner. The numerous tests and phase checks of these subsystems generate lot of data that need to be analyzed before its formal clearance for flight test or launch. This paper aims to automate this cumbersome process of data analysis and documentation using the GUI feature of Matlab®. Prior to this, the offline data were plotted using m files and figures thus obtained was inserted manually in documentation software. All these process were manual and prone to human error. Now, this entire process has been automated using the GUI feature of Matlab®. The data are plotted through a user-friendly GUI window by a single mouse click. The plots can also be exported to MS PowerPoint® and other documentation software by a single click in the required format for presentation.

Keywords Phase checks · Fault tolerant · Automation · Intelligent system

P. Gayen (✉) · P.K. Jha · P. Tandon
Centre for Advanced Systems, DRDO, Hyderabad, India
e-mail: pritsaccount@gmail.com

P.K. Jha
e-mail: pkj@cas.drdo.in

P. Tandon
e-mail: praveen.tandon@cas.drdo.in

© Springer Science+Business Media Singapore 2017
S.C. Satapathy et al. (eds.), *Proceedings of the First International Conference on Computational Intelligence and Informatics*, Advances in Intelligent Systems and Computing 507, DOI 10.1007/978-981-10-2471-9_1

1 Introduction

Aerospace vehicles are fire and forget type and need to undergo rigorous ground tests before its delivery to users. It is a complex multi-disciplinary system consisting of various systems, subsystems, and assemblies that are both intelligent and non-intelligent. They are configured around MIL-STD-1553B [1] data bus. All operations and data communications during checkout, auto launch, and test flight are managed by intelligent on board computer. The data transaction takes place over a fault- tolerant 1 Mbps serial link. The tests and phase checks generate lot of data related to system's health and it needs to monitored and analyzed both online and offline for its formal clearance. These data thus generated were earlier plotted manually using the Matlab® software and then imported into documentation software for generation of reports. This process use to take lot of time and was prone to human errors. This work is a step towards the automation of this work. It utilizes the GUI feature of Matlab® to create user-friendly screens to perform these tasks with a mouse click.

2 Related Work

During the test and phase checks, all the subsystems of aerospace vehicle are connected as per the actual configuration. The intelligent subsystem communicates with each other over MIL-STD-15553B [1] bus under the control of onboard computer which acts as a bus controller. The ongoing data transactions are captured by a bus monitor [2] computer. The captured data files are in ACE.ASF format of DDC (Data Device Corporation) [3] and have to be decoded with suitable scale factors to convert them into engineering units. After the completion of proper-decoding process, the data files are converted into text files, which are imported into Matlab® environment for plotting as per the required format. After plotting, it was converted into *.ppt/*.pptx format manually for presentation. The same can be stored in a printable format. This whole process was manually done which was both time consuming and error prone.

This work has automated this entire analysis process using the GUI feature of Matlab®. The steps of automation process are being described here with suitable screen shots.

1. The user interface (UI) is created, which contains several controls called components. These components may include toolbars, menu bars, push buttons, etc. These UIs can perform computations, read and write data files, and even communicate with other UIs.
2. After pressing a push button, the associated callback function gets executed. These functions may be a separate .m files.
3. The Guide feature of Matlab® [4] was used in the creation of UIs, and then, modifications were done programmatically to meet our requirements.

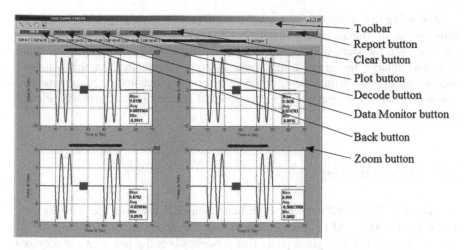

Toolbar
Report button
Clear button
Plot button
Decode button
Data Monitor button
Back button
Zoom button

Fig. 1 Screen shot of a GUI displays graphs

4. Layout Editor was used for resizing of UI windows.
5. Push buttons were added to UIs from component palette.
6. Remaining components like popup menu, axes were added to the UI subsequently.
7. Alignment tool is used for the alignment of components to add aesthetics to the interface.
8. Property Inspector tool was used for labeling the push buttons.

A user-interface design is an important aspect of this work, as the user interacts with it. It consists of various tools buttons to accomplish the required task as well as for navigation from one window to other. The UI window along with its related control buttons are as shown below. The decode button is used for the conversion of ACE.ASF data file to a text file which is then used for plotting using the "PLOT" button. The plot can be zoomed using zoom tool button.

Fig. 2 Screenshot displays
graphs separately in a GUI

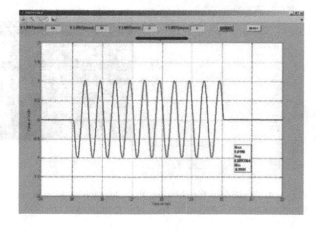

According to user-specification graphs are divided into some groups. To display those group of graphs, *uitab()* function of Matlab® is used. Some axes contain a zoom button, by which user can see that particular plot in a separate window. User can also enlarge the graph using zoom button from toolbar on this window. According to user requirement, some axes contain maximum, minimum, and average values of corresponding graph. Related screenshots are given on Figs. 1 and 2.

2.1 Data Gathering

Pressing the 'DATA MONITOR' push button on GUI window, user can store the raw data file (ACE.ASF) in his/her desired disk location. The function *uigetdir()* is used to acquire the user given path. This process saves the raw data file acquired from MILSTD1553 data bus monitor. To gather raw data, the mt2disk.exe (data gathering program by DDC) file is invoked through command prompt by this automation process. Related screenshots are given in Fig. 3.

When command prompt is running, user has to press any key on the command prompt window to stop the bus monitoring process.

To decode the data file, a decoder program is used. User has to choose the raw data file (ACE.ASF or *.ASF). The *uigetdir()* function is used to accrue the file path like the previous. To decode the raw data file, the decoder program is invoked through command prompt by this automation process. All decoded files are saved at the location provided by the user. After the completion of decode process, a message is displayed to confirm the successful completion of decode process.

Fig. 3 GUI window for choosing the folder where to save raw data (*left*). A GUI window of command prompt invoked by the automation process (*right*)

2.2 Documentation

Another GUI window is displayed, after clicking the 'REPORT' button, which has two edit text boxes, one for aerospace vehicle sequence number and another is for some specification about that particular test. A button named 'SAVE' is provided on UI window, which starts the documentation process on clicking.

After pressing the 'SAVE' button, a GUI window is displayed, in which user needs to specify the particular folder where to save the document. This is generated by the *uigetdir()* function.

Now Microsoft PowerPoint© Application is invoked using the *actxserver()* function of Matlab®. It starts the MS PowerPoint©. Then, a new slide to that PowerPoint presentation is added. Then, all axes from a UI figure window are copied to a newly created figure, and then the figure is saved as a temporary image file. After that, the picture from the temporary image file is added to the MS PowerPoint© slide. Next, another new slide to that PowerPoint presentation is added and does the same until all the axes of GUI window are covered. After covering all the axes from the GUI window, the PowerPoint presentation is saved using Matlab® [5] command. The name of that PPT is built by concatenating the aerospace vehicle name and number given by user also using a Matlab® function named '*strcat()*'. After completing the documentation process, a message is displayed to indicate that the documentation process is completed. Related screenshots are given in Fig. 4. Saved documentation (*.ppt file) is shown in Fig. 4 (Fig. 5).

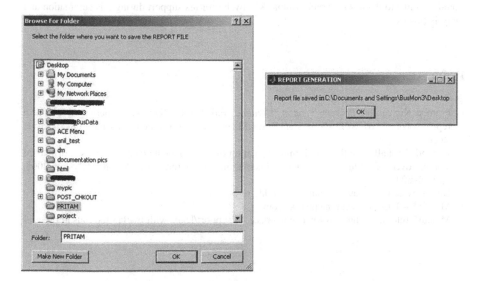

Fig. 4 GUI window for choosing the folder where to save the document file (*left*). A message box after completing the report documentation (*right*)

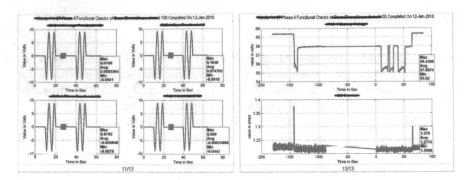

Fig. 5 Two screenshots of documentation made by the automation process (*left* and *right*)

3 Future Work

A *.ppt file is generated by this automation process, but a non-editable *.pdf can be created by this automation process after the inclusion of some Matlab® APIs. In future, the plan is to make this application run on any compatible machine without even Matlab® installation using Matlab Runtime Compiler. The automation process would become more useful if it can count the sine- and cosine-wave cycles of the test results. It is also proposed to add some intelligence into this system, so that it analyzes the data by itself and presents the result to the users.

Acknowledgments The authors would like to thank Dr. MRM Babu, Former Director, CAS, and Dr. Tessy Thomas, Director, CAS, for their constant encouragement and guidance. The authors would also like to thank Mr. Arvind Kumar Kushwaha for his support during this automation and analysis process.

References

1. Pramod Kumar Jha, Praveen Tandon, "Avionics Full Duplex Switched Ethernet(AFDX) as a replacement of MIL STD 1553B Data Bus, CSI communications, Feb 2016, page 08–10, cover story.
2. Pramod Jha, kallol Mullick et al, Implementation of real time data capture analysis, display & storage over Mil Std 1553 systems, International conference on Avionics System-2008, pp 618–624.
3. Data Device Corporation – http://www.ddc-web.com.
4. Matlab® – http://www.in.mathworks.com.
5. Matlab® tutorials – http://www.mathworks.com/support/learn-with-matlab-tutorials.html.

A Survey Study on Singular Value Decomposition and Genetic Algorithm Based on Digital Watermarking Techniques

Mokhtar Hussein and Manjula Bairam

Abstract In digital image watermarking, many techniques are used for obtaining optimal image representation; image decomposition as a standard set in the frequency domain is not necessary (DCT, DWT, and DFT). Therefore, another representation of transform was explored which is about using algebra methods and algorithms with watermarking like singular value decomposition algorithm based on watermarking. SVD algorithms have shown that they are highly strong against extensive range of attacks. In addition to that, Genetic Algorithm (GA) is used with SVD to optimize the watermarking. Many techniques and algorithms have already been proposed on the using of SVD and GA on digital watermarking. In this paper, we introduce a general survey on those techniques along with analysis for them based on the two measures, transparency and robustness.

Keywords Genetic algorithm · Singular value decomposition · Digital watermarking

1 Introduction

A digital watermarking can be defined as a sequence of special information inserted invisibly into another multimedia file; this information is related to the owner of the multimedia file and later can be extracted to be considered as copyright ownership evidence. Embedding a watermark w into the original image (cover image or host image) I is the primary conception of digital watermarking that is for achieving and obtaining copyright protection, monitoring of broadcast, access control, and etc. The watermark w could be a digital signal, a label, or a tag. The host digital media could be image, audio, video, or software. The watermarked media Iw is $I + w$.

M. Hussein (✉) · M. Bairam
Department of Computer Science, Kakatiya University, Warangal, Telengana, India
e-mail: msorori201201@gmail.com

M. Bairam
e-mail: manjulabairam@gmail.com

© Springer Science+Business Media Singapore 2017
S.C. Satapathy et al. (eds.), *Proceedings of the First International Conference on Computational Intelligence and Informatics*, Advances in Intelligent Systems and Computing 507, DOI 10.1007/978-981-10-2471-9_2

Digital watermarking algorithms and techniques are categorized into two types or domains, namely spatial and transform domains.

Spatial domain techniques are minimal robust for several types of offensives and susceptible to manipulate (to forge) by prohibited users [1]. Techniques related to transform domain are more robust and secure to several and different types of attacks, these techniques embed w into I by modulating the coefficients like Discrete Fourier Transform (DFT), Discrete Cosine Transform (DCT), Discrete Wavelets Transform (DWT), and Singular Value Decomposition (SVD). The size of w could be inserted into I is generally 1/16 of I. Robustness can be gained when considerable modifications are made to I while inserting w, regardless the domain without affecting the visual and quality of I. Those modifications are identifiable, and therefore, thus do not meet the requirements of invisibility (transparency) [2]. Designing and planning an optimal watermarking system for a specified application constantly include a trade-off in between those requirements, so that, if the mentioned issue can be addressed as an issue of optimization, then it can be resolved using one of the soft computing tools related to Artificial Intelligence (AI), viz., Genetic Algorithm (GA), Fuzzy Logic, and Neural Networks [3–5]. Watermark w could be scaled by the Scaling Factor (SF) denoted by k prior the process of embedding, this k is utilized to control the intensity and immovability of w. Watermark w can be embedded into I either via adjusting values of pixels or modifying the coefficients of transform domain. When SF is greater that means more distortion (more transparency) in the quality of I and stronger robustness. Furthermore, when SF is minimal that means much better quality of image and weaker robustness [4, 5]. Various spectral components may show diverse possibility to amendment. Thus, an individual SF may not be viable for adjusting all the values of I. Then, several SFs should be utilized for adapting to the divers spectral ingredients to minimize visual artifacts [1, 4].

For all the techniques and methods previously developed, they usually utilize predefined embedding essential rules and define their parameters of embedding, such as embedding strength and threshold. Often, it is hard to set optimal watermarking parameters, because watermarking algorithms have wider space of parameter. Consequence, those techniques and algorithms do not demonstrate desired performance [6, 7]. As a rule, digital watermarking is having two measures for performance, namely imperceptibly and robustness. Yet, inserting w inside I decrease the visual quality of image, while it is desirable that the dissolution must not be readily observed. Thence, imperceptibility indicates the concept which should be unrealizable to visual system of human. Both of the performance measures, imperceptibility and robustness, are conflicting with each other [8]. An optimal watermark system has to make equilibrium between the two. Hence, the problem of optimal watermark is from the multi objective optimization problem.

Therefore, many techniques are used with digital watermarking; they generally called soft computing tools, such as neural networks, genetic algorithms, and fuzzy logic. In addition to that, a number of transformations are used along with these techniques, such as DFT, DCT, DWT, and SVD. In this paper, we are concentrating only on highlighting the digital image watermarking techniques which used the concepts of SVD and GA.

2 Digital Watermarking Based on SVD

This section represents a general description of SVD with digital watermarking. The notion of SVD was founded for real-square matrices by Beltramin and Joda in the 1870s, for complex matrices in 1902, by Autonne, and has been expanded to rectangular matrices in 1939 by Echart and Young. Finally, SVD has been utilized in image processing applications, such as image hiding, image compression, and noise reduction.

SVD is famed as a transformation of an algorithm matrix which is relay on Eigenvector. It is an athletic tool which is applied during matrices analyzes. It is familiar that SVD is considered one of the generality sturdy techniques used in the numeric analysis along with so many number of applications, including watermarking [4, 9, 10]. SVD is also known as a mathematical method used to elicit algebraic lineaments out of image. Substance idea of SVD-based approach is to stratify the SVD to the entire I or at least to few segments of that image, and, therefore, adjust the singular values to insert the watermark. Three features are there to utilize the SVD technique with digital watermarking scheme, the first one is that the size of matrices in SVD transformation is unspecified; second, there will be no large variation of image's singular values when a slight disruption is added to that image, and the third one is that SVs exemplify the actual algebraic properties of image [11]. The above-mentioned properties of SVD are quite likeable to evolve watermarking techniques and algorithms and techniques which are particularly robust against the geometric offensives. In general, the energy or strength of w can be controlled using a scaling factor. Watermarking approach effectiveness is extremely relies on selecting an appropriate scaling factor. Ganic et al. [12] pointed that the scaling factor is set to be constant in several SVD-based studies. While that Cox et al. [13] argued that considering a unique single consonant scaling factor may not be usable in some situations, then they proposed that users can use various or multiple scaling factors instead of using only one. SVD always decompose a real matrix A with M * N to three sub matrices A = USVT, where U and V are M * N and N * N orthogonal matrices, respectively, S is considered as N * N diagonal matrix, i.e., consider a real M * N matrix A, and gather this matrix into two orthogonal matrices (U and V) and a diagonal matrix (S). The elements or entries in S are only non-zero on the diagonal and are named as singular values of A. Matrices size from SVD transformation is not fixed, so that it could be a rectangle or a square. Singular values in a digital are minimal influenced to common image processing algorithms and techniques, because they resist against attacks. When r is the rank of the matrix A and $S = \text{diag}(\lambda_i)$ where $i = 1, 2, ..., n$, then these singular values satisfy the following equation:

$$\lambda 1 \geq \lambda 2 \geq \lambda 3 \geq \cdots > \lambda \text{rank}(A) > 0.$$

When A indicates a matrix that its elements are pixel values of that image, thereafter this image can be written as:

$$A = USV^T = \sum_{i=1}^{r} \lambda_i u_i v_i^T.$$

As known, watermarking system composed of two main processes, embedding and extraction processes.

2.1 Embedding Process

The algorithm used in embedding process first applies SVD to I, then modify S with W, then apply SVD to Sm, and, finally, compute the watermarked image Iw. This algorithm can be represented mathematically as the following:

1. $I = USV^T$
2. $Sm = S + Kw$
3. $Sm = Um\ Sw\ V^T w$
4. $Iw = U\ Sw\ V^T$

2.2 Extraction Process

In the extraction process, first apply SVD to Iw which is possibly distorted, then compute possibly corrupted and, finally, extracted the watermark. This algorithm represented mathematically as follows:

1. $Iw = U\ Sw\ V^T$
2. $Smc = Uw\ Sw\ V^T$
3. $w = (Sc - S)/k.$

3 Digital Watermarking Based on Genetic Algorithm

Among all digital watermarking, imperceptibility and robustness are considered as the two inconsistent requirements required by any digital watermarking system. An effective watermarking system necessarily has to make a delicate trade-off in between these conflicting properties. This section explains how to use GA to optimize the process of watermarking embedding with respecting to the two inconsistent or conflicting required conditions transparency and robustness for various types of attacks. It presents a general information on GA and how to apply them with digital watermarking.

Genetic Algorithm is the most popular evolutionary algorithm widely used for optimization initiated by Holand in the 1960s and 1970s, as a basic principle, and attracted a great attention in several areas as a way and methodology for learning, adaptability and optimization after publishing Goldberg's book in 1984. Various techniques for digital watermarking have been optimized based on GA in spatial and frequency domains by many of the researches. GA deemed as a search and optimization technique based on the norm of natural and genetic selection. Cover image I with $M * N$ can have a number N of SVs which might detect diverse possibility to amendment. The technique or algorithm is necessary and needed to attain the optimum SFs which are output extreme transparency and robustness. Hence, an effective and powerful optimization algorithm is in needed for this objection. So for that, GA which is a familiar modernistic heuristic optimization algorithm is mostly used. The basic Genetic Algorithm is composed of five components namely: Random Number Generator (RNG), Fitness Evaluation (FE), Genetic Operator for Reproduction (GOR), Crossover (Cro.), and the last component is Mutation Operation (MO). Figure 1 represents the basic components of GA.

As stated in Fig. 1, the initial population is demanded to be at the starting of algorithm. The initial population is a group of number strings produced by GNA. Each series treated as an impersonation solution to the optimization problem being addressed.

To acquire or approach the limit of maximum performance of earlier developed watermarking algorithms, we have to locate their optimal parameters. One of the

Fig. 1 Basic components of GA

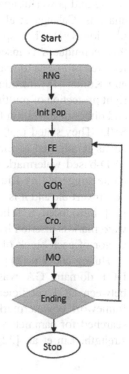

common ways of solving the problem of optimal watermarking is to treat it as a problem of optimization. Based on that, Genetic algorithms (GA) can be exercised and used to solve the problem of optimization [14].

4 Survey of Watermarking Techniques Using SVD and GA

In this section, we will introduce different proposed methods and techniques related to the using of SVD- and GA-based watermarking. First, we will present watermarking techniques based on SVD, then techniques based on GA, and finally, we will concentrate on the techniques that use both terms SVD and GA. Ganic et al. [15] suggested a dual SVD scheme based on watermarking embeds w twice. In the first layer, I is segmented to tiny segments and a part of w is embedded in each block. In the second layer, I is utilized as a single partition or block to insert the entire watermark. The intent of considering two layers to insert watermark, because the first one permits elasticity in the capacity of data, the second layer provides extra robustness and resistance to different offensive. Lee et al. [16] suggested a method for image content authentication based on SVD with enhanced security by insert or embed w into blocks which are ordered randomly, modifying and dithering the quantized greatest singular value of image block, the proposed technique is strong and powerful against VQ attack and is secure from the attacks of histogram analysis. Calanga et al. [17] presented a scheme of watermarking based on compression SVD. They split I into small blocks, then applied SVD to every block. In this technique, I is inserted into all non-zero singular values based on the topical features of I so as to make a balance in distortion with embedding capacity. Mohan and Kumar [18] introduced a robust image watermarking technique for the copyright protection of multimedia. In their suggested technique, they used SVD domain and dither quantization for inserting w in the both matrices (U and V) acquired from SVD. They stated that in their method, the greatest SV of I and U matrix coefficients is adjusted to embed w. Mohammad et al. [19] introduced a technique of SVD-based watermarking which is an amended version of the SVD-based technique suggested by Liu and Tan at [13]. The suggested technique or algorithm is non-invertible and it is mainly used for protecting legitimate ownership. Basso et al. [20] introduced a scheme of block-based watermarking based on SVD. The watermark is inserted by means of adjusting the angularity created by right singular vector of each block of I.

Huang and Wu [21] suggested a technique of image watermarking using GA in DCT domain, GA was utilized to search for suitable embedding locations in between DCT coefficients of image blocks. Kumsawat et al. [6] discussed that a framework of watermarking based on GA for performance was debated, so that GA searched for parameters of watermarking which include threshold and embedding strength. Lin et al. [22] discussed a minor searching algorithm based on GA to minimize long times of computation. Shieh et al. [23] presented a technique of watermarking optimization using GA to search optimal frequency bands for embedding watermarks. In the previous researches and presented techniques and methods, two goals for using GA, the first one is searching for an optimum parameters of watermarking (e.g., embedding strengths and threshold) [24], and the second one is to find the most appropriate embedding positions.

Veysel Aslantas presented a novel scheme of optimal watermarking based on SVD with GA, he modified SV values of I to embed w, then these amendments are optimized by employing GA to obtain the highest possible robustness without losing transparency, means without degrading the image quality. The size of I used in experiments was 256 * 256 and w with size of 32 * 32. In his suggested algorithms, GA is utilized to acquire multiple SF. Experiments outcomes show that feasibility of multiple SFs evaluated by GA and its notability over the use of a single SF. For further studies, he suggested to be executed along with the components of SVD (U and V). B. Jagdesh et al., they presented a novel scheme for optimal watermarking based on SVD and GA. This proposed technique relies on step size optimization using GA to improve the fineness of Iw and robustness of w to evaluate the effectiveness of the method, they used a cover image I with the size of 512 × 512 and w with size of 64 × 64. I is partitioned it into 8 × 8 before performing SVD on each block. They applied the embedding and extraction processes based on the suggested algorithm. They used an error matrices to test and make sure that the algorithms is normalized cross correlation NC for the robustness and peak signal-to-noise rate (PSNR) for transparency based on the following equations:

$$
PSNR = 10\log_{10}\left(\frac{\sum_{i=1}^{N}\sum_{j=1}^{N}(F(i,j))^2}{\sum_{i=1}^{N}\sum_{j=1}^{N}(f(i,j)-F(i,j))^2} \right)
$$

$$
NC = \left(\frac{\sum_{i=1}^{N}\sum_{j=1}^{N}(w(i,j)-w_{mean})(w'(i,j)-w'_{mean})}{\sqrt{\sum_{i=1}^{N}\sum_{j=1}^{N}(w(i,j)-w_{mean})^2 \sum_{i=1}^{N}\sum_{j=1}^{N}(w'(i,j)-w'_{mean})^2}} \right).
$$

Then, calculate the fitness of solution based on both the transparency PSNR and robustness $fval = -(PSNR + NC)$ based on the suggested algorithm. Different types of attacks are used to test the robustness of w. The experiments results show that the quality of Iw looks good with respect to perceptibility and PSNR, and in addition to that, this technique also presents robust to the following attacks: JPEG compression, low-pass filtering, resizing, media filtering, salt and pepper noise, and column blanking and copying. Chih-china et al., presented an adaptive amended SVD-based watermarking technique through stratify statics of the image with GA.

The proposed approach used GA to optimize the strength of w. They used I as a gray-level image (size of 512 * 512) and w (size of 32 * 32). In their experiments, the suggested approach maintains high perceptual quality of Iw. For the sake of valuation, the robustness of the suggested method, w, was examined against five attacks: Geometrical Attack (Cropping), Noising Attack (Gaussian Noise), De-noising Attack (Median filtering), Format Compression Attack (JPEG Compression), and Image Processing Attack (Sharpening, Histogram Equalization and Tempering), [CR, GN, MF, JPEG, and (SH, HE)]. So as to show the notability of their approach, they compare it with another SVD-based watermarking techniques and algorithms. Experiments results show both of the considerable amelioration in imperceptibility and robustness against various attacks and techniques of image manipulation. Shih-Chin Lai suggested an innovative technique for image watermarking through merging SVD technique with tiny GA. Appropriate scaling factors are specified using the tiny algorithm. For the purpose of testing, he used I with size of 256 $\times$ 256 and w with size of 64 $\times$ 64. The relative parameters which are used with tiny GA as follows: 5 is the size of population, 300 is the utmost generation number, and 0.95 is the probability of crossover. Lai divided the experiments into two steps, in the first step, he evaluated the robustness of the suggested approach, Iw was examined versus five sorts of attacks (CR, Rotation RO, GN, Average Filtering AF, JPEG, HE and Darken DK). In the second step: Lai compared this schema with [25, 26] to check the capability of resistance to various types of attacks, Extracted watermark quality is specified by values of NC. In [26], they apply the SVD to the whole I and insert w in the diagonal matrix of the SVD transformation and the SF is set to a single constant value. In [25] the author suggested a watermarking technique based on SVD and micro-GA (μ-GA). Singular values of I are adjusted through considering various scaling factors to insert w. The μ-GA is applied to effectively search the appropriate values of scaling factors. Result shows that in shih proposed approach, the robustness performance is superior to the other similar approach. One main disadvantage for this approach proved by Khaled that it suffers from a highly probability of detection of false positive. which means that the proposed watermarking scheme might be on-effective in the term of probability of false positive detection of I.

Poonam et al. proposed the modern technique used to get the robustness and imperceptibility much better in watermarking. In this method, the singular value of w is inserted into singular value of third-level DWT approximation matrix of I. They used the GA to optimize the scaling factor with which the watermark is embedded to I. The proposed algorithm makes use of fitness function that takes two values PSNR and correlation. They used a host image [size of 512 $\times$ 512] and watermark [size of 64 $\times$ 64], they carried out watermarking based on DWT-SVD procedure through GA on ten training samples, these procedures repeated for five attacks: Geometrical Attack, Format compression, Noising Attack, De-Noising Attack, and Image Processing Attack. They observed that after applying several types of attacks, the correlation value in between the original watermarking and the extracted one is larger if the scaling factor is high. The results shown that PSNR for the suggested algorithm is rising with the increasing value of scaling factors and their result was the best. They show that when the scaling factor increased, they got bestead performance particularly with robustness.

5 Conclusion

In this study, we have covered only the techniques of digital watermarking based on SVD and GA, we found that only a few researches have been done on the field of digital watermarking, which rely on combining SVD and GA, most of the mentioned studies in this study concentrate on some attacks, they do not cover all known attacks. Some of them concentrate on achieving higher PSNR and the other on NC. More studies need to be done in those fields to achieve more equilibrium between the two contradictory requirement perceptibility and robustness (i.e., PSNR and NC).

References

1. I.J. Cox, M.L. Miller, and J.A. Bloom, "Digital Watermarking," Morgan Kaufmann Publishers, 2002.
2. B. Jagadeesh, et al, "Image Watermarking Scheme Using Singular Value Decomposition, Quantization and Genetic Algorithm", 978-0-7695- 3960-7/10 $26.00 © 2010 IEEE.
3. Veysel Alantas, "An SVD Based Digital Image Watermarking Using Genetic Algorithm", 1-4244-0779-6/07/$20.00 ©2007 IEEE.
4. Veysel Alantas, "A singular-value decomposition-based image watermarking using genetic algorithm", 2007 Elsevier GmbH, Int. J. Electron. Commun. (AEÜ) 62 (2008) 386–394.
5. J.S. Pan, H.C. Huang, and L.C. Jain, "Intelligent Watermarking Techniques", World Scientific Publishing Company, 2004.
6. C.H. Huang, J.L. Wu, Fidelity-guaranteed robustness enhancement of blind-detection watermarking schemes, Information Sciences 179 (6) (2009) 791–808.
7. C.-C. Lai, H.-C. Huang, C.-C. Tsai, A digital watermarking scheme based on singular value decomposition and micro-genetic algorithm, International Journal of Innovative Computing, Information and Control 5 (7) (2009) 1867–1874.
8. P. Kumsawat, K. Attakitmongcol, A. Srikaew, A new approach for optimization in image watermarking by using genetic algorithm, IEEE Transactions on Signal Processing 53 (12) (2005) 4707–4719.
9. Liu R, Tan T. "An SVD-based watermarking scheme for protecting rightful ownership. IEEE Trans Multimedia 2002;4:121–8.
10. Bao P, Ma X. Image adaptive watermarking using wavelet domain singular value decomposition. IEEE Trans Circuits Systems Video Technol 2005;15:96–102.
11. R. Liu, T. Tan, An SVD-based watermarking scheme for protecting rightful ownership, IEEE Trans. Multimedia 40 (1) (2002) 121–128.
12. E. Ganic, N. Zubair, A.M. Eskicioglu, An optimal watermarking scheme based on singular value decomposition, in: Proc. IASTED Int'l Conf. on Communication, Network, and Information Security, 2003, pp. 85–90.

13. I.J. Cox, J. Kilian, F.T. Leighton, T. Shamoon, Secure spread spectrum watermarking for multimedia, IEEE Trans. Image Process. 6 (12) (1997) 1673–1687.
14. Jun Wang, Hong and Peng, "An optimal image watermarking approach based on a multi-objective genetic algorithm", Information Sciences 181 (2011) 5501–5514, Elsevier Inc., doi:10.1016/j.ins.2011.07.040.
15. E. Ganic, N. Zubair, A.M. Eskicioglu, An optimal watermarking scheme based on singular value decomposition, in: Proc. IASTED Int'l Conf. on Communication, Network, and Information Security, 2003, pp. 85–90.
16. S. Lee, D. Jang, C.D. Yoo, AN SVD-based watermarking method for image content authentication with improved security, in: Proc. ICASSP05, 2005, pp. 525–528.
17. M. Calagna, H. Guo, L.V Mancini, S. Jajodia, A robust watermarking system based on SVD compression, in: Proc. 2006 ACM Symposium on Applied Computing, 2006, pp. 1341–1347.
18. B.C. Mohan, S.S. Kumar, A robust image watermarking scheme using singular value decomposition, J. Multimedia 3 (1) (2008) 7–15.
19. A.A. Mohammad, A. Alhaj, S. Shaltaf, An improved SVD-based watermarking scheme for protecting rightful ownership, Signal Process. 88 (9) (2008) 2158–2180.
20. A. Basso, F. Bergadano, D. Cavagnino, V. Pomponiu, A. Vernone, A novel block based watermarking scheme using the SVD transform, Algorithms 2 (2009) 46–75.
21. C.H. Huang, J.L. Wu, A watermark optimization technique based on genetic algorithms, The SPIE Electronic Imaging, San Jose, CA, 2000.
22. C.H. Lin, J.L. Wu, C.H. Huang, An Efficient Genetic Algorithm for Small Search Range Problems and its Applications, Intelligent Multimedia Processing with Soft Computer, Springer, 2005.
23. S. Shieh, H.C. Huang, F.H. Wang, J.S. Pan, Genetic watermarking based on transform domain techniques, Pattern Recognition 37 (3) (2004) 555–565.
24. P. Kumsawat, K. Attakitmongcol, A. Srikaew, A new approach for optimization in image watermarking by using genetic algorithm, IEEE Transactions on Signal Processing 53 (12) (2005) 4707–4719.
25. C.-C. Lai, H.-C. Huang, C.-C. Tsai, Image watermarking scheme using singular value decomposition and micro-genetic algorithm, in: Proc. of the Fourth International Conference on Intelligent Information Hiding and Multimedia, Signal Processing, 2008, pp. 239–242.
26. F. Harung. M. Kuter, "Multimedia watermarking techniques", Proc. IEEE 87 (7) (1999) 1079–1107.

A Robust Authenticated Privacy-Preserving Attribute Matchmaking Protocol for Mobile Social Networks

M.M. Naresh Babu, A.S.N. Chakravarthy, C. Ravindranath and Chandra Sekhar Vorugunti

Abstract The advances in mobile and communication technologies lead to the advancement of mobile social networks (MSNs). MSN changed the way people communicate and exchange the private and sensitive information among the friend groups via mobile phones. Due to the involvement of private and sensitive information, MSN demands for efficient and privacy-preserving matchmaking protocols to prevent the unintended data (attribute) leakage. Many existing matchmaking protocols are based on user's private and specific data. Malicious participants may opt their attribute set arbitrarily so as to discover more information about the attributes of an honest participant. Hence, there is great chance of information leakage to a dishonest participant. In this context, Sarpong et al. had proposed a first of its kind of an authenticated hybrid matchmaking protocol that will help match-pair initiators to find an appropriate pair which satisfies the predefined threshold number of common attributes. Sarpong et al. had claimed that their protocol restricts attribute leakage to unintended participants and proved to be secure. Unfortunately, in Sarpong et al. scheme, after thorough analysis, we demonstrate that their scheme suffers from data (attribute) leakage in which the initiator and the participant can compute or achieve all the attributes of each other. Also, we show that Sarpong et al. scheme requires huge computation and communication cost. As a part of our contribution, we will propose an efficient and secure matchmaking protocol which is light weight and restricts attribute leakage to the participants.

M.M. Naresh Babu (✉) · A.S.N. Chakravarthy
Department of CSE, JNTU Kakinada, Kakinada 533003, AP, India
e-mail: itsnaresh4u@gmail.com

A.S.N. Chakravarthy
e-mail: asnchakravarthy@yahoo.com

C. Ravindranath
TITR, Bhopal 462038, MP, India
e-mail: ravindranathc@gmail.com

C.S. Vorugunti
Indian Institute of Information Technology-SriCity, Chittoor 517588, AP, India
e-mail: sekhar.iiitsricity@gmail.com

© Springer Science+Business Media Singapore 2017
S.C. Satapathy et al. (eds.), *Proceedings of the First International Conference on Computational Intelligence and Informatics*, Advances in Intelligent Systems and Computing 507, DOI 10.1007/978-981-10-2471-9_3

Keywords Mobile social network · Matchmaking · Attribute · Authentication

1 Introduction

The advances in mobile and communication technologies lead to advancement of traditional online social network to mobile social networks (MSNs). MSN facilitates real time personal and user specific data sharing and instant messaging among friend groups. Due to the exchange of private and shared information among the participants, finding a matching pair privately is a critical requirement in MSN.

A private matchmaking is a primary feature of private set intersection. Matchmaking protocol is a critical requirement for MSN in which two or more mutually mistrustful parties A and B consists of attribute sets SA and SB, desire to compute together the intersection in such a way that both A and B should not take any information particular to the other opponent. A and B must learn only the common attributes among them, i.e., $SA \cap SB$ nothing more.

In the literature, many matchmaking algorithms have been proposed based on various parameters. Few matchmaking protocols [1–3] have been proposed based on certificate authority CA, in which CA authenticates the entities attributes. Another matchmaking technique is fully distributed [4], which eliminates CA. The participants perform the distribution of attributes among themselves, computing the intersection set. The initiator and the multiparties exchange their attributes using Shamir's secret sharing scheme. The hybrid technique [5] is a commonly used technique in which the CA performs only the verification of attributes and managing the communication among the entities. The protocol participants will perform the attribute sharing and matchmaking operations. Recently, Huang et al. [6] had proposed an identity based encryption scheme for matchmaking in social networks.

However, in this context in 2015, Chiou et al. [7] and Sarpong et al. [8] had proposed matchmaking protocols in which the initiator finds the best match among multiple participants who has the maximum similar attribute as the initiator. Sarpong et al. claimed that their scheme protects user's attributes from unnecessary leakage to unintended persons. In this manuscript, after thorough analysis of Sarpong et al. scheme, we will demonstrate that in Sarpong et al. scheme, the participants can achieve the attributes of other participants and requires huge computation and communication cost.

As a part of our contribution, we will propose a secure and light weight matchmaking protocol for MSN, which resists the pitfalls in Sarpong and other related schemes.

2 Brief Review of Sarpong et al. Matchmaking Algorithm

This protocol mainly consists of a certification authority (CA) that cannot be compromised, an initiator of matchmaking algorithm, and the number of participants in the matchmaking process. This algorithm facilitates the match seekers to find the most appropriate pairs. The users of this protocol can know the actual attributes they have in common, if a pair is found and the size of the intersection (if it exits) is greater than the predefined threshold $A_{Threshold}$.

2.1 Initial Phase

Assume Alice is the initiator of the protocol to find out the closest match among 'm' participants (for brevity, we assume that Alice is communicating with a single participant Bob to find out the common attributes. The other participants also perform and exchange similar messages as Bob with Alice. Alice also exchanges same messages as it exchanges with Bob) having portable devices and can connect with each other using PAN or Bluetooth or Wifi. $A_{threshold}$ is the threshold value for the attribute matching set by the initiator Alice, i.e., to qualify as a match-pair for initiator; there should be minimum of $A_{threshold}$ number of common attributes between pairs. The initiator Alice consists of 'm' attributes, i.e., $a = \{a1, a2, am\}$ and Bob consists 'p' attributes, i.e., $b = \{b1, b2, b3.........bp\}$. In the matchmaking, if two attributes are semantically same, then only they are treated as the same.

2.2 Key Generation

K1. Alice and Bob computes RSA key pairs (e_A, d_A), (e_B, d_B), respectively, using p, q which are large prime numbers, where e_A, e_B are the public variables.
    K3. CA computes RSA key pair is (e, d), where $N = p * q$.
    K4. CA makes <e, N> public.

2.3 Attribute Certification

A1. The attributes of Alice and Bob are $a = \{a1, a2, a3...am\}$ and $b = \{b1, b2,... bk\}$.
A2. Alice exponentiates his attribute set using the public key of CA, i.e., 'e'. $a^e = \{a1^e, a2^e, ..., am^e\}$.
A3. Bob also exponentiates his/her attributes as $b^e = \{b1^e, b2^e, ...bk^e\}$.

A4. Alice to get the attributes certified by CA, forwards a message $E_e\{a^e\|ID_A\|$ $UN_A\| e, e_A\}$ to CA which contains the attribute set computed in A2, its identity, user name, its public key and CA public key. The message is encrypted with the CA public key, i.e., 'e'.

A5. Bob also to get the attributes certified by CA, forwards a message $E_e\{b^e \| ID_B$ $\| UN_B \| e \|e_B\}$ to CA which contains the attribute set computed in A3, its identity, user name, its public key and CA public key. The message is encrypted with the CA public key, i.e., 'e'.

A6. The CA certifies the Alice attributes and returns $A = \{(a1, s1), (a2, s2), ...,$ $(am, sm)\}$ to Alice, where $si = H(ID_A\|ai)^d \bmod N$ using its private key 'd'.

A7. The CA also certifies the Bob attributes and returns $B = \{(b1, \sigma1), (b2, \sigma2),$ $..., (bk, \sigma k)\}$ to Bob, where $\sigma1 = H(ID_B\|b1) \bmod N$.

2.4 Matchmaking Phase

M1. On getting the attributes certified by the CA, the private attributes of Alice and Bob becomes $A = \{(a1, s1), (a2, s2), ..., (am, sm)\}$, $B = \{(b1, \sigma1), (b2, \sigma2), ..., (bk, \sigma k)\}$, respectively.

M2. Alice picks 'm' arbitrary random numbers Ri for each attribute $i = \{1, 2, 3, ...m\}$ and computes $MAi = Si.g^{Ri} \bmod N$, i.e., $MA1 = s1.g^{R1} \bmod N$, $MA2 = s2.g^{R2} \bmod N$, $MA3 = s3.g^{R3} \bmod N$ and sends $MES1 = \{MA1, MA2, ...MAm\}$ to Bob.

M3. Bob also chooses an arbitrary numbers Pk for each attribute $k = \{1,2,3,...... k\}$ and computes $MBk = \sigma k.g^{Pk} \bmod N$, i.e., $MB1 = \sigma1.g^{P1} = H(ID_B\|b1).g^{P1} \bmod N$, $MB2 = \sigma2.g^{P2} \bmod N = H(ID_B\|b2).g^{P2} \bmod N$...and sends $MES2 = \{MB1, MB2, ..., MBk\}$...to Alice.

M4. Alice chooses an arbitrary number Ra and computes $ZA = g^{e.Ra} \bmod N$, $MBk^* = (MBk)^{e*Ra} = \{MB1^{e.Ra}, MB2^{e.Ra}, .. MBm^{e.Ra}\} = \{(H(ID_B\|b1). g^{P1})^{.e.Ra}, (H(ID_B\|b2).g^{P2})^{.e.Ra}, ...H(ID_B\|bm).g^{Pm})^{.e.Ra}.\}$.

M5. Alice performs arbitrary permutation $RPA = \zeta\{a_1, a_2, ..., a_k\}^{Ra} = \zeta\{a_1^{Ra}, a_2^{Ra}, ..., a_k^{Ra}\}$ and sends $MES3 = \{Z_A\|MBk^*\| RPA\}$ to Bob.

M6. Bob also opts an arbitrary number Rb and computes $ZB = g^{e.Rb} \bmod N$, $(MES1)^{e.Rb} = \{M1^{e.Rb}, M2^{e.Rb}, M3^{e.Rb}, ...Mk^{e.Rb}\} = \{(s1.g^{R1})^{e.Rb}, (s2.g^{R2})^{e.Rb}, ... (Sk.g^{Rm})^{e.Rb}.\}$.

M7. Bob chooses an arbitrary permutation $RPB = \zeta\{b_1^{Rb}, b_2^{Rb}, ..., bk^{Rb}\}$ and sends $MES4 = \{Z_k\|(MES1)^{e.Rb}\| RPB\}$ to Alice.

M8. Alice sends his signed message $Sig_{dA}(ID_A\|MES1\|MES2\|MES3\|MES4)$ to Bob.

M9. Bob also sends his signed message $Sig_{dB}(ID_B\|MES1\|MES2\|MES3\|MES4)$ to Alice.

M10. Now Alice and Bob verify that received MES1, MES2, MES3, MES4 values are equivalent to the received or computed values in the previous steps.

M11. Alice share his random number to Bob by sending $Sig_{dA}(ID_A \| ID_B \| R_a)$. Similarly, Bob also shares his arbitrary number by sending $Sig_{dB}(ID_B \| ID_A \| R_b)$.

M12. Alice computes a list $KA = \zeta A\{a_1^{RaRb}, a_2^{RaRb}, ..., a_m^{RaRb}\}$ and direct to Bob. Bob also computes $KB = \zeta B\{b_1^{RbRa}, b_2^{RbRa}, ..., b_k^{RbRa}\}$ and send it to Alice.

M13. To know the actual common attributes, Alice sends his random permutation by encrypting with the Bob public key, i.e., e_B, $E_{eB}(\zeta A)$. Similarly, Bob sends his random permutations to Alice by encrypting with the Alice public key, i.e., e_a, $E_{eA}(\zeta B)$.

M14. Alice already know ζB, can able to compute ζB^{-1} and retrieves $\{b_1^{RbRa}, b_2^{RbRa}, ..., b_k^{RbRa}\}$, similarly, Bob able to compute ζA^{-1} and recover $\{a_1^{RaRb}, a_2^{RaRb}, ..., a_m^{RaRb}\}$. Now both Alice and Bob know their actual common attributes.

3   Cryptanalysis of Sarpong et al. Algorithm

In this section, we deliberate security analysis of Sarpong et al. scheme.

In Sarpong et al. scheme, in M13 of matching phase, Alice sends its random permutation, i.e., $E_{eB}(\zeta A)$ by encrypting with the Bob public key. Similarly, Bob sends its random permutation $E_{eA}(\zeta B)$ by encrypting with the Alice public key. On receiving the encrypted message $E_{eA}(\zeta B)$, Alice perform the following steps as depicted below:

Step (1)   Decrypts $E_{eA}(\zeta B)$ using its private key dA, i.e., $D_{dA} E_{eA}(\zeta B) = \zeta B$.

Step (2)   Alice performs inverse operation ζB^{-1} on KB, i.e., $\zeta B^{-1}(KB) = B^{-1}\{b_1^{RbRa}, b_2^{RbRa}, ..., b_k^{RbRa}\}$ to retrieve original list, i.e., $\{b_1^{RbRa}, b_2^{RbRa}, ..., b_k^{RbRa}\}$.

Step (3)   In M11 of matching phase, Bob sends the message $Sig_{dB}(ID_B \| ID_A \| R_b)$ to Alice. Alice retrieves $\{ID_B, ID_A, R_b\}$ from the received message. Alice already know his R_a, hence Alice can perform an inverse operation on each received value in $\{b_1^{RbRa}, b_2^{RbRa}, ..., b_k^{RbRa}\}$, i.e., $\{b_1^{RbRa}, b_2^{RbRa}, ..., b_k^{RbRa}\} R_b^{-1} R_a^{-1} = \{b1, b2, b3, b4 .... bk\}$. Hence, Alice comes to know all the attributes of Bob, along with the common attributes. Similar is the case with Bob, in which Bob also comes to know all the attributes of Alice along with the common attributes by executing the above steps similar to Bob. Therefore, we can conclude that Sarpong et al. scheme fails to achieve the primary requirement of matchmaking algorithm, in which the participant and initiator must know only the common attributes.

4 Pitfalls or Anomalies in Sarpong et al. Algorithm

4.1 Requires Huge Communication Cost

In M2 and M5 steps of matchmaking process, Alice sends MES1 and MES3 to Bob, respectively. In M8, Alice again forwards MES1, MES3 to Bob in a message $Sig_{dA}(ID_A\|MES1\|MES2\|MES3\|MES4)$. Bob, on receiving the message $Sig_{dA(ID_A}\|MES1\|MES2\|MES3\|MES4)$, decrypts the message to get {ID_A, MES1, MES2, MES3, MES4} and uses MES1, MES2, MES3, MES4 to validate, whether the transferred and received values are valid or not. To validate the messages transferred, a message digest operations like hash functions, e.g., SHA-1 can be used, which outputs a fixed length data, hence reduces the need to transfer full messages.

Similar is the case with the Bob. In M3, Bob sends MES2, in M7, Bob sends MES4 to Alice. In M9, Bob again sends these messages in the form of $Sig_{dB}(ID_B\|$ MES1$\|$MES2$\|$MES3$\|$MES4) to Alice, which consumes huge communication cost.

4.2 Requires Huge Computation Cost

In M2 and M3 steps of matchmaking process, Alice and Bob selects 'm' and 'p' arbitrary numbers, respectively. Alice computes $MAi = si.gRi \bmod N$, where $1 < = i < = m$. Similarly, Bob computes $MBk = \sigma k.gPk = H(IDB\|bk).gPk \bmod N...$, where $i < = k < = p$. Totally for one participant and one initiator, the Sarpong et al. schemes $k * m$ random numbers, which requires huge computation cost.

5 Our Proposed Scheme

In this section, we present our improved scheme over Sarpong et al. scheme to remedy the security flaws as mentioned above while preserving their merits. The Key Generation and Attribute Certification phases of our proposed scheme are similar to Sarpong et al. scheme.

5.1 Matchmaking Phase

M1. On getting the attributes certified by the CA, the private attributes of Alice and Bob becomes $A = \{(a1, s1), (a2, s2), ..., (am, sm)\}$, $B = \{(b1, \sigma1), (b2, \sigma2), ..., (bk, \sigma k)\}$, respectively.

M2.　Alice picks a single arbitrary random number R1, and computes $MA1 = Si$. g^{R1}mod N, i.e., $MA1 = s1.g^{R1}$mod N, $MA2 = s2.g^{R2}$mod N, $MA3 = s3$. g^{R3}mod N and sends $MES1 = \{MA1, MA2, ..., MAm\}$ to Bob.

M3.　Each participant also chooses an arbitrary numbers P1 computes $MBk = \sigma k$. g^{P1}mod N, i.e., $MB1 = \sigma1.g^{P1=} H(ID_B\|b1).g^{P1}$ mod N, $MB2 = \sigma2.g^{P1}$ mod $N = H(ID_B\|b2).g^{P1}$mod N... and sends $MES2 = \{MB1, MB2, ..., MBk\}$... to Alice.

M4.　Alice chooses an arbitrary number Ra and computes $ZA = g^{e.Ra}$mod N, $MBk^* = (MBk)^{e*Ra} = \{MB1^{e.Ra}, MB2^{e.Ra}, ... MBm^{e.Ra}\} = \{(H(ID_B\|b1). g^{P1})^{.e.Ra}, (H(ID_B\|b2).g^{P2})^{.e.Ra}, ...(H(ID_B\|bm).g^{Pm})^{.e.Ra}.\}$.

M5.　Alice performs arbitrary permutation $RPA = \zeta\{a_1, a_2, ..., a_k\}^{Ra} = \zeta\{a_1^{Ra}, a_2^{Ra}, ..., a_k^{Ra}\}$ and sends $MES3 = \{Z_A\|MBk^*\| RPA\}$ to Bob.

M6.　Bob also opts an arbitrary number Rb and computes $ZB = g^{e.Rb}$mod N, $(MES1)^{e.Rb} = \{M1^{e.Rb}, M2^{e.Rb}, M3^{e.Rb}, ...Mk^{e.Rb}\} = \{(s1.g^{R1})^{e.Rb}, (s2.g^{R2})^{e.Rb}, ... (Sk.g^{Rm)\ e.Rb}.\}$.

M7.　Bob chooses an arbitrary permutation $RPB = \zeta\{b_1^{\ Rb}, b_2^{\ Rb}, ..., bk^{Rb}\}$ and sends $MES4 = \{Z_k\|(MES1)^{e.Rb}\| RPB\}$ to Alice.

M8.　Alice computes $M1 = ID_A \oplus h(MES1\|MES2\|MES3\|MES4)$, $M2 = h(ID_A\|MES1\|MES2\|MES3\|MES4)$ and forwards $\{M1, M2\}$ to Bob.

M9.　Bob computes $M3 = ID_B \oplus h(MES1\|MES2\|MES3\|MES4)$, $M4 = h(ID_B\|MES1\|MES2\|MES3\|MES4)$ and forwards $\{M3, M4\}$ to Alice.

M10.　On receiving $\{M3, M4\}$ from Bob, Alice achieves $ID_B^* = M3 \oplus h(MES1\|MES2\|MES3\|MES4)$, computes $M4^* = h(ID_B^*\| MES1\|MES2\|MES3\|MES4)$ and compares the computed $M4^*$ with the received M4. If both are equal, Alice authenticates Bob. Similarly, Bob achieves ID_A^* from M1, and computes $M2^* = h(ID_A^*\|MES1\|MES2\|MES3\|MES4)$. If computed $M2^*$ equals the received M2, Bob authenticates Alice.

M11.　Alice share his random number to Bob by sending an encrypted message using the Bob public key, so that the message can be decrypted only by Bob using his private key, i.e., dB where $D_{dB}(E_{eB}(ID_A\|ID_B\|R_a\|R1)) = \{ID_A, ID_B, R_a, R1\}$.

M12.　Similarly, Bob also shares his arbitrary numbers by sending an encrypted message using Alice public key, i.e., eA, where $E_{eA}(ID_B\| ID_A\|R_b\| P1) = \{ID_B, ID_A, R_b, P1\}$.

M13.　Alice computes a random permuted list $KA = \zeta A\{h(a1\|R1)^{RaRb}, h(a2\| R1)^{RaRb}, ..., h(am\|R1)^{RaRb}\}$ and direct to Bob. Bob also computes $KB = \zeta B \{h(b1\|P1)^{RbRa}, h(b2\|P1)^{RbRa}, ...,h(bk\|P1)^{RbRa}\}$ and send it to Alice.

M14.　To know the actual common attributes, Alice sends his random permutation by encrypting with the Bob public key, i.e., $e_B, E_{eB}(\zeta A)$. Similarly, Bob sends his random permutations to Alice by encrypting with the Alice public key, i.e., $e_A, E_{eA}(\zeta B)$.

M15.　Alice already know ζB, can able to compute ζB^{-1} and retrieves $\{h(b1\| P1)^{RbRa}, h(b2\|P1)^{RbRa}, ...,h(bk\|P1)^{RbRa}\}$, similarly, Bob able to compute ζA^{-1} and recover $\{h(a1\|R1)^{RaRb}, h(a2\|R1)^{RaRb}, ..., h(am\|R1)^{RaRb}\}$.

M16. For each attribute $\{a1, a2, am\}$, Alice computes $\{h(a1\|P1)^{RaRb}, h(a2\|P1)^{RaRb},$
  ....,$h(am\|P1)^{RaRb}\}$ and compares with the attribute list $\{h(b1\|P1)^{RbRa},$ h
  $(b2\|P1)^{RbRa}, ...,h(bk\|P1)^{RbRa}\}$. The comparison gives the Alice, the number of
  attributes in common and their actual values with Bob. Bob also perform same
  computations as Alice. As Alice and Bob uses hash function and session
  specific arbitrary numbers to compute $\{h(a1\|R1)^{RaRb}, ...\}$, $\{h(b1\|P1)^{RaRb},$
  ....$\}$, if an attribute sent by Bob is not matching against any value in the Alice
  attribute list, even though Alice knows P1, Ra, Rb it is computationally
  infeasible for Alice to achieve or compute the non matching attribute, this is
  due to the one way property of hash function. Similar is the case with Bob.

M17. Hence, in our scheme, there is no chance of leakage of attributes to opponent,
  in case of non-matching attributes.

6 Security Strengths of Our Proposed Scheme

Attribute Verification: In our proposed scheme, in Attribute Certification phase, the
initiator Alice and the participant Bob submit their attribute set a = $\{a1, a2, a3...$
am$\}$ and b = $\{b1, b2, ...bk\}$ to CA. The CA certifies the attributes and returns
A = $\{(a1, s1), (a2, s2), ..., (am, sm)\}$ to Alice, where si = $H(ID_A\|ai)^d$mod N.
Similarly for Bob, CA returns B = $\{(b1, \sigma1), (b2, \sigma2), ..., (bk, \sigma k)\}$ where σi = H
$(ID_B\|bi)$ modN. As CA binding the attributes with their hash value, the participants
are restricted to change their attributes later. This step restricts the attacks by
malicious and semi-honest participants.

Resists Attribute Mapping: In matchmaking phase of our scheme, i.e., M5, M7,
the initiator Alice sends the randomly permuted attribute set, i.e., RPA = $\zeta\{a_1, a_2,$
..., $a_k\}^{Ra}$ = $\zeta\{a_1^{Ra}, a_2^{Ra}, ..., a_k^{Ra}\}$ to Bob. Similarly, Bob also opts an arbitrary number
Rb and computes an arbitrary permutation RPB = $\zeta\{b_1{}^{Rb}, b_2{}^{Rb}, ..., bk^{Rb}\}$. Due to
the random permutations, even though the participant or malicious attacker
achieves a_1^{Ra}, it is impossible to map a_1^{Ra} to an entry in the list $\zeta\{a_1^{Ra}, a_2^{Ra}, ..., a_k^{Ra}\}$.
Also, in M11, M12 the Alice and Bob exchange their random numbers by
encrypting with the public key of the opponents. In M11, Alice share his random
number to Bob by sending an encrypted message using the Bob public key, so that
the message is decrypted only by Bob using his private key, i.e., dB, where
$D_{dB}(E_{eB}(ID_A\|ID_B\|R_a\|R1))$ = $\{ID_A, ID_B, R_a, R1\}$. Similar is the case with the Bob.
Hence, it is impossible for an attacker to achieve the attributes of the participants.

Dynamic Attributes: In all the previous works including Sarpong et al., the
initiator and the participants make their attribute set random by exponentiating the
attributes with random number. If the random numbers are known to the malicious
users, they can retrieve the attribute values which are static. Hence, it will leak the
attribute information. In our proposed scheme, Alice computes a random permuted
list KA = $\zeta A\{h(a1\|R1)^{RaRb}, h(a2\|R1)^{RaRb}, ..., h(am\|R1)^{RaRb}\}$ in which a hash of
an attribute is concatenated with a random number and exponentiated. In this case,

Fig. 1 The performance comparison between Sarpong et al. and our proposed algorithm

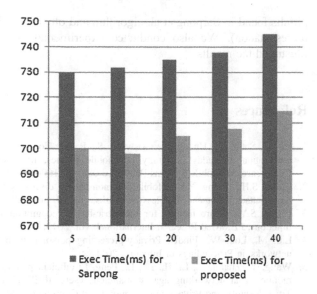

the same attribute value results in a different hash value each time it is sent. Hence, it is difficult for an attacker to achieve any information from the attribute set.

Due to space restrictions, we have discussed above attacks only. Our scheme resists all major cryptographic attacks and achieves attribute privacy.

7 Experimental Implementation

Our scheme has been tested using a simulation on Intel core i7 PC with 3.6 GHz processor and 4G RAM. We have considered the users n = 5, 10, 20, 30, 40 and each user is considered to contain varying attributes a = 5, 10, 15. We have depicted our simulation results in the Fig. 1. Figure 1 illustrates the execution time of our algorithm for the different number of users with varying number of attributes.

8 Conclusion

The involvement of user's specific and sensitive data in MSN demands for a light weight and secure matchmaking algorithm, which resists attribute leakage to participants. Sarpong et al. had proposed first of its kind of matchmaking algorithm which selects the participants that contains the threshold level of attributes matching. We have crypt analyzed Sarpong et al. scheme, and demonstrated that their scheme fails to achieve attribute privacy and requires huge storage and computation cost. We have proposed an efficient algorithm, which resists the pitfalls found in Sarpong et al. algorithm and other related schemes (static attribute representation). We also conducted experimental analysis of our scheme and illustrated the results.

References

1. Li, K., Sohn, T., Huang, S., Griswold, S.: People-Tones: A system for the detection and notification of buddy proximity on mobile phones, in Proceedings of 6th International Conference on Mobile Systems (MobiSys'08), pp. 160–173, 2008.
2. Chiou, S.H., Huang, Y.H.: Mobile common friends discovery with friendship ownership and replay-attack resistance. Wireless Networks, vol. 19, pp. 1839–1850, 2013.
3. Chiou, S.Y: Secure method for biometric-based recognition with integrated cryptographic functions, J.BioMed Research International, vol. 2013.
4. Liu. M., Lou, W.: FindU: Privacy-preserving personal profile matching in mobile social networks, in Proceedings of Infocom, 2011.
5. Wang, Y., Zhang, T., Li, H., He, L., Peng, J.: efficient privacy preserving matchmaking for mobile social networking against malicious users. IEEE 11th International Conference on Trust, Security and Privacy in Computing and Communications, pp. 609–615, 2012.
6. Lin, H., Chow, S.S.M., Xing, D., Fang, Y., Cao, Z.: Privacy preserving friend search over online social networks, Cryptology EPrint Archive, 2011. (http://eprint.iacr.org/2011/445.pdf).
7. Chiou, S.Y., Luo, C.S.: An Authenticated Privacy-Preserving Mobile Matchmaking Protocol Based on Social Connections with Friendship Ownership. J. Mathematical Problems in Engineering Volume 2015.
8. Sarpong, Xu. C., Zhang, X.: An Authenticated Privacy-preserving Attribute Matchmaking Protocol for Mobile Social Networks, International Journal of Security and Its Applications, Vol. 9, pp. 217–230, 2015.

Artificial Neural Networks Model to Improve the Performance Index of the Coil-Disc Assembly in Tube Flow

R. Kanaka Durga, C. Srinivasa Kumar, Vaka Murali Mohan, L. Praveen Kumar and P. Rajendra Prasad

Abstract An artificial neural networks model to enhance the performance index of mass transfer function in tube flow by means of entry region coil-disc assembly promoter was inserted coaxially is presented in this paper. Popular Backpropagation algorithm was utilized to test, train and normalize the network data to envisage the performance of mass transfer function. The experimental data of the study is separated into two sets one is training sets and second one is validation sets. The 248 sets of the experimental data were used in training and 106 sets for the validation of the artificial neural networks using MATLAB 7.7.0, particularly tool boxes to predict the performance index of the mass transfer in tube for faster convergence and accuracy. The weights were initialized within the range of [−1, 1]. The network limitations in all attempts taken learning rate as 0.10 and momentum term as 0.30. The finest model was selected based on the MSE, STD and R2. In this, network with 5_8_1 configuration is recommended for mass transfer training. This research work reveals that artificial neural networks with adding more number of layers and nodes in hidden layer may not increase the performance of mass transfer function.

Keywords Artificial neural network · Mass transfer · Coil-disc assembly · Tube · Backpropagation algorithm

R. Kanaka Durga · C. Srinivasa Kumar
Department of CSE, VIGNAN Institute of Technology & Sciences,
Deshmukhi, Hyderabad, TS, India

V.M. Mohan (✉) · L. Praveen Kumar
Department of CSE, TRR College of Engineering, Inole, Patancheru,
Hyderabad, TS, India
e-mail: murali_vaka@yahoo.com

P. Rajendra Prasad
Department of Chemical Engineering, AUCE, Andhra University,
Visakhapatnam, AP, India

© Springer Science+Business Media Singapore 2017   27
S.C. Satapathy et al. (eds.), *Proceedings of the First International Conference
on Computational Intelligence and Informatics*, Advances in Intelligent Systems
and Computing 507, DOI 10.1007/978-981-10-2471-9_4

Nomenclature

D	Tube Diameter, m
D_c	Coil Diameter, m
P_c	Coil Pitch, m/turn
L_c	Coil Length, m
D_d	Diameter of the disc, m
H_d	Height of the disc, m
V	Velocity of the fluid, m/s
$\bar{g}$	Mass transfer function
Re_m^+	Modified Reynolds number
P_c/d, L_c/d, d_d/d, H_d/d	Dimensionless parameter

1 Introduction

Artificial neural networks (ANN) has been well established and used for significant applications in process intensification. The enhancement in mass transfer is a popular application of artificial neural networks into engineering and technology. Applications of computer modeling systems are general in technical and scientific investigations. The advantages and the models of the neural network have been reviewed, such as capabilities of the universal approximation, flexibility and cost-cutting. The artificial neural networks model also provides fast and reliable results within a high degree of accuracy. Neural networks improved performance in terms of optimization time, scalability and robustness for most of these problems. Artificial neural networks can be used as a fault diagnostic tool in chemical and mechanical process industries. It is highly useful in natural gas dehydration, ethane extraction, LPG production and natural gas sweetening. Most importantly, it is useful to forecast the ground water level and weather updates. It is also useful to delineating rice crop management and crop production estimation, esterification process for biodiesel synthesis, pharmaceutical process development and mineral separation process. Artificial neural networks model is used to optimize and simulate the transport processes, thermal systems, radar systems, medical analysis.

The main application of artificial neural networks (ANNs) in transport processes that they recommended the impending of a generic approach of nonlinear system modeling. "Artificial Neural Network" initiates from the study which tried to recognize and suggested easy method like human brain procedure. Therefore, ANNs have behavior like general and biological process, which are consists by means of easy numerous practicing functions (neurons) connected jointly with different capacities to form the connections by means of considerable comparable and extremely consistent message passing system using multilayer perception neural network. In this study, backpropagation have been chosen because it is usually used and well recognized in this area of research and Cascade correlation have been chosen because, it solves few of the difficulties related with the backpropagation.

Transport processes have been dominated by linear control theory and technology over the last few decades. However, the use of linear system techniques is quite limiting, since a significant portion of the transport processes are inherently nonlinear. Modern developments in computer software and hardware contain advanced nonlinear control methods and algorithms. Though linear systems have the advantage of computational simplicity, in many instances it may not be practical to use them, especially for severely nonlinear processes. The impending advantages of nonlinear system have been explained and review of established linear and nonlinear modeling control techniques. The recognized neural network techniques were presented for a mass transfer was evaluated and motivated the control system.

2 Literature Review

Several investigations were made earlier to know the capacity of ANN to characterize the nonlinear methods make them a powerful implementation designed for the development of process control system and its modeling. To arrive at the above objective, several strategies have been devised and adopted in different areas using artificial neural networks, i.e., pattern recognition, sentence recognition, thermal resistivity, process identification, drying, cyclone separators, distillation column, flow in open channels, heat transfer, friction factor, mass transfer. The utilization of ANN in the process industry is relatively unusual for process control and modeling. One of the major issues is lack of set up plan procedures, which continue living used for linear system modeling, which excluding not so far for the development of the ANN process modeling, some of them are Jie Zhang and Morris [1] presented the chronological approach to construct, test and train the network with single hidden layer. Stevan et al. [2] described the modified design of a multilayer network and it is trained using backpropagation learning algorithm. Simon Haykin [3] presented the importance of the necessary component for recognizing the neural networks such as learning process. Leonard and Kramer [4] discussed the chemical engineering problems and malfunction diagnosis using artificial neural networks. Pollard et al. [5] described the applications of neural network in process identification using backpropagation. Zdaniuk et al. [6] determined the factors of Colburn "j" and "Fanning Friction" for the flow of water through circular conduit by means of helical fins as an insert promoter using "Artificial Neural Network". Vaka Murali Mohan [7] studied mass and momentum transfer in circular conduit by means of "entry region coil, disc, coil-disc assembly" as promoters. Rajendra Prasad [8] studied ionic mass and momentum transfer within standardized flow and in fluidized beds by means of "spiral coils" as promoter. Vaka Murali Mohan et al. [9] developed model and flow pattern in pipe by means of "entry region coil-disc" as insert promoter. Unal Akdag et al. [10] studied the rate of HT in annular flow by means of artificial neural networks. Jafari Nasr et al. [11] presented performance of helical wire coil as promoter in pipe using artificial neural networks. Hoskins and Himmelblau [12] described the characteristics of neural networks desirable for knowledge representation in engineering processes. Bhat and McAvoy [13] discussed the role of backpropagation for dynamic modeling and organized the methods of chemical process. Prakash Maran et al. [14] presented a comparative approach made between artificial neural network and response surface methodologies for the prediction of mass transfer parameters of osmotic dehydration of papaya. Kamble et al. [15] discussed the effect of fluidizing gas velocity on HT coefficient in immersed horizontal tube by utilizing feed forward network with backpropagation structure of Levenberg–Marquardt's learning rule. Jing Zhou et al. [16] proposed a model to detect and identify faults of K4b2 during exercise monitoring by backpropagation (BP) neural network.

3 Problem Statement

This paper described that the performance index in transport process such as mass transfer performance in tube flow by means of coaxially placed entry region Coil-Disc assembly promoter was inserted using artificial neural networks. The effects of the disc diameter, height of the disc, coil pitch and coil length on mass transfer enhancement was investigated from the experimental study. The data sets have been extracted from the experimental study and these sets were examined within the range of a geometrical parameters such as pitch of the coil $0.015 < Pc < 0.035$ m/turn; length of the coil $0.035 < Lc < 0.125$ m; diameter of the disc $0.02 < Dc < 0.045$ m; height of the disc $0 < Hd < 0.5$ m and Reynolds numbers are varied from 1200 to 14,500. The data of the experimental study is divided into two sets, one is training sets and second one is validation sets. The data of 248 sets were used in training and 106 sets for the validation of the ANN using MATLAB 7.7.0, particularly tool boxes to predict the performance index of the mass transfer enhancement in tube flow. In this mass transfer training, artificial neural networks consists of single input layer having five nodes and single output layer having single node is employed. Number of layers and nodes in hidden layer is determined through given input and it is taken as 1 and 2. For the single hidden layer case, the nodes are taken as 0–10. For two hidden layers case; in first hidden layer, five hidden nodes were taken as constant and the nodes in the second layer are varied as 1–10. The training of the network is utilized to minimize the error function with a learning rule. The generally used learning rule is gradient based such as the popular backpropagation algorithm. Neural network is trained by minimizing an error function using a learning rule such as the popular backpropagation algorithm.

4 Experimentation

In the literature, it was found that mass transfer using artificial neural networks, a most demanding method in solving the process intensification problems. Entry region coil-disc assembly insert promoter has been developed as one of the reactive augmentation techniques and is extensively used in mass transfer operations. The experimentation procedure is clearly described by Vaka Murali Mohan et al. [13]. Extensive varieties of practical applications are involved in the investigation of rate of mass transfer. A model was developed using multilayer perception neural network. It was found that mass transfer performance in tube flow with insert promoter of entry region coil-disc assembly by applying artificial neural networks has not been reported in the literature.

5   Mass Transfer

Mass transfer performance in tube flow has practical importance for the design and development of various unit operations. Entry region coil-disc assembly is an axially displaced promoter in tube, has been studied extensively for process intensification in mass transfer processes. The mass transfer dimension less function and its development was explained by Vaka Murali Mohan [13]. The geometric parameters of the turbulence promoters have a significant effect in mass transfer coefficients. For this case, the geometric dimensions are coil pitch (Pc), coil length (Lc), disc diameter (Dd), disc height (Hd) and Reynold's number are possible parameters that affect the kL values. Mass transfer coefficients were estimated using Eq. 1.

$$k_L = \frac{i_L}{n\,F\,A\,C_o} \tag{1}$$

Effect of Coil Pitch
Coil pitch is defined as gap between two consecutive turns. As the pitch increases, number of turns per unit length decreases which imparts swirl motion to the fluid leads to higher transfer coefficients. Plots are drawn for kL versus velocity with pitch as parameter and shown as Fig. 1. The figure reveals mass transfer coefficient increases with velocity. The values of the mass transfer coefficient are also found to increase up to a maximum value and then decrease with the increase in pitch. The exponents on velocity recorded marginal increase with decrease in coil pitch.
Effect of Coil Length
Coil length has an important influence as it changes axial flow of the fluid while passing through coil. This axial flow slowly changes into swirl flow, whereas the fluid passes through coil length. Maximum swirl was attained at the fluid leaving the coil, further increase records decaying swirl, finally the flow reaches to steady axial flow. All these transformations improves mass transfer coefficient by increasing turbulence. Effect of coil length is observed from Fig. 2. kL increases through increase in coil length because of induced swirl motion of the fluid. The increase continued to a maximum value at coil length of 0.095 m, beyond which a marginal decrease is observed. As the velocity of fluid increases, a marginal decrease of coefficients is noted with length. It may be attributed to a marginal decrease in velocity along the length of the coil because of drag resistance offered by the coil.

Fig. 1 Variation of k_L with velocity—Pitch of the coil as parameter

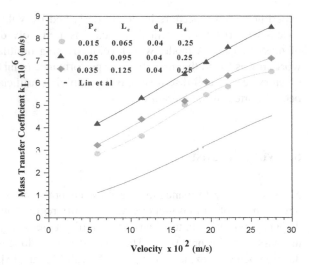

Fig. 2 Variation of k_L with velocity—length of the coil as parameter

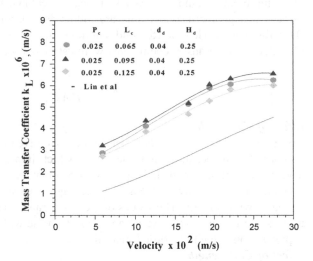

Effect of disc diameter

Disc diameter has strapping control on mass transfer. The coefficients of the mass transfer versus velocity are drawn for different disc diameters and shown as Fig. 3. The disc diameter increases mass transfer performance.

Effect of Disc Height

Disc height has strapping control on mass transfer. k_L versus velocity with disc height as parameter is drawn as Fig. 4. The plots reveal mass transfer increases with velocity. As the disc is moved away from the coil, there is scope for better swirl motion due to the coil and with good recirculatory flow may be achieved. From the above observation, one can say that there would be an optimum height for a particular set of geometric parameter combinations.

Fig. 3 Variation of k_L with velocity—disc diameter as parameter

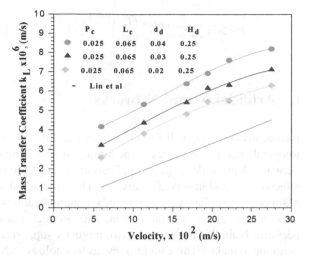

Fig. 4 Variation of k_L with velocity—height of the disc as parameter

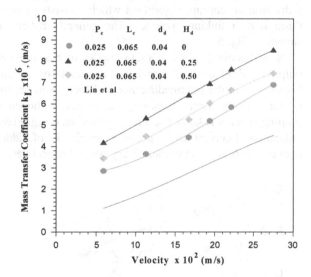

Mass Transfer function

Depending on flow conditions and entry region coil-disc assembly geometry, the mass transfer rate will increase. Flow of electrolyte during tube flow with entry region coil-disc assembly insert promoter with established momentum and concentration profiles. In turbulent core, intensity of eddies will be higher. By applying the wall similarity concept in the measurement of mass transfer coefficient in tube flow with entry region coil-disc assembly promoter in terms of dimensionless groups is presented by Vaka Murali Mohan [13] is

$$\bar{g} = 5.76[\text{Re}_m^+]^{-0.003} \left(\frac{P_c}{d}\right)^{0.003} \left(\frac{L_c}{d}\right)^{0.001} \left(\frac{d_d}{d}\right)^{0.62} \left(\frac{H_d}{d}\right)^{-0.00001} \tag{2}$$

6 Artificial Neural Networks

Neural networks are still far away from imitating the complex and difficult physiological and psychological functioning of the human nervous system. Neural networks have wide range of applications in actual world production troubles and effectively useful in several industries. The structure of a neural network bears only a superficial similarity to the brain's communications system. However, understanding the human neural system, and producing a computational system that performs brain like functions are two mutually support scientific researches that will continue to make significant progress as technology. AN artificial neural network is a division of learning algorithms which consists of several nodes that correspond through their linking synapses. The general artificial neural networks model is shown in Fig. 5.

Neural network contains a set of nodes such as input nodes receive input signals, output nodes give output signals and a potentially unlimited number of intermediate layers contain the intermediate nodes. Perceptrons are the most commonly utilized neural network forms, backpropagation and Kohonen self organizing map. This training process is known as unsupervised learning. Perceptron is a straightforward indication of current neural networks, exclusive of hidden layers. Backpropagation networks submit to an extraordinary type of neural networks that creates utilization

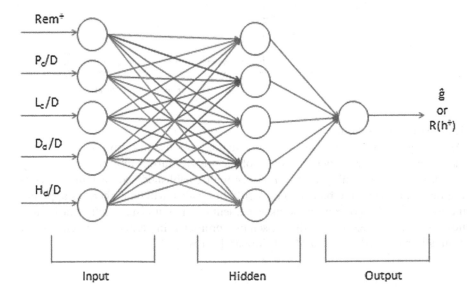

Fig. 5 Artificial neural networks model

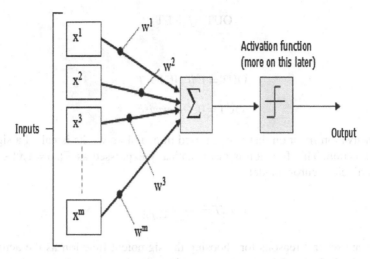

Fig. 6 Artificial neuron

of hidden layer. Backpropagation starts through random weightings scheduled its synapses. The network training is exposed to input values mutually through the right output values. The network weights are adjusted until its responses are more or less accurate.

Artificial Neurons

An artificial neuron is designed to replicate the basic functions of a biological neuron and shown in Fig. 6. The artificial neuron calculates the weighted sum of its applied inputs; each input represents the output of another neuron. The calculated sum is equivalent to the electrical potential of a biological neuron. The output is subsequently moved through an activation function, it determines whether or not it has exceeded the threshold value (T).

Here, each input to the neuron is tagged as $X_1, X_2, \ldots \ldots \ldots + X_n$ together they are referred to as the input vector X. Every input value is reproduce by its related weight $w_1, w_2, \ldots \ldots \ldots + w_n$. Weighted inputs are function of the summation block tagged as Σ. This summation block functions like a cell body, sum all the weighted inputs and produces an output called NET.

$$\text{NET} = x_1 w_1 + x_2 w_2 + \cdots + x_n w_n \tag{3}$$

This equation can be stated in a vector notation as follows:

$$\text{NET} = XW \tag{4}$$

The NET signal is then further processed by an activation function called f. This activation function can be as simple as a binary threshold unit, where,

$$OUT = f(NET) \tag{5}$$

And

$$OUT = 1, \textit{if } NET > T \tag{6}$$

$$OUT = 0, \textit{Otherwise} \tag{7}$$

The activation function that is often used in neural networks is called a sigmoid logistic function. This function is mathematically expressed as $F(x) = 1/(1 + e^{-x})$, thus in artificial neuron model,

$$OUT = \frac{1}{1 + e^{-NET}} \tag{8}$$

There are several reasons for choosing the sigmoidal function as the activation function. The S shape of this function provides suitable increase for a wide range of input levels. Besides its ability to provide automatic gain control, the squashing function also has other desirable properties; it is differentiable everywhere and its derivative are very simple to calculate and it is represented in terms of the function itself.

$$\frac{\partial OUT}{\partial NET} = OUT(1 - OUT) \tag{9}$$

Commonly used activation function is the hyperbolic tangent t function,

$$OUT = Tanh(x) \tag{10}$$

This function has a bipolar value for OUT, which is beneficial for certain network architectures. A multilayered network of these individual neurons is powerful enough to execute complicated programs in a strong manner.

7 Backpropagation (BP) Learning Algorithm

Well-organized approach used in several developments in industrial applications is "backpropagation algorithm". The BP algorithm estimated the changes in weight using "artificial neural networks", and a general approach is used as two time algorithm such as "learning rate" (LR) and a "momentum factor" (MF). The most important drawbacks of the two term backpropagation learning algorithm are used to estimate in real time system by local minima and slow meeting in real time applications. The multilayer neural networks by means of backpropagation learning algorithm are useful to compare the data of mass transfer function of the operational setting between various architectures of the ANN. The network training using "backpropagation" algorithm is developed mean square error connecting predicted and desired output. Backpropagation learning algorithm was utilized to predict mass transfer inside tube flow with insert promoter of entry region coil-disc assembly. Input and output pairs were offered the network, and adjustment of weights was adjusted to reduce error between output and actual value.

The aim of training a neural network is to update its weights through a set of inputs and outputs. It calculates suitable weight updates, mean squared error (MSE) and learning rate. A uniformly distributed initial weight set is produced within the range of $[-1, 1]$. The aim is to expose the relation among the statistical characteristics of input and learning, network parameters such as learning rate, momentum term. The parameters of network such as learning rate and momentum term were taken as 0.1 and 0.3, respectively, in all attempts.

8 Methodology

Backpropagation learning algorithm was utilized to test, train and normalize network data to envisage performance of mass and momentum transfer function. The main idea of this paper is to calculate performance index of the artificial neural networks model for the mass transfer in tube. The signal of input propagates through the network in an advanced way and "layer by layer". The major improvement is estimated several input/output and are easy to handle. The output is measured in the direction of the forward using ANN by means of "layer by layer". The first layer output layer is input of the next layer. The pass in the reverse mode are output neuron weights which are regulated since the objective of every neuron in output are adjusted by its connected weights.

"Artificial neural networks" is employed for envisage transfer rate of mass and momentum functions. The input layer consists of five input values, such as Re_m^+; P_c/d, L_c/d, D_d/d; H_d/d and output term in output layer for the mass transfer (g). A total of 354 experimental data sets are used within the range of "Reynolds number" varied from 1400 to 14,000 for mass transfer model development. Error is minimizing by varying the weights, biases and number of nodes in a hidden layer by means of output and current data. The design for converge the least error in ANN by identifying nodes and layers in the hidden layer.

9 Experimental Results

The performance mass transfer data is depending on the architecture of the ANNs. 20 types of ANN patterns are reviewed for calculating the mass transfer. The connected MSE, STD and R^2 by means of ANN patterns in the training and validation models were presented in the below mentioned Table. 248 data sets were used in training process and 106 data sets were used in the validation process. The

Fig. 7 Mass transfer training
graph for 5_8_1

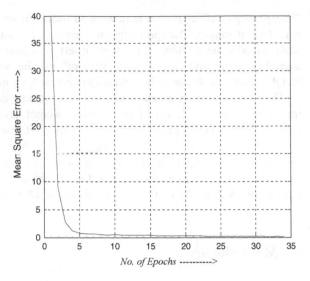

No. of Epochs ----------->

Fig. 8 Mass transfer
performance graph

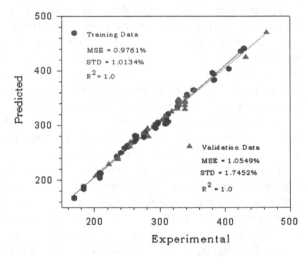

optimal ANN configuration is evaluated by predicting the results of the "training
and validation" process (Fig. 7).

The experimental work presented the Training and Validation results. It
demonstrated that SET 9 consist the configuration of 5_8_1 observed the least MSE
values and greatest accuracy compared with other networks and values are reported
in Table 1. This indicates that adding more or less number of layers and nodes in
hidden layer may not improve the efficiency of network. In this study of prediction,
ANN with configuration 5_8_1 is selected and recommended for training of the
mass transfer. It gives the least MSE = 1.0549 %, STD = 1.7452 % and R^2 = 1.

The efficiency of the mass transfer training function using "artificial neural
network" is represented in Fig. 8. The figure was drawn between experimental mass

transfer vs predicted mass transfer. It reveals the training and validation results. For the training data, MSE = 0.9761 %, standard deviation = 1.0134 %, R^2 = 1.0 and for the validation data MSE = 1.0549 %, standard deviation = 1.7452 %, R^2 = 1.0. These results show the good and significant performance of mass transfer function using backpropagation training and are recommended.

This paper reported that ANN configuration of 5_8_1 is recommended for mass transfer training for faster convergence and accuracy. This paper reveals that ANN with adding more number of layers and nodes in hidden layer may not increase mass transfer performance.

10 Conclusion

This paper presented the use of artificial neural networks to explain the mass transfer improvement in tube flow by means of coaxially placed entry region coil-disc assembly as insert promoter. The experimental data of the study is separated into two sets, one is training sets and second one is validation sets. The 248 sets of the experimental data were used in training and 106 sets for the validation of the artificial neural networks using MATLAB 7.7.0, particularly tool boxes to predict the performance index of the mass transfer in tube. Backpropagation learning algorithm was utilized to test, train and normalize network data to envisage performance of mass transfer function.

The methodology utilized, in artificial neural networks training of mass transfer functions through the network of single input layer with five nodes and single output layer of single node is employed. The number of layers and nodes in hidden layer is determined through the given input and it is taken as 1 and 2. For the single hidden layer case, the nodes are taken as 0–10. For the case of two hidden layers; in the first hidden layer, five hidden nodes were taken as constant and the nodes in the second layer are varied as 1–8. The weights were initialized within the range of [−1, 1]. The network limitations in all attempts taken, learning rate as 0.10 and momentum term as 0.30. For each experiment of the network, initial state was kept constant. Same conditions were utilized for each cycle of the input to hidden interconnection and of the hidden to output interconnection. The best model was selected based on the MSE, STD and R^2.

The results of artificial neural networks with its correlations were excellent fit through all the data. ANN model with 5_8_1 configuration is chosen for calculation of mass transfer function, with the smallest mean square error of 1.0549 %, standard deviation of 1.7452 % and R^2 is 1. The network with 5_8_1 configuration is recommended for mass transfer training for faster convergence and accuracy. This research work reveals that artificial neural networks with adding more number of the layers and nodes in hidden layer may not increase mass transfer performance.

11 Future Enhancements

The application of complex valued neural network approach is implemented for contingency analysis using the offline data for training purpose. This method can be extended for online application of the transport process. To reduce the training time, a future method of mutual information between the input and output variables is to be investigated. For larger application of the transport system having thousands of variables, input feature selection for the neural network plays an important role. As the size of the system increases the number of neurons increases, thereby increasing the training time. There is an urgent need for the future development of methods, procedures and software tools to deal with various contingencies, wide range of operating conditions. This would help in further research on accurate transfer capability computations of the transport process.

References

1. Jie Zhang., A.J. Morris "A Sequential Learning Approach for Single Hidden Layer Neural Networks" Neural Networks, Vol. 11, Issue 1, 1998, pp 65–80.
2. Stevan V. Odri., Dusan P. Petrovacki and Gordana A. Krstonosic, "Evolutional development of a multilevel neural network", Neural Networks, Vol. 6, Issue 4, 1993, pp 583–595.
3. Simon Haykin, "Neural Networks: A Comprehensive Foundation, 2nd edition" Prentice Hall PTR Upper Saddle River, NJ, USA, 1998, pages 842.
4. Leonard, J and Kramer, M. A, "Improvement of the back propagation algorithm for training ANN" Computers & Chemical Engineering, Vol. 14, Issue 3, 1990, pp 337–341.
5. Pollard, J.F., Broussard, M.R., Garrison, D.B and San, K.Y, "Process identification using neural networks" Computers & Chemical Engineering, Vol. 16, Issue 4, 1992, pp 253–270.
6. Gregory J. Zdaniuk, Louay M. Chamraa and D. Keith Waltersa, "Correlating heat transfer and friction in helically-finned tubes using Artificial Neural Networks" International Journal of Heat and Mass Transfer, Vol. 50, 2007, pp 4713–4723.
7. Vaka Murali Mohan, "Studies on mass & momentum transfer with coaxially placed entry region coil, disc, coil-disc assembly as turbulence promoter in circular conduit" Ph.D Thesis, Andhra University, Visakhapatnam, INDIA. 2008.
8. Rajendra Prasad, P, "Studies on ionic MT with coaxially placed spiral coils as turbulence promoter in homogenous flow fluidized beds" Ph.D Thesis, Andhra University, Visakhapatnam, INDIA. 1993.
9. Vaka Murali Mohan., P. R. Prasad., V. Sujatha & S. Sarveswarao "Flow pattern and model development for coaxially placed entry region coil-disc assembly as turbulence promoter in circular conduits" Expt. Thermal and Fluid Science, Vol. 32, Issue 8, 2008, pp 1748–1753.
10. Unal Akdag., M. K.Aydin., A. F.Ozguc, "Estimation of heat transfer in oscillating annular flow using artificial neural networks" Advances in Engineering Software, Vol. 40, Iss 9, 2009, pp 864–870.
11. M.R. Jafari Nasr., A.H Khalaj, S.H. Mozaffari, "Modeling of HT enhancement by wire coil inserts using artificial neural network analysis" Applied Thermal Engineering, Vol. 30, 2010, pp 143–151.
12. Hoskins, J. C and Him, D. M, "ANN models of knowledge representation in chemical engineering" Computers & Chemical Engineering, Vol.12, Iss9–10, 1988, pp 881–890.

13. Bhat, N and McAvoy, T. J, "Use of NN for dynamic modeling & control of chemical process systems" Computers & Chemical Engineering, Vol. 14, Iss 4–5, 1990, pp 573-582.
14. J. Prakash Maran., V. Sivakumar., K. Thirugnanasambandham., R. Sridhar "Artificial neural network and response surface methodology modeling in mass transfer parameters predictions during osmotic dehydration of Carica papaya L" Alexandria Engineering Journal, Volume 52, 2013, pp 507–516.
15. L.V. Kamble., D.R. Pangavhane., T.P. Singh "Experimental investigation of horizontal tube immersed in gas–solid fluidized bed of large particles using artificial neural network" International Journal of Heat and Mass Transfer, Volume 70, 2014, pp 719–724.
16. Jing Zhou., Aihuang Guo., Steven Sua "Fault detection and identification spanning multiple processes by integrating PCA with neural network" Applied Soft Computing, Vol. 14, 2014, pp 4–11.

Handling Wormhole Attacks in WSNs Using Location Based Approach

Swarna Mahesh Naidu and Vemuri Bhavani Himaja

Abstract Implementation of wireless indicator network is mainly in aggressive surroundings like army fight field, environment monitoring, atomic energy vegetation, focus on monitoring, seismic monitoring, fire and overflow recognition, etc., where continuous monitoring and real-time reaction are of innovator need. A wireless signal program has a preferable number of signal nodes that are connected mutually each distinctive easily. These signal nodes are used to sense and assess the heterogeneous factors, such as ecological stress, heat range, wetness, ground beauty products and heat, and therefore, it is very persevering to secure from the various attacks. Being restricted by sources are battery power or energy, memory potential and computational energy; these techniques are vulnerable against various types of inner and external attacks. One such challenge of attack is wormhole attack, where attackers create a postponement in between the two points in the program. In this paper, the recommended method discovers and furthermore prevents wormhole attack in wireless signal techniques. The recommended strategy uses stations information of nodes in program and uses Euclidean distance system to further recognize and restrain wormhole attack and make the connections at intervals signal nodes more secured and efficient.

Keywords Euclidean range system · Indicator node · Wireless sensor network · Wormhole strikes

S.M. Naidu (✉)
Department of CSE, JNTUACEK, Kalikiri, Andhra Pradesh, India
e-mail: mahesh.swarna1@live.com

V.B. Himaja
KL University, Vaddeswaram, Guntur, Andhra Pradesh, India
e-mail: himajavmr@gmail.com

© Springer Science+Business Media Singapore 2017
S.C. Satapathy et al. (eds.), *Proceedings of the First International Conference on Computational Intelligence and Informatics*, Advances in Intelligent Systems and Computing 507, DOI 10.1007/978-981-10-2471-9_5

1 Introduction

Efficient design and execution of wireless indicator systems has become a hot area of research in the past few decades, due to the vast potential of indicator systems to enable programs that connect the actual globe to the exclusive globe. By social media huge variety of tiny indicator nodes, it is possible to acquire data about actual phenomena that was difficult or difficult [1] to acquire in more traditional ways. In the future, as developments in micro-fabrication technology allow the cost of production indicator nodes to continue to drop, increasing deployments of wireless sensor networks are expected, with the systems gradually growing to huge variety of nodes (e.g., thousands). Implementation of wireless sensor system is mainly in aggressive surroundings like army battle field, environment monitoring, atomic energy vegetation, focus on monitoring, seismic monitoring, fire and flood recognition, etc., where continuous observance and real-time reaction are of discoverer requirement. Lack of normal information can be permitted by such programs, but they cannot tolerate the reduction of innumerable packages of critical event information. This kind of need makes the sensor nodes an important part of the network. Naturally, a radio sensor network is an interconnection among thousands, thousands or an incredible number of sensor nodes [2]. A sensor node is an embedded analogy that combines a location of microprocessor fundamentals onto a base hit processor. Whereas a sensor node having the efficient of detecting, information systems, interaction projects their limited memory potential, restricted battery power pack, less data transfer usage and less computational power makes the sensor network susceptible to many kinds of strikes (Fig. 1).

Wormhole strike is one of the denial-of-service strikes effective on the system part that can impact system redirecting, information gathering or amassing, and place-based wireless protection. The wormhole strike may be released by only one or a couple of working together nodes. In generally discovered two finished wormholes, one end overhears the packages and sends them over the pipe to the other close, where the packages are replayed to regional community. A number of techniques have been suggested for managing wormhole strike. Some techniques

Fig. 1 Wireless sensor
network application
framework

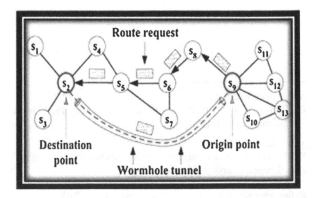

only identify the existence of wormhole in the system, while some techniques also focus on preventing or preventing the wormhole strike. Majority of the techniques provided oblige additional components back, depending on time synchronization, localization reference or make out be restrictive to consistent redirecting criteria. A strategy to preventing wormhole attack strike is provided in this document. No special components or time synchronization is needed for this method. Additionally, only self location is needed for the suggested key creation stage. The procedure utilizes depending on sending of Direction Response bundle based on the credibility of the two-hop next door neighbor sending it. The road is chosen for transmitting only if each node backwards path until the source node validates the two-hop emailer node. A wireless sensor network (WSN) is a route which includes an innovation of forecast nodes that are decidedly connected aside other. This little, less-priced, less-power, multipurpose indicator nodes can connect in quickly ranges. Each indicator node includes detecting, information systems, and interaction elements. A huge variety of this sensor nodes work together to form wireless indicator networks. A WSN usually includes 10's to thousands of a well-known node that connect through transmission programs for suspicion discussing and supportive handling. To maintain scalability and to grow the performance of the program operation, indicator nodes are regularly arranged directed toward groups. In the next section, the active of AODV redirecting approach is described in detail.

2 Related Work

Hu and Perrig [3] provided a strategy applying bundle leads, where in geographical lead and sensual lead place, higher limited on place of the recipient and highest possible time of a packet requires to travel, respectively. TIK method is suggested for protection against temporary lead, but the information of geographical place or limited time synchronism is needed. Taheri, Naderi and Barekatain used leashes strategy with customized packet transmitting technique to decrease computation expense of TIK method.

In transmitting time centered procedure (TTM), Tran, Installed and Lee bros [4] suggested an approach where each node on direction notices duration of delivering RREQ packet and getting RREP packet. Here, also time concern is the primary aspect. Singh and Vaisla customized this approach by eliminating the emailer and recipient from keeping demand and response packet timing. Various methods have been suggested by disparate writers to protect against wormhole strikes in wireless indicator networks. Buch et al. [5] suggested a strategy in accordance by the whole of the scrutinize of the probe of the sent and the obtained place of 'n' more packages by every node in system. To prohibit the harmful nodes changing the material of the packages, writers furthermore recommend a strategy to bring about an incredible key surrounded by the nodes and the base place that make out further be used in the ciphering functions. Hu et al. [6] suggested a procedure, known as bundle leads, whose objective is to limit the range journeyed by bundle system.

They explain the techniques to achieve those objectives; one is a space centered strategy, known as regional leashes which determine a high limited range that a bundle moves. At the time centered strategy known as temporary leashes, the delivering node involves in the bundle time at which it forward the bundle, ts; when getting bundle, the getting node analyzes those rate to time at which it obtained a bundle, tr. The recipient is thus able to identify if the bundle visited too far, in accordance with stated transmitting efforts and speed of light.

3 Background Approach

Let us consider a situation where wormhole end L keeps its identification revealed and node B receives Probe_Ack information as: D with 0, F with 0, H with 0 and L with 1. Here, when LK is derived considering the IDs of nodes D, F, H and L, its value does not coordinate with the Kmu value of node B, as node L is not the genuine two-hop next door neighbor and has not regarded while drawing Kmu. In common, for node N,

$$\sum_{i=1}^{N2hn} probe\_Ack\_Tag_i = 1$$

If,
then calculate

$$LK(N) = f(ID_i, 1 \leq i \leq N_{2hn})$$

Here, Kmu(N) <> LK(N), as the RREP bundle has came from a harmful node acting as a two-hop next door neighbor node. Accordingly, this situation is also frightened as an "Illegal Case".

A scenario with effective strike can be regarded where node F delivers Probe_Ack_Tag with 0 value, but it is modified to 1 by the harmful node C. In such scenario, the received Probe_Ack messages are: D with 0, F with 1 and H with 0 (Fig. 2).

This type of outcome discovers related between the produced LK value and Kmu value and so validates the two-hop next door neighbor emailer and delivers the RREP bundle. Actually, the scenario is an illegal scenario as maliciously [7] the tag value sent by one of the genuine node is modified on way. However, if the sent Probe_Ack concept by node F gets to node B for more than once via different tracks, then node B obtained different tag principles for the node F. Such scenario raises a query on the credibility of obtained tag principles and "Illegal Case" is frightened.

Fig. 2 Data transmission process over data series in application format

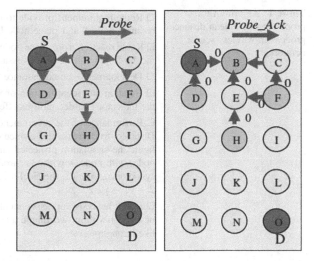

4 Proposed Methodology

The proposed technique to identify and avoid wormhole strike in Wi-fi indicator system is characterized. Suggested perform is adequate to identify and avoid wormhole strike and then interaction taken position in the seam of resource and location. The suggested plan performs for invisible and revealed to wormhole strike. The interaction in the seam of nodes begin when the resource node having to set up the interaction of the location node, it will problem the direction finding process. Resource node shows direction demand packages (RREQ) to all its available others who live nearby.

The advanced node that gets demand (RREQ) will examine the demand. If the advanced node is the location, it will respond with a direction response concept (Table 1).

(RREP). If the location node is not, resource will be sent to other next door neighbor nodes by the demanding. Since sending a bundle, each node will have to shop the transmitted identifier the past node a variety from which the demand came. Clock will be used by the advanced nodes to remove the access when no response is obtained for the demand. If there is a response, advanced nodes will liberate the transmitted identifier and the past nodes from where the response came from. The transmitted identifier, resource ID using identify the node has obtained a direction demand concept formerly. It stops repetitive demand get in dupe nodes.

By the way, AODV [8] redirecting technique performs and build interaction in the seam of nodes. When wormhole strike applied in the simulator, packages fall regularly and the wormhole strike is applied at the node. Here, a phrase "Detector" is presented some criteria which are being used to identify and avoid wormhole strike

Table 1 Algorithm for processing Euclidean distance process in wireless sensor networks

☐ Route requirement provide to all innovative nodes between resource node S and place node D
☐ Route reaction from place node D to Source node S to set up for the Route with less hops issues
☐ Development of next entrance neighbor systems at nodes
☐ Isolation and identification of wormhole attack nodes with the Indicator and decrease its effect
☐ Performance of the recommended requirements (Euclidean Distance) to decrease the chance of wormhole protect and to figure the simulation protected and effective. Thus, for two nodes with node A with synchronizes x1, y1 and node B with synchronizes x2, y2, the Euclidean variety is given by x1 − x2 2 + y1 − y2 2
☐ Result analysis and assessment which symbolizes the effect of with or without wormhole attack on the important statistics like throughput etc.

in wireless sensor network [9]. When sensor is applied and set real, the packages sending into the seam of nodes take position regularly with loss of the variety of packages falls. Also, Euclidean range system which gives the quickest distance direction between nodes will enhance bundle sending and create the transmitting of packages between nodes more protected and efficient. In Cartesian harmonizes, if p = p1, p2, ..., pn and q = q1, q2, ..., qn are two factors in Euclidean n-space, before the Euclidean distance from p to q, or from q to p is supposing by:

$$d(p,q) = d(p,q)\sqrt{(q1-p1)^2 + (q2-p2)^2 + \cdots + (qn-pn)^2} = \sqrt{\sum_{i=1}^{n}(qi-pi)^2}$$

Thus, for two nodes are mutual, node A by the whole of coordinates (x1, y1) and node B with coordinates (x2, y2), the Euclidean distance is given by x1 − x2 2 + y1 − y2 2. Locating wormhole attack nodes by providing particular location of neighbor nodes [10], the distance between them and with their x-position and y-position are this helps.

The important of the suggested plan can also be described by the cycle situation given below:

If (wormhole strike = = true)
{
If (Malicious node = = true)
{
Packets fall constantly
}
}
If (detector = = true)
{

Packets ahead regularly apply and contact Euclidean range formula
{
Packages ahead consistently with improved protection and stability
}
}

5 Experimental Results

Models are conducted in NS-2 system simulation set up on the OS Ie8 12.04 a Linux based system on wireless devices or desktop. The entire situation includes 52 nodes with two platform channels and two group leads, each group head for each group of nodes simulated in NS-2. The factors maintain for our simulation are described below.

Parameter	Value
Simulator	NS2
Simulation duration	90 s
Area	2500 m for processing
No. of nodes	50
Maximum segment node	512
Routing protocol	AODV
Interface processes or	802.11 Standard version

At time 35 a few moments, wormhole strike is applied at node 50, due to which packages fall occurs at it and no further transmitting [11] of packages from source to location occurs. Determine four given below reveals packages fall at node 50 due to wormhole strike (Fig. 3).

The factors using in our simulator are evaluated outcomes of wormhole-hitted situation [12], and then after applying suggested plan are throughput and packages missing (Fig. 4).

Throughput is determined as the common amount of effective concept distribution over an interaction route. The throughput is calculated in kilo pieces per second (kbps or kbit/s). Higher the value of throughput, indicates better the efficiency of the method [13].

Fig. 3 Packets drop in
wormhole attack in real-time
networks

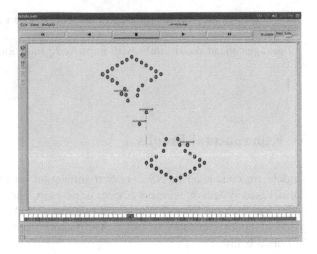

Fig. 4 Throughput analysis
with respect to time

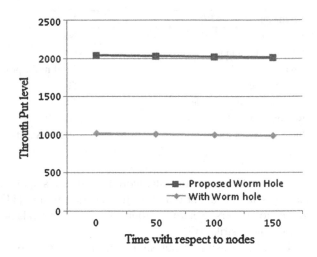

6 Conclusion

The wormhole strike is a significant problem that impacts the Wi-Fi indicator
system poorly. In this paper, the suggested stratagem has focused from one end to
the other the recognition and long lasting protection of wormhole lead in Wi-Fi
indicator system. The suggested technique is an effective and simple measure to
wormhole strike. The suggested perform clearly represents the effect of recognition
and protection using unique execution to separate wormhole and further prevent it
through Euclidean distance system. The analytics uses to confirm the suggested
perform are throughput and bundle lost over the system. And hence, the outcomes
through suggested perform are silent better in differentiate to outcomes in wormhole
strike.

References

1. "MOBIWORP: Mitigation of the wormhole attack in mobile multi-hop wireless networks," by I. Khalil, S. Bagchi, and N.B. Shroff, in *Elsevier Ad Hoc Networks*, vol. 6, no. 3, pp. 344 −62, 2008.
2. **"PREVENTION OF WORMHOLE ATTACK IN WIRELESS SENSOR NETWORK" by** Dhara Buch and Devesh Jinwala, in International Journal of Network Security & Its Applications (IJNSA), Vol.3, No.5, Sep 2011.
3. "DelPHI: Wormhole Detection Mechanism for Ad Hoc Wireless Networks", by Hon Sun Chiu King-Shan Lui, in *International Symposium on Wireless Pervasive Computing ISWPC*, 2006.
4. "Serloc: Secure range-independent localization for wireless sensor networks," by L. Lazos and R. Poovendran, in *Proceedings of the ACM Workshop on Wireless Security*, pp. 21−30, Oct. 2004.
5. "Detection of Wormhole Attacks in Wireless Sensor Network", by Dhara Buch and Devesh Jinwala in *Proc. of Int. Conference on Advances in Recent Technologies in Communication and Computing*, IEEE, 2011.
6. "Packet Leashes: A Defense against Wormhole Attacks in Wireless Networks", by Y. C. Hu, A. Perrig and D. B. Johnson, in *22nd Annual Joint Conference of the IEEE Computer and Communications Societies (INFOCOM)*, pp. 1976–1986, 2003.
7. "Using directional antennas to prevent wormhole attacks," L. Hu and D. Evans, in *Proceedings of Network and Distributed System Security Symposium*, pp. 131−41, Feb. 2004.
8. "SLAW: Secure Localization Against Wormhole Attacks Using Conflicting Sets", by Honglong Chen, WeiLou, Xice Sunand ZhiWang, in *Technical Report, The Hong Kong Polytechnic University*, pp. 1– 11, 2010.
9. "MOBIWORP: Mitigation of the Wormhole Attack" by Khalil, Saurabh Bagchiand Ness B. Shroff, in *Mobile Multihop Wireless Networks in Ad-Hoc Networks*, vol. 6, no. 3, pp. 344– 362, 2008.
10. "Detecting and Avoiding Wormhole Attacks in Wireless Ad Hoc Networks ", Na i t F a r i dNait-Abdesselam, Brahim Bensaou and TarikTaleb, in *proceeding of Wireless Communications and Networking Conference*, pp. 3117–3122, 2007.
11. "A Secure Localization Approach against Wormhole Attacks Using Distance Consistency" Honglong Chen, Wei Lou, XiceSunand and ZhiWang in *Hindawi Publishing Corporation EURASIP Journalon Wireless Communications and Networking*, vol. 2010, 11 pages, 2010.
12. "TTM: An Efficient Mechanism to Detect Wormhole Attacks in Wireless Ad-hoc Networks", Van Tran, Le Xuan Hung, Young-Koo Lee, Sungyoung Lee and Heejo Lee, in *4th IEEE conference on Consumer Communications and Networking Conference*, pp. 593–598, 2007.
13. "Visualization of wormholes in sensor networks," by W. Wang and B. Bhargava, in *WiSe 04, Proceedings of the 2004 ACM workshop on Wireless security*. ACM Press, pp. 51−60, 2004.

Video Authentication Using Watermark and Digital Signature—A Study

K.N. Sowmya and H.R. Chennamma

Abstract Widespread and easily available tools have become common for video synthesis and maneuvring in the digital era. It is therefore necessary, imperative and difficult as well to ensure the authenticity of video information. Authenticity and trustworthiness of the video is of paramount importance in a variety of areas like court of law, surveillance systems, journalism, advertisement, movie industry and medical world. Any malicious alteration or modification could affect the decisions taken based on these videos. Video authentication and tampering detection techniques due to intentional changes which are visible or invisible in a video are discussed in this paper.

Keywords Video forgery · Video watermarking · Video tampering · Blind detection · Digital signature · Video authentication · Active approach

1 Introduction

Nowadays information has a great role in our human society. Today, we can transmit information digitally over great distance in short time. This helps in connecting different societies, countries and cultures. Developments in digital technology help us to overcome many barriers in society but they also pose severe threats related to information. 'Wide usage or sharing of videos in social media like WhatsApp, Youtube, Facebook and news channels has a huge impact in our daily lives' [1]. Thus, credibility of digital information being exchanged is very

K.N. Sowmya (✉)
JSS Academy of Technical Education, Bangalore, Karnataka, India
e-mail: kn_sowmya@rediffmail.com

H.R. Chennamma
Sri Jayachamarajendra College of Engineering, Mysuru, Karnataka, India
e-mail: anuamruthesh@gmail.com

© Springer Science+Business Media Singapore 2017
S.C. Satapathy et al. (eds.), *Proceedings of the First International Conference on Computational Intelligence and Informatics*, Advances in Intelligent Systems and Computing 507, DOI 10.1007/978-981-10-2471-9_6

important. Video recordings which represent facts are used as proof in legal proceedings in the court of law and to establish content ownership. Many communication and compression techniques available, aid in sharing multimedia data such as videos, images, and audio, efficiently and feasibly. Low cost and sophisticated techniques have made digital manipulation of videos easier. Ensuring the integrity and legitimacy of video content is becoming challenging over time. To face these challenges a wide set of solutions have been proposed by researchers which rely on the fact that irreversible operations applied to a signal leave some traces that can be identified and classified to reconstruct the possible alterations that have been operated on the original source.

2 Video Authentication

Proving a given video is genuine, authentic and original has huge benefit in forensic science since it helps in establishing credibility of the video when used as electronic evidence. Video authentication requires a lot of attention due to the ease in manipulating the content of video without any visual clues. Authentication is the process or action of proving something to be true, genuine or valid [2]. Based on the objective of video authentication, the authentication system can be grouped into complete verification system and content verification system [3]. *Complete verification system* aid and ensure multimedia authentication of digital data. It considers the whole video data and do not allow any manipulations or transformations. Non manipulated data acts like messages and many existing message authentication techniques can be adopted to generate unique signature that can be used as a watermark or digital signature. In *Content verification system,* multimedia data is based on their content instead of their bit stream representations. Manipulations of the bit streams without changing or altering the meaning of content are considered as acceptable like with compression. Blind authentication approach is based on content verification system. It involves/adopts watermark and digital signature based upon the content of video data unknown or unseen by others and known only to the authenticator. In an active approach, the integrity of the data is verified by embedding an authentication signal on the content itself before the content is shared. Figure 1 shows the video authentication system in general. In this paper, we briefly discuss about the various active approaches adopted in video forensics until recently.

3 Watermark Approach

"Watermarking" is the process of hiding or embedding digital information in a carrier signal; the entrenched information should, but does not need to contain a relation to the carrier signal. Digital watermarks are used to verify the authenticity, integrity and identity of its owners. It is prominently used for tracing copyright

Fig. 1 General video
authentication system

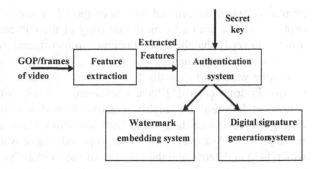

infringements in entertainment industry and for authentication in legal proceedings or court of law. Adopted watermark technique needs to be robust with respect to modifications which are unintentional and its presence should not degrade the quality of the video. One should not be able to remove or change easily.

Advantage of watermark approach in general is that they help in preserving the integrity of the multimedia data without degrading the quality. Change in the content will change the embedded watermark or affect it too which help in identifying the tampered video during authentication. Video watermarks can be classified based on factors such as their characteristics, domain as shown in Fig. 2.

3.1 Video Watermarking Based on Watermark Characteristics

Watermark in videos can be grouped into fragile and semi-fragile based on their characteristics.

Fragile Watermarking.

In fragile approach, inserted watermark is scarcely visible information that will be altered if there is any attempt to modify the video. Embedded information can be extracted to verify the authenticity of the video to find whether it is premeditated to

Fig. 2 Classification of video
watermarks

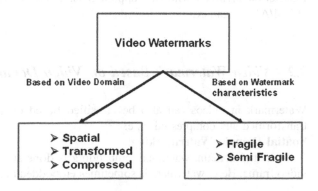

be malefic when the original data is tampered or altered. Major drawback of this approach is that if a bit by bit digital copy of the content is made then it becomes difficult to verify that the content belongs to its original owner/producers or not [4]. It is sensitive to recompression too.

Fragile watermarks usually fail to survive compressions resulting in high false alarms. Fridrich et al. [5] have considered a fragile watermarking approach for tampering detection. It is highly sensitive to modification that makes it difficult to distinguish malicious tampering from common video processing operation like recompression. Wang et al. [6] have proposed fragile watermarking for 3D models which help in determining the integrity of the model. Once the muddled watermark is generated with the help of secret key, chaotic system sequence is generated with the help of Chen-lee system which breaks the spatial continuity. It is then embedded into the uncompressed image based on steganography model. It is difficult to predict initial values and parameters chosen for generating secret key which is used for generating chaotic watermark. Blind approach adopted in their work does not require the original video for authentication.

Semi-Fragile Watermarking.

Semi-fragile watermarking is less sensitive to classical modification such as lossy compression, changing brightness, saving file in different file format with the assumption that these modifications do not affect the integrity of the video. Its main advantage is that it is sensitive to content changing alterations and robust against content preserving alterations [4]. High presence of false alarms during tampering detection is found and filters are required to reduce the false alarms. Main difficulties encountered are choice of the signature and the embedding technique [7]. Maeno et al. [8], Fei et al. [9], sang et al. [10] have all proposed semi-fragile watermarking to tolerate the common video processing operations such as recompression and at the same time detection of malicious tampering. Chen et al. [11] have also proposed a semi-fragile watermarking based on chaotic approach. Timing information of the video frames is modulated into the parameters of a chaotic system. Output of the chaotic system is used as a watermark and embedded into block based on DCT domains of video frames. Timing information for each frame is modulated into parameters of chaotic system. Mismatch between extracted and the observed timing information helps in determining temporal tampering. Restriction involved with this approach is, it cannot be applied directly to the H.264/AVC standards.

3.2   Video Watermark Based on Video Domain

Watermark in videos can also be classified based on their domain into spatial, transformed and compressed type.

Spatial Domain Watermarking.

In spatial domain, watermark embedding is done at the bit level of the original video frame. Here, watermark is sometimes embedded in the luminance value of the pixel. Raw data is available as image pixels. Watermark is applied for these pixels using color separation or conversion from RGB to YCbCr. Pixel manipulation techniques for watermark embedding like texture mapping, LSB approach resulting in low degradation of image quality with high perceptual transparency; additive watermarking can be grouped under this domain. Embedding watermark into spatial domain is easy to implement and less complex. It is fragile to image processing operations such as cropping, scaling and noise which results in lack of robustness.

Tanima et al. [12] have embedded the invisible watermark in the luminance component of the predictive p frames. The bipolar watermark sequence is embedded with the help of pseudorandom key which reduces the increase in bit rate and security is also improved. PSNR is used to evaluate the fidelity and visual quality. It is found that it is robust against temporal attacks like frame dropping, swapping, averaging, etc. Drawback of this approach is increase in block occurrence and more will help in better recovery of FDAS attacks but at the cost of decrease in randomness. Also, attacks like copy, collusion and re-encoding need to be considered.

Transformation Domain Watermarking.

In transformation domain, DCT, QDCT, DWT and DFT are the popular transformations which take advantage of the spectral coefficients of the human visual system.

DCT or discrete cosine transformations represent data in terms of frequency domain. DCT is carried out either at block level or global level based on the selection approach adopted to avoid redundancy. Though DCT's are strong against video pre-processing operations like blurring, contrast adjustment, etc., they are weak against geometrical attacks like tilting, cropping, etc. Jianfeng et al. [13] have proposed video watermarking based on DCT domain. Watermark is embedded during encoding of the video by quantization coefficients such as DCT and motion vector. Video quality may be affected. Advantage of their approach is that bit rate of video is not increased. It is easy to design and it is robust to various attacks. Kadam et al. [14] have proposed an approach which uses a 3D-DCT to obtain DC and AC values from all the blocks of the GOP for which the threshold probability is calculated and used as a step size in index modulation. Watermark is embedded using bit plane slicing method. Robustness is tested with the help of standard methods like PSNR &MSE, normalized correlation (NC) of extracted watermark and universal image quality index. Major advantage of their approach is that it can be adapted to both compressed and uncompressed domain though time required is more. Limitations include fail for geometric attacks like averaging, cropping tilting, etc. Feng et al. [15] have proposed zero watermarking technique. Watermark is embedded based on DCT quantization coefficients by intra frame prediction in real time. Features of the video media are used in creation of the watermark instead of modifying the original data or media. Intellectual property right (IPR) information database is a must to compare which at times may be target to interpretation attacks. Registration of the zero watermarks with the help of timestamp mechanism from the server is required to avoid such attacks. Tampering detection is done by matching the test video with that in the IPR database for the specified threshold. False alarm rate ratio is high which need to be handled.

Mehdi et al. [16] have proposed video authentication with the help of video codec where compression is considered for robustness. QDCT and quantization is done before embedding watermark with the help of pseudorandom number generator in the LNZ positions selected based on quantization parameters. Bit error rate (BER) is used to determine the original watermark for various errors. Spatiotemporal attacks are considered for tampering detection.

DWT divides or filters the information energy obtained into nonoverlapping multi resolution frequency bands denoted as LL, LH, HL and HH sub bands to indicate Approximation image, horizontal, vertical and diagonal components, respectively. Multiple scale wavelet decomposition can also be obtained further with the help of LL or approximation image. Watermarking application will not undergo more than four decomposition steps in general. DWT is applied on 1D, 2D and 3D images or GOP's based on application. Watermark is then embedded into the DWT quantized coefficients for each video frame. DWT based schemes are robust against noise addition in digital data. DWT schemes are more popular since they understand the human visual system better than DCT. Cox et al. [17] have proposed a secure embedding algorithm in which a watermark is constructed as a Gaussian random vector which is independent and identically distributed. It is inserted into spectral component of data which is visually most significant using spread spectrum technique. Chia-Hung et al. [18] have proposed an image authentication methodology by inserting digital watermark into the image by modifying the middle frequency component of the image. Bhargava et al. [19] have proposed wavelet based watermarking in which watermark is embedded in selected wavelet coefficient of luminance Y of the frame. The given video is first converted from RGB color space to YCbCr color space and then a 2D wavelet is applied with quantization based decomposition on the Y. Disadvantages of their approach is, it is time consuming and not all watermark can be detected in this approach. Prathik et al. [20] have proposed a watermarking technique based on discrete swarm optimization of binary particles. Initially, frame selection for embedding watermark among the GOP is done to achieve maximum PSNR. It is then converted from RGB to YCbCr domain. DWT applied to the frames generate four bands LL, LH, HL, HH, out of which other than LL remaining three bands are used for watermarking with the help of singular value decomposition to put in these bits. Their proposed scheme withstands noise addition.

Keta et al. [21] have proposed transformation algorithm for watermarking based on DCT-DWT techniques. EBCOT algorithm is used to store information and Huffman coding is used for encoding and decoding. Error correction codes like cyclic codes and hamming codes are used for error corrections. Importance of communication channel in the system design plays an important role with respect to security and robustness for the given application. Singh et al. [22] have considered or evaluated performance factors like robustness, security, etc., for various filters with the help of PSNR and normalized correlation(NC) for a specified scaling factor (k). As the 'k' value varies, quality of the image robustness is found varying and conflicting or contradicting each other which affect the applications. DFT transforms the continuous function into frequency components of sine and cosine. DFT is RST (rotation, translation and scaling) invariant. In the Fourier image, each point represents a particular frequency contained in the spatial image. The sampled transforms contain only frequencies sufficient to describe the spatial domain of the video frame/image and the resolution remains same. DFT's offer better results against geometric attacks even though computation cost is high and complex.

Compressed Domain Watermarking.

ISO and ITU-T standards like MPEG 1-4, H.264/AVC, respectively, remove the redundancies in spatial and temporal domain. They embed the information into variable length code by modifying the motion vector information.

Hartung et al. [23] have embedded watermark into compressed bit stream. Since they do not require encoding and decoding of watermark, it does not alter or effect video quality. Selection of bit pane image is controlled in a specific order. Drawback of their approach is the limitation with respect to watermark capacity and it is not so easy to design. A predetermined detector can recover the original watermark and its sequences. Liu Hong et al. [24] have designed a MPEG2 compression domain based algorithm. They use a gray scale image as a watermark and embeds it into the 'I' frames during video encoding based on MPEG2 compression standard. Watermark can be embedded into the original video sequence at two different circumstances according to them. It can be embedded into the original video before the first compression or the compressed video is decoded initially and then the watermark is embedded. Principle of the two situations remains the same. Video is regarded as a sequence of video pictures and watermark is embedded into these pictures which help in using the image watermarking algorithms. Limitations include increase in video bit rate due to embedding and video quality is also affected. Prashanth Swamy et al. [25] have used fingerprints of human as a watermarking vector which is converted into 3D form and the binary vector is generated and coded with LDPC syndrome coding technique. The generated watermark is then used in the SVD framework. Authentication is done by syndrome decoding the selected binary feature vector and the tampered one. It is a prerequisite for the correct syndrome to be available for syndrome decoding. Limitation in their approach is considering only fingerprint as a biometric trait and multimodal biometric feature can be used for effective authentication. Manoj Kumar et al. [26] have proposed an authentication scheme by adopting layered approach to video watermarking based on histogram, using sub image classification where watermark is distributed over a set of frames in the video. Robustness is verified using PSNR and MSE for temporal attacks like frame dropping, cropping, filtering, resolution modification, contrast and color enhancement. The invisible video watermark carries voluminous data by implementing a distributed secure watermark through a video file. Venugopal et al. [27] have considered temporal tampering attacks. Watermark is embedded into the quantization coefficients during encoding of the video in an uncompressed format. The blind video watermarking approach is evaluated with the help of PSNR and NC coefficients. Scene change is considered for embedding each of the bit plane obtained by converting the gray scale image. Quantization coefficients are compared for extracting watermark until all bits have been extracted, and thus robust against temporal attacks.

3.3 Limitations of Watermark Approach

Watermark is inserted precisely at the time of recording; which limits this approach to specially equipped digital cameras. Watermark based video authentication techniques only detect video forgeries in videos with watermarks, which is the minority of digital videos. Decision to embed watermark for high frequency or low frequency components is a trivial question and redundancy between frames of video also need to be considered. DCT based watermarks are not robust against geometrical transformations. Compression based watermark schemes are too sensitive and fragile.

4 Digital Signature Approach

The digital signature approach for video authentication is influenced by the signature creation environment which must correspond to international standards since they form evidence and their reliability in legal proceedings depend on their integrity or vulnerability. Digital signatures are combinations of zeros and ones based on a digital signature algorithm that serves the purpose of validation and authentication of electronic documents [28]. Validation refers to the process of certifying the contents of the document, while authentication refers to the process of certifying the sender of the document. Digital signature endorses the content of video increasing the confidence of the user. Change in the content of video due to intentional tampering would quash the digital signature. Digital signature remains distinctive for each video even though the originator/source remains the same. Most of the times, the digital signature thus generated can be embedded as watermark for authentication. The process of signature generation uses unique features to prevent forgery such that it is computationally infeasible to forge it, either by constructing a new video or by constructing a fraudulent digital signature for the tampered video. The process of signature generation is shown in Fig. 3. Feature selection varies from one video application to another depending on the domain area of usage. They need to be robust against compression, transformations and geometric translations. Features considered by various researchers for digital signature generation in video include edge [29, 30], color [31], transformations like in DCT [31–33], intensity histograms and geometric shapes and motion trajectory [34].

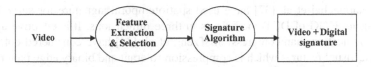

Fig. 3 Digital signature creation

Diffe and Hellman introduced digital signature in 1976 for the first time to verify the integrity of multimedia data. Ching-Yung et al. [31] have proposed digital signature based on the unique invariant properties, post editing and can identify malefic manipulations and compression. Two digital signature generators, Type 1 and Type 2 have been designed. Type-1 digital signature is used where the structure of video is unaltered after editing processes. Type 2 digital signature is used where the GOP is modified and only pixel values of pictures are preserved. DCT coefficients are used in Type 1 digital signature which is vulnerable since the value of DCT coefficients can be modified keeping their relationship unaltered. Queluz et al. [29] Ditmann et al. [30], have selected the edge/corner feature to generate digital signature. Ditmann et al. [30] have proposed content based digital signature for verifying the authenticity and integrity of video by visualizing video in the form of 3D cube and by considering edge characteristics with the help of canny edge detector. Their approach fails in detecting color manipulations when the edge characteristics remain same. When video is compressed edge features may be affected slightly affecting correctness of authentication. They are long and inconsistent at times since edge consistency is also a problem when the video is compressed.

Ramaswamy et al. [32] have designed hard authentication using cryptographic hash and digital signature via collected DC and 2 AC DCT coefficients from luminance frames in transform domain for group of pictures (GOP). A unique digital signature for every GOP helps in retransmitting only the GOP which fails authentication sequence. Computational complexity is more in their approach. Navajit et al. [33] have proposed signatures using coefficients of wavelet decomposition. Their approach helps to detect spatiotemporal tampering effectively since message authentication code is generated from group of picture frames using the transform coefficients and fixed blocks divided from the frames. Group of frames (GOF) of video is considered authentic if the distance between the original and the corresponding GOF at receiver side is below the predefined threshold else tampered. Po-chyi et al. [35] have proposed an authentication technique based on scalar/vector quantization of reliable features from the block of video frames. Authentication code thus generated is transmitted along with the video and authenticity is verified by matching the extracted feature with the transmitted code which is sensitive to malicious modification of video data. Xu et al. [36] have generated video signature using ordinal measure from average of DCT coefficients in low and middle frequencies.

Kroputaponchai et al. [37] have used spatiotemporal signature for surveillance system using HOG of DCT coefficients in three dimensions. It is effective against frame modification than compression. Tzeng et al. [34] have considered color and geometric visual features which is compression tolerant and binary edge patterns for nonoverlapping blocks. It is exposed to content modification if the attacker modifies the content keeping edges intact.

4.1 Limitations of Digital Signature Approach

Digital signature for video has its own limitations due to constraints involved in signature creation and verification process depending on applications. Techniques adopted for digital signature generation process are normally complex in nature and cannot be fitted and not feasible in surveillance cameras or CCTV's due to enormous computing and storage involved for real time data captured. In the pre-processing stage where inherent fingerprints of video is considered for signature generation efficient methods, need to be adopted for feature selection so as to check the integrity of video unlike using multiple approaches. Interception and intentional fraudulent manipulations during signature creation process or before the signature computation can compromise the authenticity of video. All countries do not follow uniform standards in judiciary for accepting video evidence and it is a matter of grave concern. Sometimes video data available from the source is unknown or from mobile cameras, with low resolution finding suitable feature vectors is a task. If the digital signature is replaced with a fraudulent one computed on the same video scene compromising of the proof of the original video happens and is deceived.

5 Conclusion and Future Scope

In this paper, we have given an overview of video authentication system. More focus is given on active approaches adopted for video authentication, i.e., watermark and digital signature. Each method comes with its own robustness and weakness. It has been noticed from our study that an exhaustive research has been carried out in the field of video authentication using watermark approach. But very few attempts have been made using digital signature for video authentication. Efficient signature scheme is today's need which can also be embedded as watermark for the given video. Signature generation which is robust to scale invariant features need to be considered. Standard dataset availability for evaluation purpose is an area which needs to be addressed by all researchers. Machine signature, metadata characteristics or inherent finger print of video—all need to be studied together in authenticating a video when its integrity is in question. Mixed mode biometric authentication where compound biometric features are considered together is an area of further interest which has immense scope.

References

1. Sowmya, K. N., and H. R. Chennamma. "A Survey On Video Forgery Detection." *International Journal of Computer Engineering and Applications* 9. (1503.00843(2015).
2. http://www.oxforddictionaries.com/definition/english/authentication.
3. Upadhyay, Saurabh, and Sanjay Kumar Singh. "Video Authentication-An Overview." *International Journal of Computer Science and Engineering Survey* 2.4 (2011).
4. Kaur, Manpreet, Sonika Jindal, and Sunny Behal. "A study of digital image watermarking." *Journal of Research in Engineering and Applied Sciences* 2, no. 2 (2012): 126–136.
5. Fridrich, Jiri, Miroslav Goljan, and Arnold C. Baldoza. "New fragile authentication watermark for images." In *Image Processing, 2000. Proceedings. 2000 International Conference on*, vol. 1, pp. 446–449. IEEE, 2000.
6. Wang, Jen Tse, Wang Hsai Yang, Peng Cheng Wang, and Yung Tseng Chang. "A Novel Chaos Sequence Based 3D Fragile Watermarking Scheme." In *Computer, Consumer and Control (IS3C), 2014 International Symposium on*, pp. 745–748. IEEE, 2014.
7. Mitrea, Mihai, and Marwen Hasnaoui. "Semi-fragile watermarking between theory and practice." *Proceedings of the Romanian Academy* (2013): 328–327.
8. Maeno, Kurato, Qibin Sun, Shih-Fu Chang, and Masayuki Suto. "New semi-fragile image authentication watermarking techniques using random bias and nonuniform quantization." *Multimedia, IEEE Transactions on* 8, no. 1 (2006): 32–45.
9. Fei, Chuhong, Deepa Kundur, and Raymond H. Kwong. "Analysis and design of secure watermark-based authentication systems." *Information Forensics and Security, IEEE Transactions on* 1, no. 1 (2006): 43–55.
10. Sang, Jun, and Mohammad S. Alam. "Fragility and robustness of binary-phase-only-filter-based fragile/semifragile digital image watermarking." *Instrumentation and Measurement, IEEE Transactions on* 57, no. 3 (2008): 595–606.
11. Chen, Siyue, and Henry Leung. "Chaotic watermarking for video authentication in surveillance applications." *Circuits and Systems for Video Technology, IEEE Transactions on* 18, no. 5 (2008): 704–709.
12. Dutta, Tanima, Arijit Sur, and Sukumar Nandi. "A robust compressed domain video watermarking in P-frames with controlled bit rate increase." In *Communications (NCC), 2013 National Conference on*, pp. 1–5. IEEE, 2013.
13. Jianfeng, Lu, Yang Zhenhua, Yang Fan, and Li Li. "A mpeg2 video watermarking algorithm based on dct domain." In *Digital Media and Digital Content Management (DMDCM), 2011 Workshop on*, pp. 194–197. IEEE, 2011.
14. Kadam, Bhakti D., and Shilpa P. Metkar. "Digital video watermarking based on dither modulation." In *India Conference (INDICON), 2014 Annual IEEE*, pp. 1–6. IEEE, 2014.
15. Feng, Gui, and Kai Huang. "H. 264 video standard based zero watermarking technology." In *Anti-Counterfeiting, Security and Identification (ASID), 2013 IEEE International Conference on*, pp. 1–4. IEEE, 2013.
16. Fallahpour, Mehdi, Shervin Shirmohammadi, Mehdi Semsarzadeh, and Jiying Zhao. "Tampering Detection in Compressed Digital Video Using Watermarking." *IEEE T. Instrumentation and Measurement* 63, no. 5 (2014): 1057–1072.
17. Cox, Ingemar J., Joe Kilian, F. Thomson Leighton, and Talal Shamoon. "Secure spread spectrum watermarking for multimedia." *Image Processing, IEEE Transactions on* 6, no. 12 (1997): 1673–1687.
18. LU, Chia-Hung, et al. "Image authentication method by combining digital signature and watermarking." *IJCSES* 1.2 (2007): 78.(Still doubt).
19. Mohanty, Saraju P., and Bharat K. Bhargava. "Invisible watermarking based on creation and robust insertion-extraction of image adaptive watermarks." *ACM Transactions on Multimedia Computing, Communications, and Applications (TOMCCAP)* 5, no. 2 (2008): 12.
20. Prathik, P., Rahul Krishna, Rahul Ajay Nafde, and K. Shreedarshan. "An Adaptive blind video watermarking technique based on SD-BPSO and DWT-SVD." In *Computer*

Communication and Informatics (ICCCI), 2013 International Conference on, pp. 1–6. IEEE, 2013.

21. Raval, Keta, and S. Zafar. "Digital Watermarking with copyright authentication for image communication." In *Intelligent Systems and Signal Processing (ISSP), 2013 International Conference on*, pp. 111–116. IEEE, 2013.

22. Singh, A. K., M. Dave, and Anand Mohan. "Performance comparison of wavelet filters against signal processing attacks." In *Image Information Processing (ICIIP), 2013 IEEE Second International Conference on*, pp. 695–698. IEEE, 2013.

23. Hartung, Frank, and Bernd Girod. "Digital watermarking of MPEG-2 coded video in the bitstream domain." In *Acoustics, Speech, and Signal Processing, 1997. ICASSP-97., 1997 IEEE International Conference on*, vol. 4, pp. 2621–2624. IEEE, 1997.

24. LIU, Hong, and Quan LIU. "Novel MPEG-2 Compression Domain-based Video Digital Watermarking Algorithm." *Journal of Wuhan University of Technology* 7 (2008): 034.

25. Swamy, Prashanth, M. Girish Chandra, and B. S. Adiga. "On incorporating biometric based watermark for HD video using SVD and error correction codes." In *Emerging Research Areas and 2013 International Conference on Microelectronics, Communications and Renewable Energy (AICERA/ICMiCR), 2013 Annual International Conference on*, pp. 1–6. IEEE, 2013.

26. Kumar, Manoj, and Arnold Hensman. "Robust digital video watermarking using reversible data hiding and visual cryptography." (2013): 54–54.

27. Venugopala, P. S., H. Sarojadevi, Niranjan N. Chiplunkar, and Vani Bhat. "Video Watermarking by Adjusting the Pixel Values and Using Scene Change Detection." In *Signal and Image Processing (ICSIP), 2014 Fifth International Conference on*, pp. 259–264. IEEE, 2014.

28. Subramanya, S. R., and Byung K. Yi. "Digital signatures." *Potentials, IEEE* 25.2 (2006):5–8.

29. Queluz, Maria Paula. "Towards robust, content based techniques for image authentication." *Multimedia Signal Processing, 1998 IEEE Second Workshop on*. IEEE, 1998.

30. Dittmann, Jana, Arnd Steinmetz, and Ralf Steinmetz. "Content-based digital signature for motion pictures authentication and content-fragile watermarking."*Multimedia Computing and Systems, 1999. IEEE International Conference on*. Vol. 2. IEEE, 1999.

31. Lin, Ching-Yung, and Shih-Fu Chang. "Issues and solutions for authenticating MPEG video." *Electronic Imaging'99*. International Society for Optics and Photonics, 1999.

32. Ramaswamy, Nandakishore, and K. R. Rao. "Video authentication for H. 264/AVC using digital signature standard and secure hash algorithm." In *Proceedings of the 2006 international workshop on Network and operating systems support for digital audio and video*, p. 21. ACM, 2006.

33. Navajit Saikia, Prabin K. Bora, Video Authentication Using Temporal Wavelet Transform, adcom, pp. 648–653, 15th International Conference on Advanced Computing and Communications (ADCOM 2007), 2007.

34. Tzeng, Chih-Hsuan, and Wen-Hsiang Tsai. "A new technique for authentication of image/video for multimedia applications." In *Proceedings of the 2001 workshop on Multimedia and security: new challenges*, pp. 23–26. ACM, 2001.

35. Po-chyi Su, Chun-chieh Chen and Hong Min Chang, Towards effective content authentication for digital videos by employing feature extraction and quantization, IEEE Transactions on Circuits and Systems for Video Technology Volume 19, Issue 5 (May 2009), p: 668–677, 2009.

36. Xu, Zhihua, Hefei Ling, Fuhao Zou, Zhengding Lu, Ping Li, and Tianjiang Wang. "Fast and robust video copy detection scheme using full DCT coefficients." In *Multimedia and Expo, 2009. ICME 2009. IEEE International Conference on*, pp. 434–437. IEEE, 2009.

37. Kroputaponchai, Teerasak, and Nikom Suvonvorn. "Video authentication using spatio-temporal signature for surveillance system." In *Computer Science and Software Engineering (JCSSE), 2015 12th International Joint Conference on*, pp. 24–29. IEEE, 2015.

A Comparative Study and Performance Analysis Using IRNSS and Hybrid Satellites

B. Kiran, N. Raghu and K.N. Manjunatha

Abstract This paper deals about the visibility position accuracies of four Indian Regional Navigation Satellite System (IRNSS). The position accuracies have been determined accurately after launching the fourth satellite with respect to standard time and it is observed by the receiver placed in School of Engineering and Technology, Jain University. The IRNSS satellite is independent of Global Position system (GPS). It is used for mapping, object tracking, navigation, surveying, and other services for users within the coverage area 1500 km.

Keywords Indian regional navigation satellite system · Global position system · Earth centered earth fixed · Indian space research organization

1 Introduction

In a communication system, the transmitter and receiver play a major role in data analysis and processing. In this connection the data flow from transmitter to receiver is taken care by satellites. From earlier days, GPS satellites also called as ground positioning system are used to provide a standard positioning service for users from respective base stations like navigation, tracking, mapping services, and other research activities [1].

In a GPS constellation, 24 Satellites and 6 spare Satellites are available in free space at different orbits to process, analyze, and transfer the data from one point of the world to another point [2]. Presently, Indian space research organization (ISRO)

B. Kiran (✉) · N. Raghu · K.N. Manjunatha
Electronics and Communication Engineering Department, School of Engineering
and Technology, Jain University, Ram Nagar, Bengaluru 562112, Karnataka, India
e-mail: b.kiran@jainuniversity.ac.in

N. Raghu
e-mail: n.raghu@jainuniversity.ac.in

K.N. Manjunatha
e-mail: kn.manjunatha@jainuniversity.ac.in

© Springer Science+Business Media Singapore 2017
S.C. Satapathy et al. (eds.), *Proceedings of the First International Conference on Computational Intelligence and Informatics*, Advances in Intelligent Systems and Computing 507, DOI 10.1007/978-981-10-2471-9_7

Fig. 1 IRNSS coverage area about 1500 km around the Indian land mass (image credit from ISRO)

has taken a step to develop a satellite-based navigation system called as Indian regional navigation satellite system (IRNSS) within Indian coverage area. IRNSS space segment consists of seven satellites [3]. In this, four satellites are available in free spaced to receive and transmit the Indian data within coverage area.

The four launched satellites are Geo synchronous satellites, they are IRNSS 1A, IRNSS 1B, IRNSS 1C, and IRNSS 1D [4]. These satellites are working in L and S band frequency, and which carry navigation payload and C-band ranging transponder used to determine the space craft range accurately [5]. Three IRNSS satellites are geo-stationary and four IRNSS satellites are geo-system and two satellites are spares in free space. The IRNSS user segment consists of the IRNSS receiver and user community. The IRNSS receiver converts IRNSS signal into position velocity and time estimates. In a tracking operation, minimum four satellites are required to find the exact position of a tracked object with an accuracy position [6]. The four satellites are required to validate for dimensions of x, y, z (position), and time arrival [7]. The IRNSS user receiver [8] is placed at SET-JU to track the IRNSS satellites, GPS satellites, and hybrid, which are available in free space. This receiver is placed at an altitude of $77.0°$ and longitude $12.130°$; it covers elevation angle $5°$ and azimuth angle $0°$ to $360°$.

The IRNSS satellite is used for tracking, mapping, navigation, and other services for users within the 1500 km coverage area Fig. 1. In GPS satellites the accuracy [9] of tracking is 20 m; when IRNSS satellites are used for analyzing the tracking accuracy [7] it is reduced to 13.04 m.

2 Methodology

The recorded data from processing unit are taken out for the analysis using a MATLAB simulator. The flow chart Fig. 2 shows the various data sets consisting coordinates, GDOP (Geometric dilution of precision), availability of number of satellites in three

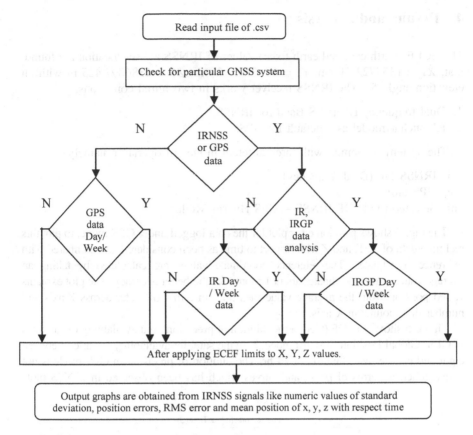

Fig. 2 Representation of the proposed design flow

different modes. Based on user's selection of a particular mode, the processed data give the minimum and maximum values with errors and plots of output as shown in results.

3 Math Modeling

The mathematical function is used in MATLAB to measure the parameters like standard deviation, RMS (Route Mean Square) error, and mean position errors for the X, Y, and Z coordinates in MATLAB and the average values of X, Y, and Z are calculated by considering the reference coordinates at the user position are used to calculate the position of X, Y, and Z. The mean and standard deviation of X, Y, and Z coordinates are computed using standard mathematical functions.

4   Results and Analysis

The ECEF (Earth centered earth fixed) values of IRNSS at user location are found, that, $X1 = 1353722.716$ m, $Y1 = 6076145.128$ m, $Z1 = 1386897.425$ m with an elevation angle 5°. The IRNSS receiver works in two initial conditions,

1. Dual frequency L5 and S Band for IRNSS.
2. Klobuchar model as a default for GPS.

The system is running with three different modes of operation namely

   i. IRNSS-DF (Dual frequency)
  ii. GPS only
 iii. Selected mode DF IRNSS + GPS Hybrid Mode

The Fig. 3 shows position error plot for the data logged under GPS mode, to plot this real time data of X, Y, and Z with respect to time as been considered and compared with reference coordinates. The reference coordinate values are calculated by taking an average value of the coordinates about 1 month for the better accuracy. The plot explains varied position from the average value measured in terms of meter across Y axis and number time counts in X axis. Fig. 4.

Figure 5 shows GDOP (Geometric dilution of precision) and availability of satellites in GPS (Global Position System) mode, here the satellite availability is varied between eight and twelve over duration of 24 hr. The dilution of precision in GPS mode is not varied much in terms of peaks and curves which has been observed in IRNSS plots

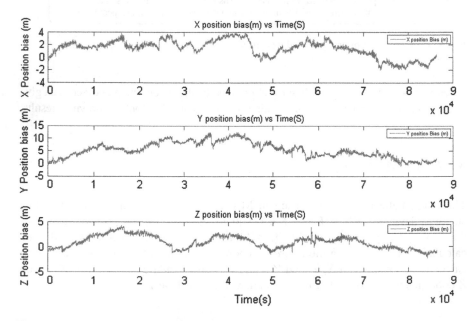

Fig. 3 X, Y, Z errors for GPS position solution

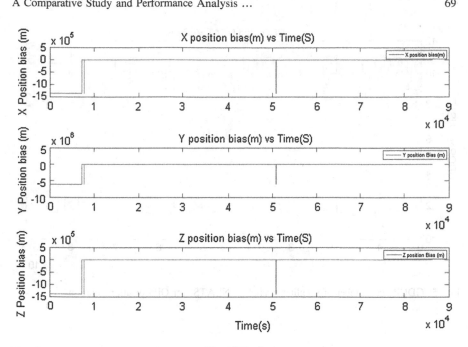

Fig. 4 X, Y, Z errors for IRNSS-2D dual frequency position solution

shown in Fig. 5. The Fig. 6 shows GDOP and NSATS for IRNSS-2D Dual Frequency Position Solution. The Fig. 7 shows the position error plots with an IRNSS and dual frequency hybrid mode of operations. The availability of satellites is extended up to 16 and 4 in hybrid and IRNSS mode respectively as plotted in Fig. 8.

4.1 Position Error with GPS Mode Only

4.2 Position Error with IRNNS Dual Frequency

4.3 Position Error for Third User Selection—DF IRNSS + GPS Hybrid Mode

The summarized comparison table of the research work covers many parameters like RMS error, Position error, Mean, and standard deviations of coordinates with

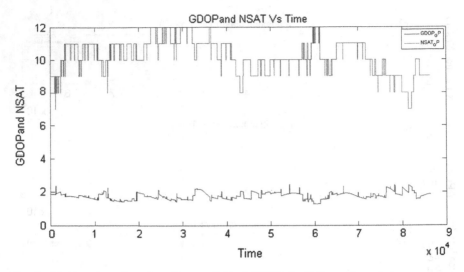

Fig. 5 GDOP and number of satellites available (NSATS) for GPS position solution

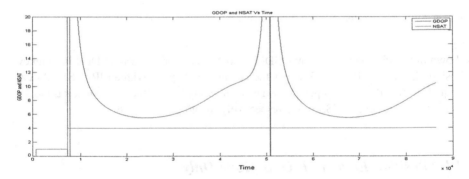

Fig. 6 GDOP and NSATS for IRNSS-2D Dual Frequency Position Solution

respect to time as shown below Table 1. The total number of processed samples is also mentioned in four and three satellites-based IRNSS operation. The result shows that there is an improved and better accuracy in error margin across all parameters. It is expected to reduce position error and other accuracies as the number of satellites increases in an orbit for the operation.

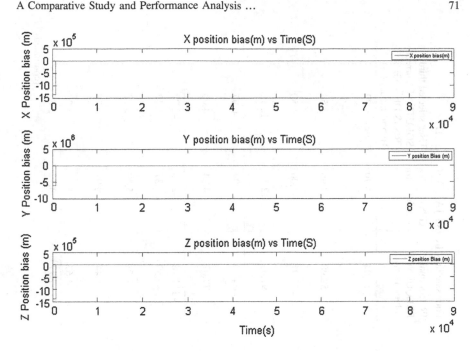

Fig. 7 X, Y, Z errors for IRNSS + GPS hybrid position solution

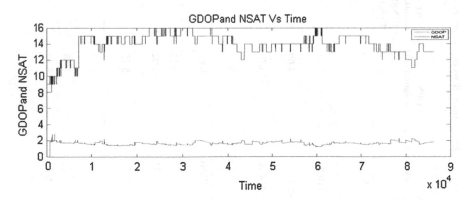

Fig. 8 GDOP and NSATS for IRNSS GPS + hybrid position solution

Table 1 Summarized comparison of three modes of week number 819

		Three satellites IRNSS 1A, 1B and 1C			Four satellites IRNSS 1A, 1B,1C and 1D		
SL no	Parameters of interest (m)	IRNSS dual frequency solution (m)	GPS solution (m)	User selected third solution (currently IRNSS DF + GPS Hybrid mode) (m)	IRNSS dual FREQUENCY solution (m)	GPS solution (m)	User selected third SOLUTION (currently IRNSS DF + GPS Hybrid mode) (m)
1	3D RMS position error	24.71	7.70	7.59	13.04	6.79	6.32
2	Max X error	510.82	7.01	6.87	52.98	3.96	3.54
3	Min X error	−500.82	−4.69	−4.31	−38.91	−2.00	−1.50
4	Max Y error	914.16	28.30	20.82	97.65	12.52	12.11
5	Min Y error	−399.444	−4.64	−3.96	−64.89	−0.94	−0.97
6	Max Z error	2258.10	8.82	8.83	79.07	4.24	4.03
7	Min Z error	−3764.58	−5.25	−4.59	−111.78	−2.25	−2.22
8	Mean X error	2.71	1.02	1.08	−0.59	1.37	1.12
9	Mean Y error	5.58	5.75	5.61	1.73	5.49	5.22
10	Mean Z error	−0.53	0.15	0.15	−1.44	0.91	0.77
11	SD X error	5.21	1.68	1.67	4.67	1.34	1.08
12	SD Y error	6.29	4.45	4.45	7.77	3.12	2.85

(continued)

Table 1 (continued)

SL no	Parameters of interest (m)	Three satellites IRNSS 1A, 1B and 1C			Four satellites IRNSS 1A, 1B,1C and 1D		
		IRNSS dual frequency solution (m)	GPS solution (m)	User selected third solution (currently IRNSS DF + GPS Hybrid mode) (m)	IRNSS dual FREQUENCY solution (m)	GPS solution (m)	User selected third SOLUTION (currently IRNSS DF + GPS Hybrid mode) (m)
13	SD Z error	22.47	1.60	1.49	9.07	1.29	1.21
14	Mean X position	1353725.43002918	1353723.744	1353723.802	1353722.123	1353724.092	1353723.837
15	Mean Y position	6076150.7138068	6076150.88	6076150.741	6076146.861	6076150.6	6076150.356
16	Mean Z position	1386896.88667261	1386897.580	1386897.576	1386895.984	1386898.344	1386898.204
17	Ref X position	1353722.716			1353722.716		
18	Ref Y position	6076145.128			6076145.128		
19	Ref Z position	1386897.425			1386897.425		

5   Conclusion

In this paper shows, the various position accuracies of GNSS system like IRNSS, GPS and hybrid are analyzed by using three and four satellites IRNSS constellation data with in fixed elevation angle of IRNSS receiver placed at school of engineering and technology, jain university and the reference coordinates are considered to evaluate the fixed position of X, Y, and Z with ECEF value. The various plots are obtained by evaluating the coordinates and the maximum, minimum errors, and RMS errors are calculated for all three and four GNSS systems.

Acknowledgment The authors would like to thank ISRO and Dr. Raju Garudachar for providing an opportunity to work and to publish this paper. This work is carried out at project laboratory department of Electrical, Electronics, and Communication Engineering, School of Engineering & Technology, Jain University.

The authors would like to thank all the faculty members of Department of Electrical, Electronics & Communication Engineering and management of Jain University for their continuous support and encouragement.

References

1. Karanpreet Kaur, "Indian Regional Navigation Satellite System (IRNSS)", scholar warrior, autumn 2013.
2. Rodrigo F. Leandro, Marcelo C. Santos, Richard B. Langley, "Analyzing GNSS data in precise point positioning software", GPS Solut, DOI 10.1007/s10291-010-0173-9.
3. Harshada Harde, Prof. M. R. Shahade, Diksha Badnore, "Indian Regional Navigation Satellite System", International Journal of Research in Science & Engineering (IJRISE), p-ISSN: 2394–8280.
4. O. Montenbruck, P. Steigenberger, "IRNSS Orbit Determination and Broadcast Ephemeris Assessment", ION International Technical Meeting (ITM), Dana Point, CA, 26–28 Jan 2015.
5. Global Indian Navigation Satellites, "Constellation studies", August ISRO-ISAC-IRNSS-TR-0887, 2009.
6. Babu R., Rethika T, and Rathnakara S. C, "Onboard Atomic Clock Frequency Offset for Indian Regional Navigation Satellite System", International Journal of Applied Physics and Mathematics, Vol. 2, No. 4, July 2012.
7. Ganeshan A. S., Rathnakara S. C, Gupta R., Jain A. K, "Indian Regional Navigation Satellite System (IRNSS) Concept", ISRO Satellite Center Journal of Spacecraft Technology, 15(2):19–23, 2005.
8. Primal majithiya, kriti khatri, J K Hota, "Indian regional navigation satellite system", Space application centre, Indian space research organization, Jan/Feb- 2011.
9. Vyasaraj Guru Rao, Gérard Lachapelle, Vijay Kumar S "Analysis of IRNSS over Indian Subcontinent B2" Position, Location And Navigation (PLAN) Group ION ITM 2011, Session B5, San Diego, CA, 24–26 January 2011.

Generalized Approach to Enhance the Shared Cache Performance in Multicore Platform

P. Pavan Kumar, Ch. Satyanarayana and A. Ananda Rao

Abstract Multicore processors have become an exemplar for high-performance computing. Designing a real-time scheduler that enhances throughput of the entire system in load scheduling in shared caches is the midpoint of the paper. Real-time scheduling schemes are appropriate for shared caches that these methods are aware of these above-mentioned issues. The existing methods work using simple memory sets; all the priorities are static and non-scalable data structure makes task priorities less flexible. Our work addresses the above issues in the soft real-time system, where task scheduling systems require functionally accurate result, but the timing constraints are softened. In order to exhibit this we use: SESC (Super Escalar Simulator) and CACTI (Cache Access Cycle Time Indicator) for investigating the efficiency of our approach on various multicore platforms.

Keywords Multicore processors · Shared cache performance · ENCAP · Re-scheduling · Load set

1 Introduction

A multicore system, which contains two or more processors on a single integrated circuit (IC) has been attached for enhanced throughput, reduced power consumption, and more powerful simultaneous processing of multiple tasks [1].

Designing an efficient scheduling method for managing available cache memory on multicore platforms is one of the main issues with which individual research

P. Pavan Kumar (✉) · A. Ananda Rao
Department of CSE, JNTU Anantapur, Anantapur, India
e-mail: paruchuripa1@gmail.com

A. Ananda Rao
e-mail: akepogu@gmail.com

Ch. Satyanarayana
Department of CSE, JNTU Kakinada, Kakinada, India
e-mail: chsatyanarayana@yahoo.com

© Springer Science+Business Media Singapore 2017
S.C. Satapathy et al. (eds.), *Proceedings of the First International Conference on Computational Intelligence and Informatics*, Advances in Intelligent Systems and Computing 507, DOI 10.1007/978-981-10-2471-9_8

scholars, research groups and chip vendors are currently tussling. This paper is about the issue of enhancing the shared cache memory in the context of soft real-time system, where task scheduling systems require functionally accurate result, but the timing constraints are softened [2]. In this work, we presume that load set τ is a set of many threaded tasks T_i, where i = 1, 2, ... n. are formulated into parallel threaded tasks (PTTs), where each PTT consists of consecutive loads, which will have various processing cost but a similar span. Each PTT has a stable storage region set (SRS) that is shared by all tasks within it. The SRS is the region of memory referenced by tasks within a PTT the size of this portion is called as the task size (TS) of the PTT. PTTs are useful for specifying groups of mutual tasks.

The idea. The idea emphasized here is inspired by the X. Xiang et al. Our focal point in this work is reassembling of discouraged tasks based on repeatedly scaling the simultaneous process, reassembling task sets with stable task size (TS) as parallel threaded tasks (PTTs) named as ENCAP (ENhancing CAche Performance) [3–7].

Other identical work. There are three related works on soft real-time multicore systems which we have an idea is one by Anderson et al. [5] which is also organized at performance based real-time applications. This paper addresses the issue of discouraging a group of task set from being co-scheduled. The second work by Calandrino [4, 18, 20–23] on shared cache memory, emphasizes offline cache profiling and heuristics set, which minimizes the run-time overheads and enhances the shared memory throughput. The third work, on trace-based program locality analysis and dynamic use, is identical to our idea [3] which has a scope, that how well the thread categorization performs when the loads are co-scheduled [2–4].

Brief contributions. we present a generalized approach for real-time work loads on multicore system, presuming a quantum-placed global periodic scheduling design. We show that our analytical method results in an enhanced shared cache utilization. We validate the accomplishment impact of our approach with simulations in SESC and CACTI 6.5.

Organization of the paper. The remaining sections of this paper are structured as follows: in Sect. 2, the background concepts to understand the multicore real-time scheduling are given, Sects. 3, 4 and 5 illustrate our generalized approach called ENCAP, the experimental evaluation phase including system configuration for our simulations on SESC and CACTI with experimental results, respectively. Finally, in Sect. 6 conclusions are drawn.

2 Background

Multicore architecture has the ability to support more than one processor and allocate tasks between them. These kinds of platforms is either loosely coupled or tightly coupled. In tightly coupled systems all the task's workload and the status information can retain the current at a low price. In contradiction, loosely coupled systems, termed as distributed systems, are very expensive because these kinds of systems need independent scheduler on different processors. Scheduling tasks in these platforms need an exceptional kind of system that is aware of timing constraint.

A real-time system is that kind of system which is attentive about the real-time constraint. The timing constraints are usually divided into two: soft as well as hard. A soft real-time system, where task scheduling systems require functionally accurate result, but the timing constraints are softened. In contrast, a hard real-time system imposes a hard deadline or time constraint.

Periodic load set. In view of this, we focus on the scheduling of a system of periodic load set (note that, end-to-end tasks in multiprocessor systems are determined as periodic loads.) $L_1, \ldots, L_n$ on M identical cores $C_1, \ldots, C_m$. Each load L_k is stated with processing time p_k and span s_k. The tth task of load L_k is indicated by L_k^t. Such a task L_k^t is ready for processing at its release time r_k^t and have to end the processing by $r_k^t + s_k$; or else it is tardy. The gap within r_k^t and r_k^{t+1} must satisfy $r_k^{t+1} \geq r_k^t + s_k$. A tardy load L_k^t may not amend r_k^{t+1}, but when L_k^t thorough its process then r_k^{t+1} allowed to thorough its processing. Formally the tardy of the load L at span s, indicated as tardy(L, s) is determined as follows from [8]:

$$tardy(L, s) = (L.p/L.s)t - assigned(L, s), \tag{1}$$

where assigned(L, s) is the amount of processor resource (time) allocated to L in [0, s). A schedule is Pfair if

$$(\forall L, s :: -1 < tardy(L, s) < 1). \tag{2}$$

The Pfair tardy constraints given in (2) have the effect of partitioning each load L into a series of unit-time sub-loads. The load utilization, p_k/s_k indicates the core own stake; the total $\Sigma_{k=1}^n \, p_k/s_k$ implies the entire utilization of the system. The system model we use for simulations is known as the unrelated processor model. In this model, each job having different speeds and processing times are unrelated but have the same size. For instance, the processing time of a data processing intensive job may be two seconds on core1 and seven seconds on core2.

Baruah et al. presented that pfair scheduler processing a periodic Job set τ on M identical cores iff

$$\Sigma_{L \in \tau} \frac{L.p}{L.s} \leq M. \tag{3}$$

We derive the k_{th} sub-load of load L as L_k, where K $\geq$ 1. Each sub-load L_k has a mock-release $r(L_k)$ and a mock-cutoff $c(L_k)$. If L_k is synchronous and periodic, then $r(L_k)$ and $c(L_k)$ are defined here.

$$r(L_k) = \lfloor \frac{k-1}{wt(L)} \rfloor \tag{4}$$

$$c(L_k) = \lceil \frac{k}{wt(L)} \rceil - 1 \tag{5}$$

From Eqs. (4) and (5), $r(L_k)$ is the initial space into which L_k could likely be scheduled, and $c(L_k)$ is the final such space. In an ER fair-scheduled approach, if L_k and L_{k-1} are part of the similar load, then L_k gets the eligibility immediately, when L_{k-1} is scheduled, so $r(L_k)$ is of less relevance [16, 17, 19]. We frequently specify to mock-cutoff and mock-releases as simply cutoffs and releases, respectively. The interval $[r(L_{k-1}), c(L_{k-1})]$ is called the frame of subtask L_k and is defined by $f(L_k)$. The length of frame $f(L_k)$, specified into $|f(L_k)|$, is defined as $c(L_k) - r(L_k) + 1$.

Subsequent-sporadic loads. The partition among successive tasks of a load is generally called as sporadic loads. The Subsequent-Sporadic (SS) load model is obtained from transmitting this a step more and letting partition among successive sub-loads of the identical load. The subsequent-sporadic (SS) load system is derived by the tuple (τ, q), where τ represents a set of loads, and q is a module that indicates when each sub-load first becomes qualified for processing. Each load may release either a definite or indefinite number of sub-loads. We presume that $p(L_k) \geq p(L_{k-1})$ for all $i \geq 2$, i.e., sub-load L_k cannot become qualified before its ancestor L_{k-1}. The terms, $r(L_k)$ and $c(L_k)$ are derived firstly by examining the alignment of the (SS)-frame of L in a periodic load system. In Subsequent-Sporadic (SS) load system, successive periodic-fair (PF) loads of L are either overlap by one span. The chunk $ch(L_k)$ is defined below.

$$ch(L_k) = \begin{cases} 0, & \text{if} \lfloor \frac{k}{wt(L)} \rfloor = \lceil \frac{k}{wt(L)} \rceil \\ 1, & \text{or else.} \end{cases}$$

From Eqs. (3) and (4) it is clear that in a periodic, synchronous task system release span of the next task in a load L is the cutoff of the current processing task, means an overlap by one slot $r(L_{k+1}) = c(L_k)$ from [8]. The overlap can be mitigated, if $ch(L_k) = 1$, and $r(r(L_{k+1})) = c(L_k) + 1$. For subsequent-sporadic loads, we define $c(L_k)$ exactly as above. Given this definition, we can define $ch(L_k)$ equal to above, which defines the initiation of the PF-frame of L_k. initiation of a task can be defined as:

$$r(L_k) = \begin{cases} p(L_k) & \text{if } k = 1 \\ max(p(L_k), c(L_k) + 1 - ch(L_{k-1})) & \text{if } k \geq 2 \end{cases} \tag{6}$$

cutoff $c(L_k)$ of load L_k is derived as $r(L_k) + |f(L_k)| - 1$. PF-frame length is equal to subsequent-sporadic loads, periodic and synchronous task systems. From Eqs. (3) and (4), we derive

$$|f(L_k)| = \lceil \frac{k}{wt(L)} \rceil - \lfloor \frac{k-1}{wt(L)} \rfloor, \tag{7}$$

and hence,

$$c(L_k) = r(L_k) + \lceil \frac{k}{wt(L)} \rceil - \lfloor \frac{k-1}{wt(L)} \rfloor - 1. \tag{8}$$

3 ENCAP

ENCAP approach is a generalized one which reschedules the loads which are neglected tasks from co-schedule. Bastoni et al. [9] presented an adaptive feedback-back framework to estimate the execution times and Calandrino et al. [2, 4] derived task re-weighting rules to encourage a load set to be co-scheduled, so that thrashing will be mitigated.

The effectiveness of scheduler. Subsequent-sporadic loads on M processors are considered to be effective at the span s if the subsequent guidelines are convinced.

G1 Initially, a load set should have been discouraged from co-scheduling and the size of the load set should not be less than the fixed storage region set (SRS) of re-partitioned load.
G2 More than one re-partitioned discouraged loads are not allowed to execute its job within a core.
G3 Time constraint of each sub-load should be the frame bounded with SS-frame span s.
G4 Tardiness of each discouraged loads should be bounded.
G5 All the partitioned discouraged load set will have a fixed storage region set (SRS) of parallel threaded tasks (PTTs).

If any discouraged load set convinced the above guidelines, then ENCAP allows to re-partition discouraged load set that should be re-scheduled based on EDF real-time scheduling algorithm.

Example Figure 1a illustrates with an example how J. Calandrino et al. discouraged the jobs N and O which are part of one PTT to co-schedule in a 3-core platform, even jobs L and M priority is lower than jobs N and O. Now with ENCAP we re-schedule jobs N and O (shown in Fig. 1b). ENCAP will have a significant performance impact not only in near terms but also in sub-sequential terms.

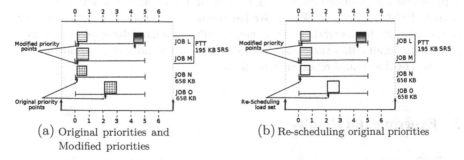

(a) Original priorities and Modified priorities

(b) Re-scheduling original priorities

Fig. 1 Original priorities, modified priorities and re-scheduling original priorities in Example-2

Theorem 1 (From *[10, 11]*) *The Load partition solution B is:*

$$m.B = m.(1 - B - B.In\frac{m-1}{B.m-1}) + In\frac{m-1}{B.m-1} \tag{9}$$

4 Experimental Setup

4.1 Tools Used

The SESC Simulator. To evaluate the efficacy of ENCAP, we have used the Super ESClar (SESC) simulator, which has a MIPS processor. It works as an event-driven emulator for multiprocessor environments [12]. We conducted our experiments on Intel sixth generation core I5 processor working at 3.5 GHz [13]. Every individual core incorporates isolated 32 k private data cache and instruction cache and also has a cooperative exclusive 256 k shared private cache. Four cores share a 6-MB on-chip shared cache and all caches run on 64-byte line size. Miss rate is taken from the results of our simulations and the miss penalty is taken as 12 clock cycles for private cache. Cycle time for private cache is 8.5 ns [12, 14].

The CACTI tool. Along with SESC we have used CACTI-6.5, to test our simulations, cache access time estimates. CACTI is an analytical model which integrates cache access time, cycle time, leakage area dynamic power and also it improves our simulations physical efficiency. SESC receives a configuration file, where CACTI-estimated cache access times are used as default attributes [12, 14, 15].

4.2 System Configuration

CACTI 6.5 deals with a config. file. The attributes (in configuration file) used in experiments are: a shared cache memory size of 14680064 bytes with each consisting of 64-byte cache block size. We have considered 90 nm chip technology for our simulations. In this work, we have intentionally considered the above-mentioned cache sizes because these are some of the shared cache memory sizes recommended by CACTI-6.5 technical report specifications [14, 15].

5 Experimental Results

In Fig. 2, x-axis represents core count on a single die and y-axis specifies the access and cycle time. We can visualize that the core count affects the access and cycle time. We have shown in our previous work, that if the no. of processor count increases on

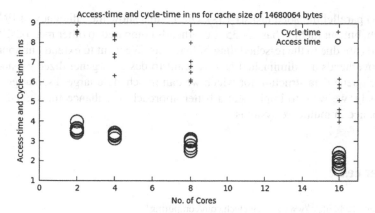

Fig. 2 Access time and cycle time of shared cache memory size of 14680064 bytes in ns

Table 1 L2 cache miss ratio per load set (Max., Min., Avg.) and cache size of 14680064 bytes in ns

Sr. no	Name of load set	Pfair	Megatask	ENCAP
1	NORMAL	89.35 % (1.65, 1.65, 1.65)	2.18 % (10.78, 10.74, 10.25)	15.36 % (3.85, 3.79, 3.53)
2	TINY_NORMAL	59.35 % (2.65, 1.65, 1.35)	2.88 % (11.68, 11.64, 11.66)	17.63 % (1.85, 1.81, 1.82)
3	ONE_COMBO	11.06 % (7.55, 4.65, 3.29)	0.88 % (6.8, 5.9, 5.9)	14.333 % (2.45, 2.41, 2.43)

a single die, then the cache access time increases, whereas the cycle time decreases (note that, even if the shared cache size is doubled, it has no effect on access time and cycle time).

The comparison of access time and cycle time of cache size 14680064 bytes is shown in Fig. 2 and this is due to ultra-speed bandwidth bus structure connected to both cache and the processor set [14] and the advancements in VLSI chip technology. Table 1 illustrates the comparison of ENCAP performance with other methods and approaches. Shared cache miss ratio per load set (Max., Min., Avg.) and we have considered the load set size of 14680064 bytes in ns for our simulations.

6 Conclusion

This paper presents a generalized approach on re-scheduling the discouraged load set to be co-scheduled, even load sets have eligibility priority as per G-EDF scheduling heuristics. Our approach (ENCAP) reorganizes the discouraged tasks based on repeatedly scaling the simultaneous process, reassembling task sets with as fixed task size (TS) parallel threaded tasks(PTTs). Our results show the efficiency of ENCAP approach, but our method has 15–35 % overheads compared to other methods. These overheads are due to the re-scheduling of load sets. We want to extend our approach so that overheads are diminished and we want to design a generalized expandable dynamic queue data structure for which we can re-schedule large discouraged load sets. Lastly, we want to implement a better approach to enhance the shared cache performance in multicore systems.

References

1. Extremetech, http://www.extremetech.com/computing/
2. Calandrino, J., Anderson, J.: On the Design and Implementation of a Cache-Aware Multi-core Real-Time Schedule. In: 21st Euromicro Conference on Real-Time Systems, pp. 194–204, IEEE, Londan (2009).
3. Xiang, X., Bin Bao, Chen Ding, Shen, K.: Cache Conscious Task Regrouping on Multicore Processors. In: 12th IEEE/ACM International Symposium on Cluster, Cloud and Grid Computing, pp. 603–611. IEEE, China (2012).
4. Calandrino, J., Anderson, J.: Cache-Aware Real-Time Scheduling on Multicore Platforms: Heuristics and a Case Study. In: 20th Euromicro Conference on Real-Time Systems, pp. 299–308, IEEE, USA (2008).
5. Anderson, J., Calandrino, J., Devi, U.: Real-Time Scheduling on Multicore Platforms. In: Euromicro Conference on Real-Time Systems, pp. 299–308, IEEE, Paris (2008).
6. Burchard, A., Liebeherr, J., Oh, Y., Sang, H. Son.: New Strategies for Assigning Real-Time Tasks to Multimocessor Systems. IEEE Transactions on computers, Vol. 44, No. 12, USA (1995).
7. Jia Xu.: Multiprocessor Scheduling of Processes with Release Times, Deadlines, Precedence, and Exclusion Relations. IEEE Transactions on computers, Vol. 19, No. 2, USA (1993).
8. Anderson, J., Anand Srinivasan.: Pfair Scheduling: Beyond Periodic Task Systems. In: 7th International Conference on Real-Time Computing Systems and Applications, pp. 297–306, IEEE, London (2000).
9. Block, A., Brandenburg, B., Anderson, J., Quint, S.: An Adaptive Framework for multiprocessor Real-Time Syatems. In: Euromicro Conference on Real-Time Systems (ECRTS), pp. 23–33, IEEE, China (2008).
10. Devi, U. PhD: Soft Real-Time Scheduling on Multiprocessors. Chalmers University of Technology, Goteborg, Sweden (2003).
11. Andersson, B., Abdelzaher, T., Jonsson, J.: Partitioned Aperiodic Scheduling on Multiprocessors. In: International Parallel and Distributed Processing Symposium. Nice, pp. 22–26, France (2003).
12. SESC, SESC Simulator. SESC is a microprocessor architectural simulator. http://iacoma.cs.uiuc.edu/~paulsack/sescdoc/.
13. INTEL, http://www.intel.com/pressroom/archive/releases/2010/20100531comp.html,0531comp.html
14. Ramasubramaniam, N., Srinivas, V.V., Pavan Kumar, P.: Understanding the Impact of Cache Performance on multicore Architectures. In: 4th International conference on Advances in Information Technology and Mobile Communication (AIM-2011), pp. 403–406, Springer-LNCS, Nagapur (2011).
15. CACTI6.0, CACTI-6.0 (Cache Access Cycle Time Indicator), http://www.cs.utah.edu/~rajeev/cacti6/
16. Bastoni, A., Brandenburg, B., Anderson, J.: An Empirical Comparison of Global, Partitioned, and Clustered Multiprocessor EDF Schedulers. In: 31st Real-Time Systems Symposium (RTSS), pp. 14–24, IEEE, Italy (2010).

17. J. Anderson, J., Anand Srinivasan.: Mixed Pfair/ERfair scheduling of asynchronous periodic tasks. J. of Comp. and Sys. Sciences, Vol. 68, No. 1, PP. 157–204, Springer, Heidelberg (2004).
18. UNC Real-Time Group. LITMUS RT project, http://www.cs.unc.edu/anderson/litmus-rt/.
19. AMD, http://www.amd.com
20. Andersson, B. Static-priority scheduling on multiprocessors. University of North Carolina, Chapel Hill, NC (2006).
21. Calandrino J. PhD: On The Design and Implementation of a Cache-Aware Soft Real-Time Scheduling for Multicore Platforms. University of North Carolina, Chapel Hill, NC (2009).
22. Calandrino, J., Leontyev, H., Block, A., Devi, U., Anderson, J.: *LITMUS*^{RT}: A testbed for empirically comparing real-time multiprocessor schedulers. In: 27th IEEE Real-Time Sys. Symp., pp. 111–123, China (2006).
23. UNC Real-Time Group. *LITMUS*^{RT} homepage. http://www.cs.unc.edu/anderson/LITMUSRT.

A Novel Framework for Road Traffic Data Management for Cloud-Based Storage

Md. Rafeeq, C. Sunil Kumar and N. Subhash Chandra

Abstract With the tremendous growth of population and the consequent road traffic increase, the demand for optimized traffic data collection and management framework is demanding extensive research. The collection of traffic data using multiple sensors and other capture devices have been addressed in multiple researches deploying the mechanism using geodetically static sensor agents. However, the sheath factors for the parallel research outcomes have significantly ignored the fact of data replication control during processing. This work proposes a novel framework for capturing and storing traffic data. During the multi-node traffic data analysis, controlling the replication in order to reduce the cost hasalso been a challenge. Recent research outcomes demonstrate the use of agent-based sensor networks to accumulate road traffic data. However, a multipurpose framework for accumulating and managing the traffic data is still a demand. The outcomes of this research is also to consider the most effective cloud-based storage for the traffic data with the knowledge of most popular cloud-based storage service providers. The accumulation of the data is also followed by the predictive system for road traffic data analysis. Hence in this work we also explore the use of standard machine learning techniques to identify the most suitable technique with performance consideration. Also this work proposes a performance evaluation matrix for comparing the traffic frameworks.

Keywords Road traffic management · Performance evaluation · Cloud storage · Framework components · Framework functions

Md.Rafeeq
CSE, CMR Technical Campus-Hyd, Secunderabad, India
e-mail: mdrafeeqcse@gmail.com
URL: http://www.cmrtc.acin

C. Sunil Kumar (✉)
ECM, SNIST, Hyderabad, India
e-mail: r9949043145@gmail.com
URL: http://www.snist.ac.in

N. Subhash Chandra
HITS Hyd, Keesara, India

© Springer Science+Business Media Singapore 2017
S.C. Satapathy et al. (eds.), *Proceedings of the First International Conference on Computational Intelligence and Informatics*, Advances in Intelligent Systems and Computing 507, DOI 10.1007/978-981-10-2471-9_9

1 Introduction

In the face of exponential growth of traffic and road links, it is the need of research to explore new directions for managing and predicting traffic situations in order to gain road safety, better traffic managements and finally gain higher productivity during peak hours by reducing the traffic burdens. Across the world ranging from city to urban the tremendous growth of traffic is leading towards a major problem. The highly populated cities around the world are facing the problem of better traffic management. Technologically advanced cities deploy agent-based multiple multimedia censor-based networks to collect and analyse the traffic data to provide better solutions for management and prediction of the road conditions [1]. Data collection methods include automatic and manual collections of high amount of data and then the analysis are often done in legacy system or manually. However, the collected data reach a high volume and became highly difficult to manage. Moreover, the predictive analysis also demands a high computational power to run the predictive analysis algorithms. The continuous efforts from parallel researches have demonstrated various monitoring agents and software management stacks deployed for accumulation and management of traffic data. The approaches are classified into two major categories as actively deployed agents into the network for accumulation with management and in contrast the passively deployed agents into the network for traffic data management only. Their merits and demerits are discussed in this work. This understanding will certainly help to propose a new novel framework for traffic data management [2]. The deployment of sensor based agents for accumulating the traffic data is also been proposed by multiple parallel research outcomes. The agents are deployed to collect timely data from the network as spatial and temporal. The conditional data collection from the network is also been proposed by various frameworks [3]. Generic approaches have proposed the collection of sensor-based data from the network based on Time Division Multiple Access or TDMA for collecting dynamic pattern data from the network and also the continuous data collection methodologies. In this work, we also analyse the data collection methods in order to obtain a better understanding of the network data flow for the prosed framework. In order to store and manage the high load data collected from the proposed framework, it is important to find the most suitable storage option for the big road traffic data [4]. Recent research progresses have demonstrated the benefits of using the cloud storage services for higher efficient data management. However, competitive studies have not been conducted in the recent researches to propose the most cost-effective storage solution and solution provider benefits. Here in this work we also study the cost and performance comparisons for various non-traditional data storage solutions from various cloud storage service providers such as Dropbox, Google Drive, High Tail, One Drive and Sugar Sync. This elevated understanding will help to define the most efficient framework for road traffic data management framework. This work also analyses the implementation abilities of machine learning for road traffic data prediction [5]. In this work, we analyse the predictive systems for traffic data analysis and understand the most suitable machine learning approach for the same. This understanding will certainly help to propose anovel framework for predictive system for avoidance of road traffics in the further continuation of this work. The rest of the paper is organized as follows: in Sect. 2, we understand the outcomes from parallel researches; in Sect. 3, we understand the generic components of the road traffic monitoring system; in Sect. 4, we propose a novel framework for road traffic data management framework with the components; in Sect. 5, we realize the results from the proposed framework and in Sect. 6 we present the conclusion of this work.

2 Outcomes of the Parallel Researches

In this part of the research we study the outcomes from the parallel researches conducted to achieve prediction systems based on the road traffic data.

Notable works by Porikli et al. have demonstrated a congestion-free network for traffic data transmission system for video and image-based data. The prediction system is based on the hidden Markov Model. The work defines five level of congestion control for the traffic data flow [6, 7]. Another benchmarking work by the Atikom et al. has demonstrated the effect of on-road vehicle velocity for the predictive model of road traffic density [6, 7]. The deployed model is also capable of estimating the congestions on the road for a defined vehicle velocity and has been widely tested. Different popular outcomes from the work of Krause et al. have demonstrated the prediction system based on the vehicle speed and density of the vehicles at a given instance. The system deploys fuzzy logic of first order to predict the road conditions focusing on traffic density. The use of fuzzy logic has also been analysed and deployed in the work of Jia et al. for predictive analysis of traffic data. However, the significant difference demonstrated in this work is the use of multiple approaches to predict the results [7]. The work demonstrates the use of manually interred logic and the use of machine learning based on neural networks. With the understanding of the above listed outcomes, we also understand the other parallel research outcomes based on data accumulation and management frameworks. Hence considering the situations, we identify the following problems to be addressed in the current traffic management situations: (1) management of agent-based censor network to implement a low-cost infrastructure and normalization of the data under pre-processing; (2) comparison and identification of most suitable cloud storage architecture for replication of data considering the low-cost Erasure models; (3) comparison and identification of most suitable artificial neural networks for processing the data in high speed to achieve reliable and timely solutions in the light of Elastic Cloud Computing properties; (4) identification of most suitable algorithm and proposal of a novel neural network solution for a timely and reliable predictive model for the road traffic analysis.

In this work we propose the relevant solutions to the problems as: (1) agent-based network implementation for multimedia road traffic data accumulation; (2) Cloud Storage Models considering the low-cost replication for high data reliability during processing at multiple nodes with the comparative study of cloud solutions; (3) artificial neural networks most suitable for road traffic data processing.

3 Components of Traffic Monitoring System

The general architecture for traditional road traffic prediction and management systems consists of multiple layer [8, 9] like the sensor layer, monitoring layer and server network layer (Fig. 1).

Here in this work, we understand the functionalities of each component:

A. **Sensor Network**: The sensor networks, generally supported by the wireless network for data communication in the form of wireless sensor network are deployed on any frame work for accumulation of data. The general-purpose sensor network demonstrates the workability without human efforts. The network also demonstrates the efficiency to work in any weather condition and also in various visibility conditions. The implementation is very much cost effective as this runs on the low power consumption. Also the integration of the sensor network is possible with video and other management systems. This includes inclusion and removal of any nodes dynamically in the network for timely network management. Also the data collectors in the same layer are programmable.

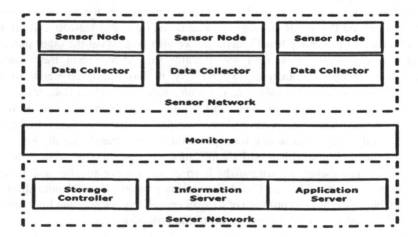

Fig. 1 Traditional road traffic prediction system

B. *Monitor*: The second most important component is the monitoring agents into the network. The management components are multi-functional. The functions of the agents generally include managing the connectivity of the sensor nodes, data transmission control, hardware control, bandwidth monitoring and also latency control.

C. *Server Network*: To analyse the traffic data collected from the sensor layer by the sensor network, the servers run the applications for data management and also the predictive system is deployed in the same servers. The generic process of predictive system is learning the pattern from the initial data and then processing the data for further prediction. Hence in this section we understand the generic architecture of any road traffic data accumulation and management framework. This understanding helps us in defining the proposed methodology in Sect. 4.

4 Proposed Road Traffic Data Management Framework

Henceforth with the detailed understanding of the generic traffic data monitoring and management framework, we propose the novel framework for road traffic data management control with replication control on cloud storage. In the proposed framework, we have considered the layer-based approach for better control and management of the agent-based components. The agents in the wireless network are single-function oriented, but the collective network is multipurpose. In this study, we propose the framework consisting of deployed network layer, monitoring layer, application management layer, storage layer and finally the server-based server layer (Fig. 2).

Henceforth, we demonstrate the components of all layers in the network here:

A. *Network Layer*: The first and onsite layer of this proposed framework is the network layer. The layer includes the agents required to collect the traffic data, monitor the data collection with connectivity and the significantly small amount of local storage. The deployed layer also includes the on-vehicle sensor agents. Here we understand the components of network layer.

- **Image Sensor Node**: The imaging agents installed on the site are connected to the network with high-performance capturing capabilities. The imaging agents are also capable of functioning in various weather conditions. Here we propose the optimal configuration considering the power and performance balance (Table 1). The capture devices must satisfy the same configuration scheme to reproduce the same performance.
- **Data Accumulator Agent**: The programmable data accumulator agents are configured to capture the image and video data. The optimizer components into these agents are configured to capture the video or image data while the motion in the object is detected. This optimizes the captured data in size.

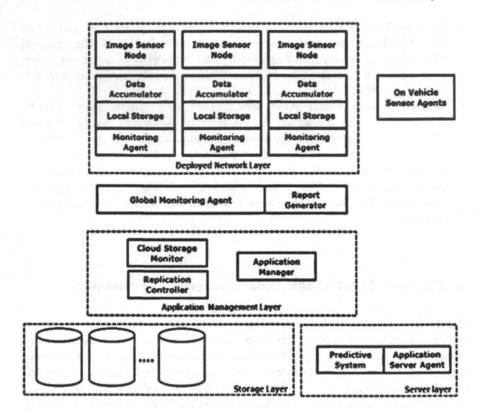

Fig. 2 Proposed framework for road traffic data management with replication control

Table 1 Image sensor node configuration

Parameter type	Proposed optimal value
Exposure responsiveness	1/120–1/160 s
Focal ratio	F/5 to F/4
ISO film speed rating	100
Lance focal length	18–20 mm
Resolution	300–350 DPI (horizontal and vertical)
Rotation and movement positioning	Co-sited
On-board compression mode	3
APEX brightness	6–7.5
Colour space	sRGB (recommended)
Digital zoom	5–8

The algorithm used for motion detection for multiple objects consider each object as collective vector object as O_t at any given time instance t. The collection of the detectable features based on the time instance are noted as $F_1, F_2 \ldots F_t$. The algorithm is deploying a probabilistic function to estimate the next state of the specified object denoted as $P(O_t|F_1, F_2, \ldots, F_t)$. The outcome of this probabilistic function is the next possible coordinates for the object under motion tracking. Further the probabilistic function at any given time is considered to function on a weighted and time-dependent sample set of video data. The sample set is defined as $\{S_t^{(v)}, v = 1, \ldots, V\}$ with the considered weight for corrected predictive location is $\pi_t^{(v)}$. The V parameter denotes the number of video data in the selected data set. Henceforth, we understand the relation between the learning features and the probabilistic prediction as:

- Firstly we consider the sample data set with V number of video data as

$$\left\{S_0^{(v)}, v = 1, \ldots, V\right\} \tag{1}$$

- The weight component for enhanced learning is considered as

$$\left\{\pi_0^{(v)}, v = 1, \ldots, V\right\} \tag{2}$$

- Hence, the final probabilistic function for next location determination is

$$P(O_t|F_1, F_2, \ldots, F_t) \tag{3}$$

- Henceforth we consider the change in enhanced learning factor as

$$P(O_t|O_{t-1}) \tag{4}$$

- The next step is to apply the Eq. 4 on Eq. 1 and obtain a new dataset for prediction and obtain $\left\{S_t^{(v)}\right\}$
- Hence the final step is to consider the updated feature tracking for iterative calculation of the motion as

$$\pi_t^{(v)} = \frac{P(F|S^{(v)})}{\sum\limits_{i=1}^{V} P(F_t|S^i)} \tag{5}$$

Table 2 Image sensor node configuration

Parameter type	Proposed optimal value
Technology	ZIP or JAZ or USB
Preferred mode of operation	No or less mechanical parts
Read speed	20–200 MBPS
Write speed	30–280 MBPS
Form factor	2.5–3.5
Storage capacity	5–12 GB
Supported storage format	JPEG

- **Local Storage**: The local storage container is designed to store data locally from the capture agents in case of loss of connectivity between the network layer and other layers. The local storage containers are desired to perform on-board compression and store small amount of data. However, the storage containers are expected to deliver low latency. Here we propose the factors to be considered for a local storage to balance the performance and cost implications (Table 2).

- **Monitoring and Connectivity Agent**: The monitoring and connectivity agents are deployed to monitor inter- and intra-connectivity of the network layer.

- **On-Vehicle Sensor Agents (Optional)**: The optional components of the network layer are the °F sensor components for better performance of this framework. However, these components are not compulsory in this proposed framework. Here we list the sensors:

  (a) Positional sensor for latitude and longitude information; (b) speed sensor for acceleration and deceleration information, (c) temperature sensor, (d) obstacle detection sensor.

B. *Monitoring Layer*: The monitoring layer consists of two major components, global monitoring agent and report generator. The global monitoring agent works same as the local monitoring agent with the enhanced performance to manage and detect the connectivity and data transmission rates. The report generator is a programmable software packet on the network to filter the data and generate report-based data for the queries generated from the server components.

C. *Application Management Layer*: The application management layer is equipped with three components such as cloud storage monitor, replication controller and application protocol manager for higher management and performance enhancement of the complete framework. The cloud storage monitor is the software agent to analyse and keep track of the data storage meter for cost and scalability control for the framework data. However, the framework data are also minimized in the previous layers by data accumulator and global monitoring agents.

D. *Storage Layer*: The storage layer is the third part stack of the storage solutions based on cloud service providers. This work also considers the selection parameters of a cloud storage solution for cost-effective storage. The details of

Table 3 System functionalities

Parameter type	Proposed optimal value
UI – 1	The system providers the feature to graphically provide information from the user interface
UI – 2	The system provides the user interface to manage the system view for complete observation
UI – 3	The system provides the visual interpretation of the system on click for every node in the traffic model
STAT – 1	The system provides the statistics related to the traffic monitoring
STAT – 2	The system also provides the statistics related to each vehicle registered into the framework
ST – 1	The system provides the in-detail view of the storage system overall estimation
PRED – 1	The system connects with the predictive analysis component of the framework
PRED – 2	The system manages all predictive query and reports
FUNC – 1	The performance of the system is at least equal or greater to the real-time data capture speed
FUNC – 2	The system is in sync always with the traffic lights deployed on the road
FUNC – 3	The system provides the on road vehicle detail statistics like travel time and all
FUNC – 4	The system is also capable of providing multi-lane road traffic prediction

the storage solutions comparative study are discussed in the later part of this work.

E. **Server Layer**: The server layer comprises two major components: application server agent and the predictive system for traffic prediction. Here we discuss the components:

- **Application Server Agent**: The application server agent is an application for analysing the traffic data for making the data ready for predictive system module.

The features of the system are also described in this work (Table 3).

5 Proposed Performance Evaluation Matrix

The evaluation of the road traffic monitoring systems is continuous. The outcomes from the parallel researches have demonstrated availability of the number of application and frameworks for the same purpose. The performance of the road traffic monitoring and management systems needs to be measured on a common platform as performance evaluation matrix; however, the availability of the performance evaluation matrix is still thrive for the present research trends. Hence in this work we also propose the performance evaluation matrix with various properties ranging from dash board or monitoring to I/O device performances (Table 4).

Table 4 Performance evaluation matrix

Type	Name	Description
	Name of node	Unique name of the node
	Type	Architecture type
Dash board parameters	Overall health indicator	Running, stopped, critical
	Last accessed	Last accessed date and time
(Overall monitoring)	Total availability	Time of total availability
	Memory utilization	Total local sensor memory utilization
	Disk utilization	Total local disk utilization in MB
	Network utilization	Total Network Utilization in time
	Active memory	Amount of active memory in KB
Memory parameters	Over heading memory	Amount of over heading memory in KB
	Swappable memory	Amount of swappable memory in KB
	Total shared memory	Amount of total shared memory
	Memory temperature	TEMPERATURE OF THE MEMORY UNITS
	Container name	Unique name of the storage container
Storage parameters	Container size	Container size in GB
	Container utilization	Container utilization in GB
	Network device ID	Unique id for the network devices
	Up time	Total up time
Networking parameters	Down time	Total down time
	IP address	Unique assigned IP address
	MAC address	Unique assigned MAC address
	Data transfer rate	Data transfer rate in megabytes per second
Peripheral parameters (I/O Device)	Device ID	Unique device ID
	Type	Read or write
	Read count	Number of read operations

6   Results

The results of simulation testing for the proposed framework under the NetSIM tool are satisfactory. The framework is tested on the parameters proposed in the performance evaluation matrix (Table 5).

The results demonstrates the most effective and sustainable nature of the framework. However, due to network congestion during data transmission, it has been observed that the parameter values seem not to be available for longer runs. Hence this issue will be addressed in the further research for this work.

Table 5 Performance evaluation matrix

Parameter observation	Availability of the properties		
	Test process—1 (duration—60 min)	Test process—2 (duration—90 min)	Test process—3 (duration—200 min)
Name of node	Available	Available	Available
Type	Available	Available	Available
Overall health indicator	Available	Available	Available
Last accessed	Available	No continuous availability	Available
Total availability	Available	Available	No continuous availability
Memory utilization	Available	Available	Available
Disk utilization	Available	Available	Available
Network utilization	Available	No continuous availability	No continuous availability
Active memory	Available	No continuous availability	No continuous availability
Over heading memory	Available	Available	No continuous availability
Swappable memory	Available	Available	Available
Total shared memory	Available	No continuous availability	No continuous availability
Memory temperature	Available	Available	Available
Container name	Available	Available	Available
Container size	Available	Available	Available
Container utilization	Available	Available	Available
Network device ID	Available	Available	Available
Up time	Available	Available	No continuous availability
Down time	Available	Available	No continuous availability
IP address	Available	Available	Available
MAC address	Available	Available	Available
Data transfer rate	Available	Available	Available
Device ID	Available	Available	Available
Type	Available	Available	Available
Read count	Available	Available	No continuous availability

7   Conclusion

These works consider the parallel research outcomes of recent time and find multiple approaches to build a scalable and sustainable framework for traffic monitoring system. Based on the parallel research outcomes, we understand the focus area of this research. We also understand the generic components of any road traffic management and monitoring system consisting of Sensor Nodes, Data Collector, and Monitoring agents, Storage Controller, Information Server and Application Server. Based on this understanding of the application framework, we propose a novel proposed framework consisting of Image Sensor Nodes, Data Accumulator, Local Storage, Local Monitoring Agent, Global Monitoring agent, Report generator, cloud storage monitor, replication controller, application manager, cloud storage solutions and Application server program. For the proposed framework, we also built the application program for monitoring. The functionalities are listed in this work. During the framework building process, we have also demonstrated the optimal data collection process for the proposed framework. This work also proposes the performance evaluation matrix for evaluating the performance of any generic road traffic management and monitoring system. For the storage of the traffic data, we propose the cloud storage solutions with Erasure replication control and the comparative study of major cloud storage specialist service providers are also been carried out in this work. With the consideration of the results, we understand the proposed framework is highly sustainable and reliable. However, the network data congestion demands further research. This work does not consider the predictive analysis of the traffic data and will be considering it for the future research on the same framework.

References

1. R. Yu and M. Abdel-Aty, "Utilizing support vector machine in real time crash risk evaluation," Accid. Anal. Prev., vol. 51, pp. 252–259, Mar. 2013.
2. R. Yu, M. Abdel-Aty, and M. Ahmed, "Bayesian random effect models incorporating real-time weather and traffic data to investigate mountainous freeway hazardous factors," Accid. Anal. Prev., vol. 50, pp. 371–376, Jan. 2013.
3. R. Yu and M. Abdel-Aty, "Utilizing support vector machine in real time crash risk evaluation," Accid. Anal. Prev., vol. 51, pp. 252–259, Mar. 2013.
4. C. Xu, A. Tarko, W. Wang, and P. Liu, "Predicting crash likelihood and severity on freeways with real-time loop detector data," Accid. Anal. Prev., vol. 57, pp. 30–39, Aug. 2013.
5. M. Fazeen, B. Gozick, R. Dantu, M. Bhukhiya and M. Gonzalez "Safe driving using mobile phones", IEEE Trans. Intell. Transp. Syst., vol. 13, no. 3, pp. 1462–1468 2012.
6. B. Hardjono, A. Wibowo, M. Rachmadi and W. Jatmiko "Mobile phones as traffic sensors with map matching and privacy considerations", Proc. Int. Symp. MHS, pp. 450–455 2012.
7. K. Jasper, S. Miller, C. Armstrong and G. Golembiewski National evaluation of the safetrip-21 initiative: California connected traveler test bed final evaluation report: Mobile Millennium, 2011.

8. W. Zhao and X. Tang, "Scheduling Data Collection with Dynamic Traffic Patterns in Wireless Sensor Networks," Proc. IEEE INFOCOM'11, pp. 286–290, Apr. 2011.
9. J. C. Herrera, D. B. Work, R. Herring, X. J. Ban, Q. Jacobson and A. M. Bayen "Evaluation of traffic data obtained via GPS-enabled mobile phones: The Mobile Century field experiment", Transp. Res. C, Emerging Technol., vol. 18, no. 4, pp. 568–583 2010.

Minimum Description Length (MDL) Based Graph Analytics

Sirisha Velampalli and V.R. Murthy Jonnalagedda

Abstract Graph is a way of representing data as a web of relationships. Graphs consist of nodes and edges where nodes are concepts and edges are the relationships. Graph analytics is the science of solving problems expressed on graphs. Graph analytics is useful for finding hidden patterns, relationships, similarities and anomalies in graphs. These tasks are useful in many application areas like protein analysis, fraud detection, health care, computer security, financial data analysis, etc. Minimum description length (MDL) comes from information theory, which can be used for universal coding or universal modeling. SUBstructure Discovery Using Examples (SUBDUE) algorithm uses MDL to discover substructures. In this paper, the use of MDL for graph analytics is shown by applying MDL encoding to various graph datasets and in particular graph matching is solved using MDL. Further, comparative analysis is done to show how MDL value changes w.r.t. varying graph properties. subgen tool is used to generate graph datasets. Statistical tests are applied and we came to know in which cases MDL value changes significantly.

Keywords Analytics · Graphs · MDL · Graph matching · SUBDUE

1 Introduction to MDL and Motivation

Nowadays the requirement for graph analytics is growing rapidly in diverse fields including large network systems, semantic search and knowledge discovery, natural language processing, cybersecurity, social networks, chemical compounds. Graph analytics provides deeper understanding of data. Relational analytics can explore "one-to-one" and "one-to-many", whereas by using graphs we can compare "many-to-many". Graph matching, finding frequent graph patterns, graph partitioning, graph

S. Velampalli (✉) · V.R. Murthy Jonnalagedda
Department of CSE, University College of Engineering, JNTUK, Kakinada 533003, India
e-mail: sirisha.velampalli@gmail.com

V.R. Murthy Jonnalagedda
e-mail: mjonnalagedda@gmail.com

© Springer Science+Business Media Singapore 2017
S.C. Satapathy et al. (eds.), *Proceedings of the First International Conference on Computational Intelligence and Informatics*, Advances in Intelligent Systems and Computing 507, DOI 10.1007/978-981-10-2471-9_10

coloring, graph-based anomaly detection, etc., are some of graph-related problems. MDL of a graph is defined in [1] as the number of bits needed to describe a graph completely. The concept of MDL [2] was first introduced by Jorma Rissanen in 1978. MDL principle says the relation between the regularity in data and compression of data. It states that whenever we are able to compress the data well, then it implies there is much regularity in data. In particular, MDL is developed and suited for [3] model selection problems. MDL can do prediction and estimation as well. MDL principle is used in [1, 4–6] for substructure discovery, decision tree induction, genetic programming and image processing. The concept of MDL is employed in many data mining tasks [7] including outlier detection, clustering, feature selection, discretization etc.

Section 2 explains minimum description encoding of graphs. Section 3 explains how MDL can be calculated, by taking a small example graph. Section 4 explains how graph matching problem can be solved using MDL. Section 5 shows how MDL value changes with respect to varying graph properties. Section 6 concludes the paper with future work.

2 Applying MDL to graphs

The following steps are given in [1] for MDL encoding of graphs.
First the entire graph should be represented using adjacency matrix, say A.
$A[i, j] = 1$, when there is edge between vertex i and vertex j.
$A[i, j] = 0$, when there is no edge between vertex i to vertex j.

Step 1: For encoding vertex labels, vbits are needed.

$$vbits = lgV + Vlgl_u \tag{1}$$

where
V is the number of vertices,
l_u is the number of unique labels.

Step 2: rbits are needed to encode the rows in the adjacency matrix.

$$rbits = lg(b + 1) + \sum_{i=1}^{V} lg(b + 1) + \sum_{i=1}^{V} lg\binom{V}{K_i} \tag{2}$$

$$= lg(b + 1) + Vlg(b + 1) + \sum_{i=1}^{V} lg\binom{V}{K_i} \tag{3}$$

$$= (V + 1)lg(b + 1) + \sum_{i=1}^{V} lg\binom{V}{K_i} \tag{4}$$

where
K_i is the number of 1s in the ith row
$b = max_i K_i$

Step 3: For encoding edges, ebits are needed.

$$ebits = e(1 + lgl_u) + (K + 1)lgm \qquad (5)$$

where
K is the number of 1s in adjacency matrix.
$m = max_{i,j}\, e(i,j)$ $e(i, j)$ = number of edges that are present in between i and j.

Step 4: Total number of bits for the entire graph

$$Total\ bits = vbits + rbits + ebits \qquad (6)$$

3 MDL-Sample Example

MDL encoding for graphs is explained in this section by taking a very small graph (Fig. 1).

First we need to write the adjacency matrix for the graph as:

$$\text{Adjacency matrix} = \begin{matrix} & A\ A \\ A \\ A \end{matrix}\begin{pmatrix} 0 & 1 \\ 0 & 0 \end{pmatrix}$$

Encoding of graph consists of the following steps:
Step 1: First the vertex bits are encoded as:

$$vbits = lgV + Vlgl_u$$

number of vertices, V = 2.
The number of unique labels, $l_u = 2$ (i.e. label A, label edge)

$$\therefore vbits = lg2 + 2lg2 = 1 + 2 = 3$$

Fig. 1 Sample graph

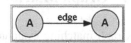

Step 2: Second we need to encode rows as:

$$rbits = (V+1)lg(b+1) + \sum_{i=1}^{V} lg\binom{V}{K_i}$$

b = maximum number of 1's in rows = 2
K_1 = number of 1s in first row = 1
K_2 = number of 1s in second row = 0

$$\therefore rbits = (2+1)lg(2+1) + \sum_{i=1}^{V} lg\binom{V}{K_i}$$

$$= 3lg2 + lg\binom{2}{K_1} + lg\binom{2}{K_2}$$

$$= 3lg2 + lg\binom{2}{1} + lg\binom{2}{0}$$

$$= 3(1) + 1 + 0 = 4$$

Step 3: Finally, we encode edges as:

$$ebits = e(1+lgl_u) + (K+1)lgm$$

e = number of edges in the graph = 1
K = number of 1s in adjacency matrix = 1
$m = max_{i,j}\, e(i,j)$ = maximum number of edges between 'i' and 'j' = 1

$$\therefore ebits = 1(1+lg2) + (1+1)lg1 = 1(1+1) + 2(0) = 2$$

Step 4: Total number of bits:

$$Total = vbits + rbits + ebits$$

$$\therefore Totalbits = 3+4+2 = 9$$

4 Graph Matching

Graph matching evaluates the similarity between two graphs. Graph matching [8]
can be exact and inexact. Graph matching has numerous application areas including
image processing, mobile computing, bioinformatics, multilevel graph algorithms,
etc. Nauty [9] is one of the popular algorithm for exact graph matching. Inexact

Fig. 2 Sample graphs (i, ii, iii)

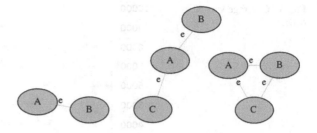

graph matching is also useful in certain domains. In molecular compounds, partial matching in substructures preserve important chemical properties. [8] used distance metrics to solve inexact graph matching problem.

4.1 Graph Matching—Our Approach

Graph matching through MDL is explained in this section using 3 sample graphs shown in Fig. 2. MDL values for graphs shown in Fig. 2 are 11 bits, 22 bits and 26 bits respectively.

MDL encoding difference between the first 2 graphs in Fig. 2 = 11 bits.
MDL encoding difference between second and third graphs in Fig. 2 = 4 bits.
MDl encoding difference between first and third graphs in Fig. 2 = 15 bits.

4.2 Interpretation

From the MDL values we can interpret that, whenever the difference between two graph encoding values is low, then we can match those graphs with lower transformation cost.

We can exactly match second and third graphs in Fig. 2 by adding only 1 edge (MDL difference = only 4 bits), but to match first and third graph we need to add 1 vertex and 2 edges (MDL difference = 15 bits).

5 Variation of MDL Value w.r.t. Varying Graph Properties

We used [10] subgen tool to generate artificial datasets with varying graph connectivity and graph coverage values and checked how the value of MDL changes w.r.t. graph properties and size. Connectivity [1] is the number of external connections on each instance of the substructure. Coverage [1] is the percentage of the final graph to

Fig. 3 Coverage versus MDL

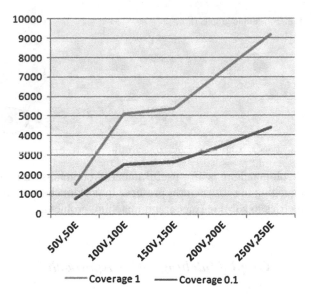

Fig. 4 Connectivity versus MDL

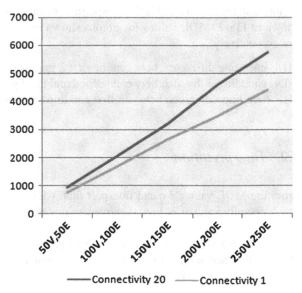

be covered by the instances of the substructure. MDL value w.r.t. increasing graph size and coverage value is shown in Fig. 3. MDL value w.r.t. increasing graph size and connectivity value is shown in Fig. 4. The x-axis shows varying graph(V,E) sizes with an increment of 100 nodes. Y-axis shows the corresponding MDL values with varying coverage value in Fig. 3 and varying connectivity value in Fig. 4.

5.1 t-Test Statistics

We performed t-test [11] to know how the value of MDL changes w.r.t. increase in coverage value. t value is 0.1687 and p value is 0.8702.

We performed t-test to know how the value of MDL changes w.r.t. increase in connectivity value. t value is 3.2446 and p value is 0.0315.

5.2 Interpretation

From the obtained t and p values, we interpreted that there is no significant difference in MDL value w.r.t. increase in coverage value and we also interpreted that there is significant difference in MDL value w.r.t. change in connectivity value.

6 Conclusion

In this paper minimum description length (MDL) is applied to graphs and the graph matching problem is solved through MDL. We studied how the value of MDL changes w.r.t. varying graph properties. Further, we want to solve more graph problems like anomaly detection, clustering, classification using MDL.

References

1. D.J. Cook, L.B. Holder, D.J. Cook, and S. Djoko "Substructure discovery in the SUBDUE system.," in Proc. of the AI Workshop on Knowledge Discovery in Databases, pages 169–180, (1994)
2. Minimum description length, https://en.wikipedia.org/wiki/Minimum_description_length
3. Minimum Description Length (MDL), http://www.modelselection.org/mdl/
4. Quinlan, J.R., Rivest, R.L., Inferring decision trees using the minimum description length principle, Information and Computation, 227–248, (1989)
5. Hitoshi. I.B.A., H.DE and S.TAISUKE, Genetic Programming using a Minimum Description Length Principle. Advances in Genetic Programming, (1994)
6. Leclerc. Y.G. Constructing simple stable descriptions for image partitioning. International Journal of Computer Vision, 73–102, (1989)
7. Matthijs van Leeuwen and Jilles Vreeken, Mining and Using Sets of Patterns through Compression, In: Charu C. Aggarwal. Jiawei Han, Frequent Pattern Mining-Springer US (2014)
8. Kaspar Riesen, Xiaoyi Jiang, Horst Bunke, Exact and Inexact Graph Matching: Methodology and Applications, In: Charu C. Aggarwal, Haixun Wang (eds.)-Managing and Mining Graph Data- Springer US (2010)
9. The Stony Brook Algorithm Repository, http://www3.cs.stonybrook.edu/~algorith/implement/nauty/implement.shtml

10. C. Noble and D. Cook, Graph-Based Anomaly Detection. Proceedings of the 9th ACM
    SIGKDD International Conference on Knowledge Discovery and Data Mining., 631–636,
    (2003)
11. The T-Test, http://www.socialresearchmethods.net/kb/stat_t.php

Implementation of Intrusion Detection System Using Artificial Bee Colony with Correlation-Based Feature Selection

K. Kanaka Vardhini and T. Sitamahalakshmi

Abstract In any organization, information can be transmitted over the Internet. Maintaining confidentiality of the data during transmission plays a vital role. Many existing approaches like firewalls, antivirus, encryption and decryption techniques are available to provide security. But still these approaches suffer due to the sophisticated nature of the attackers. So we are moving towards swarm intelligence approaches to build intrusion detection system. In this paper, we use a swarm intelligence approach namely the artificial bee colony to implement classifier. It will generate classification rules to detect the intruder. For this KDD dataset was used. Before classification rule generation subsets were generated depending on the correlation existing between attributes and target class label to reduce complexity. The results show that ABC effectively identified the different types of attacks compared to the existing ones.

Keywords Data transmission · Intrusion detection system · Classifier · Artificial bee colony · KDD cup 1999

1 Introduction

In communication over the Internet, different services, namely authentication, confidentiality, integrity has to be provided. Intrusion detection system, will be installed to provide confidentiality during data transmission. It performs analysis of data traffic and identifies unauthorized activity if it exists. Firewalls are the best example for this. It will identify different attacks which are discussed below either at the host or in the network.

K. Kanaka Vardhini (✉)
Department of CSE, ASCET, Gudur, India
e-mail: kanakavardhini@gmail.com

T. Sitamahalakshmi
Department of CSE, GITAM University, Visakhapatnam, India
e-mail: tsm@gitam.edu

© Springer Science+Business Media Singapore 2017   107
S.C. Satapathy et al. (eds.), *Proceedings of the First International Conference on Computational Intelligence and Informatics*, Advances in Intelligent Systems and Computing 507, DOI 10.1007/978-981-10-2471-9_11

DOS attack: In this attacker tries to make the server unnecessarily busy. The attacker will send a huge number of sync packets to server and server becomes unnecessarily busy to respond. So that it cannot provide services to the authorized people.

Probe: This is one type of passive attack, in which the attacker listens to the channel and performs the traffic analysis to find the vulnerabilities. The scanning can happen either at the network devices or during the transmission.

User to login: This is one type of active attack. In this attacker is authorized person, but trying to access data having access control. He does not have privileges to access the data.

Remote to login: This is also one type of active attack. In this the attacker enter into the network by pretending to be an authorized person.

In general, the intrusion detection system (IDS) was used to avoid or to detect above discussed types of attacks. By analyzing the traffic over the network the IDS will detect the attack and gives the alarm. Some of the IDS not only detect it, but also try to prevent the attacks by performing action like shutting down, logging-off, etc. [1]. In this paper, we used swarm intelligence approach to implement IDS.

2 Literature Review

In 2012, Mohammad Sazzadul Hoque et al. developed intrusion detection system by applying genetic algorithm [2]. They used standard deviation to measure fitness of the chromosomes and proved that the proposed one effectively detecting the attacks.

In 2013, G.V. Nadiammai, M. Hemalatha [3] proposed EDADT algorithm for generating classification rules to detect attacks. The results proved that the proposed algorithm detects attacks more effectively compared to existing algorithms.

In 2014, by using visualization strategy Luo et al. developed a novel intrusion detection system [4]. They evaluated the 5-class classification problem with the proposed algorithm and obtained high accuracy in detecting the intruder.

In 2014, Chen et al. proposed a parallel and scalable compressed model [5]. They also analyzed its application on intrusion detection. They used KDD99 and CMDC2012 datasets to demonstrate the efficiency of the proposed model.

In 2010, Karegowda et al., implemented classifier along with the genetic algorithms and they used medical datasets for the experiments [6].

In 2011, Lavanya and Usha Rani generated the classification rules on cancer dataset with decision tree algorithm [7]. They proved that selection of relevant features before classification rule generation increased the efficiency of classifier.

3   Methodology

From the literature survey, it could be observed that most of the current intrusion detection systems suffer from high false alarms rates. The Artificial Bee Colony (ABC) optimization is applied to reduce the false alarm considerably. KDD cup 99 was used as an input dataset. Before applying ABC, authors have followed a typical feature selection process to reduce number of attributes for getting efficient results. The proposed approach is as follows:

In this paper we have taken KDD dataset as input for IDS. The artificial bee colony was used to generate classification rules to detect the different types of attacks. Before generating the classification rules, subsets were generated using the CFS method to get more effective results. After getting subsets some of the attributes have string data. So we converted the string data into numerical data for calculating the fitness of solution and called it as data mapping. Later authors applied an artificial bee colony on these subsets to generate classification rules for IDS. The whole methodology is shown in Fig. 1.

3.1   Description of KDD99 Dataset

In this KDD cup-99 dataset was used as input [8]. In 1998, MIT Lincoln labs delivered this dataset which describes different types of attacks namely smurf, satan, Pod, backtrap, imap, buffer overflow, perl, root kit, load module, etc. KDD dataset discuss about 24 types of attacks which are categorized as denial of service, remote to login, probe, user to login attacks depending on their nature.

Every record of the KDD cup 99 consists of 41 attributes along with the class label. In this some of the attributes have categorical data and some of them numerical data. Table 1 shows all the 41 attributes along with their data type.

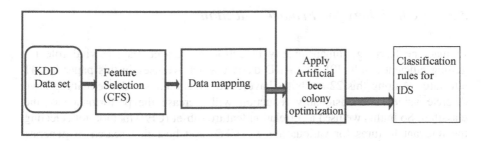

Fig. 1 Architecture of the proposed IDS

Table 1 Features of KDD

Feature name	Type
duration	Continuous
protocol_type	Discrete
service	Discrete
src_bytes	Continuous
dst_bytes	Continuous
flag	Discrete
land	Discrete
wrong_fragment	Continuous
hot	Continuous
num_failed_logins	Continuous
logged_in	Discrete
num_compromised	Continuous
root_shell	Discrete
su_attempted	Discrete
num_root	Continuous
num_file_creations	Continuous
num_shells	Continuous
num_access_files	Continuous
num_outbound_cmds	Continuous
is_hot_login	Discrete
is_guest_login	Discrete
count	Continuous
serror_rate	Continuous
rerror_rate	Continuous
same_srv_rate	Continuous
diff_srv_rate	Continuous
srv_count	Continuous
srv_serror_rate	Continuous
srv_rerror_rate	Continuous

3.2 Methodology for Feature Selection

In machine learning, dimensionality reduction of dataset plays a vital role for getting efficient results [9]. The KDD dataset which we used in this paper has 41 attributes. Among this 22 attributes have continuous data and the remaining have discrete values. Dimensionality reduction will increase the performance of the classifier. So in this we used a correlation feature subset (CFS) method for selecting the relevant features for various attacks. CFS used best first search to generate subsets. The subsets are generated depending on the correlation existing among attribute/target classes and the correlation between attributes [10, 11]. Equation 1 is used to generate the subsets.

Table 2 Selected attribute set from KDDCup 99 using CFS

S. no	Feature number	Feature name
DOS	3, 5, 23, 27, 31, 34, 36	Service, src_bytes, Count, rerror-rate, srv-diff-host-rate, dst-host_same_srv_rate, dst_host_same-src_port_rate
R2L	2, 3, 4, 5, 6, 11, 31	protocol_type, service, flag, src_bytes, dst_bytes, num_failed_logins
U2R	2, 3, 4, 12, 33, 32, 30	protocol_type, service, flag, logged_in, dst_host_srv_countdst_host_srv_rate, dst_host_count
Probe	2, 3, 5, 27, 29, 31, 33	protocol_type, service, src_bytes, srv-diff-host-rate, num_failed_logins, srv_diff_host_rate, dst_host_srv_count

$$Merit_s = \frac{N\overline{r_{ca}}}{\sqrt{N + N(N-1)\overline{r_{aa}}}}, \qquad (1)$$

where N is the NUMBER of attributes,
r_{ca} the correlation between attributes/target classes and
r_{aa} the Correlation among the attributes.

3.3 Data Mapping Procedure

The subsets with different features for all types of attacks are shown in Table 2. In these subsets some of the attributes hold the string values. The authors changed this string values to numeric as shown in Table 3.

Table 3 Mapped values for service attribute

No.	String	No.	String	No.	String	No.	String
1	auth	16	hostnames	31	Netstat	46	systat
2	bgp	17	http	32	Nnsp	47	telnet
3	courier	18	http_443	33	nntp	48	tim_i
4	csnet_ns	19	imap4	34	ntp_u	49	time
5	ctf	20	iso_tsap	35	Other	50	urh_i
6	daytime	21	Klogin	36	pm_dump	51	urp_i
7	discard	22	Kshell	37	pop_2	52	uucp
8	domain	23	Ldap	38	pop_3	53	uucp_path
9	domain_u	24	Link	39	Printer	54	vmnet
10	eco_i	25	Login	40	Private	55	whois
11	efs	26	Mtp	41	Smtp	56	Z39_50
12	exec	27	Name	42	sql_net		
13	ftp	28	netbios_dgm	43	Ssh		
14	ftp_data	29	netbios_ns	44	Sunrpc		
15	gopher	30	netbios_ssn	45	Supdup		

3.4 Artificial Bee Colony (ABC)

In general we have different types of bees in nature. These bees are classified as employee, onlooker, and scout bees. The employee bees initiate its food search from the nearest places to its hive. Employee bees perform the waggle dance depends on the nectar amount. While the waggle dance is on, the onlooker bees would locate the area in the hive where the highest amount of nectar is present. They start transferring the nectar/food to their hive depriving the employee bees of their food source. Later, employee bees will search for new source. The above discussed process of bees simulated as bee colony approach [12]. This optimization starts with initialization of number of bees (Solutions), number of iterations and stopping criteria. It has two passes namely forward and backward passes. In forward pass, a set of solutions is chosen from the available options randomly [13]. The quality of solutions is derived in backward pass. The best solutions could be kept and bad ones abandoned based on the quality of selected solutions. And the fitness can be calculated accordingly Eq. 2

$$\text{Fit}_i = \frac{1}{1 + \text{object function} f_i} \tag{2}$$

The selection of employee bees (NS) and onlooker bees is directly proportional to available food sources. Initially the position of food sources generated randomly. The nectar amount for new source can be calculated by using following Eq. 3

$$V_{ij}: = X_{ij} + \emptyset \left(x_{ij} - X_{kj} \right) \tag{3}$$

where i is 1 ... NS, j is 1 ... N and $\emptyset$, any number in [0, 1].

The selection of new food source by an employee bee ends if it founds new source is better, stays back if not. And all employee bees share such information about food sources helping onlooker bees in their search for best food source. The probability at which onlooker bee search for the best food source is as follows:

$$Pi = \frac{\text{Fit}_i}{\sum_{j=1}^{NS} \text{Fit}_j} \tag{4}$$

where NS is the number of food sources and Fit_j the fitness of new solution.

If the fitness of the food source selected by the onlooker is not improved, the selected solutions will be discarded. And then, the corresponding employee bee would become a scout bee and repeats the cycle till it finds the next best solutions using the following:

$$X_i^j = x_{\min}^j + \text{rand}[0, 1] \left(x_{\max}^j - x_{\min}^j \right) \dots, \tag{5}$$

where $x_{\max}^j$ = upper bound of population and $X_{\min}^j$ = lower bound of population.

4 Experiments and Results

In this section, the author discusses how artificial bee colony is used for generating the classification rules. Each record of KDD 99 has 41 feature values along with one class label.

Initially the subsets are generated by selecting relevant features for various attacks. Based on the existing correlation between the class labels and features and between the features subsets are generated. The subsets generated by CFS are shown in Table 2. Table 3 explains the mapping procedure.

Finally, authors obtained subsets with different features along with class label depending on the type of attack. Rules for the classification are generated by using ABC subsets. In this regard, a modified ABC algorithm has been used for appropriate result within a short time using the following equation

$$V_{ij} := X_{ij} \tag{6}$$

Some of the classification rules generated by the above proposed approach are shown in Table 4.

The following performance measures are used for calculating the accuracy of generated rules.

Table 4 Classification rule generated from ABC

Rule no.	Rule description	Attack type	Accuracy
1	src_bytes(i, 2) ≤ 2964 & srv_diff_host_rate(i, 5) ≤ 1	DOS	97.955
2.	count(i, 3) > 0 & src_bytes(i, 2) ≤ 2964 & srv_diff_host_rate(i, 5) ≤ 1	DOS	96.7543
3.	count(i, 3) > 270 & src_bytes(i, 2) ≤ 2964 & srv_diff_host_rate(i, 5) ≤ 6.700000e-001	DOS	81.7916
4.	count(i, 3) > 290 & reerror_rate(i, 4) ≤ 0	DOS	78.335
5.	duration = 26 or 134, protocol = tcp, service = FTP or login, flag = SF	R2L	82.781
6.	protocol = tcp, service = telnet, flag = RSTO, src_byte = 125 or 126, dst_byte = 179, hot = 1, num_failed_login = 1	R2L	75.198
7.	protocol = tcp, service = ftp, flag = SF, src_byte > 980, dst_host_count = 255, dst_host_count	R2L	90.955
8.	protocol = tcp, service = telnet or ftp_data, flag = SF, loggin_in = 1, dst_host_same_srv_rate = 1	U2R	96.7543
9.	protocol = tcp, service = telnet, flag = SF, logged_in = 1, dst_host_srv_count6 2, dst_host_diff_srv_rate6 0.07	U2R	81.1979
10.	protocol = tcp, service = telnet or ftp, flag = SF, dst_host_count = 255, dst_host_diff_srv_rate = 0.02	U2R	75.5553
11.	protocol = ICMP, service = SF or SH, src_byte = 8, same_srv_rate = 1, srv_diff_host_rate = 1	Probe	73.915

$$\text{True positive rate} = \frac{TP}{TP + FN} \tag{7}$$

$$\text{True negative rate} = \frac{TN}{TN + FP} \tag{8}$$

$$\text{Total accuracy} = \frac{TP + TN}{TP + FN + TN + FP} \tag{9}$$

5 Conclusion

This paper presents the method to detect intruders within and between the networks obstructing data transmission, finding solutions for the recurring security challenges. The latest swarm intelligence technique has been adapted which is based on a bee colony approach. It was found that the rate of false alarms has been reduced and the accuracy of classifier was enhanced. It was also found that the used algorithm is adaptable and flexible in finding the intruders effectively.

References

1. Usman Asghar Sandhu et al.: A survey of intrusion detection & prevention techniques, IPCSIT vol. 16 (2011).
2. Mohammad Sazzadul Hoque et al.: An Implementation Of Intrusion Detection System Using Genetic Algorithm, (IJNSA), Vol. 4, No. 2, March 2012.
3. G.V. Nadiammai, M. Hemalatha.: Effective approach toward Intrusion Detection System using data mining techniques, Cairo University Egyptian Informatics Journal.
4. Bin Luo, Jingbo Xia.: Indian Journal of Computer Science and Engineering (IJCSE) A novel intrusion detection system based on feature generation with visualization strategy published in Expert Systems with Applications, Elsevier, 2014.
5. Tieming Chen, Xu Zhang, Shichao Jin, Okhee Kim.: Efficient classification using parallel and scalable compressed model and its application on intrusion detection published in Expert Systems with Applications, Elsevier, 2014.
6. AshaGowda Karegowda, M.A. Jayaram, A.S. Manjunath.: Feature Subset Selection Problem using Wrapper Approach in Supervised Learning, International Journal of Computer Applications 1(7):13–17, February 2010.
7. D. Lavanya, Dr. K. Usha Rani.: Performance Evaluation of Decision Tree Classifiers on Medical Datasets, published in IJCA Volume 26– No. 4, July 2011.
8. Mahbod Tavallaee, Ebrahim Bagheri, Wei Lu, and Ali A. Ghorbani.: A Detailed Analysis of the KDD CUP 99 Data Set, CISDA 2009.
9. Abd-Alsabour N et al.: Feature Selection for Classification Using an Ant Colony System Published in e-Science Workshops, 2010 Sixth IEEE International Conference.
10. Rajdev Tiwari, Manu Pratap Singh.: Correlation-based Attribute Selection using Genetic Algorithm, IJCA Volume 4– No. 8, August 2010.

11. Mital Doshi1, Dr. Setu K Chaturvedi.: Correlation Based Feature Selection (CFS) Technique To Predict Student Performance, (IJCNC) Vol. 6, No. 3, May 2014.
12. D. Karaboga, B. Basturk.: On the performance of artificial bee colony (ABC) algorithm, Applied Soft Computing 8 687–697, Elsevier (2008).
13. Balwant Kumar, Dr. Dharmender Kumar.: A review on Artificial Bee Colony algorithm, IJET, 2(3) (2013) 175–186.

Virtualization Security Issues and Mitigations in Cloud Computing

S. Rama Krishna and B. Padmaja Rani

Abstract This paper presents various security issues related to hypervisor in cloud. This paper also brings issues possible with a malicious virtual machine running over hypervisor such as exploiting more resources than allocated by VM, stealing sensitive data by bypassing isolation of VM through side channel attacks, allowing attacks to compromise hypervisor. In this paper, we also bring security measures or requirements to be taken and architectures that are needed by hypervisor to handle various security concerns.

Keywords Security · Hypervisors · Cloud computing

1 Introduction

Several enterprises believed cloud to be a platform to fulfil their requirements such as increased scalability, availability, and upfront setup cost, etc. Though web2.0, Internet, distributed computing are technologies that enable cloud computing, in reality virtualization is the key technology to extract the exact sense of utilization maximization of resources. Sharing the resources is possible with virtualization is called to be multi-tenancy. Where the physical resources are virtualized and provided for multiple users to share them. Hypervisor looks at resolving issues of provisioning, de provisioning of virtual machines, their migration and isolated use to share a common physical space by multiple tenants (Fig. 1).

Security is the major concern in the cloud because several users share their data in cloud without noticing their co tenants in the same physical space. In this case

S. Rama Krishna (✉)
Department of Computer Science & Engineering, VRS & YRN College
of Engineering & Technology, Chirala 523155, India
e-mail: ccvy.ram@gmail.com

B. Padmaja Rani
Department of Computer Science & Engineering, JNTUCEH, Hyderabad, India
e-mail: padmaja_jntuh@yahoo.co

© Springer Science+Business Media Singapore 2017   117
S.C. Satapathy et al. (eds.), *Proceedings of the First International Conference
on Computational Intelligence and Informatics*, Advances in Intelligent Systems
and Computing 507, DOI 10.1007/978-981-10-2471-9_12

Fig. 1 To show the functionality of virtual machine

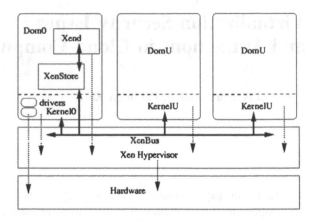

hypervisor should ensure a strong VM isolation mechanism because if a VM is vulnerable with its compromised security cause danger to the remaining who share the common spaces. Even denial of service attack is also possible with the compromised VM in hypervisor where the vulnerable VM gets hold of shared resource and cause data leakage also. If the shared resources were hijacked, then co-tenant VMS may be slowed down also.

Remaining of this paper is organized as: Sect. 2 brings Side Channel attacks in Hypervisor. Section 3 brings performance-based attacks that cause target machine slow down. In Sect. 4 security measures needed to protect hypervisor are mentioned. Section 5 presents VM Isolation security issues and counter measures. Section 6 contains conclusions of the derived paper.

2 Side Channel Attacks and Defences

Side channel attack creates opportunity for a co-resident VM to gain access data of other VM without their intervention. It creates a bypassing method to access data. CPU cache, memory, power consumption and network used in extraction of data in side channel attack. Software happenings will be traced by observing behaviour in hardware [1]. Yu et al. [2] takes CPU cache response time to check whether target VM co-resident or not. Cache behaviour is analysed using linear regression of the values collected by load pre-process with cubic spline and load predictor. A malicious VM occupies a major part of CPU cache then targets co-resident by simple data request to it. Then it executes load measuring program over malicious VM for measuring access time of cache. Literature [2] observes and proves that higher cache access time implies more activities by co-resident. The experiment proposal also verified with three VMS sharing resource. Vulnerable VM not only analyses CPU cache access time, but also can get data of the target machine. Literature [3] describe data hijacking of co-resident VM by infecting malware into the software.

This attack targets at obtaining information from the target VM without its notice covertly and it will not leave any trace. For doing this the attacker uses memory bus. Vulnerable VM locks memory bus by sending 1 for issuing atomic CPU instruction, memory latency will increase with execution of CPU atomic instruction. If the memory bus released latency will decrease and it transfers bit 0. Similarly, it also build other side channels.

For exploiting cache contention for manipulating latency times [3], this attack also calculates bandwidth by finding the length of overlapped execution time of malicious VM with target VM. This attack can be defended by making some changes to the scheduler in the hypervisor. To successfully defend, the scheduler can try to limit the overlapping execution times of any two VMS on the system while maintaining an acceptable level of performance. The scheduler should still maintain fairness because it does not know which VM is malicious. In order to maintain an acceptable level of performance, the scheduler should limit the frequency of VM switching which reduces performance [3]. System performance may decrease due to limiting of the overlapped execution time. Pumping noise to the side channel proposed to prevent attack due to which error rate increases and bandwidth reduced. Created noise due to random atomic memory access defends the side channel attack that happens over memory bus contention.

Literature [1, 4] proposes other methods for defending side channel attacks such as Xenpump. This method limits effectiveness in timing channels. Bandwidth of the timing channel is limited by adding some random latencies by Xenpump. Hence confusion is created to vulnerable VM that receives channel bandwidth. That unpredictability created in the receiver VM in the generated latency information is because of VM or hypervisor. This proposed model also decreases the system performance. Literature [1] presents another kind of side channel attack called as a Cache-based side channel attack that uses the prime trigger probing method for attack. Like the previous case attacker VM occupies the cache by accessing many lines and records. Then triggering is done while target VM is running a message is encoded by accessing parts of cache. Once target VM finishes its job VM used for probing starts accessing the cache, where each line used to access cache causes cache miss. It has a higher access time when compared with the base line.

A diagram is shown (Fig. 2) to describe Prime trigger probing.

Fig. 2 Prime trigger probe method

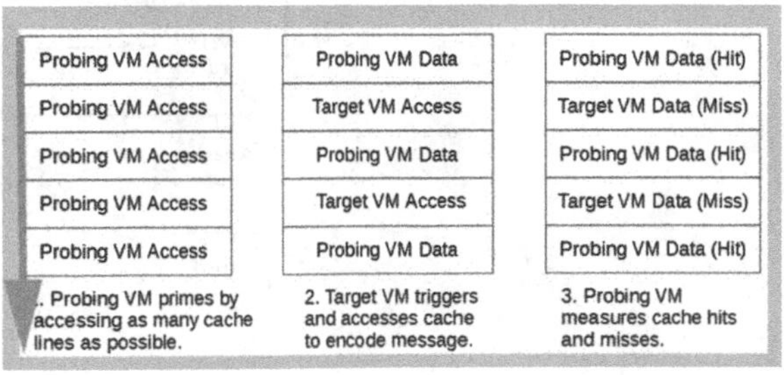

Probing VM Access	Probing VM Data	Probing VM Data (Hit)
Probing VM Access	Target VM Access	Target VM Data (Miss)
Probing VM Access	Probing VM Data	Probing VM Data (Hit)
Probing VM Access	Target VM Access	Target VM Data (Miss)
Probing VM Access	Probing VM Data	Probing VM Data (Hit)
1. Probing VM primes by accessing as many cache lines as possible.	2. Target VM triggers and accesses cache to encode message.	3. Probing VM measures cache hits and misses.

Flushing is done between time of switching of target VM and probing VM to defend this attack, but this method creates additional overhead of 15 % [1].

3 Performance-Based Attacks and Defences

An attack is called performance-based attack if the target VM slows down because of resource hijacking by attacker VM. Literature [5] mentioned an attack which cheats scheduler to achieve 98 % of CPU usage over a physical machine. Credit-based scheduling is done in Xen hypervisor which uses token bucket scheduling. Every VM is allowed to get a credit at most 300 to store. VMs which are running will be debited its credits by one on every scheduler tick. An attacker VM blindfolds the scheduler enter before scheduler tics when a co-resident VM is in execution at scheduler which will not reflect in change of credits of attacker VM. The scheduler works in two modes, a one boost mode and a non-boost mode. In boost mode hypervisor could not differentiate among VMS waking by deliberately yielding and scheduler executed VMS. In the other mode attacker will never be debited his CPU credits. This problem can be eliminated in two ways. By using a high-precision clock and randomized scheduler ticks. In the first method the scheduler uses a clock with high precision and measures CPU usage even when VM yields or it is idle. Other method clock ticks at random intervals 10 or 30 ms. The Amazon EC2 instance will be allowed to access from 40 to 85 % of cap. An attacker VM shares resources, and calculates CPU usage if it is 85 % no more co-resident available in location otherwise there exist a co-resident [5] (Fig. 3).

I/O based attacks were discussed in literature [6]. Specially designed I/O workloads are deployed over shared queues of I/O for reducing the performance of target VM. First scheduling characteristics of hypervisor observed and extracted from targeted VM by attacking. Using this information I/O resources are hijacked and hence resulting in slow down of target I/O performance and access. Using

Fig. 3 Randomized schedule tick rate prevents a VM from being preempted

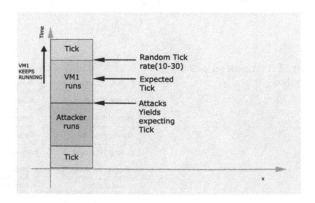

multiple disks to store data and accessing and the random schedule ticker can protect from I/O performance reducing attacks [5].

4 Hypervisor Security Issues and Defences

Along with side channel attacks there is one more attack possible called as virtual machine escape is an attempt of VM compromising hypervisor. Hypervisor is a layer that provides separate VMS for isolated tenants. This section presents security issues associated with VMM (Virtual Machine Manager) and VM (Virtual Machine).

A. VMM Security

In order to say that hypervisor is secured it should have a trusted VM. Still there are several security concerns with VMM.

(1) VMM Vulnerabilities:

Literature [7] mentioned, six major security vulnerabilities of VMM and emulators using auditing of the source code, fuzzing techniques. Further vulnerabilities were presented in [8] issues including VMware, Xen and other Softwares.

(2) VM-Based Rootkits:

As traditional Rootkit kernel is available in OS kernel, VM-based Rootkits are available in VMM. Kernel rootkit detector could not identify these VM-based rootkits. VMware and Virtual PC were used to develop these Proof of concept VM-based rootkits. They were used to observe the stealthiness of environment by considering parameters such as installation time, boot time and memory footprints [9]. Literature [10] mentioned Blue pill a VM-based rootkit. It can be installed dynamically without rebooting the system on the fly. Blue Pill leverages nested virtualization features in AMD svm. While a hypervisor is running Blue pill enable itself VMM layer. Kernel Guard [11] uses VMM policies to monitor kernel memory access for root kit attack protection. It blocks dynamic code and prevents root kit attack.

(3) VMM Transparency:

VMM detectability is one more major concern. In order to protect from VM detection threats it is obvious to host potential hostile codes like honepots. Literature [12] states that VMM transparency is not feasible because of discrepancies in between physical machine and VM. Several clues like time sources, hardware abstraction, and overhead were left to make VM detection possible. Four essentials for detecting VMM were mentioned in literature [13]. In same literature they mentioned an experiment which is run for detecting remotely different VMM types. Remote verifier can detect P IV architecture and VMM type (Xen, Linux,

VMware). VMM version and type can be revealed, so this may be an attack possible over VM transparency.

(4) **Platform Integrity**:

Users of the cloud has to blindly believe in trust of VMM because there is possibility of co tenants modifying data. VMM should ensure the trust for each VM that is in execution in each layer of software stack. Literature [14] proposes a model called Terra. Terra is a model built as a prototype for trusted VMM. For every application it assigns a different VM. Integrity of data can be ensured by deploying each application with an optimized OS. Literature [15] suggests Trusted Platform Module (TPM) which is a Trusted Computing Group specified security definition. Trusted Computing Group extends in accommodating virtualization techniques. vTPM embedded in TPM and can run on over external co-processor and VM. TPM 1.2 extends command set for accommodating vTPM, which enables TPM to access every VM. Hypervisor deserves tools to measure integrity of VM that is running. HIMA is proposed in literature [16] which is based on hypervisor agent. Isolating measurement agent and target is desired for tool to measure integrity. HIMA makes sure that VMS that pass integrity check only run on VMM. So it ensures healthy program in execution.

(5) **Hypervisor Control Flow Integrity**:

Literature [17] proposed a method for providing hypervisor control flow integrity called as hypersafe. Literature [18] suggested Trusted Platform Module (TPM) based on hardware to provide secure attestation, crypto graphic hashes, signatures and secure storage. Second method ensures load time integrity where as the first provides run time integrity which is very crucial. For implementing run time integrity checking Hyper safe uses two techniques. Those are restricted pointer indexing method and non-bypassable memory lock down method. Unauthorized page writes are prevented by locking down memory pages. The designed unlocking process ensures no modification is done to code or data of hypervisor. Malicious code injection for flow control in hypervisor can be prevented using memory page locking system. Literature [17] figures out Hypersafe implementation as an extension in Xen hypervisor. A new layer is created for indirecting all operators in restricted pointer indexing method in Hypersafe. This works as previous technique and control flow targets are pre-computed and stored in a table. This approach provides call target and return target to follow control flow graphs. Without any change to existing hypervisor this method can be added as an extension to compiler [17]. Use of protected hooks is suggested for monitoring untrusted VM execution to get control over applications running on it in Lares [19] framework. Applications running over untrusted VM will be monitored by VMI and security policies when control is transferred to security VM by hooks. Customized OS may not support Lares framework because change is needed on the guest OS on the fly. A state-based control flow comprising static and dynamic control flow provides kernel integrity [20]. Static control flow checking uses hashing whereas dynamic control flow checking uses control flow graphs generated of source code. Wei et al. Addressed risks in managing security of virtual images such as publishers risk, retriever risk and repository administrator risk. The suggested solutions for access control of images and filtering and scanning images proved the better result than treating images independently. But filters may not give 100 % accurate results; virus scanning may not guarantee identifying malware in vmimage [21].

(6) **Hypervisor Integrity Checking**:

Hypersentry is a method suggested in Literature [22] for providing hypervisor security. Hypervisor is added with this new software component Hyper sentry to ensure integrity and stealth protection. For isolation of hypervisor with TPM Hypersentry uses existing hardware and also provides software integrity. Scrubbing is an attack used to remove evidence of attack without detection of higher software layer. Hypersentry acts as stealth and out-of-band channels were used to trigger this. Intelligent platform management interface (IMPI), Baseboard Management controller and System Management Mode (SMM) are used as out-of-band channels. IMPI is implemented in hardware of the hypervisor and functions independently to the CPU and other softwares of the system. BMC is a component installed over mother board for providing interface among hardware management component to remote verifier. SMM is triggered by IMPI call for providing secure environment and prevents manipulations over software which is running on machine. Interaction among these components is given in Fig. 4.

A verifiable, non-interruptible and deterministic measurement is provided in Hypersentry. It saves CPU current state after checking it thoroughly restores it. Hypersentry also provides integrity, authentication and attestation as output.

(7) **Return Oriented Programming Attack on Hypervisors**:

Return oriented programming (ROP) is one more attack mentioned in Literature [23] over Xen hypervisor which is very successful attack. It uses existing code for attack. Turing language was created by sequence chaining which ends in return statement. This is an extension of DEP (Data Execution Prevention) which is security measure implemented in most of systems today. ROP attack modifies hypervisor data which are used for control level of VM privilege level. An attacker can modify their VM level from normal level to privileged. Literature [24] suggests a defence method for ROP problem. In this solution stack is analysed continuously looking for possibilities in occurrence of ROP attack and quarantined for the use of

Fig. 4 HyperSentry
architecture [22]

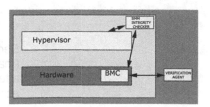

further investigations. As ROP requires many address that range in program this key feature is used to search ROP attacks using libraries.

(8) **Modifying Non-control Data**:

Literature [25] attempts to bring forward attacks over non-control data in hypervisor. There is a possibility of vulnerability in three different types of non-control data: privilege-level data, resource utilization and security policy data. An attacker uses this privilege-level data for escalation of VM privilege level. Resource utilization data helps attacker in gaining access to shared physical resources. Using security policy data attacker can attempt for side channel attacks for stealing sensitive data from target VM. Hypervisor version number helps in execution of attack over non-control data attack. Memory offsets are calculated with the help of version number of hypervisor for modifying non-control data. Writing in non-controlled data memory locations have to be limited by hardware can prevent attacks. Hypersafe [25] can be used in preventing these attacks by using non-bypassable memory lockdown.

(9) **VM Rollback Attack**:

Literature [26] mentions VM rollback attack. It assumes hypervisor is compromised already. This compromised hypervisor tries to execute VM from its older snapshot without owner's awareness. This attack damages target VM's execution history and undo security patches and updates make it vulnerable target VM. This lets attacker to bypass security system (Fig. 5).

By roll back VM state that attacks an attacker gets a chance to execute brute force password attack. Actually this will occur as when there is brute force attack occurred target VM raises security alert but compromised hypervisor brings its previous snapshot by roll back and allowing brute force attack to be possible. Using suspend/resume function we can prevent this roll back attack. But it makes developing solution more complex because it cannot distinguish between normal suspend/resume and an attack of roll back. One more requirement is this solution should not create burden for users. By securely logging all rollback actions and

Fig. 5 Demonstrates this attack

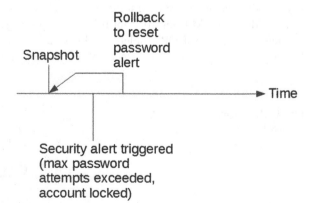

auditing them can prevent roll back attack. Even TPM can be used in protection of log integrity. VM boot, VM suspend, VM resume and VM resume are four hyper calls used in logging information. Isolating and encrypting the VM'S memory in hypervisor helps in protecting memory hence creates solution to the rollback attack. This solution also prevents hypervisor to modify or read memory pages [26].

B. **VM Security**

Virtual Machine gives opportunity to interact with hardware in a multi-tenant and shared mode over VMM. VMS running like this should be secured and they should make sure proper security measures need to be taken. Introspection and Secure resource allocation are issues related to VM security.

(1) **Introspection**: VM Introspection is a process to track data flow inside guest VMS; it has many challenges. Moonsols livecloudkd is one such implementation presented in Literature [27]. It debugs a guest OS running over Hypervisor by allowing KD and windbg to inspect Windows Virtual Machine delivered from Microsoft Hyper-V R2 hypervisor. VIX is a tool suite presented in Literature [28] used for introspection of Xen. It tracks guest VMS Process by mapping domU's virtual memory to Dom0's memory address in VMM.

(2) **Secure resource allocation**: complete isolation of VMS in hypervisor will reduce performance and efficiency in resource utilization hence it may not be a deserved solution. Efficiency in utilizing resources with security is needed. So literature [29] suggested resource sharing in Hypervisor through shype [30]. It uses MAC-based policy for security to share resources without compromise in security with minimized overhead.

5 VM Isolation Techniques

This section brings four different approaches used in isolation of VM. Literature [31] suggests security Turtles an architecture based on nested virtualization for protection of guest VM.

Even though attacks possible in Level 1 Hypervisor for VMS running Level 2, Level 0 Hypervisor is the highest privileged which protects.

(1) Lifetime kernel code integrity of Level 1 hypervisor.
(2) Code-data integrity in QEMU-KVM daemons.
(3) Data integrity in Level 2 execution of Guest VM.
(4) Guest VMS running in Level 2 need to be aware that weather there is any violations in above 3 requirements.

Secure Turtles believe that outside attacks are not possible over hypervisor of Level 0. Literature [32] eliminates hypervisor almost and proposed a new approach. Minimizing code base area will reduce vulnerabilities in virtualization software also reduces its functionality. Multi-tenancy is a needed requirement to be provided in cloud for provisioning resources to customers on demand leveraging economies. In this literature they propose a solution temporary Hypervisor; a temporary hypervisor that runs at initialization. It is consists of pre-allocated hardware resources (process cores, memory, etc.). It sets virtualized I/O to avoid indirection for bringing the VM a more direct interaction with hardware.

Provisioning each VM with individual I/O devices is not much practical so virtualized I/O devices are encouraged to be used. Once virtual I/O devices are allocated, hard will make sure isolation in between virtual machines without hypervisor. Literature [33] describes a mandatory access control policy (MAC) which allows VMS to share resources like disk, networks, etc., this also supports multiple hypervisors to share these resources. Bind time authorization is used to obtain high performance and Chinese wall is also included. Chinese wall assigns every VM a type. It will not allow to run two VMS concurrently that there is conflict with type. Hence it prevents attacker VM not to use covert channels for accessing target VM sensitive data. Type-enforcement is done to specify access of VMS to resources. Literature [34] suggests a multilevel set of security. There are two different types of hypervisors known as pure isolation and shared hypervisor. In hypervisor with pure isolation partitioning is done in machine and it will not allow resource sharing except memory and CPU. Hypervisor with sharing will allow file sharing. There will be two partitions low level and high level. Low level is allowed to read/ write in same level of security where as higher security partition has read only permission to access security data of low level partition. One-way network implementation is done. Secure shared file store is suggested in another implementation. Hypervisor uses cross-ring call for access control to sub system. It also implemented as separate partition which uses message passing mechanism.

6 Conclusion

Virtualization enable cloud computing facilitated several guest VMS to share common physical hardware. Hypervisor is the key component in virtualization. Hence it must resist attack effectively by isolating VMS. But in reality it is vulnerable and exposed to several security flaws such as VM escape. It is said to be the most serious among several attacks said above. An escaped VM will compromise several co-resident VMS. Several architectural and design changes are needed in Hypervisor for potential resistance of VM escape attack. Side channel attacks hijack system resources and steal sensitive data of co-resident VMS. Some solutions were discussed in mitigation of side channel attacks as adding noise, etc. Hypervisor security enables security to the cloud environment which results in trust building and enterprises motivation of migration to cloud.

References

1. M. Godfrey and M. Zulkernine, "A Server-Side Solution to Cache-Based Side-Channel Attacks in the Cloud," Proc. Of 6th IEEE International Conference on Cloud Computing, 2013, pp. 163–170.
2. S. Yu, X. Gui, J. Lin, X. Zhang, and J. Wang, "Detecting vms Co-residency in the Cloud: Using Cache-based Side Channel Attacks," Elektronika Ir Elektrotechnika, 19(5), 2013, pp. 73–78.
3. F. Liu, L. Ren, and H. Bai, "Mitigating Cross-VM Side Channel Attack on Multiple Tenants Cloud Platform," Journal of Computers, 9(4), 2014, pp. 1005–1013.
4. J. Wu, L. Ding, Y. Lin, N. Min-Allah, and Y. Wang, "xcnpump: A New Method to Mitigate Timing Channel in Cloud Computing," Proc. Of 5th IEEE International Conference On Cloud Computing, 2012, pp. 678–685.
5. F. Zhou, M. Goel, P. Desnoyers, and R. Sundaram, "Scheduler Vulnerabilities and Coordinated Attacks in Cloud Computing," Journal of Computer Security, 21(4), 2013, pp. 533–559.
6. Z. Yang, H. Fang, Y. Wu, C. Li, B. Zhao, and H. Huang, "Understanding the Effects of Hypervisor I/O Scheduling for Virtual Machine Performance Interference," Proc. Of 4th IEEE International Conference on Cloud Computing Technology and Science (cloudcom 2012), 2012, pp. 34–41.
7. T. Ormandy, "An Empirical Study into the Security Exposure to Hosts of Hostile Virtualized Environments," in cansecwest, 2007.
8. The MITRE Corporation, "Common Vulnerability and Exposures (CVE)," http://cve.mitre.org/, Mar. 2011.
9. S. King and P. Chen, "Subvit: implementing malware with virtual machines," in IEEE Symposium on Security and Privacy, May 2006.
10. J. Rutkowska, "Subverting Vista kernel for fun and profit," 2006.
11. J. Rhee, R. Riley, D. Xu and X. Jiang "Defeating dynamic data kernel Root-kit attacks via VMM based guest transparent monitoring". In proceedings of ARES 2009, conference 2009, To appear.
12. T. Garfinkel, et al., "Compatibility is not transparency: Vmm detection myths and realities," in hotos, 2007.
13. J. Franklin, et al., "Remote detection of virtual machine monitors with fuzzy benchmarking," SIGOPS Oper. Syst. Rev., April 2008.
14. T. Garfinkel, et al., "Terra: a virtual machine-based platform for trusted computing," in SOSP, 2003.
15. Trusted Computing Group, http://www.trustedcomputinggroup.org/, June 2011.
16. A. Azab, et al., "Hima: A hypervisor-based integrity measurement agent," in ACSAC, dec. 2009.
17. Z. Wang and X. Jiang, "hypersafe: A Lightweight Approach to Provide Lifetime Hypervisor Control-Flow Integrity," Proc. Of IEEE Symposium on Security and Privacy, 2010, pp. 380–395.
18. M. Kim, H. Ju, Y. Kim, J. Park, and Y. Park, "Design and Implementation of Mobile Trusted Module for Trusted Mobile Computing," IEEE Transactions on Consumer Electronics, 56(1), 2010, pp. 134–140.`.
19. B.D. Payne, Macaroni, M. Sharif and W. Lee." Lares: an architecture for secure active monitoring using virtualization." Security and privacy IEEE Symposium ON, 0:233–347.
20. N.L. Petroni, Jr and M. Hicks, " automated detection of persistent kernel control flow attacks". In CCS'07: proceedings of the 14th ACM conference on Computer and communications security, pages 103–115, New York NY, USA 2007, ACM.
21. Jinpeg Wei, Xiaolan Zhang, Glenn Ammons, Vasantha Bala, Peng nns, "Managing security of virtual machine images in a cloud environment", in CCW'09 proceedings, Chicago, Illinios, USA, ACM 978-1-60558-78-4/09/11.

22. A. Azab, P. Ning, Z. Wang, X. Jiang, X. Zhang, and N. Skalsky, "hypersentry: Enabling Stealthy In-context measurement of Hypervisor Integrity," Proc. Of 17th ACM Conference on Computer and Communications Security, 2010, pp. 38–49.
23. B. Ding, Y. Wu, Y. He, S. Tian, B. Guan, and G. Wu, "Return- Oriented Programming Attack on the Xen Hypervisor," Proc. Of 7th International Conference on Availability, Reliability and Security, 2012, pp. 479–484.
24. X. Jia, R. Wang, J. Jiang, S. Zhang, and P. Liu, "Defending Return-oriented Programming Based on Virtualization Techniques," Security and Communication Networks, 6(10), 2013, pp. 1236–1249.
25. B. Ding, Y. He, Y. Wu, and J. Yu, "Systemic Threats to Hypervisor Non-control Data," Information Security, 7(4), 2013, pp. 349–354.
26. Y. Xia, Y. Liu, H. Chen, and B. Zang, "Defending against VM Rollback Attack," Proc. Of 2nd International Workshop on Dependability of Clouds, Data Centers and Virtual Machine Technology (DCDV 2012), 2012.
27. Moonsols, "livecloudkd," http://www.moonsols.com/2010/08/12/livecloudkd/, Aug. 2011.
28. B. Hay and K. Nance, "Forensics examination of volatile system data using virtual introspection," SIGOPS Oper. Syst. Rev., April 2008.
29. R. Sailer, et al., "Building a mac-based security architecture for the xen open-source hypervisor," in ACSAC, 2005.
30. S. Berger, et al., "vtpm: virtualizing the trusted platform module," in USENIX Security Symposium, 2006.
31. F. Liu, L. Ren, and H. Bai, "Secure-Turtles: Building a Secure Execution Environment for Guest vms on Turtles System," Journal of Computers, 9(3), 2014, pp. 741–749.
32. J. Szefer, E. Keller, R. Lee, and J. Rexford, "Eliminating the Hypervisor Attack Surface for a More Secure Cloud," Proc. Of 18th ACM Conference on Computer and Communications Security, 2011, pp. 401–412.
33. R. Sailer, T. Jaeger, E. Valdez, R. Caceres, R. Perez, S. Berger, J. Griffin, and L. Van Doorn, "Building a MAC-based Security Architecture for the Xen Open-source Hypervisor," Proc. Of 21st Annual Computer Security Applications Conference (ACSAC 2005), 2005, pp. 276–285.
34. P. Karger, "Multi-level Security Requirements for Hypervisors," Proc. Of 21st Annual Computer Security Applications Conference (ACSAC 2005), 2005, pp. 267–275.

Secured Three-Tier Architecture for Wireless Sensor Networks Using Chaotic Neural Network

C.N.S. Vinoth Kumar and A. Suhasini

Abstract Security in wireless sensor framework is a noteworthy test for all researchers now-a-days. Reinforcing the verification framework before association foundation is the right way to upgrade the structural planning furthermore to give the secured correspondence from busybody. This paper proposes a mixture security instrument utilizing chaotic neural network while setting up the association between the two hubs. In this paper, an arbitrary pair savvy key created is scrambled utilizing chaotic neural network. The proposed calculation incredibly enhances the security of the key shared between the hubs. The CPU time for the proposed calculation is assessed for different key generation calculation.

Keywords Wireless sensor networks · Chaotic neural network · Key encryption

1 Introduction

Transmission of information in the wireless sensor networks has been enormously expanded which pulls in the examined information to add to the advancement of WSN. The WSN comprises widely disseminated sensor hubs joined as a network. The WSN utilizes numerous sensor hubs inside or neighboring zone to transmit the information over the hubs. Before transmission of information, association between the hubs ought to be set up before the trading of data.

WSN system is inclined to various assaults. The most widely recognized assault confronted by WSN is the replication assault in which the busybody hacks the security key required to build up the association and make a clone of the beneficiary hub to get the transmitted data. To guarantee the insurance of the system from

C.N.S. Vinoth Kumar (✉) · A. Suhasini
Annamalai University, Chidambaram, Tamil Nadu, India
e-mail: vinothcns@gmail.com

A. Suhasini
e-mail: suha_babu@yahoo.com

© Springer Science+Business Media Singapore 2017
S.C. Satapathy et al. (eds.), *Proceedings of the First International Conference on Computational Intelligence and Informatics*, Advances in Intelligent Systems and Computing 507, DOI 10.1007/978-981-10-2471-9_13

busybody, the key required to set up the association ought to be reinforced and more secured. In this paper, the insurance of the random pair shrewd key is upgraded utilizing the chaotic neural system encryption to shield the system from replication assault. Different writing has been concentrated on replication assaults confronted by WSN and other security issues found in static and portable WSN. The study on writing gave different numerical and operational strategies on securing the system. Eschenauer et al. [1] proposed a probabilistic key dispersion plan which helps in making the introductory association between the sensor hubs. An irregular arrangement of keys was chosen by the sensor hubs from the key pool which has the likelihood of at least one normal key between the sensor hubs.

Chan et al. [2] proposed a two-key pre-conveyance plan which comprises q-composite key pre-appropriation and irregular pair savvy key plan. In this building design, two sensor hubs are required to figure the pairwise key from the q circulated keys.

Liu et al. [3] proposed a secured correspondence between the sensor hubs empowering in order to use the cryptographic frameworks with the bi-variate key polynomial. This strategy guarantees the immediate key foundation between the two sensor hubs. Better trade-off and security is achieved between the pre-distribution schemes.

Rasheed et al. [4] proposed a gathering key pre-appropriation plan with three-level security plans. This framework is more secure and t-arrangement safe. In this framework, the stationary hub helps us to confirm and build up the association between the sensor hub and portable sink. The information will be transmitted from versatile sink to sensor hub through the stationary access hub. The vast majority of the written works focuses on key era systems and neglected to focus on keeping up the key quality of the created key. This paper proposes a high-secure WSN information transmission utilizing three-level security plans and chaotic neural network. Whatever is left of the paper is sorted out as takes after. In Sect. 2, three levels of security plans is discussed and in Sect. 3, high-secure key era is examined. In Sect. 4, test results are examined and the paper is finished up in Sect. 5.

2 Three-Tier Security Scheme

Remote sensor network comprises vast number of sensor hubs which transmit the information to the versatile sinks through the stationary access hubs. In three-level security plans, two arrangements of polynomial pool, for example, mobile polynomial pool and static polynomial pool were utilized. To acquire the data from the versatile sink, the key from the portable polynomial pool is utilized. To set up the association between the sensor hub and stationary access hub, the key from the static polynomial pool is utilized. A sensor hub called stationary access point is arbitrarily chosen to transmit the information from the sensor hub to portable sink. The stationary access point will contain a typical polynomial from the portable polynomial pool. The basic key is confirmed, and after that, the stationary access point will go about as a transmission span between the sensor hub and the portable sink. Each stationary hub contains its own particular polynomial pool; furthermore, there will be a typical polynomial with the versatile polynomial pool which validates the association for information transmission.

2.1 Data Transmission Between the Mobile Node and Sensor Node and Its Key Discovery

The association between the portable node (M) and sensor node (S) can be set up utilizing direct key establishment or indirect key establishment. The association between the versatile node (M) and sensor node (A) can be built up utilizing the middle stationary access nodes (A).

In the direct key establishment, the portable node (M) finds a neighboring stationary access hub (An) and sends the pair astute key and builds up the association in the middle of M and A. At that point, the stationary access hub (A) sends the pair astute key to sensor node (S), and after that, sets up the association in the middle of An and S. Hence, the association between the mobile hub and sensor hub will be set up utilizing the stationary access hub.

In the indirect key establishment, the versatile node (M) finds a neighboring stationary access node (A) and sends the pair savvy key and builds up the association in the middle of M and A. At that point, the stationary access node (A) sends the pair shrewd key to the sensor node (S), and afterward, if the stationary access node (A) did not validate with the polynomial keys between the sensor hub, then the stationary access hub needs to discover another stationary access hub to set up the association with the sensor hub.

3 High-Secure Key Generation

3.1 Random PairWise Key Pre-distribution

This paper proposes a high secured key era instrument utilizing chaotic neural network. In this segment, the irregular key pre-appropriation is performed first, and after that, the arbitrary key is encoded utilizing Hopfield chaotic neural network.

In this segment, the pair savvy key setup in random key pre-distribution is examined. This setup contains four stages.

(i) *Key Pre-appropriation stage*: In this stage, the polynomial key pool is created and the produced key pool is circulated among the sensors.
(ii) *Sensor arrangement stage*: In this stage, the sensors are consistently scattered in a substantial zone.

(iii) *Key revelation stage*: In this key disclosure stage, every sensor will recognize its common key with its neighboring sensors and make a guide, taking into account the key-imparted connection to neighboring sensors.

(iv) *Pair shrewd key foundation stage*: If the sensor imparts a key to its given neighbor, then the key can be utilized as pair insightful key.

3.2 Encryption of Key Using Hopfield Chaotic Neural Networks

Disordered system has numerous fascinating properties of good cryptosystem, for example, periodicity, blending and affectability to introductory conditions. [5] Yu et al. planned a postponed confused neural system based cryptosystem. This cryptosystem makes utilization of the disordered directions of two neurons to create essential double successions for encoding plaintext. The cryptosystem utilizing Hopfield neural network is talked about as follows:

$$\begin{pmatrix} \frac{dx1(t)}{dt} \\ \frac{dx2(t)}{dt} \end{pmatrix} = -A\begin{pmatrix} x1(t) \\ x2(t) \end{pmatrix} + W\begin{pmatrix} \tan h(x1(t)) \\ \tan h(x2(t)) \end{pmatrix} + B\begin{pmatrix} \tan h(x1(t-\tau(t))) \\ \tan h(x2(t-\tau(t))) \end{pmatrix} \quad (1)$$

where $\tau(t) = 1 + 0.1 \sin(t)$, the initial condition of differential equation (Eq. 1) is given $x_i(t) = \Phi_i(t)$ when $-r \le t \le 0$, where

$$r = \max_t\{\tau(t)\}, \Phi_i(t) = (0.4, 0.6)^T$$

The arrangement of deferred differential comparisons is fathomed by the fourth-request Runge–Kutta system with time-step size $h = 0.01$. Assume that $x1$ (t) and $x2(t)$ are the directions of deferred neural systems. The ith emphases of the disorderly neural systems are $x1_i = x1(ih)$, and $x2_i = x2(ih)$.

A methodology proposed in [6] was embraced to produce an arrangement of free and indistinguishable (i.i.d.) paired arbitrary variables from a class of ergodic disordered maps. For any x characterized in the interim $I = [d, e]$, we can express the estimation of $(x - d)/(e - d)$ that has a place with [0, 1] in the accompanying twofold representation:

$$\frac{x-d}{e-d} = 0. \quad b_1(x)b_2(x), \ldots, b_i(x), \ldots, x \in [d, e], b_i(x) \in \{0, 1\} \quad (2)$$

The ith bit $b_i(x)$ can be expressed as:

$$b_i(x) = a_0 + \sum_{r=1}^{2i-1} (-1)^{r-1} \Phi_{(e-d)(r/2i)+d}(x) \qquad (3)$$

where $\Phi_t(x)$ is a threshold function defined by

$$\Phi_t(x) = \begin{cases} 0, & x < t \\ 1, & x \geq t \end{cases} \qquad (4)$$

The abovesaid four mathematical statements empower to produce the essential paired succession. These double successions are utilized for encryption as expressed by Yu et al. [5]. Disorganized neural systems offer extraordinarily build memory limit. Every memory is encoded by an unstable periodic orbit (UPO) on the riotous attractor. A turbulent attractor is an arrangement of states in a framework's state space with exceptionally uncommon property that the set is a drawing in set. So, the framework starting with its initial condition in the appropriate basin, eventually ends up in the set.

In the proposed framework, the irregular key is encoded utilizing Hopfield chaotic neural network

1. Calculate the length of the key and divide the key into subsequences of 8 bytes.
2. Set the parameters, μ and the initial point $x(0)$.
3. The chaotic sequence $x(1)$, $x(2)$, $x(3)$, ..., $X(M)$ is evolved using the formula:

$$X(n+1) = \mu x(n)(1 - x(n))$$

4. Create $b(0)$, $b(1)$, ..., $b(8M-1)$ from $x(1)$, $x(2)$, ... $x(M)$ by the generating scheme that $b(8m-8)$, $b(8m-7)$, ... $b(8m-1)$ ... is the binary representation of $x(m)$ for $m = 1, 2, ... M$.
5. For $n = 0$ to $(M-1)$,
   For $i = 0$ to 7
   $j = \{0, 1, 2, 3, 4, 5, 6, 7\}$

   $W_{ji} = 1$ if $j = i$ and $b(8n+i) = 0$
   $\qquad -1$ if $j = i$ and $b(8n+i) = 1$
   0 if $j \neq i$

   $\theta_i = -1/2$ if $b(8n+i) = 0$
   $\qquad 1/2$ if $b(8n+i) = 1$

End

For $i = 0$ to 7, d_i is calculated using:

$$d_i = f\left(\sum_{i=0}^{i=7} w_{ij}d_i + \theta_i\right),$$

where $f(x)$ is 1 if $x \geq 0$.
End

$$g(n) = \sum_{i=0}^{i=7} d_i 2^i$$

End

In this manner, the scrambled key g is gotten.

The key quality of the scrambled key g is more contrasted with different methods.

4 Experimental Analysis

In this segment, the proposed encryption calculation is assessed utilizing MATLAB 7.14. The created arbitrary key is encoded utilizing chaotic neural network. The CPU time for encoding the irregular key created utilizing different calculations was concentrated on.

Table 1 shows the comparison of the evaluation of CPU time for the proposed algorithm for various polynomial key generation algorithms (Figs. 1 and 2).

The simulated graphical result shows the various time spans in seconds taken by CPU for encrypting and decrypting the polynomial key generation algorithms. It shows that sensed information will be encrypted/decrypted in a faster manner using key generation algorithm respectively. (Figs. 1 and 2).

Table 1 Evaluation of CPU time for the proposed algorithm

S. no	Polynomial key generation algorithm	CPU time for encryption	CPU time for decryption
1	SHA160	0.4992	0.1716
2	SHA224	0.6240	0.1872
3	SHA256	0.5928	0.1872
4	SHA384	0.7176	0.1716
5	SHA512	0.8736	0.1560

Fig. 1 CPU time for
encryption

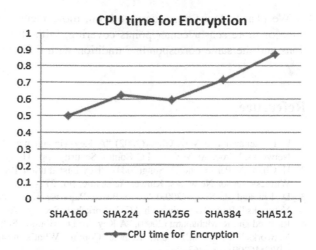

Fig. 2 CPU time for
decryption

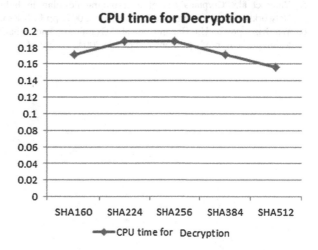

5 Conclusion and Future Work

This paper proposes a high-secure three-level structural engineering utilizing
Chaotic neural network. The proposed calculation scrambles the irregular pair
insightful polynomial key produced utilizing SHA calculation. The CPU time of the
proposed calculation is examined for different polynomial key era calculations. The
quality of the key is incredibly enhanced by encoding the key utilizing disordered
neural system. A renovated secure three level plans going for the establishment of
remote sensor frameworks with polynomial pool based adaptable sink. A recreation
test under MATLAB demonstrates that the calculations exhibited in this paper
outflank the encryption and decoding time with results.

We plan to focus our future work on, more identified boggling application by three of more sensor center points covering, with strengthened grouping calculations and the same corresponding tradition with different portable sinks.

References

1. L. Eschenauer and V.D. Gligor(2002) "A Key-Management Scheme for Distributed Sensor Networks," Proc. ACM Conf. PC Comm. Security, pp. 41–47.
2. H. Chan, A. Perrig, and D. Song(2004), "Key Distribution Techniques for Sensor Networks," Wireless Sensor Networks, Kluwer Academic, pp. 277–303.
3. D. Liu and P. Ning, (2003) "Area Based Pairwise Key Establishments for Static Sensor Networks," Proc. To begin with ACM Workshop Security Ad Hoc and Sensor Networks.
4. Rasheed and R. Mahapatra (2009), "A Key Pre-Distribution Scheme for Heterogeneous Sensor Networks," Proc. Int'l Conf. Remote Comm. What's more, Mobile Computing Conf. (IWCMC'09), pp. 263–268.
5. Yang et al. "Cryptanalysis of a cryptographic plan in light of deferred Chaotic Neural Networks". Arrangements and Fractals 40 - 2009; pp 821–825.
6. Wu Xiao posse, Hu Hanping, et al. "Examining and enhancing a tumultuous encryption technique. Disarray", Solitons & Fractals 2004; 22: pp 367.

A Regularized Constrained Least-Squares De-blurring Method for SPECT Images

Neethu Sasi and V.K. Jayasree

Abstract Nuclear medicine images suffer from blur due to the scattering of emitted radiations. An image processing technique is proposed in this paper to reduce blur in nuclear images. This is achieved in two main stages. A maximum likelihood estimate of the distortion operator or the point spread function is computed from the image itself. Then, regularized least-squares filtering is performed constrained to the noise power computed from the image. Pre-filtering is also done to avoid unwanted high frequency drops. The algorithm is tested on real cardiac single-photon emission computed tomography images. Quantitative and qualitative evaluations of the algorithm show the potential of proposed algorithm in reducing blur while maintaining high peak signal-to-noise ratio.

Keywords Least-squares filtering · Maximum likelihood estimate · Nuclear medicine imaging · Single-photon emission computed tomography imaging

1 Introduction

Heart disease is a major health problem in the developed world [1]. According to the statistics in 2008, about 17.3 million people died from heart diseases, representing 30 % of all global deaths [2]. This statistic marks urgency for the early diagnosis of heart diseases. Nuclear medicine imaging is becoming common nowadays in the diagnosis of heart diseases [3]. Nuclear medicine imaging is a non-invasive technique in which a radioactive tracer is injected into the human body. This tracer is carried to the heart through blood flow. Using a rotating camera, multiple images in different directions are obtained which show the uptake of tracer

N. Sasi (✉)
Government Model Engineering College, Ernakuam, Kerala, India
e-mail: neethumsasi@gmail.com

V.K. Jayasree
College of Engineering, Cherthala, Kerala, India

© Springer Science+Business Media Singapore 2017   137
S.C. Satapathy et al. (eds.), *Proceedings of the First International Conference on Computational Intelligence and Informatics*, Advances in Intelligent Systems and Computing 507, DOI 10.1007/978-981-10-2471-9_14

by the heart. Different heart slices, that is, short axis, horizontal long-axis and vertical long-axis slices, are then reconstructed from these multiple images [4].

Single-photon emission computed tomography (SPECT) image is a popular nuclear medicine imaging tool in the diagnosis of cardiac diseases. But its diagnostic accuracy is reduced by patient motion and photon scattering. This introduces blur in such type of images. This work proposes a method to improve the visual quality of cardiac SPECT images by reducing blur. Methods to improve the visual quality of nuclear images fall into two categories: algorithms performed during the reconstruction process and algorithms performed on the reconstructed images. The first category methods are based on the system point spread function (PSF) modeling [5] and the second category makes use of image processing techniques. Our method falls under the second category. A two-stage image de-convolution technique is employed using maximum likelihood estimate and regularized least-squares filtering.

Many authors had addressed the problem of improving the quality of blurred images. A non-negativity and support constraints recursive inverse filtering (NAS-RIF) algorithm was proposed in [6] and its extension in three-dimensional domains is given in [7]. Bayesian tomography reconstruction methods are considered in [8]. In [9], the blur transfer function for a SPECT image is approximated with a two-dimensional symmetric Gaussian function. An analysis of blind de-convolution algorithms is presented in [10]. In [11], a two-step iterative shrinkage and thresholding algorithm has been proposed, exhibiting a faster convergence rate.

2 Methods

2.1 Image Blurring Model

A basic image blurring model in SPECT imaging is shown in Fig. 1.

The original image is represented by $o(x, y)$; $b(x, y)$ represents the blurred image; $h(x, y)$ represents the point spread function or the distortion operator; and $n(x, y)$ represents the noise. The blurred image is obtained by the convolution between the original image and the point spread function. In a noisy system, noise also gets added to it.

Fig. 1 Image blurring model

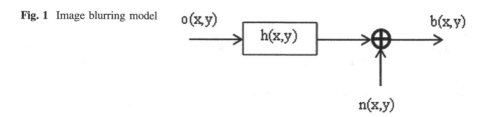

2.2 Proposed Algorithm

This paper proposes a method to de-blur cardiac SPECT images. As depicted in Fig. 1, getting back $o(x, y)$ from $b(x, y)$ becomes a de-convolution problem. In this paper, this is achieved using a two-stage de-convolution algorithm.

Different slices or tiles in a cardiac SPECT image represent the blood flow through the heart at different instances of time. A slice or a tile is a 2D image of a portion of the heart at an instant of the heart cycle. As an initial step, each tile, $b_i(x, y)$, is extracted from the image, and further processing is done on the tiles taken individually. In the first stage, an estimate of the PSF is obtained using maximum likelihood estimation principles. In the second stage, constrained least-squares filtering is done to get a better estimate of the original image tile using the ML estimate of the PSF. The main contribution of this paper is that the least-squares filtering is constrained to the noise power estimated from each tile.

Maximum likelihood estimate. From Fig. 1, getting back $o(x, y)$ from $b(x, y)$ becomes a de-convolution problem. In maximum likelihood approach, original image, blurred image, point spread function and noise are treated as stochastic quantities with probability density functions $p(o)$, $p(b)$, $p(h)$ and $p(n)$, respectively. The likelihood function $p(b/o, h, n)$ defines the random process defined by Fig. 1. Maximizing the likelihood function over o, h, n gives the maximum likelihood solution.

$$\{o_{ML}, h_{ML}, n_{ML}\} = \arg_{o,h,n} \max p(b/o, h, n). \tag{1}$$

In the case of cardiac SPECT images, the distortion operator for each tile is unknown. A maximum likelihood estimate of the original image tile and the distortion operator or the point spread function is made as in Eq. (2). We consider noise in the next step.

$$\{o_{ML}, h_{ML}\} = \arg_{o,h} \max p(b/o, h) \tag{2}$$

Discrete Fourier transform is used to implement maximum likelihood estimation. This may introduce ringing effect. Pre-filtering is done to reduce this ringing effect. Maximum likelihood estimate of the original image and the point spread function are obtained using Bayesian approach. The detailed algorithm for estimating the original image and the PSF is given below.

Step 1: An initial value is assigned for the PSF:

$$\hat{h}_i^{(initial)}(x, y) = \text{ones(arbitrary size)} \tag{3}$$

Step 2: An initial value is assigned for the original image:

$$\hat{o}_i^{(\text{initial})}(x, y) = b_i(x, y) \tag{4}$$

Step 3: Since this is an iterative algorithm, the maximum number of iterations to be performed is set as P.

Step 4: The ML estimate is computed with non-negativity constraints applied in each step.

$$\hat{o}_i^{p+1}(x, y) = \hat{o}_i^p(x, y) \left[\hat{h}_i^p(x, y)^* * \frac{b_i(x, y)}{\hat{h}_i^p(x, y) * \hat{o}_i^p(x, y)} \right] \tag{5}$$

$$\hat{h}_i^{p+1}(x, y) = \hat{h}_i^p(x, y) \left[\frac{b_i(x, y)}{\hat{h}_i^p(x, y) * \hat{o}_i^p(x, y)} * b_i(x, y)^* \right] \tag{6}$$

Step 5: When the number of iterations reaches P, $\hat{h}_i^P(x, y)$ gives the ML estimate of PSF.

Constrained least-squares filtering. As shown in Fig. 1, $b(x, y)$ is obtained by the convolution between $h(x, y)$ and $o(x, y)$.

$$b(x, y) = h(x, y) * o(x, y) + n(x, y) \tag{7}$$

$$b = Ho + n \tag{8}$$

Restoration using least-squares filtering is done subjected to the constraint:

$$\|b - Ho\|^2 = \|n\|^2 \tag{9}$$

The frequency domain solution to this problem is:

$$\hat{O}(u, v) = \left[\frac{H^*(u, v)}{|H(u, v)|^2 + \gamma |P(u, v)|^2} \right] B(u, v) \tag{10}$$

In our algorithm the noise power is computed from the mean and variance estimated from each image tile and the parameter γ is selected proportional to the noise power. $P(u, v)$ is the regularization operator, and a smoothening function is used as the regularization operator. ML estimates obtained from the first stage act as inputs to the least-squares restoration operation.

3 Results

3.1 Simulation Setup

Simulations have been carried out in MATLAB to evaluate the performance of the proposed method both qualitatively and quantitatively. 35 real cardiac SPECT images taken from 35 different patients are used as the database. The patients are in the age group 40–60 years. 14 patients have normal functioning of the heart and 21 patients have myocardial perfusion defects. The images are in gray scale with size 952×510 and 40 tiles in each image.

Visual quality is used as the qualitative measure for evaluating the performance of the proposed method. The difference between the blurred image and the de-blurred image can be seen by the visual analysis of the images.

Peak signal-to-noise ratio (PSNR) and blur metric are used to analyze the images quantitatively.

PSNR is defined as:

$$PSNR = 10 * \log\left(\frac{Int_{max}^2}{MSE}\right) \tag{11}$$

The maximum intensity in the image is given by Int_{max} and the mean squared error is given by MSE. The higher the value of PSNR, the better is the quality of the image in terms of signal-to-noise ratio.

Blur metric gives the amount of difference in blur between the blurred image and the de-blurred image. This measure is based on the discrimination between different levels of blur perceptible on the same image [12]. The lower the value of blur metric, the better is the quality of the image.

3.2 De-blurring Results

In this section, we present the results of applying the proposed method on cardiac SPECT images. The proposed method is evaluated qualitatively by the visual analysis of the images. Figure 2 shows four tiles from the input blurred image. The image de-blurred using the proposed method is shown in Fig. 3.

We quantitatively evaluated the performance of the proposed method in terms of blur metric and peak signal-to-noise ratio (PSNR).

Table 1 compares the performance of the proposed algorithm with conventional image de-blurring techniques. The use of the proposed method yields a lower blur as indicated by the blur metric and lower noise as indicated by the value of PSNR.

Fig. 2 Blurred image

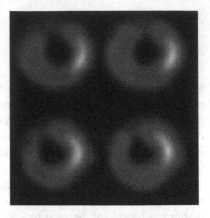

Fig. 3 De-blurred image
using proposed method

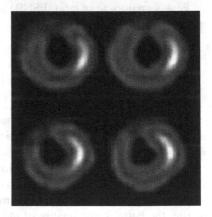

Table 1 Comparison results
of applying the proposed
method on cardiac SPECT
images

Method	Blur metric	PSNR
Blurred image	0.8427	25.0656
Blind de-convolution	0.8151	25.4652
Lucy Richardson	0.8168	17.4167
Proposed method	0.7485	26.0954

Figure 4 shows the ability of the proposed method in reducing blur. The lower the
value of blur metric, the lower is the amount of blur present in the image. The
proposed method gives better PSNR value as shown in Fig. 5.

Fig. 4 Comparing the proposed method with other state-of-the-art methods in terms of blur

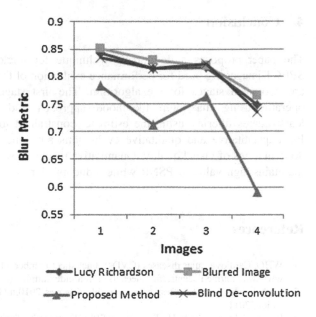

Fig. 5 Comparing the proposed method with other state-of-the-art methods in terms of PSNR

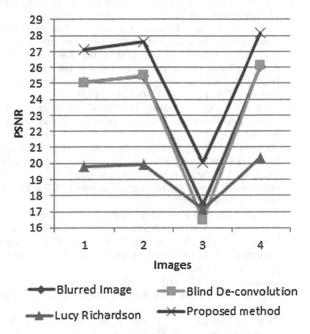

4 Conclusion

The paper proposes a de-blurring technique for nuclear images. Real cardiac SPECT images are used for performance evaluation of the proposed method. There are two main stages for the algorithm. The first stage estimates the distortion operator using maximum likelihood approach and the second stage does least-squares filtering using the estimate, constrained to the image noise power. Both quantitative and qualitative evaluations are done. Qualitative evaluation is done in terms of visual quality. Quantitative results show that the proposed method maintains high value of PSNR while reducing blur.

References

1. WHO, Cardiovascular diseases (CVDs), Fact sheet number 317 Updated March 2013 http://www.who.int/mediacentre/factsheets/fs317/en/index.html.
2. Global status report on non-communicable diseases 2010.mGeneva, World Health Organization, 2011.
3. Rahmim, A., and Zaidi, H.: PET versus SPECT: strengths, limitations and challenges. Nuclear Medicine Communications 2008, 29, pp. 193–207 (2008).
4. Camici, P., G. and Rimoldi, O., E.: The Clinical Value of Myocardial Blood Flow Measurement. The Journal of Nuclear Medicine, 50(7), pp. 1076–1087 (2009).
5. Mikhno, A., Elsa, D., Angelini, Bing Bai, Laine, A.F.: Locally weighted total variation denoising for ringing artifact suppression in pet reconstruction using PSF modeling, 2013 IEEE 10th International Symposium on Biomedical Imaging: From Nano to Macro, San Francisco, CA, USA, pp. 1240–1243 (2013).
6. Kundur, D, and Hatzinakos, D.: "Blind image restoration via recursive filtering using deterministic constraints," in Proc. Int. Conf. Acoustics Speech, and Signal Processing, 4, pp. 547–549, (1996).
7. Mignotte M., and Meunier, J.: "Three-dimensional blind deconvolution of SPECT images," IEEE Transactions on Biomedical Engineering, 4(2), pp. 274–281, (2000).
8. Gindi, G., Lee, M., Rangarajan, A. and Zubal, I.G.: "Bayesian reconstruction of functional images using anatomical information as priors," IEEE Transactions on Medical Imaging, 12 (4), pp. 670–680, (1993).
9. Madsen, M.T. and Park, C.H.: "Enhancement of SPECT images by Fourier filtering the projection set," Journal of Nuclear Science, vol. 26, pp. 2687–2690, (1979).
10. Levin, A., Weiss, Y., Durand, F., Freeman W.T.: "Understanding blind deconvolution algorithms", IEEE Transactions on Pattern Analysis and Machine Intelligence, 33(12), pp. 2354–2367 (2011).
11. Bioucas-Dias, J.M. and Figueiredo, M.A.T.: "A new TwIST: Two-step iterative shrinkage/thresholding algorithms for image restoration," IEEE Trans. on Image Processing, 16, pp. 2992–3004, (2007).
12. Crete, F., Dolmiere, T., Ladret, P. and Nicolas, M.: "The Blur Effect: Perception and Estimation with a New No-Reference Perceptual Blur Metric", SPIE Electronic Imaging Symposium Conf Human Vision and Electronic Imaging, San Jose, (2007).

Effective Video Data Retrieval Using Image Key Frame Selection

D. Saravanan

Abstract Owing to the rapid growth of multimedia technology, multimedia information is easily accessed by any user and the same information construction and distribution are also very easy. Due to technology development, the multimedia information increases due to variety of factors: it can be uploaded by unprofessional users nowadays. Due to the low quality and the large number of duplicated video files available, this leads video extraction more and more complex. The general method of representing each video segment is shot that consists of series of frames. Among this series, the input frame based shot method is specifically assisted for searching the video content as clients provided image query/search where an image will be matched with the indexed key frames with assist of resemblance distance. As a result, the key frames selection is most significant, and several methods are used to automate the process. This paper proposes a new technique for key frame selection. The proposed method shows significantly good and the experiments prove the above statement.

Keywords Video data mining · Data mining · Key frame extraction · Knowledge extraction · Multimedia data · Data extraction · Video dates · Frames

1 Introduction

Data mining is a process of detecting knowledge from a given huge set of data. Of the available huge data set, multimedia is the one which contains diverse data such as audio, video, image, text and motion, and video data play a vital role in the field of video data mining. In short, the application of video data is called video data mining. Data mining helps the users to retrieve the efficient content using data mining preprocessing operations. Increasing the quantity of video content reduces

D. Saravanan (✉)
Faculty of Operations & IT, IFHE University, Hyderabad 501 203, Telangana, India
e-mail: Sa_roin@yahoo.com

© Springer Science+Business Media Singapore 2017   145
S.C. Satapathy et al. (eds.), *Proceedings of the First International Conference on Computational Intelligence and Informatics*, Advances in Intelligent Systems and Computing 507, DOI 10.1007/978-981-10-2471-9_15

the quality of the content; flexibility of the network produces duplicated information. Technology brings the huge amount of multimedia data sets. From this data set it is very difficult to the user, extract the needed informaton. For reduce the burden (Searching) data mining helps remove the unwanted infromation from the given data set. Nowadays, information is shared in the form of images instead of text information. Technology allows the users to share and upload this information in an easy manner. For retrieving these multimedia content such as image and motion, pixel values are quite difficult. This brings today most challenging operations for both users and researchers. While some basic forms of multimedia retrieval are available on the Internet, these tend to be inflexible and have significant limitations. The key frame selection is having two main issues: 1. The number of key frame(s) utilized, (The first issue is tackled by where the amount of key frames for every shot will be decided arbitrarily using the shot length). 2. The significant representative frame(s) selection in a shot. (The second issue is generally complicated for choosing the frames automatically with maximum semantic value. This issue is handled through minimizing the redundant frames with the help of the methods, for example, relevance ranking). Existing methods for searching video to identify co-derivatives have substantial limitations: they are sensitive to degradation of the video; they are expensive to compute; and checking the whole video files is quite complex, and also comparing the entire video content is not possible. Existing techniques perform direct comparison of video features between the query clip and the data being searched, which is computationally expensive.

2 Existing System

- There is no proper indexing and retrieval process available. Existing indexing techniques are suitable for only few sets of video files.
- Vast amount of video files are currently available on the web. There is not proper mechanism to arrange these contents. Proper arrangement reduces the searching time of the user.
- Current technique focuses on text-to-image retrieval; it never produces good result. Every search engine never returns the same type of information even if the users' query may be the same.

2.1 Drawback of Existing System

- This approach does not consider about the false positives and false negatives in the given video.
- The fault tolerance value is not reduced.

- Incorporating genetic variations into the design will affect the accuracy of this work.
- It considers only the action-based video mining.
- Human motion detection specifies only a particular region.

2.2 Advantage of Proposed System

- Compared to other search method, this method can also reduce the detection time.
- It first pre-processes the query image and extracts the features of that image.
- Trained videos are stored in the database, and the features of the trained videos are clustered using the extracted features of the query input image.
- Finally, features matching procedure is implemented to identify the similar features and to retrieve the relevant video.
- This method provides an efficient video retrieval using an image as input.
- Efficient clustering process is implemented.
- Features matching provides an efficient and accurate similar video retrieval

3 Literature Survey

Effective Multimedia Content Retrieval [1]. This paper brings the effective multimedia content retrieval using hierarchical clustering algorithm. Clustering provides grouping the data set effectively; it also reduces the searching time. Real-Time Human Pose Recognition in Parts from Single Depth Images [2]. The problem of predicting the human pose recognition in parts in a single depth image is discussed here. A new method should be proposed to quickly and accurately predict the position of the body joints from a single depth image. In Video Mining with Frequent Item set Configurations, [3] a new method for mining frequently occurring objects and scenes from videos is proposed. Object candidates are detected by finding recurring spatial arrangements of affine covariant regions. Content Based Image Retrieval using Color Histogram; [4] this paper brings the information retrieved based on the content; image can be extracted using image features such as text, pixel value, motion, frame value. Here, they proposed image feature vector technique, construct color histogram. An enhanced technique based on content-based image retrieval and video indexing [5]; this paper brings the new indexing technique for video data file using one of the image features like color. With the help of global color histogram (GCH) and histogram analysis, they create an indexing operation.

4 Experimental Setup

Experimental setup consists of a four-step operation. First, creation of an admin database; here, video frames are converted as shots, and those values are stored in the database [6]. Followed by users' image query, here, user input frames are taken, whose values are compared with existing stored value, and then, creation of key frame indexing; it reduces the searching time. The last phase called video retrieval is the process of user input frame matched with stored value, and similar value returned to the user. This experiment is conducted for various video files such as song, game, debate, news, and animated sets of video files. The entire process is constructed and tested using JAVA coding. Experiment is conducted as: for each video file, one input frame is selected, and then, based on the input, output extractions are done.

4.1 Creation of Admin Database

Key frame selection process is done with a two-step process. First phase called training phase or user side phase. The second phase is input query phase or server side phase. In the first phase, video files are converted as frames; after successful elimination of duplication frames, frames histogram values are calculated and stored in the admin database. Using file handling method, the duplicate files are eliminated. Key frame is determined by the frame that is the most same to the average frame. In the second phase, user input query; here, user input frame is compared with our stored frame value. Frames those are matching with input frame are extracted and returned to the user (Fig. 1).

4.1.1 Video Training Set Algorithm

Step1: Select input video file.
Step2: Segment the input file extract frames.
Step3: Eliminate duplication using image histogram technique.
Step4: Select key frames.
Step5: Per from Client and Servicer side operations.
Step 6: Based on users input, do video frame matching.
Step7: Retrieve and send matched result.

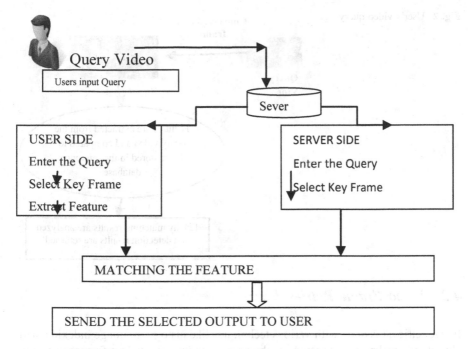

Fig. 1 Proposed architecture

4.2 User's Image Query

Fine matching stage is specially designed to choose which image in the image database is the most relevant one with that of the query image. Also, the measure to signify the degree of similarity has to be described. In this phase, input query image features are extracted using image feature extraction [7]. Extracted value is compared with admin-stored database values. The matching frames are returned to the user (Fig. 2).

Pseudo code for duplication removal:

```
Grey value = lngGrayScaleValue = (0.299 * clrPixel.R) +
(0.587 * clrPixel.G) + (0.1114 * clrPixel.B)
Grey = Σ Grey value
'Grey' gave the value of the grey value of the whole image.
Grey value = Image1.Greyvalue - Image2. Grey Value
If Grey value < threshold then Duplicate Image
```

Fig. 2 User's video query

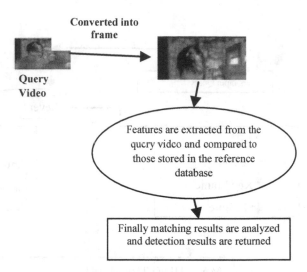

4.3 Video Frame Retrieval

Image retrieval is done with various techniques like histogram, image indexing, and algorithmic techniques. Indexing is the process to improve the performance and to reduce the searching time. Frame retrieval consists of the following steps:

Step1: Select the input frame.
Step2: Calculate the pixel value of input frame.
Step3: Convert the value obtained in Step2 into histogram value.
Step4: Compare this value with the existing value stored in the database (Fig. 3).

Fig. 3 Steps of video frame retrieval process

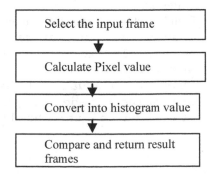

4.4   Creation of Key Frame Indexing

Due to the availability of video content in the web, it is very difficult to organize and retrieve the content. Increasing the usage of video data file, today, it is necessary for video data indexing. Many number of indexing techniques are available today, but the experimental result shows that each technique supports a particular type of video files only. For this reason, improvement is needed in each methodology. Here, we proposed key frame indexing technique for image retrieval, and the experimental results prove that the proposed technique provided better results (Fig. 4).

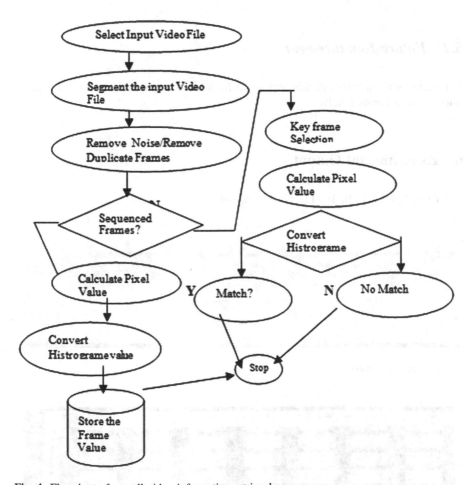

Fig. 4 Flowchart of overall video information retrieval process

5   Conclusion and Future Enhancement

This work brings out an efficient technique for video information retrieval. Because of the network development and its flexibility, the usage of image information has increased in recent years. This is because of the increasing demand for image information, and also, it provides very easy access to the video contents. Every day, the amount of these image files is increasing due to various factors like usage of mobile phones, twitter, facebook, YouTube and more. Creation and distribution of these files are very easy. Still, we suffer to manage and retrieve these contents There are no efficient techniques or tools presently available. This paper brings the key frame image indexing technique for retrieving and efficiently indexing the video files. Results proved that the proposed method is more efficient.

5.1   *Future Enhancement*

The future enhancement of this technique by adding additional features will produce more accurate results.

6   Experimental Output

See Figs. 5, 6, 7, 8, 9, 10, 11, 12, 13 and 14.

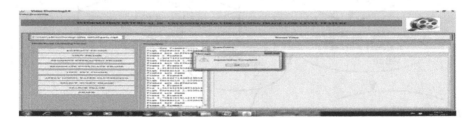

Fig. 5 Video segmentation

Fig. 6 Video shots (Frames)

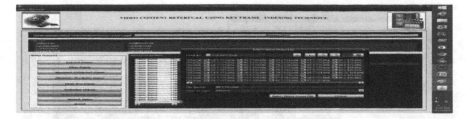

Fig. 7 Duplication removal operations

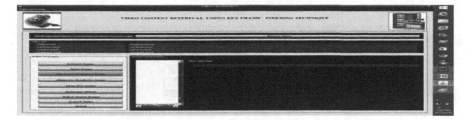

Fig. 8 Training the input frames

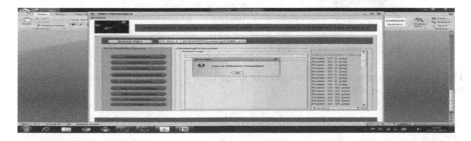

Fig. 9 Clustering the frames

Fig. 10 Key frame selection

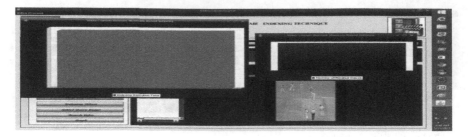

Fig. 11 Output for selecting input key frame

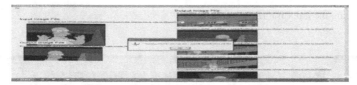

Case 1: 1 input and 8 output files

Case 2: 1 input and 15 output files

Fig. 12 Image comparison based on user input image Case 1: 1 input and 8 output files. Case 2: 1 input and 15 output files

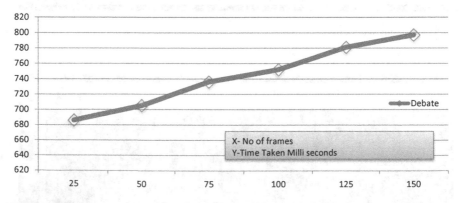

Fig. 13 Performance graph for news video file

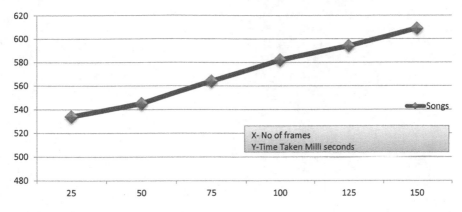

Fig. 14 Performance graph for song video file

References

1. D. Saravanan.: Effective Multimedia Content Retrieval. International Journal of Applied Environmental Sciences, Volume 10, Number 5, pp. 1771–17783, (2015).
2. Jamie Shotton, Alex Kipman, Andrew Fitzgibbon, Mark Finocchio,: Time Human Pose Recognition in Parts from Single Depth Images, Communications of the ACM, Volume 56 Issue 1, Pages 116–124, (January 2013).
3. H. Sundaram et. al., "Video mining with frequent item set configuaration", CIVR 2006, LNCS4071, Pages 360-369, 2006.
4. D. Saravanan, A. Ramesh Kumar, 2013 "Content Based Image Retrieval using Color Histogram", International journal of computer science and information technology (IJCSIT), Volume 4(2), 2013, Pages 242–245.
5. Ghorpade Krishna Bhimraj, Swapnalini Pattanaik, "An enhanced Technique Based on Content Based Image Retrieval and video Indexing" Proceedings of 19th IRF International Conference, 1st February 2015, Pune, India, ISBN: 978-93-84209-85-8, Pages 45–50.
6. D. Saravanan, V. Somasundaram 2014. Matrix Based Sequential Indexing Technique for Video Data Mining, Journal of Theoretical and Applied Information Technology 30th September 2014. Vol. 67 No. 3.
7. D. Saravanan, Performance Anlaysis of video data image using Clustering Technique, Indian journal of science and technology,Vol 9(10), DOI:10.17485/ijst/2016/v9i10/79731, March 2016, ISSN (Print): 0974-6846. Pp 01-06

A Review of Joint Channel Assignment and Routing Protocols for Wireless Mesh Networks

Satish S. Bhojannawar and Shrinivas R. Mangalwede

Abstract It is envisaged that Quality-of-Service (QoS) support for multimedia-rich applications is the key to the success of next-generation wireless mesh networks (WMNs). However, QoS support in WMN is challenging, because of the limited network capacity and intensive resource requirements. In the recent past, multi-radio multi-channel (MRMC) networking is used as a promising approach to boost network capacity. By assigning non-overlapping channels to radios, MRMC-WMN can reduce interference and, hence, increase the network capacity. The performance of MRMC-WMNs is very much dependent on the working of routing and channel assignment. Routing and channel assignment are tightly coupled and should be jointly optimized. Over the years, different research works have been done to address the issues in a joint channel assignment and routing (JCAR). This paper critically reviews the existing JCAR approaches taking different parameters, such as routing metrics used, interference model, and methodology used. The authors also present the pros and cons of these approaches. Future research directions in JCAR are also discussed.

Keywords Multi-radio multi-channel wireless mesh network · Quality-of-service · Channel assignment · Routing

1 Introduction

A wireless mesh networking is a promising technology for ubiquitous last-mile Internet access. WMN is a self organizing, self-configuring, and self-healing technology used for the implementation of various applications, such as home

S.S. Bhojannawar (✉) · S.R. Mangalwede
Department of Computer Science and Engineering, Gogte Institute of Technology, Belagavi, India
e-mail: satishsb2007@gmail.com

S.R. Mangalwede
e-mail: mangalwede@git.edu

© Springer Science+Business Media Singapore 2017   157
S.C. Satapathy et al. (eds.), *Proceedings of the First International Conference on Computational Intelligence and Informatics*, Advances in Intelligent Systems and Computing 507, DOI 10.1007/978-981-10-2471-9_16

networking, military communication system, industry monitoring and control, public service communication metropolitan area networks, transportation systems, building automation, environment monitoring, healthcare systems, security surveillance systems, and spontaneous networking [1].

Communication in WMNs can create bottlenecks for the successful implementations of various applications. To deploy different applications, WMNs need to support efficient communication that handles QoS requirements of applications without disturbing the on-going operation of network [2, 3]. Routing severely affects the communication quality, as it provides end-to-end path between source and destination. It is essential for the routing algorithm to provide feasible path that provides minimum assured level QoS. The bandwidth available on the links and network data transmission capacity determines the existence of feasible paths which are specified by the channel assignment. As physical channel capacity of link is shared with other interfering links; channel assignment tries to assign non-overlapping channels to interfering links. On the other hand, channels are assigned based on the traffic load, which is specified by the routing. Hence, channel assignment and routing both need to be jointly optimized. Since joint channel assignment and routing is a fundamental issue for the performance optimization in multi-radio multi-channel wireless mesh networks; in this paper, we focus only on the existing JCAR approaches for WMNs.

The paper is organized as follows. In Sect. 2, we describe main idea, and the basic steps of every JCAR approach individually along with their pros and cons. In Sect. 3, we present future direction for the JCAR for WMNs. Finally, the conclusion is drawn in Sect. 4.

2 Joint Channel Assignment and Routing Approaches for WMNs

In this section, review the existing JCAR approaches for multi-radio multi-channel WMNs.

2.1 Robust Joint Channel Assignment and Routing with Time Partitioning (RCART) [4]

RCART performs routing under highly extremely traffic load. Depending on the different traffic profiles, distinct routing strategies are used at different times of the day. The protocol model is used to measure the interference. This approach has implicit focus on QoS.

It uses following three steps to solve joint routing and channel assignment problem.

Step1. **Time partitioning and construction of traffic profile**: This step divides the time into periodic intervals and constructs traffic profiles within each interval. If particular routing scheme is applicable to similar traffic profiles, these profiles are grouped using the hill climbing algorithm. Traffic across different access points within each interval is then represented as convex region, which then acts as a input for the next two steps.

Step 2. **Robust routing**: Depending upon the traffic requirements that fall within convex region, this steps finds robust routing strategy. This is done using flow, path, and channel capacity constraints.

Step 3. **Channel assignment**: During the time interval determined by the step 1, based on the traffic distribution knowledge learnt in the step 2, radios are assigned to static channels.

Pros:

(i) It uses traffic-based routing.
(ii) Node switches channel assignment and routing at pre-assigned times.
(iii) To improve the performance, this approach identifies fitting time intervals for the forthcoming traffic profiles.

Cons:

(i) Hill climbing will not necessarily find the global maximum.
(ii) Congestion ratio may vary depending upon the number of channels.
(iii) This approach has constrained-based performance (trade-off required).
(iv) To reduce the time complexity, the heuristic algorithm is used to determine the number of intervals.

2.2 Channel Assignment and Routing Using Ant Colony Optimization in Multi-radio Wireless Mesh Networks [5]

This approach solves JCAR problem using extended ant colony optimization (ACO) framework. We expand ACO framework by executing the channel assignment procedure and routing procedure with the aim of reducing the network interference. This approach has implicit focus on QoS. The routing procedure uses link estimation routine to estimate the quality/cost of transmission link in terms of average packet transmission delay and path estimation routine to reduce the intra- and inter-interference to meet the demands and properties of WMNs. On particular node, the distributed channel assignment procedure is used to send an ant agent to switch channels/interfaces. Channel assignment procedure dynamically assigns channels based on the current traffic on node as indicated by the routing.

Pros:

(i)   No a-prior traffic flow information is required
(ii) It has inherent parallelism
(iii) It handles the network traffic dynamics
(iv) It discovers high-throughput paths with less inter and intra—flow interference.

Cons:

(i)   It has implicit focus on QoS
(ii) The time to convergence is uncertain
(iii) ACO-based algorithms crucially rely on repeated path sampling, thereby leading to a significant overhead

2.3 Generalized Partitioned Mesh Network Traffic and Interference Aware Channel Assignment (G-PaMeLA) for WMN [6]

Divide and conquer methodology is used to solve JCAR problem by dividing the problem into number of small sub-problems. These sub-problems are then locally optimized and solved sequentially (Phase-I). Each sub-problem covers nodes which are characterized by the similar attribute, such as number of hops to gateway. A post-processing procedure is used to combine the results of sub-problems to improve the network connectivity (Phase-II). The sub-problems are solved in sequential order as given by the ranking function. Each of the sub-problems is expressed as anteger liner programming optimization problem which is solved using the branch and cut method.

This approach has two different versions: (i) generalized and (ii) customized. Generalized version is used for the unknown topology, where the number of sub-problems is equivalent to network width to the gateway. With known topology, customized version solves JCAR problem by customizing the number of ranking functions and imposing various routing constraints. Physical interference is model used. Load and inference aware routing metric is used.

Pros:

(i)   This approach is easily scalable.
(ii) This approach has traffic and interference aware channel assignment.
(iii) It makes efficient usage of radio interface by defining logical links.

Cons:

(i) It considers stable channel condition and traffic loads, and does not consider online traffic.

(ii) It uses centralized approach to solve the JCAR problem.

(iii) It gives optimal solution only for known network topologies like grid topology.

2.4 Joint Routing and Channel Assignment Protocol (JRCAP) for the IEEE 802.11s Mesh Networks [7]

JRCAP is distributed, on-demand routing protocol to solve JCAR problem. It merges routing and channel assignment to provide path and to allocate channels for links constructing the path. It has two phases, configuration phase and routing phase. In configuration phase, density-based clustering algorithm divides the network into equitable clusters. To each cluster, a permanent channel is assigned using DSATUR (Degree of Saturation) graph-coloring algorithm. Load-aware channel selection scheme is used to assign channels. In routing phase, on-demand routing protocol inspired by hybrid wireless mesh protocol is used to find route to destination. A routing metric called maximal residual capacity (MRC) is used. MRC is based on the data transmission rate, channel diversity, and channel load. A protocol interference model is used.

Pros:

(i) It uses on-demand routing that is used to find routes.

(ii) It considers channel diversity.

(iii) Balanced clustering is used to restrict the number of nodes in each cluster to minimize the interference.

Cons:

(i) Dedicating a interface on each node in each cluster poses heavy overhead.

(ii) Static channel assignment is used.

(iii) It does not respond well with varying densities specially in hierarchical Topologies.

(iv) Messaging overhead is more.

2.5 Joint Routing and Channel Assignment (JRCA) Scheme for WMN [8]

JRCA is centralized approach to solve JCAR problem. A quality-aware route selection process is used to find end-to-end communication path between source and destination in multi-radio multi-gateway wireless mesh networks. In this approach, any casting routing model is used. The objective of this approach is to find. (i) the optimal gateway among the set of gateways for each source (ii) best route from source selected gateway to source with channel being assigned on each link of that route. A combination of genetic algorithm and backtracking is used to assign channel for each link of the route. On-demand routing algorithm is used to find the routes. A combination of end-to-end probability of success and routing delay is used as routing metric. Co-channel interference is modeled using conflict graph. Genetic algorithm along with backtracking is applied to reduce the convergence time.

Pros:

(i) Performance characteristics of the data packet transmission are used for estimating the route quality metric
(ii) It captures the effects of inter-flow interference and intra-flow interference
(iii) This approach has increased throughput and delivery ratio.

Cons:

(i) Because of centralized approach—difficult to get complete network connectivity information
(ii) It has large control overhead in channel negotiation
(iii) Equal transmission load is considered to calculate POS and delay, but in practice, loads may differ from node to node

2.6 Joint QoS Routing and Channel Assignment (JQRCA) for WMN [9]

It is on-line, iterative approach which attempts to maximize acceptance rates of user demands. It is QoS-based routing along with channel assignment. It uses two phases to find path. Based on the end user's QoS demand, the routing algorithm first finds end-to-end path. If this path does not fulfill the demands of end user, then QoS-driven dynamic channel assignment algorithm is used to discover the violated links and tries to reassign channels to make the path to fulfill demands of end user. If this algorithm fails, a new alternative path is found and the same attempt is made. This iterative process continues for k paths, until a path is found that meets end-user demand. A protocol interference model is used.

Pros:

(i) It has explicit focus on QoS.
(ii) Dynamic channel assignment is used to adjust network resources as per the end-user demands.
(iii) On-demand routing algorithm.

Cons:

(i) It is centralized algorithm.
(ii) It uses greedy call admission control.
(iii) It uses static interference sets.
(iv) Rerouting of flows is not allowed.
(v) Response time is more.

2.7 Channel and Routing Assignment with Traffic Flow (CRAFT) for WMN [10]

It is adaptive, distributed, cooperative easy to implement solution to JCAR problem. It jointly optimizes channel assignment and routing to enlarge an objective function that prototypes interference and traffic demands of all nodes. CRAFT tries to improve network performance in terms of throughput and end-to-end delay. Each node maintains routing and channel assignment (RCA) decision table to make routing decision. A physical interference model is used. In each adaption, CRAFT makes each node's routing and channel assignment decision to enhance the objective function. CRAFT has three phases.

Start phase: Builds a RCA decision of whole network by exchanging RCA decision tables among nodes. Link-state routing protocol is used.

Improvement phase: Each node tries to improve the objective function by computing RCA decision table. Each node distributes its newly computed RCA decision table to network and then waits for random period of time to re-enter this phase.

End phase: If there is no improvement in the objective function, CRAFT ends.

The CRAFT-random decision approach can execute CRAFT several times to finalize RCA decision. CRAFT-traffic-prioritized decision approach assigns high priority to higher traffic and routes it using better path.

Pros:

(i) Event drive updates are used.
(ii) It improves throughput by reassigning channels for small number of nodes.
(iii) Destinations with high traffic are given high priority.

Cons:

(i) Processing overhead is more, as RCA decision tables are frequently distributed among nodes.
(ii) Latency increase with large network.
(iii) Link-state routing sees link as either working and failed which is not always true in WMNs.

2.8 Joint Temporal-Spatial Multi-channel and Routing Assignment for Resource Constrained WMN [11]

It is distributed heuristic scheme. Every node is assumed to have single radio interface. Different channel assignment methods are used for gateway nodes and non-gateway nodes. Gateway node rotationally utilizes its channels to provide fair access to other nodes. Temporal scheme makes sure that all the non-gateway and non-sender nodes have fair access to the gateway. Spatial scheme assigns channels to non-gateway nodes based on their neighbor channel usage by avoiding channel interference among nodes. A transmitter–receiver conflict avoidance model is adapted [12], with each of the node maintaining its own clocks independently. Many different routing factors, such as channel usage of neighbors, node's hop count, memory size, transmission history, and memory usage ratio, are used for routing.

Pros:

(i) To maximize the channel bandwidth, utilization separate control and data packets are transmitted.
(ii) Channel assignments are done on per node basis.
(iii) All non-gateway nodes have fair access to gateway node.
(iv) Multiple routing factors are considered.

Cons:

(i) Network collision ratio can be high if pair of node's neighbors is unable to update their channel usage information due to asynchrony among nodes
(ii) Spatial/temporal reuse are good for fixed topologies, where position and types of nodes are known
(iii) Fairness may be sacrificed, as channel for a link is assigned in locally optimized fashion.

2.9 QoS Aware Joint Design for WMN [13]

It is joint routing, channel assignment, and scheduling algorithm for MRMC-WMNs to maximize the gateway throughput with QoS differentiation for end users. It is semi-distributed approach, where in it has centralized routing and channel assignment and distributed scheduling. This approach has three steps.

Topology Building: A minimum expected delay routing is used to build tree topology from an original network topology.

Channel Assignment: A weight aware channel assignment is used to assign channel and number of time slots to different links based on collected relative weights. Based on transmission history, channel and time slots are adjusted periodically.

QoS Aware Traffic Scheduling: In its given time slot, each parent router dynamically distributes transmission time to its children based on the transmission requirements.

This approach makes differentiation among users and provides control for fair partition of resources, such as channel and transmission time slots, between different classes of services (CoS). A protocol interference model is used. This approach considers only outgoing traffic from router to gateway. Expected minimum delay based on aggregated traffic load in each router used as routing metric.

Pros:

(i)   Semi-distributed approach is used that has advantages of centralized and distributed approach.
(ii) It provides real-time traffic differentiation by partitioning the resources between different classes of services.
(iii) Global traffic balance is achieved.

Cons:

(i)   The position of gateway is important.
(ii) It has more channel switching delay due to frequent channel switching.
(iii) Because of strict priority rule, there is need to set predefined transmission resources for each CoS.
(iv) Aggregate effect of interference of links transmitting in concurrent timeslots is not considered explicitly.
(v)   Distributed scheduling may cause network overloading.

2.10 QoS–Aware Routing Protocol on Optimized Link Source Routing (CLQ-OLSR) for IEEE 802.11 MRMC-WMN [14]

It is OLSR-based cross-layer QoS conscious routing protocol to support mission critical multimedia applications. This approach constructs multi-layer virtual logical mapping over physical topology. CLQ-OLSR implements pair of routing protocols.

(i) **Modified OLSR (M-OLSR)**: Using best-effort radio interface, it constructs routing table and estimates bandwidth.
(ii) **Logical routing**: Using real-time radio interfaces, it finds optimized logical path.
    At each node, based on channel utilization on each associated channel, bandwidth is estimated using the passive listening method. Once the construction of routing table and bandwidth estimation is done, based on topology and bandwidth estimation, logical routing creates logical full mesh and finds optimized path. The path is established to acclimatize QoS demands of real-time traffic, to balance network load, and to avoid congested route.

Pros:

(i) Path establishment is on the basis of current network topology and available bandwidth.
(ii) As bandwidth is estimated using passive listening, it reduces the network overhead.
(iii) Channel assignment is on the basis of channel selection index.

Cons:

(i) Generating a logical full mesh topology is complex and expensive process.
(ii) Failure of node in logical full mesh topology leads to network partitions.
(iii) Best-effort interface used by M-OLSR may lead to congestion.
(iv) Frequent transmission of topology control and bandwidth information result in higher messaging overhead.

2.11 Joint Multi-radio Multi-channel Assignment, Scheduling and Routing in WMN [15]

This approach is a Latin squares-based JCAR approach with multi-access scheduling for MRMC-WMNs. Nodal interference information is used to form different inter-cluster cliques and intra-cluster cliques. Latin squares are then applied to map resulting clustering structure to channels and radios. Within each cluster, Latin squares schedule the channel access among nodes in a collision-free manner. This approach guarantees network connectivity using interference cluster and bridge clusters. On-demand coloring-based multi-access scheduling is used. Node coloring and cluster coloring are used to schedule channel access time for each node within intra-clique and inter-clique, respectively. Forwarding speed is used as routing metric. Using two-hop interference information, the co-channel interference is avoided by isolating collision. Location-aware and link-adaptive route discovery mechanism is used with forwarding speed as routing metric. It exploits and integrates physical propagation models, location information, and radio utilization efficiency to provide near optimal routing paths to guarantee QoS for applications. Protocol interference model is used.

Pros:

(i) Guarantees fair allocation of radio and channel to clusters.
(ii) Alternative route is provided to mitigate the influence of forwarding node selection.
(iii) It reduces transmission failure.

Cons:

(i) Common channel is used for exchanging control information.
(ii) Static channel assignment is used.
(iii) Graph-coloring-based multi-access scheduling using protocol model for interference may not result in correct and realizable schedules.
(iv) The accuracy of the clique-based clustering result may be degraded at the expense of simplicity of the method.

2.12 Joint Channel Assignment and Routing in Rate-Variable WMN [16]

This approach is based on joint linear programming and genetic algorithm (LPGA) to solve JCAR problem. Rate-variable model is used to increase the network throughput. Physical interference model is used to evaluate the network capacity. Physical interference model computes the signal-to-interference ratio at each active node and compares it with an appropriate threshold. It is joint solution for the problem, in that genetic algorithm models channel assignment (CA) which is then used to get the network topological structure. For the resulting network topology, the routing problem is solved using LP formulations. For the interplay between routing and CA, the fitness value of chromosome is defined as value of linear objective function. Different constraints, such as radio, interference, link connection, and flow and link utilization constraints, are used in this approach.

Pros:

(i) Rate-variable model is used.
(ii) It attempts to maximize network capacity with granted connectivity.
(iii) The data rates of nodes are based on interference and channel assignment.

Table 1 Comparison of JCAR approaches

Approach	Routing metric used	Interference model used	Methodology used
[4]	Network congestion	Protocol	Time partitioning
[5]	Delay	Protocol	Ant colony optimization
[6]	Interference load aware	Physical	Divide and conquer
[7]	Maximum residual capacity	Protocol	Density-based clustering
[8]	Interference delay aware	Protocol	GA and backtracking
[9]	Minimum interference	Protocol	Greedy
[10]	Interference load aware	Physical	Adaptive
[11]	Multi-factor	Protocol	Temporal-spatial reuse
[13]	Minimum expected delay	Protocol	Differentiated service
[14]	Available bandwidth	Protocol	Passive listening
[15]	Forwarding speed	Protocol	Latin square clique-based clustering
[16]	Interference aware transmission rate	Physical	LPGA

Cons:

(i)   Computational complexity is more.
(ii) Parameters of GA are to be selected very carefully.
(iii) It has sub-optimal solution.

The comparison of JCAR approaches is given in Table 1. The [6–11, 15, 16] have used interference aware routing metrics. The [4, 5, 7–11, 13–15] use the protocol interference model which is simple but does not measure interference accurately. The [6, 10, 15] use physical interference model which accurately measures interference, but the model is complex and expensive. The [5, 9, 13] have frequent channel reassignment, which results in more channel switching delay. [9, 13, 14] have explicit focus on QoS, where [9] is centralized approach, and [13, 14] have much overhead.

3   Future Direction

Despite the fact that several research work, to address JCAR problem, the major issue is how to improve the network performance under varying traffic conditions without causing too many overheads remains unanswered. In this section, we provide a few directions for further research in JCAR for WMNs.

- Protocol model used to characterize interference, may be unrealistic in practice, thus the results connected with protocol model can be misleading. To curb this issue, a new mechanism like reality check can be used to set interference range in the protocol model to decrease the solution gap between protocol model and physical model.
- Designing a cross-layer routing metric to efficiently handle QoS provisioning in MRMC-WMNs.
- During channel assignment, due to non-cooperative environment in competitive WMNs, some node may misbehave which in-turn affects the working of some protocols. Therefore, it is necessary to implement a mechanism to deal with selfish behavior of nodes to have proper working of WMN.
- Channel switching delay is considered as an overhead for overall end-to-end delay. On the other hand, using a static channel assignment approach to avail the benefits of reduced overhead and stable topology will lack from the capacity improvement gained by MRMC environment. Therefore, a well-estimated trade-off is necessary to overcome the problem arising from switching overhead.
- A multi-rate adaptation capability improves the performance of WMNs. No MRMC protocol exploits the multi-rate adaptation capability of the IEEE 802.11 wireless network interface cards. Hence, developing multi-rate adaptation schemes is to provide the optimal solution to JCAR problem.

4   Conclusion

In this paper, we have identified the research approaches associated with joint routing and channel assignment algorithms in multi-radio wireless mesh networks. The paper also discussed each of the approachs along with their pros and cons. In the end, we outlined important future research directions in JCAR.

References

1. Akyildiz, I. F., Wang, X., Wang W: Wireless Mesh Networks: A survey. In: Computer Networks, 47(4), 445–487 (2005).
2. I. Hou, V. Borkar, P. R. Kumar: A Theory of QoS for Wireless. In: Proc. of INFOCOM 2009, pp. 486–494, (2009).
3. Shah, Ibrar Ali.: Channel assignment and routing in cooperative and competitive wireless mesh networks, Brunel University School of Engineering and Design PhD Theses (2012).
4. Jonathan Wellons, Yuan Xue: The robust joint solution for channel assignment and routing for wireless mesh networks with time partitioning. In: Elsevier, Ad Hoc Networks (2011).
5. Fawaz S. Bokhari,: Channel assignment and routing in multi-radio wireless mesh networks using smart ants. In: PerCom Workshops (2011).
6. Vanessa Gardellin, Sajal K. Das, Luciano Lenzini, Claudio Cicconetti, Mingozzi: G-PaMeLA: A divide-and-conquer approach for joint channel assignment and routing in

multi-radio multi-channel wireless mesh networks. In: Elsevier J. Parallel Distributed Computing. Voulume-71 (2011).

7. Sana Ghannay and Sonia Mettali Gammar: Joint routing and channel assignment protocol for multi-radio multi-channel IEEE 802.11s mesh networks. In: IFIP WMNC (2011).

8. Amitangshu Pal and Asis Nasipuri. In: JRCA: A Joint Routing and Channel Assignment Scheme for Wireless Mesh Networks. In: IEEE International Performance Computing and Communication Conference (PCCC) (2011).

9. Bahador Bakhshi, Siavash Khorsandi: On-line joint QoS routing and channel assignment in multi-channel multi-radio wireless mesh networks. In: Elsevier, Computer Communications 34 (2011).

10. W.L. Warner Hong Fei Long Pengye Xia S.-H. Gary Chan: Distributed Joint Channel and Routing Assignment for Multimedia Wireless Mesh Networks. In: IEEE International Conf. on Multimedia and Expo (2012).

11. Yan Jin, Weiping Wanga, Yingtao Jiang, Mei Yang.: On a joint temporal–spatial multi-channel assignment and routing scheme in resource-constrained wireless mesh networks In: Elsevier Ad Hoc Networks 10 (2012).

12. M. Kordialam and T. Nandagopal: The effect of interference on the capacity of multi-hop wirelessnetworks. In: IEEE Symposium on Information Theory 470, June (2004).

13. Peng Sun, Nancy Samaan: A QoS aware joint design for wireless mesh networks. In: Springer Science+Business Media New York (2013).

14. Yuhuai Peng, YaoYu n, LeiGuo, DingdeJiang, QimingGai: An efficient joint channel assignment and QoS routing protocol for IEEE 802.11 multi-radio multi-channel wireless mesh networks. In: Elsevier Journal of Network and Computer Applications (2013).

15. Di Wu, Shih-Hsien Yang, Lichun Bao: Joint multi-radio multi-channel assignment, scheduling, and routing in wireless mesh networks. In: Springer Science+Business Media New York (2013).

16. Chang-Sheng Yin, Xui Xiong, Xu Zhang.: Joint Channel Assignment and Routing in Rate-variable Wireless Mesh Network. In: IEEE ICSPCC (2013).

Correlated Cluster-Based Imputation for Treatment of Missing Values

Madhu Bala Myneni, Y. Srividya and Akhil Dandamudi

Abstract Improved imputation has a major role in the research of data pre-process for data analysis. The missing value treatment is implemented with many of the traditional approaches, such as attribute mean/mode, cluster-based mean/mode substitution. In these approaches, the major concentration is missing valued attribute. This paper presents a framework for correlated cluster-based imputation to improve the quality of data for data mining applications. We make use the correlation analysis on data set with respect to missing data attributes. Based on highly correlated attributes, the data set is divided into clusters using suitable clustering techniques and imputes the missing content with respect to cluster mean value. This correlated cluster-based imputation improves the quality of data. The imputed data are analyzed with K-Nearest Neighbor (KNN) and J48 Decision Tree multi-class classifiers. The efficiency of imputation is ascertaining 100 % accuracy with correlated cluster mean imputed data compared with attribute mean imputed data.

Keywords Missing values · Imputation methods · Clustering · Correlation

1 Introduction

Missing value data are a real, yet difficult issue went up against in data quality. Missing qualities may create inclination and influence the nature of the administered learning procedure or the execution of grouping calculations [1]. In any case, most learning calculations are not very much adjusted to some application spaces,

M.B. Myneni (✉) · Y. Srividya
Institute of Aeronautical Engineering, Hyderabad, India
e-mail: baladandamudi@gmail.com

Y. Srividya
e-mail: svidya.y@gmail.com

A. Dandamudi
NIIT University, Neemrana, Rajasthan, India
e-mail: akmail1289@gmail.com

© Springer Science+Business Media Singapore 2017   171
S.C. Satapathy et al. (eds.), *Proceedings of the First International Conference on Computational Intelligence and Informatics*, Advances in Intelligent Systems and Computing 507, DOI 10.1007/978-981-10-2471-9_17

because of the trouble with missing qualities, as most existed calculations are outlined under the suspicion that there are no missing values in data sets, which infer that a dependable strategy for managing those missing qualities is vital [2]. The different approaches are available to deal with missing value attributes in data mining [3]. For instance, in a database, if the known values for a property are: 2 in 60 % of cases, 6 in 20 % of cases, and 10 in 10 % of cases, it is sensible to expect that each attribute with missing estimations 2 or 3.4 will be loaded [4]. These strategies are not totally agreeable approaches to handle missing worth issues. In the first place, these strategies just are intended to manage the discrete qualities and the constant ones are discretized before imputing the missing values, which may lose the originality of the data.

As of late, the numerous analysts concentrated on the theme of attributing missing values. The machine learning systems additionally incorporate auto cooperative neural system, decision tree imputation, etc. All of these are pre-replacing methods. Inserted strategies incorporate case-wise erasure, a pathetic decision tree, dynamic way era, and some famous strategies, for example: CN2, C4.5, and CART [5–7]. Among missing quality attribution techniques that we consider in this work, there are likewise numerous current measurable techniques. [2, 8] For imputation, the recommended clustering algorithms are EM, KNN, and SKNN with minimum absolute error. Measurements-based systems incorporate direct replace substitution under standard deviation and mean/mode system. Be that as it may these methods are not totally agreeable approaches to handle missing worth issues.

2 Methodology

This paper addresses the novel method correlated cluster-based imputation which is implemented using R packages [9] and analysis is done on WEKA. In this experiment, the data used are ecoli data set collected from uci data repository. In this data set, total of 336 instances of 8 variables are observed. The characteristics of this data set are given below.

Data set Characteristics

mcg: num 0.49 0.07 0.56 0.59 0.23 0.67 0.29 0.21 0.2 …
gvh: num 0.29 0.4 0.4 0.49 0.32 0.39 0.28 0.34 0.44 0.4 …
lip: num 0.48 0.48 0.48 0.48 0.48 0.48 0.48 0.48 0 0.48 …
chg: num 0.5 0.5 0.5 0.5 0.5 0.5 0.5 0.5 0.5 0.5 …
aac: num 0.56 0.54 0.49 0.52 0.55 0.36 0.44 0.51 0.46 …
alm1: num NA NA NA NA 0.25 0.38 0.23 0.28 0.51 0.18 …
alm2: num 0.35 0.44 0.46 0.36 0.35 0.46 0.34 0.39 0.57 …
class: Factor w/8 levels "cp", "im", "imL", …

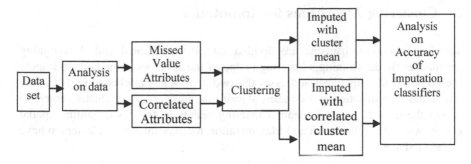

Fig. 1 Block diagram for correlated cluster-based imputation

The data set contains seven numerical attributes and one factor attribute. The methodology adopted for experimental work is shown in Fig. 1. This work proposes the novel method correlated cluster-based imputation for missing value treatment.

To impute the missing values, the following steps are adopted:

Analysis on data: The nature of the data is observed and found that missing values are present in the data set. Correlation is found between missed data attribute and remaining attributes in data set. The correlation analysis estimates the strength of association between two attributes. The correlation between attributes can be positive or negative. The positive value gives a strong association between variables and the negative value gives weak association between variables. After completion of the correlation analysis, the data set is divided into clusters.

Clustering: This is implemented using Expectation and Maximization (EM) and k-means clustering algorithms. In existing methods, the clustering is applied on missing value attribute and imputing with cluster mean. In the proposed correlated clustering approach, instead of considering missing value attribute, the highly correlated attribute is considered for clustering. The clustering is performed on ecoli data based on correlated attributes by specifying the number of clusters as 8 which are the number of distinct class values. After applying the clustering on missed data set, each attribute cluster center mean values are implanted with missed data. The imputed data are analyzed with classifiers.

Accuracy of Imputation using classifiers: The data set contains 8 class labels. Therefore, the classification is implemented with suitable classifiers. The imputation effect is tested with chosen classifiers. Here, the classifiers are tested with three cases:

(1) Imputation with cluster mean/mode value of missed value attributes (ecoli1).
(2) Imputation with EM cluster mean/mode center value of highly correlated attribute with missed value attribute (ecoli2).
(3) Imputation with k-means cluster mean/mode center value of highly correlated attribute with missed value attribute (ecoli3).

The results of the proposed method are examined in terms of exactness, kappa statistic, and error rate.

3 Clustering Algorithms for Imputation

The Imputation mechanisms are divided into the traditional and data mining approaches. In data mining, clustering is one of the appropriate methods to impute the missed data. This paper addresses the two approaches in clustering as Expectation and Maximization and k-means algorithms. Both are using cluster centers to model the data; however, k-means clustering results in clusters of similar spatial extent, while the Expectation and Maximization mechanism allows clusters to have different shapes.

3.1 Expectation Maximization (EM)

Expectation maximization algorithm is a natural generalization of maximum-likelihood estimation to the incomplete data case. When data are partially missing or hidden, the appropriate method is finding the data distribution using maximum-likelihood estimate [10]. The algorithm has two major steps:

1. Exception (E) step—In this step, it estimates the probability of each attribute in respective cluster—$P(C_j|x_k)$. Each element is composed by an attribute vector (x_k). The likelihood of each element is given by relevance degree of the observations of each cluster element in comparison with the attributes of the other elements of cluster C_j. Here, x is considered as input data, M is the total number of clusters, t is an instance, and initial instance is zero.
2. Maximization (M) step—In this, the next step is based on the parameters of the probability distribution of every class of the previous executed step. The computation starts with calculating the mean (μ_j) of class j obtained through the mean of all observations in relevant function. The covariance matrix at each iteration using Bayes' theorem is calculated. The each class probability is computed through the mean of probabilities (C_j) in function of the relevance degree of each observation from the class.

3.2 k Means

The k-means algorithm is a partition-based algorithm in clustering. It is highly suitable for the imputation of missing values [11]. One of the drawbacks of k means is the number of clusters which should be specified as the first step, but it is overcome in this application, because data contain the predefined classes. The algorithm has few iterative steps, which are as follows:

1. Arbitrarily choose k objects from D as the initial correlated cluster centers
2. Repeat
3. Based on the mean value of the objects in the cluster (re)assign each object to the cluster to which the object is most similar.
4. Update the cluster means, i.e., calculate the mean value of the objects for each cluster.
5. Repeat until no change.

4 Accuracy of Imputation Using Classifiers

The imputation effect is tested with J48 Multi-class Classifier and KNN Classification methods [12].

4.1 J48 Multi-class Classifier

J48 algorithm is supervised classification algorithm. It generates a class using a pruned or unpruned C4.5 decision tree. Decision tree is constructed using divide and conquer based on the gain ratio. The decision tree models are understandable and accurate. The attribute with highest info gain is root node and the process is repeated to construct the tree. Splitting of nodes is based on information gain. The time to construct the tree is less. For this data set, multi-class decision tree has used.

4.2 k-Nearest Neighbor (KNN)

KNN can predict both qualitative attributes (the most repeated data within the k-nearest neighbors) and quantitative attributes (the mean among the k-nearest neighbors). It is not required to create a predictive model for each attribute with missing values. Actually, it does not create explicit models as decision tree or rules. Thus, the KNN algorithm can be implemented to work with any attribute as class, by just changing the attributes which are considered in the distance metric. In addition, this approach can easily impute on multiple missing data.

5 Results

In the correlated cluster, imputation approach is applied on ecoli data. The summary of missing data is found from the existing R function is .na (data set). The found missing values are 40 %. The data are explored and correlation is found between all

Table 1 Correlation coefficients of all attributes with respect to class values

	im	imL	ims	imU	Om	omL	pp
mcg	0.12	0.35	0.37	0.36	0.30	0.33	0.28
gvh	0.08	0.03	0.1	0.07	0.27	0.10	−0.2
lip	0.03	0.05	0	0.001	0.02	0.05	0
chg	−0.001	−0.001	−0.001	−0.001	−0.001	−0.001	−0.001
aac	0.08	0	0.08	0.1	0.2	0.2	0.02
alm 1	0.4	0.3	0.3	0.4	0.1	0.2	0.1
alm 2	0.3	0.1	0.1	0.3	−0.09	−0.17	−0.02

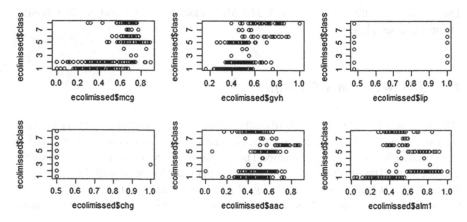

Fig. 2 Correlation plots

attributes. In this experiment, the class attribute is correlated with all other attributes. The class attributes contain 8 class labels. The correlation coefficient values of all class labels with respect to other attributes are shown in Table 1.

The correlation analysis on all attributes is plotted in Fig. 2. In this correlation analysis, the attributes mcg and alm1 are highly correlated in all the classes; hence, these attributes are selected for clustering.

On selected attributes, EM and k-Means clustering algorithms are used to find imputed values. The cluster mean values are used to impute the missed values. Now, two imputed data sets are formed with imputed EM cluster center mean/mode values referred as ecoli2 and imputed with k-Means cluster center mean/mode values referred as ecoli3.

After imputing the missed values, the imputed data accuracy is tested with classification using selected classifiers. The main motto behind the work is to evaluate the accuracy of the imputation methods to treat missing data. The performance is compared with the performance obtained by the internal algorithms used by J48 and KNN to learn with missing data, and by the mean or mode imputation method.

Table 2 Accuracy of imputed data sets with J48 and KNN classifiers

	J48_ecoli1	J48_ecoli2	J48_ecoli3	KNN_ecoli1	KNN_ecoli2	KNN_ecoli3
Correctly classified instances (%)	91	92.8	91	78.869	93.4	100
Error	0.0357	0.0316	0.0357	0.0571	0.0051	0.0051
Correctly classified instances (%)	91	92.8	91	78.869	93.4	100
Kappa statistic	0.8	0.9	0.8	0.7	0.9	1

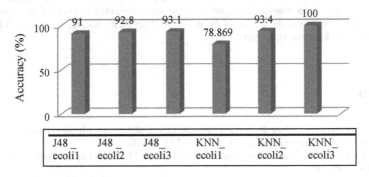

Fig. 3 Accuracy improvement with cluster-based imputation

In these experiments, missing values were implanted, in different rates and attributes into the data sets. The performance of all missing data treatments was compared using correctly classified instances, kappa statistic, and error rate. In particular, we are interested in studying the response of these imputations when the implanted data are from cluster center mean values.

Table 2 shows the accuracy of imputed data sets which are tested with J48 and KNN classifiers. In this analysis, three data sets are considered as default imputation with missed attribute mean/mode, imputation with EM cluster center, and imputation with k-Means cluster center.

Figure 3 shows the accuracy improvement with correlated cluster-based imputation while comparing with default classifier mean/mode imputation. Among the EM and k-Means clustering algorithm, the k-Means algorithm is giving high accuracy for imputation and error is also minimized with K-means.

6 Conclusion

This experimental treatment of missing values is implanted with the combination of correlation and clustering. Instead of clustering on missed attribute, the highly correlated attributes-based clustering is giving accurate result. In this work, two correlated cluster-based algorithms Expectation and Maximization and k-Means algorithm are used. The imputed data are tested by two multi-class classifiers as J48 and KNN. The results are ascertaining 100 % accuracy with KNN Classifier on k-means attribute center imputed data. This work can be extended with hybrid methods on different types of missing data.

Acknowledgments Authors gratefully acknowledge the computational facility created in the college under DST's FIST Programme (SR/FST/College-2009/2014(C)) which helped them to carry out the work. The authors are also grateful to the anonymous reviewers for their constructive comments which improved the quality of the paper. Authors thank the management of VBIT for their support and kind encouragement.

References

1. Lakshminarayan K, Harp S. A., and Samad T: Imputation of Missing Data in Industrial Databases: Applied Intelligence, 11: pp. 259–275 (1999).
2. Pearson, R. K.: The problem of disguised missing data: SIGKDD Explor. Newsl. Vol.8, no.1, pp. 83–92 (2006).
3. Grzymala-Busse J. W. and Hu. M.: A Comparison of Several Approaches to Missing Attribute Values in Data Mining: In RSCTC'2000, pp. 340–347 (2000).
4. Hua, M. and Pei, J.: Cleaning disguised missing data: a heuristic approach: In Proceedings of the 13th ACM SIGKDD international Conference on Knowledge Discovery and Data Mining, KDD '07. ACM, New York, NY, pp. 950–958, (2007).
5. Calders, T., Goethals, B., and Mampaey, M.: Mining itemsets in the presence of missing values: In Proceedings of the 2007 ACM Symposium on Applied Computin (2007).
6. Clark and Niblett . T.: The CN2 Induction Algorithm: Machine Learning, 3(4):261–283 (1989).
7. Quinlan J. R.: C4.5 Programs for Machine Learning: Morgan Kaufmann, CA (1988).
8. Batista G. E. A. P. A. and Monard M. C.: K-Nearest Neighbour as Imputation Method: Experimental Results. Technical report, ICMC-USP (2002).
9. Nuryazmin Ahmat Zainuri, et al.: A Comparison of Various Methods for Missing Values in Air Quality Data: Sains Malaysiana 44(3), :449–456; (2015).
10. Peng Shangu, Wang Xiwu, Zhong: The study of EM algorithm based on forward sampling: Qigen Electronics, Communications and Control (ICECC), pp. 4597 (2011).
11. Wagstaff, K., Cardie, C., Rogers, S. and Schroedl, S.: Constrained k-means clustering with background knowledge: In Proc. of the 18th Intl. Conf. on Machine Learning, pp. 577–584 (2001).
12. Madhu Bala Myneni, M. Seetha: Comparative Analysis on Scene Image Classification using Selected Hybrid Features: International Journal of Computer Applications (0975–8887) vol. 63, no.2, pp. 44–47 (2013).

Application of Pseudo 2-D Hidden Markov Model for Hand Gesture Recognition

K. Martin Sagayam and D. Jude Hemanth

Abstract Hand gesture recognition is the temporal pattern analysis with mathematical interpretation. It provides the means for the non-verbal communication among the people, more natural and powerful means of human–computer interaction (HCI) for the virtual reality application. The development of human-computer stochastic processes has led to a 1-D hidden Markov models (1DHMMs) and training algorithms to find the high recognition rate and low computational complexity. Due to their dimensionality and computational efficiency, Pseudo 2-D HMMs (P2DHMMs) are often favored for a flexible way of presenting events with temporal and dynamic variations. Both 1-D HMM and 2-D HMM are present in hand gestures, which are of increasing interest in the research of hand gesture recognition (HGR). The main issue of 1-D HMM is the fact that the recursiveness in the forward and backward procedures typically multiply probability values between themselves. Hence, this product quickly tends to zero and goes beyond any machine storage capabilities. This work presents an application of Pseudo 2-D HMM to classify the hand gestures from measured values of an accelerating image. Comparing an experimental result between 1-D HMM and Pseudo 2-D HMM with respect to recognition rate and accuracy, it shows a prominent result for the proposed approach.

Keywords Human–computer interface · 1-D HMM · Pseudo 2-D HMM · HGR

K. Martin Sagayam (✉) · D. Jude Hemanth
Department of ECE, Karunya University, Coimbatore, India
e-mail: martinsagayam.k@gmail.com

D. Jude Hemanth
e-mail: jude_hemanth@rediffmail.com

© Springer Science+Business Media Singapore 2017   179
S.C. Satapathy et al. (eds.), *Proceedings of the First International Conference on Computational Intelligence and Informatics*, Advances in Intelligent Systems and Computing 507, DOI 10.1007/978-981-10-2471-9_18

1 Introduction

Gesture recognition is the mathematical modeling of three-dimensional perspective by an electronic gadget. It provides the way for the non-verbal correspondence among the general population. Human communication comes in many modalities, including speech, gestures, facial, and bodily expressions. Among various gestures, hand motion plays an effective method for human–computer interaction (HCI), such as praying, legal and business transactions (handshake, judge hammering), traffic control, counting, game playing, appreciative gestures, etc. These data encode information by their temporal trajectories. Gesture classification is a critical component of such gesture recognition system (GRS). Literature survey reveals the various possible research works on gesture recognition.

Zhou et al. [1] have proposed a hand gesture recognition using kinect sensor. A novel distance metric approach is used in this work. Juan et al. [2] have proposed a parametric search for an image processing Fuzzy C-means hand gesture recognition system. This work describes parametric search simultaneously that is found using a neighborhood parameter search routine, which distance weighting for static hand gesture recognition. Chih-Ming Fu et al. [3] have proposed a hand gesture recognition using a real-time tracking method and hidden Markov models. This paper describes the modeling and recognition of actions through motor primitives. C. Li et al. [4] have proposed a segmentation and recognition of multi-attribute motion sequences, to determine various possible communications with the intelligent space by simple hand gestures. Bashir et al. [5] have proposed an object trajectory-based activity classification and recognition using the hidden Markov models.

This work presents an application of Pseudo 2-D HMM to recognize a collection of hand gestures from a measured values of an accelerating image, Comparing HMM approaches (1-D and 2-D) against with Pseudo 2-D HMM, to find the best solution by hybridize with the neural network coefficient [6]. This leads to increase the reliability and performance of the system.

2 Proposed Methodology

The proposed methodology used for hand gesture recognition based on HMM approaches is shown in Fig. 1. The various stages of the proposed techniques are collection of data set (i.e., kinect sensors, camera, etc.), pre-processing, feature extraction, and classification.

At first, the collection of data set with good quality has to be preprocessed using suitable thresholding and edge detection technique, and, the second, to find the maximum possible features from the previous stage using geometry-based feature extraction. The extracted output gives as input to the classifier, to train and manage the classification with a good recognition rate and accuracy rate. The various

Fig. 1 Proposed
methodology

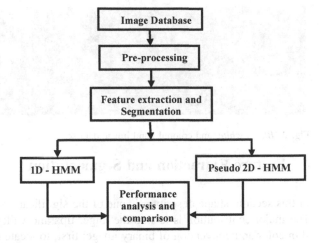

classifiers with Pseudo 2D-HMM classifier are compared to prove that the proposed
method shows the better result in terms of recognition rate and accuracy rate.

3 Image Database and Pre-Processing

The database contains 600 images with 352 × 288 gray-level images with intensity
value ranges from 0 to 255. It contains 5 gestures taken from 10 subjects, and each
gesture is 10 variations. This data set is available at the website of NUS University
of Singapore [7]. The data set with three categories, such as black, white, and
colored images. Each represents different notations for non-verbal communication.
Figure 2 shows that the sample black, white, and colored hand image data set.

Pre-processing is done by detecting the edges which involves identifying and
locating sharp discontinuities in the hand image which, in turn, characterises
boundaries in the image. There is an extremely large number of edge detection
operators frequently used in the gradient function. The gradient function f and its
direction θ are given below, respectively.

$$\nabla f = \left[\frac{\partial f}{\partial x}, \frac{\partial f}{\partial y}\right] \tag{1}$$

$$\theta = \tan^{-1}\left(\frac{\partial f}{\partial y} \Big/ \frac{\partial f}{\partial x}\right). \tag{2}$$

Using Eqs. 1 and 2 is used to detect horizontal, vertical, and diagonal edges.
Sobel operators were efficient in detecting the objects with low noise [8].

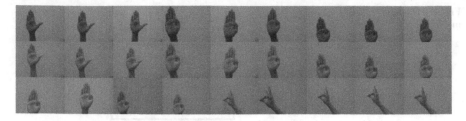

Fig. 2 *Black, white,* and colored hand image data set

4 Feature Extraction and Segmentation

In this section, shape descriptor is one of the significant parts of image processing. The major contribution is to find the finger tips and valley points, segmenting the skin color and conversion of binary image: first, to locate the finger tips and valley point from the shape-based features, and then, to determine the orientation of an individual feature with respect to the reference points. Figure 3 shows the process involved in locating and extracting features from the hand contour.

The idea of one-dimensional Savitzky-Golay channels can be effectively stretched out to two-dimensional polynomial fitting [9]. The 2D polynomial $f(x, y)$ utilized in this work has the accompanying structure. It is most common for detecting meaningful color feature.

$$f(x, y) = C_{nm} + C_{(n+1)m}x + C_{n(m+1)}y + C_{(n+1)(m+1)}xy + C_{(n+2)m}x^2 + C_{02}y^2 \quad (3)$$

where,
x and y　—2D coordinates of data point
C　　　　—Column vector coefficient
n and m　—rows and columns in column vector matrix

Consider the two points (x_1, y_1) and (x_2, y_2) in the 2D image plane, the distance 'D' between these points is expressed as,

$$D = \sqrt{(x_2 - x_1)^2 + (y_2 - y_1)^2}. \quad (4)$$

At origin (0, 0) Eq. 4 becomes,

$$D = \text{Round} \left(\sqrt{(x^2 + y^2)} \right). \quad (5)$$

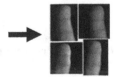

Fig. 3 Pre-processing and hand contour extraction

From Eq. 4, feature matching between two points in 2D image plane has been determined with respect to height, width, valley, and tip points. Section 5 discusses the classification of the hand image using the recognition based on HMM approach.

5 Classification Based on P2D HMM

Hidden Markov models are statistical model used for creating the building block of an observation sequence, with a high level of flexibility. They leave the observed sequence for recovering the hidden layer by pairing with each observed hidden states. Figure 4 shows the system with the same state or switch to the next state but could not go back to a previous state. This is referred to as left–right model or Bakis model.

[10] describes the extension of 1D HMM by adding robustness of the feature detection against in changing of position and size. Such a model is called Pseudo 2D hidden Markov model (P2D HMM) or otherwise known as planar HMMs that are stochastic automata with a 2D plan of the states. A planar HMM can be seen as a vertical 1D HMM that connection together an uncertain number of super-states, as shown in Fig. 5. The three major steps are followed, such as evaluation, decoding, and training, using the Forward algorithm, Viterbi algorithm, and Baum–Welch algorithm, respectively [11].

The number of the states is proportional to the complexity of the shape of the gestures. The general representation of a P2D HMM can be given by $\Lambda = \{\lambda, A, B, \pi\}$, where

- $\lambda = \{\lambda^{i+1}; i = 0, 1,.......N - 1\}$ is the $N - 1$ possible super-states in the structure
- λ^{i+1} is a 1D HMM super-state in the structure
- $s = \{s_{i+1}; i = 0, 1,........N - 1\}$ is the $N - 1$ possible super-state λ^i
- $v = \{vi + 1; i = 0, 1,.....N - 1\}$ is the transition state output to the next state
- $A^i = \{\alpha^i(l)\}^i_{l=1,....N}$ is the transition probabilities within super-state λ^i
- $B^i = \{b_m(l)\}^i_{m=1,...N}$ is the set of super-state output probabilities λ^i
- $\pi^i = \{\pi_{i+1}; i = 0, 1,......N - 1\}$ is the initial state probabilities λ^i
- $A = \{a_{ij}\}_{ij=1,......N}$ is the transition probabilities within the P2D HMM

Fig. 4 *Left—right* 1D HMM

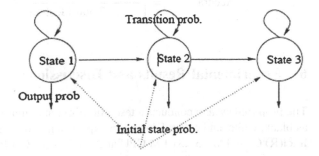

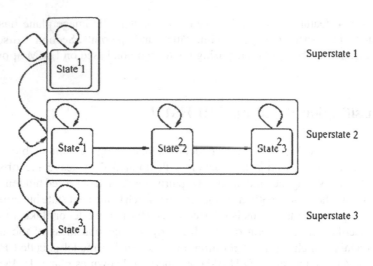

Fig. 5 Schematics of pseudo 2D HMM

- $\pi = \{\pi_{i+1}; i = 0, 1, \ldots N - 1\}$ the initial probabilities of super-state of the P2D HMM

Correspondingly to the one-dimensional model, the P2D HMM will relate a state grouping Q to a perception succession $O = \{o_{xy}\}_{x=1,\ldots X, \; y=1,\ldots Y}$. The state sequence of Q will consist of two sub-sets to link with the successive super-state position at (x, y).

More formally, the parameters of a P2D HMM can be communicated as follows:

- Transition probability of super-state: $a_{ij} = P[q_y = \lambda^j \mid q_{y-1} = \lambda^i]$
- Super-state probability at initial: $\pi_i = P[Q = \lambda^i \mid \Lambda]$
- State transition probability of super-state: $\alpha_1 = P[q_{xy} = s_1 \mid q_{x-1y} = s_k]$
- Super-state output probability: $bm = P[o_{xy} = v_i \mid q_{xy} = s_j]$
- State probability at initial: $\pi_j = P[q_{1y} = s_j \mid \lambda^i]$

Following on the structure of a P2D HMM, model evaluation has to perform using technique in [12]. After the classification of the hand gesture from the data base, calculate the accuracy rate using Eq. 6.

$$\text{Accuracy rate} = \left\{ \frac{\text{Gesture classified correctly}}{\text{Total Gesture}} \times 100\% \right\}. \tag{6}$$

6   Experimental Results and Discussions

The proposed system conducts a test with 600 hand images of three categories, such as black, white, and colored datas. An experimentation has conducted on a Sony PC Intel(R) Core(TM) i3-2310 M CPU at 2.10 GHz, 4 GB RAM, and 64-bit operating system. The tool used for the implementation is MATLAB R2014b. In pre-processing stage, sobel operator is used to return the edges, at where the gradient value is high. This avoids the false edge detection rather than the other edge detection technique, as shown in Fig. 6.

Fig. 6 Pre-processed output of sample hand image

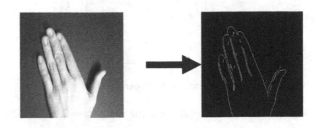

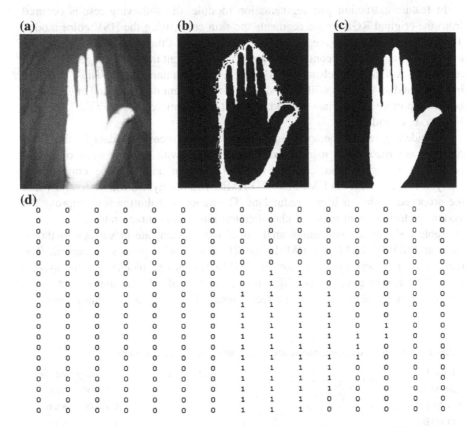

(d)

```
0  0  0  0  0  0  0  0  0  0  0  0  0  0  0
0  0  0  0  0  0  0  0  0  0  0  0  0  0  0
0  0  0  0  0  0  0  0  0  0  0  0  0  0  0
0  0  0  0  0  0  0  0  0  0  0  0  0  0  0
0  0  0  0  0  0  0  0  0  0  0  0  0  0  0
0  0  0  0  0  0  0  0  0  0  0  0  0  0  0
0  0  0  0  0  0  0  0  0  0  0  0  0  0  0
0  0  0  0  0  0  0  0  1  1  0  0  0  0  0
0  0  0  0  0  0  0  0  1  1  0  0  0  0  0
0  0  0  0  0  0  0  1  1  1  1  0  0  0  0
0  0  0  0  0  0  0  1  1  1  1  0  0  0  0
0  0  0  0  0  0  0  1  1  1  1  0  1  0  0
0  0  0  0  0  0  0  1  1  1  1  1  1  0  0
0  0  0  0  0  0  0  1  1  1  1  1  0  0  0
0  0  0  0  0  0  0  1  1  1  1  1  0  0  0
0  0  0  0  0  0  0  1  1  1  1  0  0  0  0
0  0  0  0  0  0  0  1  1  1  1  0  0  0  0
0  0  0  0  0  0  0  1  1  1  1  0  0  0  0
0  0  0  0  0  0  0  1  1  1  0  0  0  0  0
0  0  0  0  0  0  0  1  1  1  0  0  0  0  0
```

Fig. 7 **a** RGB image. **b** Detection of skin color pixel. **c** Binary image conversion. **d** Binary matrix of hand image

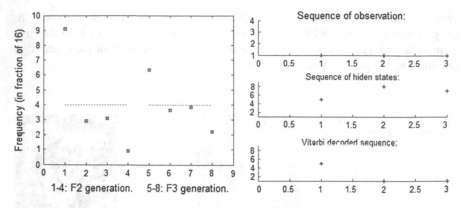

Fig. 8 Heredity factor = 2; phenotype = 4; data = 2; size = 529 × 3

In feature extraction and segmentation module, the following results occured from the original RGB image segments the skin color using the HSV color model. Then, it is converted into the binary image using Otsu's threshold [13]. This results into binary matrix which contains '1' and '0' to represent the hand image, where '1' represents skin color pixels and '0' represents the remaining parts. Thus, the hand image from data set shows different shapes with different positions and situations to be develop more appropriate method to detect a hand more accurately, as shown in Fig. 7a, b, c, and d.

Considering feature vector point from the two-dimensional data in the environment are recognized using Pseudo 2-D HMM with hybridization of neural coefficient and gray values, as shown in Fig. 8. In addition, the comparative analysis is made with 1D HMM and 2D HMM without hybridization, to show that the proposed method is highly redundant. The major contribution is to improve the system performance in terms of classification rate and accuracy rate.

Table 1 shows a comparative analysis of recognition rate in various methods, such as ESOM, HMM, 1D HMM, and 2D HMM with proposed method. The recognition rate was satisfying for all HMM-based tested blocks, but the system using P2D HMM gave an identification percentage of 99.3 %, that is, 600 hand images from the data set were perfectly recognized with a minimum tolerance

Table 1 Comparison of classification rate and accuracy rate—hand gesture

Method	Classification rate (%)	Accuracy rate (%)
ESOM [14]	84	550/600 = 91.6
HMM [15]	49	520/600 = 86.6
1D HMM	81	545/600 = 90.8
2D HMM	83	548/600 = 91.3
P2D HMM: gray values [12]	94.5	589/600 = 98.1
P2D HMM + Neural network coefficients	96.5	596/600 = 99.3

factor. In addition with the hybrid system of P2D HMM also compared against with P2D HMM + ANN result obtained for P2D HMM + ANN produced the better classification rate and accuracy rate.

7 Conclusions

A low-complexity 2D HMM structure is proposed for hand gesture recognition for human–computer interaction. This work is experimented with various gestures and postures. As an extension, a hybrid P2D HMM is also proposed in this work, to find the optimized decoding path in both training and testing phase using Viterbi algorithm. This tends to improve the system performance measures in terms of accuracy rate and recognition rate that are promising for virtual reality applications.

References

1. Zhou Ren, Junsong Yuan, Member, IEEE, Jingjing Meng, Member, IEEE, and Zhengyou Zhang, Robust Part-Based Hand Gesture Recognition Using Kinect Sensor, IEEE Trans. Multimedia, vol. 15, no. 5, August 2013.
2. Juan Wachs, Helman Stern, Yael Edan, Parameter Search For an Image Processing Fuzzy C - Means Hand Gesture Recognition System, IEEE transaction on Image Processing, pp. 341–344, 2003.
3. Chih-Ming Fu et.al, Hand gesture recognition using a real-time tracking method and hidden Markov models, Science Direct – Image and Vision Computing, vol. 21, Issue 8, pp. 745–758, 1 August 2003.
4. C. Li, P. Zhai, S. Zheng, B. Prabhakaran, Segmentation and recognition of multi-attribute motion sequences, in: Proceedings of the ACM Multimedia Conference, pp. 836–843, 2004.
5. F.Bashir, A.Khokhar, D.Schonfeld, Object trajectory-based activity classification and recognition using hidden Markov models, IEEE Trans. Image Processing 16(7), pp. 1912–1919, 2007.
6. Pawel Sokolski, Tomasz Rutkowsi, Hybrid of Neural Networks and Hidden Markov Models as a modern approach to Speech Recognition Systems, Pomiary – Automatyka – Robotyka, vol. 17, no. 2, pp. 449 – 455, 2013.
7. http://eeeweba.ntu.edu.sg/computervision/people/home/renzhou
8. Raman Maini, Himanshu Aggarwal, Study and Comparison of Various Edge Detection Techniques, International Journal of Image Processing (IJIP), vol. 3, Issue. 1, pp. 1–11, 2009.
9. N. Otsu, A threshold selection method from gray-level histograms, IEEE Trans. Systems, Man and Cybernetics, 9(1):62–66, 1979.
10. S.-S. Kuo and O.E. Agazzi, Keyword spotting in poorly printed documents using Pseudo 2D Hidden Markov models, IEEE Trans. Pattern Analysis and Machine Intelligence, PAMI-16 (8): 842–848, 1994.
11. S. Chen, H. Tong, Z. Wang, S. Liu, M. Li and B. Zhang, Improved generalized belief propagation for vision processing, Mathematical Problems in Engineering, vol. 2011, Article ID 416963, pp. 12, 2011.
12. Samaria. F., Face Recognition Using Hidden Markov Models, Ph.D. Thesis, University of Cambridge, 1994.

13. Fatma Mohammed, A.S.M. Kayes, Walid Mohammad, Attila Poya, Geometrical Feature Extraction of Human Hand, International Journal of Computer and Information Technology (ISSN: 2279 – 0764), vol. 02 – Issue 04, 2013.
14. Nguwi, Y. Y., Cho, S. Y., Emergent self-organizing feature map for recognizing road sign images, Neural Computing and Application, 19(4), 601–615, 2010.
15. Hsien, J. C., Liou, Y. S., Chen, S. Y., Road Sign Detection and Recognition Using Hidden Markov Model, Asian Journal of Health and Information Sciences, 1(1), 85–100, 2006.

Cleaning and Sentiment Tasks for News Transcript Data

Vidyashankar Lakshman, Sukruth Ananth, Rabindra Chhanchan
and K. Chandrasekaran

Abstract Today, vast amount of news in various forms is hosted on the web. They include news articles, digital newspapers, news clips, podcasts, and other sources. Traditionally, news articles and writings have been used to carry out sentiment analysis for topics. However, news channels and their transcripts represent vast data that have not been examined for business aspects. In this light, we have charted out a methodology to gather transcripts and process them for sentiment tasks by building a system to crawl Webpages for documents, index them, and aggregate them for topic analysis. Vector space model has been used for document indexing with predetermined set of topics and sentiment analysis carried out through the SentiWordNet data set, a lexical resource used for opinion mining. The areas of insight are mainly the polarity index (degree of polarity or subjectivity) of the news presented as well as their coverage. This research shows insights that can used by businesses to assess the content and quality of their content …

Keywords Sentiment analysis · News transcript · News sentiment · Topic analysis

1 Introduction

For organizations and services, knowing the customer requirements and opinions is very critical for improving businesses. To obtain these opinions, there are numerous sources of data available on the web, which run to terabytes of data. These

V. Lakshman (✉) · S. Ananth · R. Chhanchan · K. Chandrasekaran
National Institute of Technology Karnataka, Surathkal, India
e-mail: vidyashankar.bl@gmail.com

S. Ananth
e-mail: ananth.sukruth@gmail.com

R. Chhanchan
e-mail: rabindrachhanchan@gmail.com

K. Chandrasekaran
e-mail: kchnitk@gmail.com

© Springer Science+Business Media Singapore 2017   189
S.C. Satapathy et al. (eds.), *Proceedings of the First International Conference
on Computational Intelligence and Informatics*, Advances in Intelligent Systems
and Computing 507, DOI 10.1007/978-981-10-2471-9_19

include news articles, blogs, tweets, etc. In the past, a lot of tools and services have been developed to provide companies with sentiments regarding their products using online reviews places like Amazon, eBay, etc. Apart from these, the applications have been found in many places like ImDB reviews and ratings analysis, movie review analysis [1], and news and editorials [2, 3]. Some unique research [4] has been carried out in the areas of stock market prediction and election results whilst using tweets and a few other opinionated sources like blogs and editorials [5]. The marketing reaches of companies have also been measured using the sentiment analysis on various places, where the product may be discussed about. This gives companies an insight into the effectiveness of their strategies.

Sentiment analysis (also known as opinion mining) refers to the use of natural language processing, text analysis, and computational linguistics to identify and extract subjective information in source materials [6–8]. A basic task in sentiment analysis is classifying the polarity of a given text at the document, sentence, or feature/aspect level, whether the expressed opinion in a document, a sentence, or an entity feature/aspect is positive, negative, or neutral [7]. Advanced, "beyond polarity" sentiment classification looks, for instance, at emotional states, such as "angry," "sad," and "happy."

Tweets from twitter help in providing time series data and opinion regarding the product or the topic in question. Apart from this, there are a number of publicly available APIs which add much more number of sources of information to provide opinion. A large number of works have mostly concentrated on reviews, online articles, and forums. However, sentiment analysis has been used to examine news as well. News articles have commonly been used to carry out this task. While being a good source, they must not be the only sources we limit ourselves to. We wish to examine to news programs and podcasts for the sentiment analysis. We can use publicly available transcripts for this purpose. News channels provide transcripts of their programs that are well maintained and provide a wealth of information about the way they reach out to the audience and the influence of their news.

The rest of the paper is organized as follows. Section 2 presents related work. Section 3 describes the system overview. Section 4 provides an explanation of the results. Section 5 concludes the paper with future work.

2   Related Work

The earliest work on sentiment analysis involved using online product reviews and more general documents that included webpages and news articles [9]. Sentiment orientation is a measure of subjectivity in words and documents [10, 11]. Sentiment orientation of each word determines the polarity of the entire document [12]. Opinion lexicon can be automatically created with supervised or unsupervised learning.

In recent years, a lot of research has been done to find sentiment through online news articles and other media. In this paper by Raina et al. a method has been proposed to mine opinions using common sense extracted from ConceptNet and SentiWordNet to perform opinion analysis on online news articles [13]. A large corpus of news articles has been used. Their results include 71 % accuracy in classification, with 91 % in neutral sentences. ConceptNet is a semantic network containing all the relevant data that a computer should know to understand text-based sources written by people. It is made up of concepts which are expressed in the form of words and phrases of many of the languages spoken. This helps in imbibing a common sense to the computer while trying to understand text-based sources produced by languages. ConceptNet is an open source API, developed with GPLv3 code hosted by GitHub. SentiWordNet is a publicly available lexical resource for opinion mining. SentiWordNet assigns to each synset of WordNet three sentiment scores: positivity, negativity, and objectivity [14, 15].

News articles usually tend to have a neutral vocabulary as opposed to emotionally charged, such as reviews and editorials [16]. The paper by Fong et al. discusses the use of MALLET (Machine Language for Learning Toolkit) to implement and train many algorithms that are available on the toolkit, compare, and contrast them. The opinion mining on online news provides just the subjective analysis of the topic, Kevin et al. address the reader aspect instead [17]. The paper discusses ways to retrieve documents and news articles that have relevant content and a degree of emotion. The news collected is classified into reader-emotion categories and examined under different features and test the feasibility of emotion ranking. Text segmentation is an important task in while analyzing documents that cover multiple topics, a lot of methods have been developed for classifying document text, transcribed speech, etc. Co-occurence and clustering have long been used to segregate documents. In a research by Freddy et.al. various algorithms have been discussed for segmentation and determining boundaries. It covers the usage of similarity matrix, ranking, and clustering to segment topics and consequently determine boundaries for topics [18–20].

Recent research has involved using audio-based text sources for opinion mining. Chloe et al. address the issue of analyzing spontaneous speech and their transcripts [21]. The problem of mining of telephonic conversations effectively involves Transcripting and detection of opinions. One module is for the speech transcription for the call-center conversations. The next is the information extraction module which is based on semantic modeling of semantics and sentiment while imbibing many language rules and linguistics. They also discuss the feasibility of opinion detection based on call-center transcripts by comparing its outputs on manual transcripts and automatic transcripts.

3 System Overview

3.1 Data Set Source

The experiments mentioned in the paper have been carried out in Ubuntu 12.04. The data set for the experiments have been aggregated from CNN Transcripts available online. Transcripts are periodically updated based on the day of delivery of the news and also categorized based on the individual programs they are presented in. These transcripts present an exact picture of the news as presented by the reporters. To draw a brief picture, the transcripts contain all the metadata that is representative of the program like the airing date, time, and presenters, including sometimes the transcripts of the video clips played to support the news. The structure of the transcript will be briefly described in the subsequent sections.

3.2 Understanding the Data

A typical transcript begins with the show title and the headlines covered under the same. This is follow by the aired date and time. The headlines have recently been updated to add the air timing as well. In a conversation, the speakers are mentioned in capital letters and often with their designation. Few of the designations are CNN ANCHOR, CNN CONTRIBUTOR, UNIDENTIFIED MALE/FEMALE, or other designations as may be relevant. (COMMERCIAL BREAK), (BEGIN VIDEO-TAPE), (END VIDEOTAPE) are other metadata used in the transcripts to understand the flow. These can be useful for a few cleaning tasks which we will be examined in the subsequent sections.

> NATO to Hold Military Exercises in Ukraine; Obama: U.S. to Downgrade, Destroy ISIS; Search for Answers in Foley, Sotloff Deaths; Remembering Steven Sotloff.

The headlines indicate the news being spoken about. Therefore, the headlines and the corresponding part of the transcript can be mapped to the topic. Topics can be extracted from the headlines as opposed to detecting them from the documents. The news is also presented in the same order as mentioned in the headlines.

> REZA SAYAH, CNN CORRESPONDENT: Yes, and I think that could be interpreted by Moscow as a provocative move by Washington and NATO. Indeed, this is an annual exercise that takes place every year in western Ukraine. It was scheduled to take place a couple of months ago, NATO postponed it, because of the conflict. It was agreed it would happen later this month in September. That means American troops will be on the ground in western Ukraine outside of the conflict zone. We are eager to see what Moscow's reaction is to that.

> JOHN BERMAN, CNN ANCHOR: Reza Sayah for us. Thanks so much for clearing it up.
>
> It sounds like as far as a cease-fire goes, it may be. Check back in Friday.
>
> MICHAELA PEREIRA, CNN ANCHOR: Check back in Friday, and maybe, the framework is there for something concrete. That is the hope.
>
> BERMAN: Let us hope.
>
> From Ukraine now to the U.S. strategy in the fight against ISIS, President Obama vows that he wants to destroy and degrade the group. He also later said that he wants to make them a manageable threat. Are his words strong enough?

3.3 Building the Data Set

This task is categorized into two important submodules: the first module where the Webpages are crawled for transcripts and second module for storage of the transcripts. For crawling the Webpages, we have made use of python with beautiful soup python package to parse the HTML pages. The Webpages are categorised mainly on the date, on which the news or show was hosted. Therefore, a set of transcripts belonging to a day can be obtained by modifying the date in the URL. A sample URL would be of the form http://transcripts.cnn.com/TRANSCRIPTS/2014.11.04. html. By programmatically modifying the date, transcripts over a few days, weeks, or even months can be obtained.

3.4 Data Storage

MongoDB is an open source document-based database. Being a No-SQL database, it supports JSON structures and dynamic schemas. MongoDB has widely been used to store data crawled off the Web, especially where scaling is necessary. In our application, transcripts can be from more than just one news source, therefore, in circumstances, where the data are bound to grow that MongoDB would be a good choice for a DB. In our work for deciding the schema, we have considered only the CNN transcripts. Due to this, the schema used may not be stable. MongoDB being schemaless, addition of new rows will not affect the existing documents. With these factors in mind, we have chosen MongoDB as our storage database.

```
(
¡d : ObjectId(7df78ad8902c)
source : CNN
datetime : 2014.11.04 − 14 : 00
headlines : ElectionCoverage; PollsAbouttoClose
inHawaii, Oregon, Idaho; InterviewwithTedCruz
transcript :
```

< *Thecorrespondingtranscriptfortheheadline* >
)

Continuing from the previous, the transcripts that are parsed are store into the MongoDB in the transcript field of the table. A pre-processing is done on this transcript to fetch the aired time which is stored along the datetime field. The date can be provided programmatically, while the headlines too can be parsed from the Webpages and stored under headlines field of the table.

3.5   Cleaning of Transcript

Since our goal is to prepare for sentiment tasks, elements of the document that do not contribute to the headline or topic must be eliminated. The cleaning of transcript here refers to the removal of unnecessary details from the documents. These include the speakers involved in the transcript, headlines, designations, and other metadata involving commercials and video clips. From the conversation itself, sentences containing salutations or wishes can be removed as they do not contribute to the headline or topic under discussion. This is achieved with basic string processing on the transcript.

3.6   Text Segmentation

For sentiment tasks specific to a certain topic or an area of interest, it is necessary to segment the transcript based on topics. For this task, we have used the gensim python package. It is a library designed to extract semantic topics from documents automatically. Gensim is mainly aimed at processing plain texts and can, hence, be used for our experiment. Gensims implementations [22] of vector space algorithms can be used to categorize our documents based on the topic.

Vector space model is a model that is used to capture the relative importance of terms in a document [23]. Documents and queries are represented as vectors:

$$d_j = (w_{1,j}, w_{2,j}, \ldots, w_{t,j})$$
$$q = (w_{1,q}, w_{2,q}, \ldots, w_{n,q}).$$

For obtaining the vector space, we use a predetermined set of words that are extracted from the headlines. This helps us further narrow down the index terms that have to be used in classifying a document. For a document on the database, we fetch the headlines and prepare a list of index terms. The transcript, however, contains a discussion of multiple topics. For simplification, the transcript is further separated into multiple documents while using the commercials and videotape metadata. These documents are then indexed and allotted a topic. The entire corpus is tokenised and stop words are removed prior to experiment.

While using the gensim package, we focus on two major concepts, the corpus and the vectors. The corpus here would be the collection of transcripts. These transcripts are further divided into a document, on which a basic indexing task is performed. A vector space model represents a document by an array of features. The feature of importance here is inferred by checking the number of times, a certain word appears in a document.

The gensim subpackages corpora, models, and similarities are used to carry out the experiment. The outcome of this task would be a document tagged to a topic. Similarly, tagged documents can be further aggregated and used for sentiment tasks. The index or tag for a document will be extracted from the vector space based on the highest ranking from the given list of index terms.

3.7 Sentiment Timeline

Procedure

In this section, we propose a method to obtain useful analytics that can used to examine various business aspects for news channels. As observed in the previous experimental results, the transcripts have been cleaned and indexed based on the topics derived from the headlines. A few news programs only covered a single topic, in which case, the transcript can be used directly for the topic under analysis. Now, we present a case study to understand the purpose and some of the results that can be derived from the available data (Fig. 1).

To understand this, we have chosen the mid-term elections of 2014 as the basis for our experiment. Given the topic, we estimate dates between September 2nd week and October 2nd week which can provide us useful transcripts that are relevant to the topic under discussion. The mid-term election and Barack Obama mas a topic mostly gained momentum a fortnight before the actual elections. Hence, it is important to choose more number days before the actual event. The elections were completed on Nov 4th 2014. The keywords here would be, "Barack Obama", Obama, mid-term, and elections.

The transcripts for the given time line are stored in MongoDB in the order of date and time. The transcript are queried and segmented based on topics. For the given use case, we have indexed documents into two main categories, "Barack Obama" (or

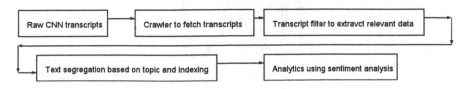

Fig. 1 Block diagram for the proposed sentiment framework

"Obama"), "election" (or "mid-term election"). Documents are aggregated day wise and topic wise and stored in a text file for further processing.

To understand the business perspectives of the news, we have covered two main measurements, one being the extent of coverage, the particular has received over a day or the time line in consideration and the polarity of the news. Here, in our experiment, the polarity is not further divided into positive, negative, or neutral. It is merely calculated to show the extent of polarity of the news. To measure this factor, we have used "Polarity Index" varying between 0 to 100. Polarity index can be obtained using sentiWordNet scores for individual words. sentiWordNet can be used to calculate aggregated score over the entire data set for a particular topic. The index varies between 0 and 1 in steps of 0.1. A low index indicates that the news is mostly objective in nature and a higher polarity index indicates that the news carries a certain degree of positive/negative sentiment with it.

4 Results and Analysis

From the observations, from the previous experiments on, we observed that most of the news tend to be neutral in nature. The primary analysis involves measuring the sentiment over time. For example, "Barack Obama" and "mid-term elections" will have received continuous coverage both prior and post the actual elections. A higher value of sentiment would indicate a polarizing effect of news and on the audience. Objective news would indicate that the news is mostly neutral in nature and will minimal effect on the opinions of audience.

The information gives news channels an opportunity to gauge the polarity of their news. The sentiment carried by the news indirectly affects their quality, viewership, and approval ratings. With multiple document sources, that might include other news channels as well as podcasts, the experiment can be carried and compared across sources. Competing news channels can measure the coverage and polarity of each other, which might be useful for business features (Fig. 2).

Fig. 2 Polarity values on aggregating news transcripts over a fortnight

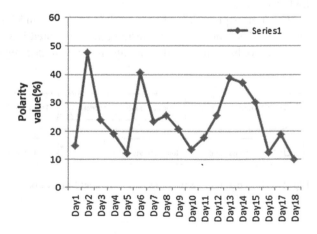

Fig. 3 Coverage values on aggregating headlines for "Obama"

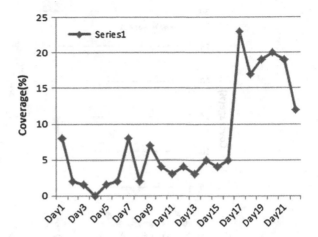

For the next part of the analysis, we examine the coverage of the news. Indicating the polarity of the news will not give the overall impact of the topic being investigated. Hence, we measure "coverage" as an additional parameter. Along with the polarity, the coverage values will be provided to assess the impact. A topic might have high polarity, but its impact is limited to the duration for which it is aired. Combining both these analysis will help us analyze better. The coverage for midterm elections is given in the graph. The values indicate the percentage of coverage it received for the given time quantum. 0 % would indicate no coverage and 100 % indicate only a single topic being discussed the entire time. The coverage for "midterm elections" is given below in the graph.

The news coverage is measured using the "datetime" field of the database. The "datetime" field stores the time of airing of the program. We can use this to calculate the coverage each topic has received in a transcript. Here, the assumption is that each news item receives equal coverage in the allotted air time. By aggregating time, over the document set for a day, we can get coverage for individual topics over a day. The values can subsequently be studied over a timeline (Fig. 3).

Given the graphs for two metrics of measurement, i.e., the polarity and coverage, it is possible to asses these across news channels and also with in the news channels. The increasing polarity values indicate an increasingly polarizing news which peaks around the election time and reduces again. Increasing values indicate the news tending towards a sentiment and hence influencing the public opinion on the topic being assessed. The coverage provides another metric for influence, a higher coverage indicates a higher reach for the news be polarized or subjective news (Fig. 4).

Across news channels, polarity for the same topic can be compared across time line, a higher one would indicate a news channel providing a more polarizing news for the same topic at the same interval. Unusually, higher coverage and unusual polarity values could indicate an affinity for certain topics.

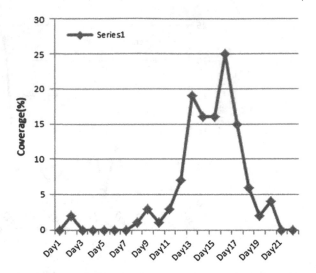

Fig. 4 Coverage values on aggregating headlines for "Election"

5 Conclusion and Future Work

A basic method to clean, segment, and perform sentiment tasks for news transcript
data has been addressed in this paper. The method though leaves scope for improvement on tasks like document segmentation as well as the premise and extent to which
sentiment analysis can be done with this data. In addition, we would like to build an
analysis tool that would inculcate all such methods to provide usable analytics for
businesses. The data sets used for our experiments have only been from a single
news source; however, we would like to expand this analysis to other media sources
to bring in new insights in the future

References

1. Li Zhuang, Feng Jing, Xiao-yan Zhu, and Lei Zhang. Movie review mining and summarization.
   In *Proceedings of the ACM SIGIR Conference on Information and Knowledge Management
   (CIKM)*, 2006.
2. Suman Basuroy, Subimal Chatterjee, and S. Abraham Ravid. How critical are critical reviews?
   the box office effects of film critics, star power and budgets. *Journal of Marketing*, 67(4):103–
   117, 2003.
3. Mikhail Bautin, Lohit Vijayarenu, and Steven Skiena. International sentiment analysis for
   news and blogs. In *Proceedings of the International Conference on Weblogs and Social Media
   (ICWSM)*, 2008.
4. Levon Lloyd, Dimitrios Kechagias, and Steven Skiena. Lydia: A system for large-scale news
   analysis. In *Proceedings of String Processing and Information Retrieval (SPIRE)*, number 3772
   in Lecture Notes in Computer Science, pages 161–166, 2005.

5. Tomohiro Fukuhara, Hiroshi Nakagawa, and Toyoaki Nishida. Understanding sentiment of people from news articles: Temporal sentiment analysis of social events. In *Proceedings of the International Conference on Weblogs and Social Media (ICWSM)*, 2007.

6. Philip Beineke, Trevor Hastie, Christopher Manning, and Shivakumar Vaithyanathan. Exploring sentiment summarization. In *Proceedings of the AAAI Spring Symposium on Exploring Attitude and Affect in Text: Theories and Applications*, 2004.

7. M.S. Neethu and R. Rajasree. Sentiment analysis in twitter using machine learning techniques. In *Computing, Communications and Networking Technologies (ICCCNT), 2013 Fourth International Conference on*, pages 1–5, July 2013.

8. Bo Pang, Lillian Lee, and Shivakumar Vaithyanathan. Thumbs up? sentiment classification using machine learning techniques. In *Proceedings of EMNLP*, pages 79–86, 2002.

9. Jeonghee Yi, T. Nasukawa, R. Bunescu, and W. Niblack. Sentiment analyzer: extracting sentiments about a given topic using natural language processing techniques. In *Data Mining, 2003. ICDM 2003. Third IEEE International Conference on*, pages 427–434, Nov 2003.

10. Bing Liu and Lei Zhang. A survey of opinion mining and sentiment analysis. In Charu C. Aggarwal and ChengXiang Zhai, editors, *Mining Text Data*, pages 415–463. Springer US, 2012.

11. Walaa Medhat, Ahmed Hassan, and Hoda Korashy. Sentiment analysis algorithms and applications: A survey. *Ain Shams Engineering Journal*, 5(4):1093–1113, 2014.

12. Peter D. Turney. Thumbs up or thumbs down?: Semantic orientation applied to unsupervised classification of reviews. In *Proceedings of the 40th Annual Meeting on Association for Computational Linguistics*, ACL '02, pages 417–424, Stroudsburg, PA, USA, 2002. Association for Computational Linguistics.

13. P. Raina. Sentiment analysis in news articles using sentic computing. In *Data Mining Workshops (ICDMW), 2013 IEEE 13th International Conference on*, pages 959–962, Dec 2013.

14. Stefano Baccianella, Andrea Esuli, and Fabrizio Sebastiani. SentiWordNet 3.0: An Enhanced Lexical Resource for Sentiment Analysis and Opinion Mining. In *Proceedings of the Seventh Conference on International Language Resources and Evaluation (LREC'10)*, Valletta, Malta, May 2010. European Language Resources Association (ELRA).

15. Andrea Esuli and Fabrizio Sebastiani. Sentiwordnet: A publicly available lexical resource for opinion mining. In *In Proceedings of the 5th Conference on Language Resources and Evaluation (LREC06*, pages 417–422, 2006.

16. S. Fong, Yan Zhuang, Jinyan Li, and R. Khoury. Sentiment analysis of online news using mallet. In *Computational and Business Intelligence (ISCBI), 2013 International Symposium on*, pages 301–304, Aug 2013.

17. Kevin Hsin-Yih Lin, Changhua Yang, and Hsin-Hsi Chen. Emotion classification of online news articles from the reader's perspective. In *Proceedings of the 2008 IEEE/WIC/ACM International Conference on Web Intelligence and Intelligent Agent Technology - Volume 01*, WI-IAT '08, pages 220–226, Washington, DC, USA, 2008. IEEE Computer Society.

18. Alekh Agarwal and Pushpak Bhattacharyya. Sentiment analysis: A new approach for effective use of linguistic knowledge and exploiting similarities in a set of documents to be classified. In *Proceedings of the International Conference on Natural Language Processing (ICON)*, 2005.

19. Freddy Y. Y. Choi. Advances in domain independent linear text segmentation. In *Proceedings of the 1st North American Chapter of the Association for Computational Linguistics Conference*, NAACL 2000, pages 26–33, Stroudsburg, PA, USA, 2000. Association for Computational Linguistics.

20. Lun-Wei Ku, Li-Ying Li, Tung-Ho Wu, and Hsin-Hsi Chen. Major topic detection and its application to opinion summarization. In *Proceedings of the ACM Special Interest Group on Information Retrieval (SIGIR)*, pages 627–628, 2005. Poster paper.

21. Chloé Clavel, Gilles Adda, Frederik Cailliau, Martine Garnier-Rizet, Ariane Cavet, Géraldine Chapuis, Sandrine Courcinous, Charlotte Danesi, Anne-Laure Daquo, Myrtille Deldossi, Sylvie Guillemin-Lanne, Marjorie Seizou, and Philippe Suignard. Spontaneous speech and opinion detection: Mining call-centre transcripts. *Lang. Resour. Eval.*, 47(4):1089–1125, December 2013.

22. Radim Řehůřek and Petr Sojka. Software Framework for Topic Modelling with Large Corpora. In *Proceedings of the LREC 2010 Workshop on New Challenges for NLP Frameworks*, pages 45–50, Valletta, Malta, May 2010. ELRA.
23. G. Salton, A. Wong, and C. S. Yang. A vector space model for automatic indexing. *Commun. ACM*, 18(11):613–620, November 1975.

Inventory Management System Using IOT

S. Jayanth, M.B. Poorvi and M.P. Sunil

Abstract In this paper, we are presenting an efficient system for managing the inventory for various applications dealing with solid or liquid assets. By implementing the inventory management based on IOT, we eliminate the unnecessary man power and make it automated between the measurement and order placement stages, thereby improving the efficiency of inventory management. The idea utilizes the ultrasonic transducer and a processing device with capability to connect to the Internet, such as a Raspberry Pi, to measure the inventory and send a mail to the supplier and/or to the company personnel for order placement, as well as display the present stock availability on a Web page hosted by our system.

Keywords IOT · Raspberry pi · Ultrasonic transducer

1 Introduction

Inventory management refers to the calculation of the available stocks. Inventory management is implemented in the various stages of the production line. It is a necessary tool for the ever-growing industries, whose demands for products are growing and delays cause a major problem [1–4]. This system is not only limited to industries, but are extended to medical and other fields. The presented design is cost-effective and can be implemented to address simpler or complex issues. The complex issues involve the stock management of wheat, barley, and other cereal manufacturing plants which require high efficiency. Our design can also be used for

S. Jayanth (✉) · M.B. Poorvi · M.P. Sunil
Department of Electronics and Communication Engineering,
School of Engineering and Technology, Jain University, Bangalore, India
e-mail: Jayanthsuresh09@gmail.com

M.B. Poorvi
e-mail: poorvimb83@gmail.com

M.P. Sunil
e-mail: mp.sunil@jainuniversity.ac.in

© Springer Science+Business Media Singapore 2017 201
S.C. Satapathy et al. (eds.), *Proceedings of the First International Conference
on Computational Intelligence and Informatics*, Advances in Intelligent Systems
and Computing 507, DOI 10.1007/978-981-10-2471-9_20

stock management of liquids as well, such as in detergents, soft drinks, and more. In medical applications, they are used in efficient storage and stock updating of drugs.

The technology used for this efficient stock management is based on Internet of Things. IOT is basically a physical network of components with embedded software, electronic concepts, and components. It allows various components to be sensed and remotely maintained by a network infrastructure. It also facilitates integration amidst the physical space and computer-like systems.

2 Literature Review

A typical inventory management system available in the market is built around software, which monitors the stocks. The inventory details are updated by the workers of the industry by various methods. One such method includes barcode scanners. Such type of Inventory management system is suitable for industries dealing with packaged cartons. However, when it comes for liquid stocks, it is impossible to keep a track of its inventory using barcodes. Another method is using RFID integrated along with ZIGBEE [5, 6]. The RFID readers themselves have a limited detection range. If multiple consignments are on the same palette, it is difficult for the RFID reader to read the signal precisely. The cost increases if there is a lot of stock and they need to be labelled using the RFID, and multiple node reflects to multiple RFID readers and cost [6]. The disadvantage of RFID is that it can to used only to monitor packed goods and can be implemented only at the

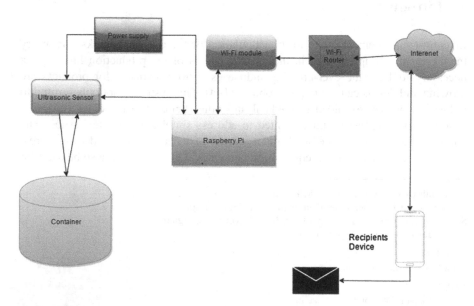

Fig. 1 Functional block diagram of the proposed design

packaging stage. By establishing a shared inventory for vivid departments of the enterprise, it allowed them to timely and efficiently maintain and control their inventory [7, 8]. While measuring inventory, there are several measuring sensors that are available in the market, such as a load cell.

3 Functional Block Diagram of the Proposed Design

In our design, we strive to improve the performance, yet maintain the simplicity of the system. Here, we use the power of IOT to simplify and make it efficient by eliminating any unnecessary human interference and automating the entire network responsible for inventory management. The dividing line between the pre-existing design and the proposed design is the use of dedicated hardware for Inventory management. The other ingenious idea incorporated in our design that sets it apart is the effective utilization of the ultrasound transducer to measure the inventory. Since our design has a dedicated hardware, it can run on batteries. This is effective when the system is installed in industries that do not depend mainly on electric power for its operation. In the presented design, an ultrasonic transducer is implemented to measure the stocks available. The design is generalized as shown in Fig. 1. There is no need for modification for change in inventory type. The same transducer can be used to manage both solid and liquid stocks with no changes. This is achieved using the ultrasonic transducer. The transducer is used to measure the time taken for a pulse to travel from the top of the container to the surface of the filled container and return back. This time is used to determine the distance from the top of the container to the surface of the inventory. By assuming two values, i.e., max and threshold, where max is the distance for full inventory and the threshold is the distance for acceptable minimum inventory. It is clear that max < threshold, since the distance increases with decrease in stocks. The threshold value to be so chosen that the industry should be capable of functioning till the new goods arrive. The stock measurement occurs as shown in Fig. 2. The heart of the system is a Raspberry Pi, which is used for two purposes. First, it is interfaced to the ultrasonic sensor to determine the time (Fig. 3).

Speed = Distance/Time.
where speed refers to the speed of sound in air
Speed = 330 m/s

Hence, the above equation becomes

34000 = Distance/(Time/2)
17000 = Distance/Time
Distance = Time * 17000

Fig. 2 Ultrasonic distance
measurement

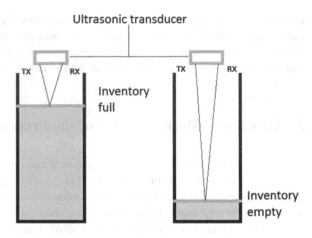

Second, the system is connected to the Internet through an LAN cable or using a suitable wireless Wi-Fi module. Figure 1 shows the system connected to the network via Wi-Fi. The presented design lets that the system sends an e-mail to the supplier as well the company's inventory manager. The inventory management based on IOT eliminates any human interaction with the system, thereby automating it with high efficiency.

As we can see, when the stocks reach the threshold value determined in the first stage, the system sends a mail directly to the supplier with the required amount of stocks as shown in Fig. 4. It also alerts the inventory manager by sending him a mail about the new order.

Using the multitasking ability of the Raspberry Pi, we run two programs necessary for our design. First, the code required for measuring the distance and sending the mail is executed. Second, a Web server is installed and made to run in the background to host a Web page. The Web page contains information, regarding the inventory and whether new stocks are added. This is achieved by installing Apache Web server on the Raspberry Pi. The apache Web server runs on boot, thereby one can access the Web page by entering the local IP address of the raspberry Pi in their browser. The IP address of the Raspberry PI can be made static, so as to fix the Web page address, else every time, the Raspberry Pi boots the IP address changes.

The distance is calculated every 10 min and compared with the threshold value to decide whether to generate a mail or not. After the mail is sent, the system waits until the inventory crosses the threshold value. This is necessary, because the system would continuously send a mail every 10 min if not.

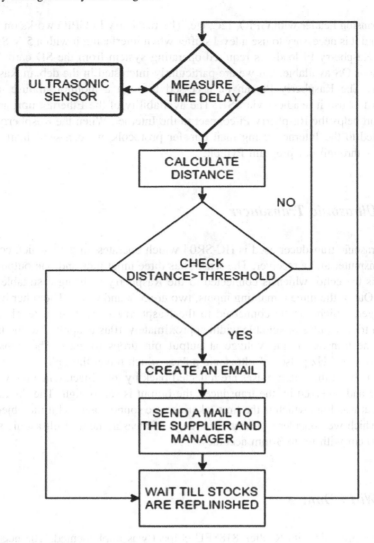

Fig. 3 Flow chart of the proposed design

4 Hardware Implementation and Its Components

4.1 Raspberry Pi

The heart of the system is based on the infamous Raspberry Pi. There are various models of Raspberry Pi available in today's market. This design was implemented on the Raspberry Pi Model B. The Raspberry Pi Model B boosts a 1 GHz processor with a 512MB RAM. It also has two on board USB ports, an Ethernet port and a 26 pin expansion header with GPIO, I2C, etc. The raspberry Pi GPIO works on 3.3 V logic, and it is necessary to use a level shifter when interfacing it with a 5 V Sensor.

The Raspberry Pi loads its required operating system from the SD card. There are various OS available, but we are particularly interested in the debian Raspbian Wheezy. The Raspberry Pi can be accessed through a monitor, mouse, and a keyboard or use it headless via SSH. The availability of the Ethernet port and the USB port helps the Raspberry Pi connect to the Internet. When the Raspberry Pi is connected to the Internet, using mail transfer protocols, we can send mails automatically through our program Python.

4.2 Ultrasonic Transducer

The ultrasonic transducer used is HC-SR04 which operates on 5 V, which consists of a transmitter and a receiver. The sensor has three input pins and one output. The output is the echo, which is connected to the Raspberry Pi using a suitable logic shifter. Out of the three remaining inputs, two are 5 V and GND. The other input is the trigger, which can be connected to the Raspberry Pi without a level shifter. When a trigger pulse of suitable width approximately 10us is applied to the trigger pin of the transducer, the voltage at output pin drops to zero. The transducer transmits 8 40 kHz pulses of ultrasonic pulses which travel through the air until it meets an object that these pulses are reflected back by the object. Once the wave is reflected and received by the transducer, the output is set to high. The duration of the output was low, which is the time taken for the sound wave to hit the object and return which we consider as delay. The range of measurement of ultrasonic sensor is 2–400 cm with up to 3 mm accuracy.

4.3 Wi-Fi Dongle

The Wi-Fi module with Realtek 8188EU chipset was implemented. The necessary drivers should be added to the Raspberry Pi. The Wi-Fi router can be any 2.4 GHz router. In our design, we have used the Wi-Fi dongle instead of the LAN connection, because of the mobility criteria. Hence, we used Leoxsys Wi-Fi dongle to enable wireless connectivity [9].

5 Software Implementation and Its Tools

The Raspberry Pi requires an operating system (OS) to boot. The Raspbian Wheezy was our choice in the presented system, because of its simplicity. In order for the Raspberry Pi to run on this powerful OS, we require a suitable capacity micro SD

Fig. 4 A typical mail received by the supplier from the system

Fig. 5 Prototype of the presented design

card with an adapter. To install the OS, we require software capable of creating a bootable disk from an OS image. This is achieved using the win32disk imager software which runs on a windows PC. Once the memory card is prepared, it is inserted to the Raspberry Pi and is booted.

Once the Raspberry Pi boots, it is necessary to tweak certain functions. One such function is to expand filesystem, which is necessary to completely utilize the memory of the SD card. All the packages should be updated and upgraded using the terminal. This can be achieved by connecting a monitor, keyboard, and a mouse, as shown in Fig. 5 or install an SSH application, such as on a PC running windows, and do necessary changes to the raspberry pi, as shown in Fig. 6.

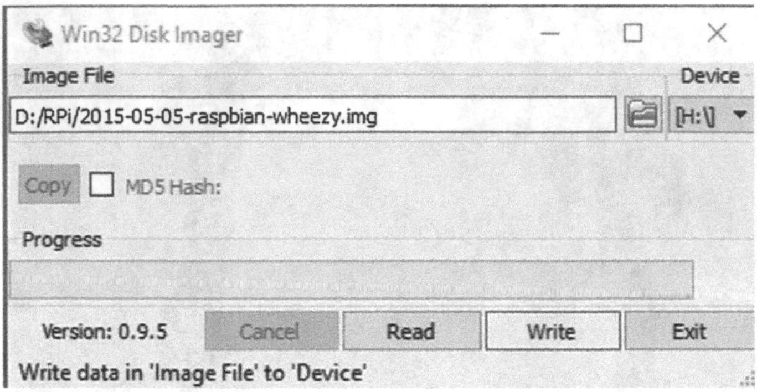

Fig. 6 Win32disk imager software window

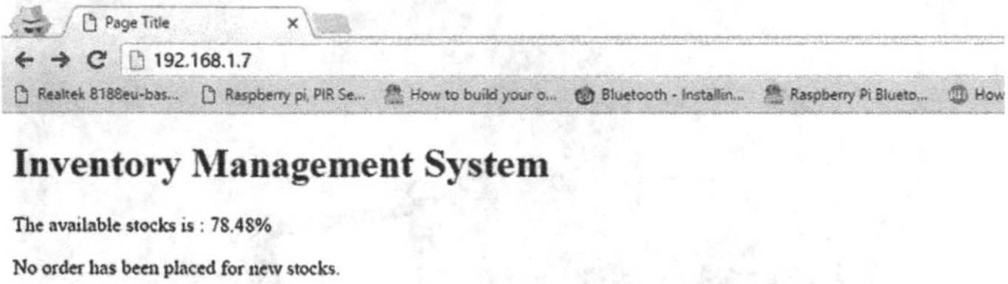

Fig. 7 Web page hosted by the system

Raspbian Wheezy comes along with Python. The program for the presented system is written using python and is added to execute at the boot time. Hence, when the system boots, the application automatically runs. To send an e-mail to the supplier and the inventory manager, it is necessary to install mail transfer protocol. We have used SMPT in our design due to its ease of configuration. The SMPT and GPIO functions can be accessed and controlled using python libraries.

6 Results and Discussions

With the above-described hardware and software implementation, the system was realized and tested. The graphical user interface can be enhanced if necessary to visually display the estimated amount. The presented design is meant to be compact, cost-effective, and simple to implement. Hence, a lot of effort was not concentrated on the GUI. This system is not limited to a single sensor. Multiple sensors can be connected to the same Raspberry Pi due to the availability of multiple GPIO pins and I2C. If storage containers are separated by great distance and multiple sensors are required, multiple systems can be installed due to their low cost. Our system detects the amount of inventory every 10 min, which can be varied according to the rate of inventory consumption. When an e-mail is sent to the supplier, the mail is delivered within 15 min. This delay is caused due to the refresh rate of new mails.

The Web page is hosted by the inventory management system is illustrated in Fig. 7. The Web page displays the inventory percentage for information and also shows if new stocks are allowed. One can access the inventory through any computer connected to the same network. The Web page can be developed further to add more details and display stocks of various storage containers in the same facility. The Web page can be made available on the Internet by acquiring a static IP from the ISP, which can help people monitor the stocks from remote location.

7 Conclusion

It is evident that this system is cost-effective. The Raspberry Pi is a 35$ computer, which is the fractional cost of the pre-existing software to track the inventory. Using ultrasonic sensors to measure the stocks, we have simplified the system which can be used for both solid and liquid stocks. Since the ultrasonic sensor needs to be placed on the top of the container, it is simpler and easier to mount. As we can clearly see, the system directly sends a mail to the supplier, thereby reducing human errors. Since the threshold value is so chosen that the stocks is sufficient for operation till the new stocks arrive, the system is self-sustained and there is no delay caused due to insufficient inventory. Because of the low cost, easy implementation, and efficient design, it can be implemented in hospitals, small-scale industries, and large-scale industries, where the limitations are our imagination.

References

1. EunSu Lee, Kambiz farahmand "Simulation Of A Base Stock Inventory Management System Integrated With Transportation Strategies Of A Logistics Network," Proceedings of the 2010 Winter Simulation Conference.
2. O. Jukic, I. Hedi, "Inventory management system for water supply network," MIPRO 2014, 26–30 May 2014, Opatija, Croatia.
3. Liling Xia, "The Design and Implementation of Distributed Inventory Management System Based on the Intranet Architecture," Proceedings of the IEEE, International Conference on Information and Automation, Shenzen, China June 2011.
4. Zheng Li, Li Jialing, Supply chain management, Beijing," China Central Radio and TV University Press, 2006. 3:129–132.
5. Xiaojun Jing, Peng Tang, "Research and Design of The Intelligent Inventory Management System Based on RFID," 2013 Sixth International Symposium on Computational Intelligence and Design.
6. Ding Long-gang, "Based on RFID, Wi-Fi, Bluetooth, ZigBee of things of electromagnetic compatibility and interference coordination," Internet of things Technology. 2011. 1, 59–61.

7. Ding Yan, "Automobile manufactures in the supply logistics warehouse management system optimization studies," Liaoning Institute of Technology, 2007.
8. Xia0-Rong Lei, Da-Xi Wang, "Inventory management system for auto parts enterprises based on the agile supply chain," 2012 Second International Conference on Instrumentation & Measurement, Computer, Communication and Control.
9. LI Zhong-cheng, "Design and Implementation of the Internet of Things in Intelligent Warehouse Management [J]," Computer systems & applications, 2011.

A Personalized Recommender System Using Conceptual Dynamics

P. Sammulal and M. Venu Gopalachari

Abstract E-commerce applications are popular as a requirement of emerging information and are becoming everyone's choice for seeking information and expressing opinions through reviews. Recommender systems plays a key role in serving the user with the best Web services by suggesting probable liked items or pages that keeps user out of the information overload problem. Past research of the recommenders mostly focused on improving the quality of suggestions by the user's navigational patterns in history, but not much emphasis has been given on the concept drift of the user in the current session. In this paper, a new recommender model is proposed that not only identifies the access sequence of the user according to the domain knowledge, but also identifies the concept drift of the user and recommends it. The proposed approach is evaluated by comparing with existing algorithms and perhaps does not sacrifice the accuracy of the quality of the recommendations.

Keywords Recommender system · Ontology · Usage patterns · Conceptual dynamics

1 Introduction

Internet has left a significant mark in all fields, such as e-commerce, science and technology, education and research, and telecommunication. From the past couple of decades, the research and development of the Web services hasbecome exponential and accelerated by many cutting edge technologies such as big data and

P. Sammulal (✉)
Jawaharlal Nehru Technological University, Hyderabad College of Engineering, Jagtial, Telangana, India
e-mail: sam@jntuh.ac.in

M. Venu Gopalachari
Chaitanya Bharathi Institute of Technology, Hyderabad, India
e-mail: venugopal.m07@gmail.com

© Springer Science+Business Media Singapore 2017   211
S.C. Satapathy et al. (eds.), *Proceedings of the First International Conference on Computational Intelligence and Informatics*, Advances in Intelligent Systems and Computing 507, DOI 10.1007/978-981-10-2471-9_21

cloud computing. Popular service providers on the Web such as Netflix, Last.fm music and Amazon are trying to promise satisfaction to their customers by predicting their interests toward the domain by means of recommender systems.

Recommender systems are the providers of personalized recommendations that exist in various types with respect to the strategy used, out of which the first one is content based (CB), in which recommenders try to analyze the users' access sequence; the second one is collaborative filtering (CF) that tries to aggregate the interests of the neighbors of a user. Over a period, hybrid recommenders evolved that combined the features of CB and CF to make suggestions better. The recommenders generally focus on the patterns of the navigation sequence of the customer by means of the user's past history. The log file in the server can be the source of finding the access patterns of a user under various types of criteria.

Broadly, there are two issues identified in this scenario where the first one is if these patterns do not consider the true semantics behind the access patterns, then the outcome will limits the quality of prediction. That means the recommender must have domain knowledge to provide meaningful suggestions, so that the user can be satisfied. For instance, if the user is accessing a movie portal such as Netflix, then the access patterns must be simplified to the genre of the movie rather than the title of the movie, assuming that the genres will say the semantic of a movie page. To achieve this, one has to construct and incorporate the knowledge using the ontology of that particular domain, and the real challenge is in constructing the knowledge with reasonable efficiency. The second issue is recommending according to the dynamics in the user's interest that drifts from one concept to another.

The patterns identified for a customer even with knowledge hardly gives the user profile little historic. If the user is with another concept which is not in the pattern, then it definitely leads to dissatisfaction of the user. For example, assume that the recommender stored the access pattern concept for a user as "romantic movies" and suggests accordingly, and assume that the user is currently accessing a set of "action movies"; then it definitely leads to the dissatisfaction of the user proving lag in the predictive accuracy.

In this paper, a recommender system is proposed that is focused on resolving the two above-mentioned issues. To accomplish this, the proposed system includes the following tasks:

- Develop a methodology to construct the domain knowledge to identify the concepts.
- Develop a model to find the sequence patterns by integrating the knowledge.
- Propose a recommendation strategy that also identifies the concept drift in the access pattern and suggests accordingly.

The experimental results carried on benchmark data sets clearly show the improvement in the performance of the proposed framework when compared with the popular existing models and also prove the importance of analyzing the interest drift of the user by evaluating with appropriate measures.

The rest of the paper is organized as follows. In Sect. 2, related work is presented. In Sect. 3, the architecture and design details of the proposed recommender strategy are described. Section 4 analyzes the implementation and experimentation part. Finally, Sect. 5 gives the conclusion followed by references.

2 Related Work

Adomavicius and Tuzhilin, in [1], discussed the classification of the recommenders as content based [2, 3], collaborative filtering [4] and hybrid methods [5, 6]. Although many of the researchers kept their efforts in improving the accuracy of the recommender through the technique used [7], some focused on metrics such as "diversity" (average dissimilarity among all recommendation pairs) [8], individual diversity (average dissimilarity of recommendation pairs limiting to a user) [9] and aggregate diversity (average of dissimilarity of all users) [10] as important as accuracy to satisfy thebuser. Some works [11] proposed recommenders considering a new kind of metric "novelty" (amounts to the user's surprise w.r.t. the time for searching a page or item) in the evaluation of recommendations and act accordingly. But all these studies did not consider the concept of the item or the user and did not try to identify the drift of concept.

In general, the access sequence patterns can be learned by probabilistic algorithms and association analysis [12]. Ezifie and Y. Lu proposed a sequence pattern mining algorithm using a tree structure called PLWAP-Mine [13] which showed better results than other pattern mining techniques. In [14], Nguyen proved that PLWAPMine integrated with the Markov model can enhance the performance of mining. However, these algorithms consider only the usage history, but not the semantics at all which lags the quality of recommendations.

L. Wei and S. Lei made a model by integrating the ontology with usage mining, so that the patterns are subject oriented rather than item or page oriented to improve the performance of the recommender [15]. There were several ontologies constructed on different domains such as personalized e-learning and software to generate the recommendations of the ontology with the significant terms in the web site used [3]. S. Salin and P. Senkul proposed applying these domain concepts even on access sequence instances and then tried to make accurate suggestions [16]. These studies did not consider the dynamics of the ontology instance and also made no focus on the efficiency of constructing the ontology.

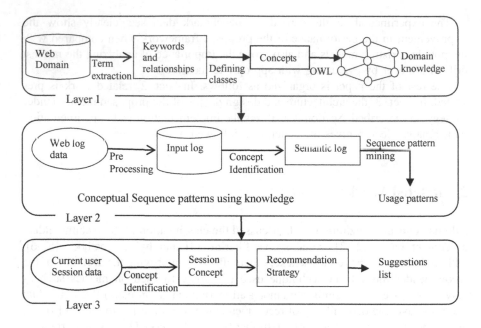

Fig. 1 Layered architecture of the proposed recommender system

3 Recommender Model with Conceptual Semantics

The proposed model for recommenders based on concept and its dynamics can be defined in three layers as shown in Fig. 1. In the first layer, the construction of ontology for the domain will take place by manual, automatic or semi automatic approaches. Though several techniques are available in constructing the ontology, there still exists demand for the customized domains depending on the purposes. To accomplish this task, the informal information provided is Web content; somehow it seems tedious to process the huge provided content. Instead, one can make use of the informal information available in or associated with a Web page such as title, tags or URL of the page to construct the domain knowledge. Typically, the second layer of the model is to find the patterns of the user's access sequence. Generally, the sequence pattern modeling uses the Web usage log after some traditional pre-processing techniques to find patterns. However, the proposed model extends preprocessing in its way to get semantic log, so that concept-oriented patterns can be extracted accordingly. In the model, the recommendation strategy is defined in the third layer that uses not only the patterns provided in the second layer to assess the user's navigation, but also the domain ontology constructed in the first layer to identify the existence of the concept drift if any. Finally the recommendation strategy in this layer gives the suggestions of the items or Web pages to the user.

3.1 Conceptual Knowledge Construction

Here, the title of the page and the tags provided for a Web page are used as sources to derive concepts to generate the ontology. Generally, the tags and title of the Web page contains key terms that represent the content of the page. The idea is to define the concepts from these key terms depending on the number of occurrences and combinations of the terms. This task can be accomplished by following the steps that involve defining concepts and relationships among concepts.

3.1.1 Defining Concepts

Let $\{T_1, T_2...T_n\}$ be the titles and tags of m number of pages or items in the domain.

Step 1: The stop word removal technique is applied to the titles and tags of the pages, so that only a set of raw terms $\{w_1, w_2...w_k\}$ will be derived in each page title.

Step 2: Find frequent terms by means of the association analysis technique which gives the significance of a set of terms that can be assumed as concept C_i.

Step 3: From the derived concept set $C = \{C_1, C_2...C_m\}$, identify the most generalized and specialized concepts.; 2

Step 4: Identify a relationship from all the concepts to at least one of the mentioned generalized concepts.

3.2 Sequence Pattern Mining with Concepts

This layer of model gives the access patterns of the user as per the domain ontology constructed and stored with concepts and relationships among concepts. This task can be of two parts, where the first one is about preprocessing and the second one is pattern mining.

3.2.1 Preprocessing

The literature provides many data preprocessing techniques as four different classes (data cleaning, reduction, integration and transformation) to make mining qualitative in terms of accuracy. Here, the Web log contains the access records of all the users of all sessions for a particular period. This log information will act as

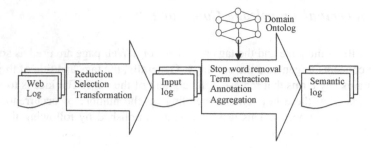

Fig. 2 Preprocessing of the log to extract the semantic log

raw input for extracting the navigational sequences of the user. Generally, the record in a log contains the IP address, time stamp of access, title of the page, URL of the page, protocol used, session id and typically the tags of the page. Data reduction is the first step that is applied to remove unnecessary fields such as protocol name in the log table. Information such as time stamps or session ids must be transformed to the format that can be processed. This log will be given for data selection so that only the records of that particular user will be extracted with its session ids (Fig. 2).

After making traditional preprocessing, the log is applied with advanced techniques to transform it as semantic log. In this model, the titles of the page are applied with stop word removal techniques to avoid the terms such as 'the', 'are' and 'is'. Thereafter, the term extraction technique will consider the raw terms and their combinations by means of the frequency measure. Then, annotation is made to make the records in the user's input log represented with the concepts for which the domain ontology is used to annotate the relevant concept label for each record. Finally, the aggregation step shows the access records with conceptual information.

3.2.2 Mining Patterns

Once the semantic log is constructed, the sequential learning method will be applied to get patterns. To get patterns, the proposed model uses the basic theme of association patterns extraction by the TITANIC algorithm [14], which outperforms in constructing the lattice of the concept for a user.

The above algorithm gives the access sequences as a set of combinations preserving the order of a particular user results the personalized lattice of concepts. The personalized concept lattice is the hierarchy of concepts in a tree structure constructed using the support and confidence measures.

Algorithm to find conceptual patterns:

Input: Set of Users U = {U₁,U₂ ... Uₙ}
           Set of Concepts C = {C₁, C₂...Cₘ}
           Set of access sequences AS of all users initialized as AS[Uᵢ]={φ}for 1≤i≤n
Output: Returns top p number of records in AS[Uᵢ]
           // p is predefined as per requirement of application

1.   Start
2.   For all users $U_i \in U$
3.   For all concepts C_k exists in User's log where $C_k \in C$
4. $X[U_i] \leftarrow C_k \cup X[U_i]$ //Add C_k to $X[U_i]$.
5.   Calculate $f(x_k)$ average frequency of each concept set x_k in all web sessions.

$$f(x_k) = \frac{\sum_{j=1}^{N} frequency\ of\ x_k\ session\ j}{N}$$

     //Where N is number of sessions participated by the user U_i
6.   If x_k satisfies the minimum support threshold then
7. $AS[U_i] \leftarrow x_k \cup AS[U_i]$ //Add x_k to $AS[U_i]$
8.   Make a set S with combinations of concepts in AS[Uᵢ]
9.   For each combination set x_k in S where $S \neq \{φ\}$
10.   Repeat steps 4 to 6
11. Sort all sequences in AS[Uᵢ] according to $f(x_k)$
12. For all sequences x_k in AS[Uᵢ]
13.   Calculate confidence of x_k
14. $AS[Ui] \leftarrow AS[U_i] - x_k$ //remove x_k from AS[Uᵢ] s
15. Stop

3.3   Recommendation Strategy

The final layer in the architecture deals with the recommendation strategy to suggest pages by means of the knowledge gained in the first layer and the patterns from the second layer. The primary task in this step is to identify the recent concept pattern of the current session. Thereafter, if the current concept matches any part of any of the existing patterns saved for the user, then suggestions has to be made accordingly. If the pattern matches part of the existing pattern but not with the current concept, then the suggestion will switch away from the traditional recommendation path to the current concept pages dynamically. Thus, the concept drift of the user's interest identified and the suggestions will be changed dynamically by the proposed method.

Table 1 Experimentation values for the two recommenders

Technique/measure	Hit ratio(%)		MAE	
	100 K dataset	1 M dataset	100 K dataset	1 M dataset
PLWAP	45.52	49.06	0.7869	0.7641
Proposed recommender	52.48	68.22	0.7199	0.6901

4 Experimentation Results

The experimental setup to evaluate the proposed model is kept on a benchmark dataset Movielens of two variants with 100 k and 1 M ratings. The one with 100 k ratings is provided for 1682 movies by 943 users, whereas the other one is provided for 3900 movies by 6040 users. The ratings are on the scale from 1 to 5, defining 1 for low quality and 5 for high quality. To evaluate the methodology, the mean absolute error (MAE) measure is used, and to evaluate the efficiency of recommendations, hit ratio [6] measure is used:

$$MAE = \frac{\sum_{i=1}^{n}|ar_i - pr_i|}{n}, \tag{1}$$

$$Hit\ ratio = \frac{Ar}{Re} \tag{2}$$

where $\{ar_1, ar_2...ar_n\}$ are the actual ratings, $\{pr_1, pr_2...pr_n\}$ are the predicted ratings, 'A_r' is the total number of recommendations accessed by the user and 'R_e' is the total number of recommendations. This experimentation compared the proposed model with one of the popular existing usage model PLWAP for the two variants of the data set for the top ten recommendations.

Table 1 shows the summary of the hit ratio and MAE of the proposed model as well as the existing model. It clearly says that the proposed model outperforms the PLWAP model in terms of the number of suggestions that are accessed by the user that is relevant to the user interest.

5 Conclusion

The typical recommenders based on usage history cannot use the semantics and will not consider the concept drift of user interest. This paper made study on the recommenders with concepts as well to find interest drift of the user on top of concepts in the sense to make user satisfied. The proposed model constructs the ontology, mines the patterns and applies on the current Web access sequence of the user. The proposed model was evaluated by comparing with popular existing methods and the results showed that the model outperformed in terms of performance.

References

1. G. Adomavicius and A. Tuzhilin, "Toward the next generation of recommender systems: A survey of the state-of-the-art and possible extensions," *IEEE Trans. Knowledge Data Eng.*, vol. 17, no. 6, pp. 734–749, June 2005.
2. M. Pazzani and D. Billsus, "Content-based recommendation systems," in *The Adaptive Web*, P. Brusilovsky, A. Kobsa, and W. Nejdl, Eds. Berlin Heidelberg, Germany: Springer-Verlag, 2007, pp. 325–341.
3. M. Venu Gopalachari, P. Sammulal, "Personalized Web Page Recommender System using integrated Usage and Content Knowledge", in the proceedings of 2014 IEEE ICACCCT, 2014. pp. 1066–1071.
4. J. Schafer, D. Frankowski, J. Herlocker, and S. Sen, "Collaborative filtering recommender systems," in *The Adaptive Web*, P. Brusilovsky, A. Kobsa, and W. Nejdl, Eds. Berlin Heidelberg, Germany: Springer-Verlag, 2007, pp. 291–324.
5. E. Amolochitis, I. T. Christou, and Z. H. Tan, "Implementing a commercial-strength parallel hybrid movie recommendation engine," *IEEE Intell. Syst.*, vol. 29, no. 2, pp. 92–96, Mar. 2014.
6. M. Venu Gopalachari, P. Sammulal, "Hybrid Recommender System with Conceptualization and Temporal Preferences", Proceedings of the Second International Conference on Computer and Communication Technologies, AISC 380, pp. 811–819, Springer, 2015.
7. M. Jahrer, A. Töscher, and R. Legenstein, "Combining predictions for accurate recommender systems," in *Proc. 16th ACM SIGKDD Int. Conf. Knowledge Discovery Data Mining*, New York, 2010, pp. 693–702.
8. S. Vargas and P. Castells, "Rank and relevance in novelty and diversity metrics for recommender systems," in *Proc. 5th ACM Conf. Recommender System*, New York, 2011, pp. 109–116.
9. M. Zhang and N. Hurley, "Avoiding monotony: Improving the diversity of recommendation lists," in *Proc. ACM Conf. Recommender Systems*, New York, 2008, pp. 123–130.
10. G. Adomavicius and Y. Kwon, "Improving aggregate recommendation diversity using ranking-based techniques," *IEEE Trans. Knowledge Data Eng.*, vol. 24, no. 5, pp. 896–911, May 2012.
11. F. Fouss and M. Saerens, "Evaluating performance of recommender systems: An experimental comparison," in *Proc. IEEE/WIC/ACM Int. Conf. Web Intelligent Agent Technology*, Washington, D.C.: IEEE Computer Society, 2008, pp. 735–738.
12. B. Mobasher, "Data Mining for Web Personalization," in *The Adaptive Web*. vol. 4321, P. Brusilovsky, A. Kobsa, and W. Nejdl, Eds.: Springer-Verlag Berlin, Heidelberg, 2007, pp. 90–135.
13. C. I. Ezeife and Y. Lu, "Mining Web Log Sequential Patterns with Position Coded Pre-Order Linked WAP-Tree," *Data Mining and Knowledge Discovery*, vol. 10, pp. 5–38, 2005.
14. S. T. T. Nguyen, "Efficient Web Usage Mining Process for Sequential Patterns," in *Proceedings of the 11th International Conference on Information Integration and Web-based Applications and Services*, Kuala Lumpur, Malaysia 2009, pp. 465–469.
15. L. Wei and S. Lei, "Integrated Recommender Systems Based on Ontology and Usage Mining," in *Active Media Technology*. vol. 5820, J. Liu, J. Wu, Y. Yao, and T. Nishida, Eds.: Springer-Verlag Berlin Heidelberg, 2009, pp. 114–125.
16. S. Salin and P. Senkul, "Using Semantic Information for Web Usage Mining based Recommendation," in *24th International Symposium on Computer and Information Sciences, 2009.*, 2009, pp. 236–241.

Mitigating Replay Attack in Wireless Sensor Network Through Assortment of Packets

Vandana Sharma and Muzzammil Hussain

Abstract There are many attacks possible on wireless sensor network (WSN) and replay attack is a major one among them and, moreover, very easy to execute. A replay attack is carried out by continuously keeping track of the messages exchanged between entities and replayed later to either bring down the target entity or affect the performance of the target network. Many mechanisms have been designed to mitigate the replay attack in WSN, but most of the mechanisms are either complex or insecure. In this paper, we propose a mechanism to mitigate the replay attack in WSN. In the proposed work at each node, assorted value of a received packet is maintained in a table and the reply attack is detected or mitigated using their assorted value of the already received packets. The proposed mechanism was simulated and its performance evaluated and it was found that the proposed mechanism mitigates the replay attack, secures the network and takes less time for processing.

Keywords Replay attack · Sensor nodes · Hash · MD5 · Occurrence

1 Introduction

WSN was introduced first time in military and heavy industrial applications. Governments and universities began using WSN in many applications such as natural disaster prevention, weather stations and fire detection. With the increased use of the WSN mission in critical environments such as military and health-care applications, these environments need to be secured.

V. Sharma (✉) · M. Hussain
Department of Computer Science and Engineering, Central University of Rajasthan,
Bandarsindri, Kishangarh, Ajmer 305817, Rajasthan, India
e-mail: 2014mtcse022@curaj.ac.in

M. Hussain
e-mail: mhussain@curaj.ac.in

© Springer Science+Business Media Singapore 2017   221
S.C. Satapathy et al. (eds.), *Proceedings of the First International Conference on Computational Intelligence and Informatics*, Advances in Intelligent Systems and Computing 507, DOI 10.1007/978-981-10-2471-9_22

The sensor node is composed of the sensor, processor, transceiver, ADC, memory antenna and power generator. The sensor produces a considerable response whenever it detects any change in its external environment. The main component of the sensor node is the sensor; it senses the data and converts it from analog to digital through ADC. The data are then processed and the processed data are then sent through the transceiver to the base station. The transceiver does the task of both transmitting and receiving [1, 2]. For location finding and mobility handling, a location finding system and mobilizer are introduced. The power generator supplies power. WSNs are placed in a location where humans cannot reach easily, so that they should be provided with a sufficient amount of energy to power the system.

The major challenges of WSN are ad hoc deployment, dynamic nature and unattended operation. Due to the dynamic environment, sensor nodes should be developed so that they can cope with environmental challenges. Due to this, unattended operation sensor nodes should be configured so that they can adapt to the environment changes automatically. WSNs are deployed in places where humans cannot go physically, so physical security becomes a major concern. Confidentiality, authentication, integrity and availability are the major security requirements of WSN [2, 3].

WSNs have limited storage capacity and computational power. The battery of WSN should be used efficiently. All attacks on WSN are to deplete the resources of WSN. There are many attacks possible, such as selective forwarding, sinkhole, Sybil, impersonation and eavesdropping [4]. Replay attack can be carried out easily by keeping track of all the messages being sent between nodes and sending them back later, to waste the resources of the target node in processing the message. The target node gets drained and will be unable to perform its actual task, leading to denial-of-service attack [5, 6]. In this paper, we have studied the impact of replay attack and have proposed a mechanism to mitigate it.

2 Related Work

So far, mechanisms have been designed to detect and mitigate replay attack. Here, we summarize the effective mechanisms among them, but most of them suffer from problems such as security, complexity and synchronization.

2.1 Lamports Password Authentication

This method implements a onetime password, due to which the attacker cannot eavesdrop on messages exchanged. This method is implemented between the user and the server. A system which uses this method will never use the same password.

In this, a password table is used to verify the legality of the user's identity. In this mechanism, there is a risk of losing the password table; if it is used, the attacker can misuse it. To overcome the stolen table attack of the Lamport Password authentication scheme, nonce-based and timestamp-based schemes were introduced.

2.2 Nonce

In password-based authentication protocol, [7] the server sends a challenge or a random value to the client and the client will respond by sending h(c ‖ p), where h is the hash, c is the challenge and p is the password. ‖ denotes concatenation. The server checks the password in its database and whether it is correct; if yes, then the client is validated.

With a passive attack, the attacker eavesdrops, but does not alter the message. The attacker can see c and h(c ‖ p), so he can use this cluster to create passwords until a match is found. Random challenge is used to avoid this attack.

If the attacker is active, then he can change the messages; the attacker can send his own challenge c' and wait for the client's response. The client will send the response h(c' ‖ p). Using the same challenge c', he can get the precomputed table. Now, the attacker can attack several passwords. A client nonce avoids this by the following:

The server sends a random challenge. The client chooses a nonce n.
The client sends n ‖ h(n ‖ c ‖ p).

The server re-computes h(c ‖ n ‖ p) and sees if this value matches the one the client sends.

2.3 Timestamps

It is a method to ensure the freshness of the message. This is a very old method to find out that the message which is received is original or replayed. In this, every message is sent along with a timestamp [8]. The receiver after receiving the message checks whether the timestamp is within the acceptable range; if yes, then the message is accepted otherwise dropped. Through Network Time Protocol every computer maintains accurate time.

To implement this method, clock synchronization of the two entities is a must. Synchronization is required to maintain the accuracy and precision of the timestamp. Here, we need to make a list of old timestamps, large storage is required and due verification overhead is suffered.

2.4 Sequence Number

Sequence number is assigned to be monotonically either increasing or decreasing to each transmitted message. If a message is replayed, it will have a very small or very old sequence number and can be discarded [9]. In this mechanism, forced replayed messages cannot be detected and also it is difficult to maintain the sequence number.

2.5 Receiver Authentication Protocol

RAP [7] can be used for two purposes, detection and prevention. Detection is a scheme which aims to detect an intruder that replays beacons without preventing it from doing so. In the prevention mode, the challenge–response message exchange takes place before data transmission. In this, the sender transmits the data only when the receiver is authenticated. Energy efficiency is a major requirement; so in normal condition, detection is done and the system switches to the expensive prevention mode only if required.

2.5.1 Detection Mode

RAP-D aims to detect the replayed packets. As we can see in Fig. 1, if a sender node A wants to transmit data to receiver B, B broadcasts a beacon and A answers backs with data and a challenge Cp. On the following beacon, B acknowledges the reception of data packets and sends encrypted version of challenge Cp using shared

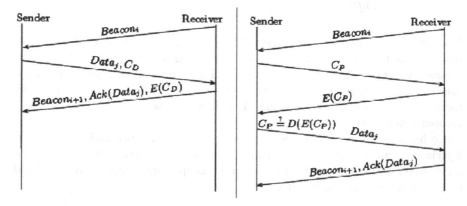

Fig. 1 Detection of the replayed beacon shown in the first figure and prevention of beacon replayed shown in the second figure

key Krap. B validates the response by decrypting it and comparing it with the original value. If they are the same, then the message is not replayed.

2.5.2 Prevention Mode

RAP-P aims to prevent the beacon replay at the cost of increased overhead. It can be seen in Fig. 1. In this, data is not sent right after the beacon. A sends a challenge Cp and waits for the encrypted challenge from B. Only if the received value is correct, then the data is exchanged. This is more costly because two more messages are being sent.

3 The Proposed Algorithm

In this work, we have designed a mechanism to mitigate replay attack. In the proposed protocol, for each received message, the node calculates its hash value and stores it in the table. The computed hash is searched in the occurrence table; if not found, then the message is treated as fresh and it is entered into the occurrence table with occurrence set to one. Otherwise, the message is replayed once and is discarded by the node.

Algorithm
```
Create table()
{
    Struct table
    {      char hp;
           Int occur;
    }
Struct table t[100];
}

For each received packet M at any node
For I from 1 till n
    If (table[1].id == hp &&  table[1].occur > 0) then
        Print " Packet is replayed"
    Else
           Make entry in table
           Table[n+1].id = hp;
           Table[n+1].occur = 1;
           Print "packet is fresh"
           Break;
    End
```
where n is the maximum number of entries in the table, Table[i].id is the hash of the ith entry and hp is the hash value.

Whenever a message is received at any node, it is checked that a table is created or not; if yes, then the hash of the received message is compared with the already existing hashes. If a match is found, then the message is declared as a replayed message; otherwise, it is a new message and entry is created in the table for hp and occur. Here, hp is hash of the received message and the occur value is set to one. Loop 'n' is the total number of entries in the table. This loop will run for every message which is received. The table is a structure, which consists of two members table[i].id for hash of message and table[i].occur for occurrence value.

If table[i].id is equal to the recently calculated hash and table[i].occur > 0, then the message is replayed; otherwise, a new entry is created for the hash and occur value and the packet is declared as fresh.

4 Implementation

The proposed protocol was implemented in Python database. For enabling communication, three entities are defined, two clients and one server. Clients communicate through the server. The proposed protocol is implemented at the server, while replayed messages are discarded by the server and fresh messages forwarded to clients. The server is defined as a local host with IP address 127.0.0.1 at port 5000. Clients are at port 0. Socket API is used for communication between client and server.

$$ipc\ s = socket.socket(socket.AF\ INET, socket.SOCK\ DGRAM).$$

Fig. 2 client1 sending a hello message to client2 through the server

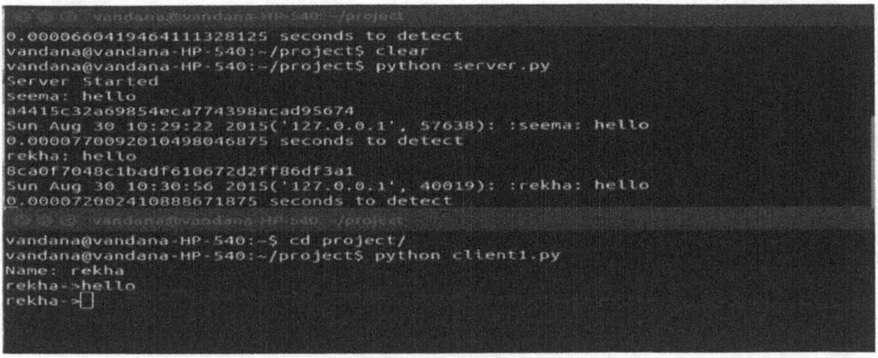
0.0000660419464111328125 seconds to detect
vandana@vandana-HP-540:~/project$ clear
vandana@vandana-HP-540:~/project$ python server.py
Server Started
seema: hello
a4415c32a69854eca774398acad95674
Sun Aug 30 10:29:22 2015('127.0.0.1', 57638): :seema: hello
0.0000770092010498046875 seconds to detect
rekha: hello
8ca0f7048c1badf610672d2ff86df3a1
Sun Aug 30 10:30:56 2015('127.0.0.1', 40019): :rekha: hello
0.000072002410888671875 seconds to detect
vandana@vandana-HP-540:~$ cd project/
vandana@vandana-HP-540:~/project$ python client1.py
Name: rekha
rekha->hello
rekha->

Fig. 3 client2 sending a hello message to client1 through the server

The messages are exchanged between communicating entities as datagram. The hash of the received message is calculated using MD5 as g = hashlib.md5(str (data))hashdigest().

5 Result and Analysis

The proposed algorithm was simulated in Python. One server and two clients (client1 and client2) were created. client 1 will send a hello message to client 2 and client2 responds with a hello message. If client1 sends a hello message again, the

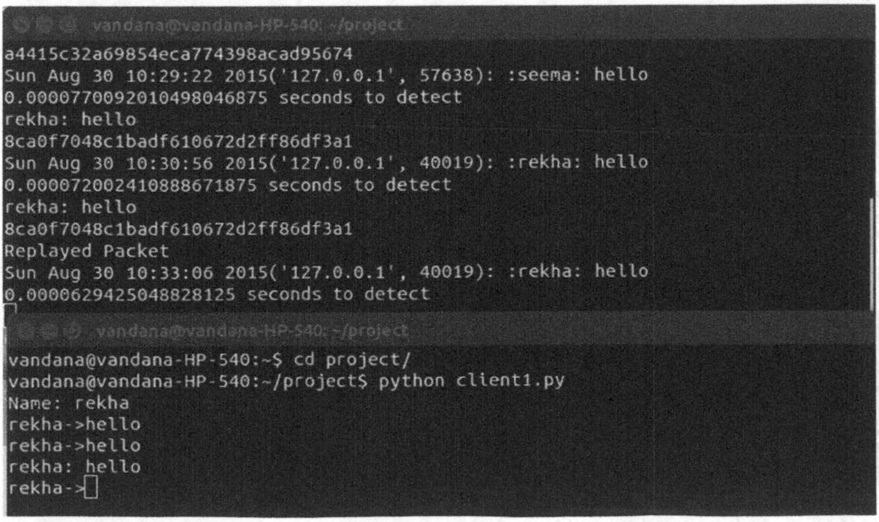
vandana@vandana-HP-540: ~/project
a4415c32a69854eca774398acad95674
Sun Aug 30 10:29:22 2015('127.0.0.1', 57638): :seema: hello
0.0000770092010498046875 seconds to detect
rekha: hello
8ca0f7048c1badf610672d2ff86df3a1
Sun Aug 30 10:30:56 2015('127.0.0.1', 40019): :rekha: hello
0.000072002410888671875 seconds to detect
rekha: hello
8ca0f7048c1badf610672d2ff86df3a1
Replayed Packet
Sun Aug 30 10:33:06 2015('127.0.0.1', 40019): :rekha: hello
0.0000629425048828125 seconds to detect
vandana@vandana-HP-540: ~/project
vandana@vandana-HP-540:~$ cd project/
vandana@vandana-HP-540:~/project$ python client1.py
Name: rekha
rekha->hello
rekha->hello
rekha: hello
rekha->

Fig. 4 The server detecting a replayed hello message sent by client1

server detects it as a replayed one. The proposed mechanism is simple as the nodes just need to compute the hash of a received packet and compare it with the existing values in the table. The mechanism is secure, as it is impossible for any node to mislead the receiver node by resending packet as a fresh one by changing a few parameters because the hash is computed for the whole packet (Figs. 2, 3 and 4).

6 Performance Comparison of Bloom Filter with the Proposed Protocol

The performance of the proposed protocol is compared to that of the existing protocol (Bloom filters). It has been observed that the proposed protocol has a less number of hash computations and the probability of detecting a replayed packet is the same. However, the proposed protocol takes less time than that taken by Bloom

```
 ●●●  vandana@vandana-HP-540: ~/project
vandana@vandana-HP-540:~/project$ python bloom_filter.py
hello everyone
Filter size 1049 bytes
False
0.0001938343048095703125 seconds to detect
vandana@vandana-HP-540:~/project$ █
```

Fig. 5 Time taken to detect the replay message through the Bloom filter is 0.0001938343 s for a particular input

```
 ●●●  vandana@vandana-HP-540: ~/project
vandana@vandana-HP-540:~/project$ python server.py
Server Started
vandana: hello everyone
ff3c76ec6e8c87eb220bd3b37092b737
Fri Aug 28 16:02:28 2015('127.0.0.1', 48239): :vandana: hello everyone
0.0000698566436767578125 seconds to detect
```

Fig. 6 Time taken to detect the replay message through the proposed protocol is 0.000069856643 s for the same input as in the Bloom filter

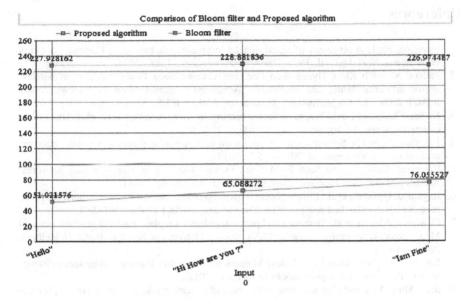

Fig. 7 The comparison of the time taken by the Bloom filter and the proposed algorithm for various inputs. Inputs are mapped in the X-axis and time in milliseconds is on the Y-axis. It shows that the time taken by the proposed algorithm is very less compared to that by the Bloom filter

filters mechanism. The computation time of the proposed protocol was found to be 0.0006985 s and that of the bloom filters mechanism 0.0001950 s for the same and under a similar environment (Figs. 5, 6 and 7).

7 Conclusion

Security is a major issue in WSN. The network has to be protected against various possible attacks so as to extend the lifetime of sensor nodes and also to avoid its malfunctioning. Replay attack is very common in WSN and may lead to many more attacks like DoS. Hence, it is very much needed to design a mechanism to mitigate any replay attack in WSN.

We have designed an algorithm to detect and mitigate any replay attack. The receiving entity calculates the hash of the received packet and stores in its database along with a parameter occurrence. Any packet with occurrence one is detected as replayed. The proposed algorithm is implemented in Python and its performance is studied and verified. It has been found that the proposed protocol is successful in mitigating the replay attack in WSN. The time taken by the proposed protocol is found to be less then that compared to the existing mechanism.

References

1. Syverson, Paul.: A taxonomy of replay attacks [cryptographic protocols] Computer Security Foundations Workshop VII, 1994. CSFW 7. Proceedings. IEEE, 1994.
2. Pathan, Al-Sakib Khan, Hyung-Woo Lee, and Choong Seon Hong.: Security in wireless sensor networks: issues and challenges. Advanced Communication Technology, 2006. ICACT 2006. The 8th International Conference. Vol. 2. IEEE, 2006.
3. Karlof, Chris, and David Wagner.: Secure routing in wireless sensor networks: Attacks and countermeasures. Ad hoc networks 1.2 (2003): 293–315.
4. Jesudoss, A., and N. Subramaniam.: A survey on authentication attacks and countermeasures in distributed environment. IJCSE, vol 5.2 (2014).
5. Raymond, David R., and Scott F. Midkiff.: Denial-of-service in wireless sensor networks: Attacks and defenses. Pervasive Computing, IEEE 7.1 (2008): 74–81.
6. Raymond, David R., et al.: Effects of denial-of-sleep attacks on wireless sensor network MAC protocols. Vehicular Technology, IEEE Transactions on 58.1 (2009): 367–380.
7. Di Mauro, Alessio, et al.: Detecting and preventing beacon replay attacks in receiver-initiated MAC protocols for energy efficient WSNs. Secure IT Systems. Springer Berlin Heidelberg, 2013. 1–16.
8. Baayer, Aziz, Nourddine Enneya, and Mohammed Elkoutbi.: Enhanced timestamp discrepancy to limit impact of replay attacks in manets. (2012).
9. Sung-Ming, Yen, and Liao Kuo-Hong.: Shared authentication token secure against replay and weak key attacks. Information Processing Letters 62.2 (1997): 77–80.

Progressive Genetic Evolutions-Based Join Cost Optimization (PGE-JCO) for Distributed RDF Chain Queries

K. Shailaja, P.V. Kumar and S. Durga Bhavani

Abstract The finest way of semantic Web representation for further indexing and querying becomes robust due to the RDF structure. The magnified growth in semantic Web data, RDF query processing, emerged as complex due to numerous joins. Henceforth, to achieve scalability and robustness toward search space and time, the query joins must be optimized. The majority of existing benchmarking models have been evaluated on a single source. These methods utterly failed to optimize the queries with nested loop join, bind join, and AGJoin, which is due to the search cost only being considered as the optimization factor by all of the existing models. Hence to optimize the distribute chain queries, here in this paper we propose a novel evolutionary approach, which is based on progressive genetic evolutions to identify the optimized chain queries even for distributed triple stores. The experimental results show that the significance of the proposed model over the existing evolutionary approaches is optimal. Also, it is obvious to confirm that the metrics and evolution process introduced here in this paper motivates future research to identify the new dimensions of chain query optimization to search in distributed triple stores.

Keywords Query optimization · RDF · Triple store · SPARQL · Join cost optimization · PGE-JCO · Query chain · Distributed RDF · Search space · Access cost

K. Shailaja (✉)
MCET, Hyderabad, Telangana, India
e-mail: shailajamtech2006@yahoo.co.in

P.V. Kumar
UCE, OU, Hyderabad, Telangana, India
e-mail: pvkumar58@gmail.com

S. Durga Bhavani
JNTUH, Hyderabad, Telangana, India
e-mail: sdurga.bhavani@gmail.com

© Springer Science+Business Media Singapore 2017   231
S.C. Satapathy et al. (eds.), *Proceedings of the First International Conference on Computational Intelligence and Informatics*, Advances in Intelligent Systems and Computing 507, DOI 10.1007/978-981-10-2471-9_23

1   Introduction

Web is the present scenario of data sharing. The magnified speed in increase of data volumes in Web leads to the information processing complexity. To handle this issue, one significantly acceptable solution is the semantic Web. More precisely, semantic Web is the structure of relating the people and data through similarity of data content, context, concept or the semantic scope [1]. Storing data of the scmantic Web is another dimension of the data storage and many of the storage models have emerged in the last two decades [2]. RDF is one of the fundamental data models [3], which is a language of forming semantic relations between Web data. The triple-store format that follows to relate the data in RDF is the most successful structure. Despite the facility of Web page storage under the semantic relations, there are significant challenges such as optimal scan of the RDF store to extract the desired information, indexing the RDF store and balancing the distributed triple stores to store and search the data. SPARQL [4], which emerged with the motivation of SQL, is one optimal solution to deal with these challenges.

The SPARQL is capable to query-distributed triple stores. To acquire the information that links with many combinations of any of the triples, an SPARQL query restraints the set of queries using joins. The complex select query of SPARQL that aimed to extract information from the RDF store needs to be optimal in search space usage [5]. Hence, the optimizing query chain to scan the RDF store is a significant research area and, moreover, this research issue is more complex if distributed RDF stores are to be scanned against the given query. The scan time, search space usage and store access cost are together influenced by the order of joins involved in the given query.

The reserach objective of many scholars is to analyze and deliver divergent query optimization strategies, which is in the context of semantic Web. The significant query optimization solutions [6] are based on evolutionary strategies, such as genetic algorithm [7], iterative improvement (II) [8], simulated annealing (SA) [9] also called two-phase optimization (2PO) algorithm were used to address the problem of query path optimization. The de facto issue is that though these many solutions emerged, none of them are capable of optimizing the queries in all contexts, such as loop back joins, left deep tree, right deep tree and bushy tree representation of the queries. Another active objective is optimizing queries against distributed RDF stores. Hence it is obvious to concude that, there is a significant need to define query optimization strategies.

The article is organized into multiple sections. Section 2 explores the review of the work related to our proposed model. Section 3 contains the detailed description of the model proposed called PGE-JCO, which is followed by Sect. 4 that elaborates the experimental setup and performance analysis of the proposal. Section 5 concludes the model devised in this article.

2 Related Work

The elementary concepts associated with efficient processing [5] of SPARQL queries has been studied in literature. The study was performed on (i) the complexity analysis of all operators in SPARQL query language, (ii) equivalences of SPARQL algebra and (iii) algorithm for optimizing semantic SPARQL queries. The complexity analysis shows that all fragments of SPARQL fall into the category of NP.

To handle the large amount of RDF data, there is a need to optimize the join of the partial query results. An ant colony system (ACS) [10] was proposed in literature to query effectively in the semantic Web environment. The improvement in solution costs was compared with the existing algorithms such as genetic algorithm (GA) [11] and two-phase optimization (2PO) [12]. It was proved that the ACS approach outperformed the existing approaches.

To optimize a special class of SPARQL queries called RDF chain queries, a new genetic algorithm was devised called the RCQ-GA [11]. The algorithm optimizes the chain queries by finding the order in which joins are to be performed. The performance of the algorithm was compared with two-phase optimization and the results show the quality of the solution and consistency of the solution.

Cardinality estimation-centric join ordering algorithm [13] is one that aimed to optimize the join order of SPARQL query to achieve minimal execution time. The performance analysis of this model was done on star and arbitrary queries. The search optimization in this model is considered to be only metric and is also specific to scan the single RDF store.

The particle swarm optimization is another evolution strategy that was used [14] to optimize the queries of distributed database context. Since the particle movement is in the context of the probability distribution, the computational complexity of this model is also not linear. The model is not described as specific to distributed RDF and optimality varies according to the parameters set.

A parallel join algorithm [15] is the ensemble of three algorithms that joins manifold queries in an enclosed manner. The access cost is not considered to optimize the join cost which is a considerable constraint of the model, but this model is robust toward data size.

Querying graphs called the G-SPARQL [16] is another benchmark model in the literature. G-SPARQL partitions the query into subqueries and executes each subquery in the context of optimal memory usage. The computational complexity of the model is not linear and also lacks the optimization of access cost and search space.

Another SPARQL query processing model [17] that uses an index structure to optimize the search space toward RDF triples scan was proposed. This model relied on a tree structure to optimize query joins. This model is robust if SPARQL query is aimed at a single RDF store, since the access cost is not the factor of join cost optimization.

A Map Reduce framework [18] was devised in the context of minimizing the response time. The algorithm that was devised to support this framework is All-Possible-Join tree (APJ-tree) algorithm. To optimize the search space, the bloom filter is used by the APJ-Tree. Similar to the other benchmarking models discussed, this model also does not consider the access cost as a factor of join cost optimization and seems to be robust to scan the query on a single RDF store.

The constraints observed in all of these existing models are that they were introduced to optimize the queries in single source environments and do not evince optimal performance and robustness in query chains with a large number of joins. Moreover, the access cost is not considered as the factor of join cost optimization.

Henceforth, here in this paper we defined a novel evolutionary computational model called "Progressive Genetic Evolutions-based Join Cost Optimization (PGE-JCO) for Distributed RDF Chain Queries", which is based on genetic evolution approach. Two metrics called "Search of Minimal Combinations (SMC)" and "Search by Minimal Access Cost (SMAC)" are defined and used to explore the cost function of the genetic algorithm.

3 Progressive Genetic Evolutions-Based Query Chain Join Optimization

Progressive genetic evolution strategy [19] restricts the number of evolutions to a minimum. To do this, the progressive evolution strategy considers the new generations with fitness better than any of the pair of parents. The other significance of this proposal is the strategy of distributed RDF query chain cost assessment, which is done by assessing the optimality of multiple joins toward access cost and search cost. The objective function (see Sect. 3.2) estimates the cost of a query chain through the aggregation of the cost observed for each individual join involved in the query chain.

3.1 Querying RDF

The semantic Web search on the RDF structure through a SPARQL queries is done by representing in the form of a tree structure. The given query is formed in three different formats called left deep tree, right deep tree and bidirectional deep tree, also known as bushy tree. Any of these trees retains the subjects, predicates or objects as leafs and joins as nodes. Hence, the formation of the tree is done as depth first (leaves first). Connecting the query join as tree node to other query join exists in the tree as a node forms a bushy structure, where as a query join connected to a subject, predicate or an object delivers right deep tree and a subject, predicate or an object connected to other join under other join node concludes left deep tree.

The context of the proposed model is to optimize the join cost of a query that is aimed at distributed RDF, and hence the query structure adapted during the exploration of the proposal is bidirectional or bushy tree.

3.2 Objective or Cost Function

To identify the optimality of a query formed as a bushy tree, we defined two metrics called Search of Minimal Combinations and Search Minimal Access Cost. These two metrics are proposed, since the join cost optimization is always assessed through the search space used and the search time taken. The search space used is primarily proportionate to the number of combinations traversed due to the two leaves connected under a join node. To reflect this, here we proposed a metric called Search of Minimal Combinations. The other key factor is access time, which is the same if all join nodes of the tree connecting the leafs target a single RDF store, but if the target triple stores are different for each leaf of the join nodes involved, then the access time is different for different leafs. To do this, we proposed another significant metric called search by minimal access cost, which is dependent on the metric called search by minimal combinations. The approach of assessing these metrics is explored below.

Search of minimal Combinations (SMC) is the possible number of combinations between the count of subjects and count of objects under a given predicate. The maximum possible combinations can be measured as follows:

$$mpc(J_k) = 1 - \left(\sum_{i=1}^{|S|} \sum_{j=1}^{|O|} \{1 \exists p(s_i \to o_j) \equiv P\} \right)^{-1},$$

// where the above equation counts all the possible combinations of the subject s_i and object o_j with the expected predicate P; $p(s_i \to o_j)$ is the actual predicate of the subject s_i and object o_j. "$mpc(J_k)$" indicates the max possible combinations under the kth join, which is normalized to the value between 0 and 1.

Search by Minimal Access Cost (SMAC) is the aggregate cost to access the distributed RDFs under different subjects to obtain the objects under the expected predicate. This can be assessed as follows:

$$ac(J_k) = 1 - \left(\sum_{i=1}^{|S|} \sum_{j=1}^{|DRDF|} \{ac(s_i \Rightarrow rdf_j) \otimes vc\} \right)^{-1},$$

// where in the above-defined equation, $|S|$ is the total number of subjects under a query, $|DRDF|$ the number of RDFs considered under distributed architecture, $ac(s_i \Rightarrow rdf_j)$ the access cost to rdf_j under the selected subject s_i and vc represents the number of visits. The normalized aggregate access cost of the kth join is represented by $ac(J_k)$.

The cost observed at the kth join can be measured as follows:

$$c(J_k) = 1 - (mpc(j_k) \otimes ac(J_k)).$$

Cost Estimation of a Query Chain
For a given query chain qc,

$$c(qc) = \frac{\sum_{k=1}^{|qc|} \{c(J_k) \exists J_k \in qc\}}{|qc|},$$

//where $c(qc)$ is the average cost of the joins found in qc.

3.2.1 Root Mean Square Distance of the Join Costs

The root mean square distance of query chain qc is measured as follows:

$$RMSD(qc) = \sqrt{\frac{\sum_{k=1}^{|qc|} \left(min(\{c(j_1), c(j_2), \ldots\ldots, c(j_{|qc|})\}) - \{c(J_k) \exists J_k \in qc\} \right)^2}{|qc|}},$$

//where in the above equation, $RMSD(qc)$ is the root mean square distance of the join costs observed in qc.

3.3 Progressive Genetic Evolution Strategy

To deliver a cost optimal query chain for distributed RDF, the complex permutations and combinations should be checked between all possible query joins. With regard to this, here we devised Progressive Genetic Evolutions Strategy to identify the set of cost optimal query chains. The exploration of the algorithm is as follows:

Input:

- Set of queries $Q = \{q_1, q_2, q_3, \ldots, q_{|Q|}\}$ involved in the possible query chains to extract the desired information from a distributed RDF.
- Set of possible joins $\begin{cases} S = \{jq_1 = \{j_{11}, j_{12}, \ldots \ldots j_{1k}\}, \\ jq_2 = \{j_{21}, j_{22}, \ldots j_{2l}\}, \ldots \ldots \\ jq_{|Q|} = \{j_{|Q|1}, j_{|Q|2}, \ldots \ldots j_{|Q|p}\}\} \end{cases}$ available for each query in Q.
- Approximate number of possible query chains set $QC = \{qc_1, qc_2, qc_3, \ldots \ldots, qc_{|QC|}\}$
- Max evolutions threshold *met*

Progressive Genetic Evolutions:

a. Find joins that are common for any two given query chains such that the queries that are in the successor part of the selected join or queries that are in the predecessor part of the selected join be the same (order of queries need not be the same) in the given pair of query chains. For all joins found under this condition in the given pair of query chains,

   i. Split the first and the second query chains given as lqc_1, rqc_1 from first query chain and lqc_2, rqc_2 from second query chain at a join that is common for both. Then, reform the two new query chains by connecting lqc_1 and rqc_2 that forms the first one and by connecting lqc_2 and rqc_1 that forms the second.

   ii. Consider any of newly formed query chains, if their cost is less than the cost of any of the parent query chains.

b. If any new query chains are formed from the above process (step a), then add all those to the query chain set QC and discard the parent query chains that cost more than the resultant query chains.

c. Continue the progressive genetic evolutions on QC, till QC is not updated with new query chains.

d. Assess the total cost of the each query chain and root mean square distance of the join costs.
e. Order the resultant query chains in ascending order of their total cost and select the n best query chains. Then order these n best query chains in ascending order based on their RMSD and select the first query chain as the best for the process.

3.4 The Pseudo Code Representation of the Model Devised

$pc \leftarrow true$
$\quad$ While (pc) Begin
$\qquad nQC \leftarrow \phi$
For each query chain $\{qc_i \forall qc_i \in QC\}$ Begin // find crossover points (common joins)
$\quad$ For each query chain $\{qc_j \exists (qc_j \in QC \wedge j \neq i\}$ Begin
$\quad nQC \leftarrow \text{PGE-JCO}\,(qc_i, qc_j)$

//Progressive Genetic Evolutions, leading to evolution of an improved form in successive generations
$\qquad$ End
$\quad$ End
$\overline{QC} \leftarrow QC$ //clone the QC as $\overline{QC}$

$\overline{QC} \leftarrow nQC$ //adding new query chains to $\overline{C}$

$if\left(QC \triangleq \overline{QC}\right)$ Begin // equaling by definition

$\qquad pc \leftarrow false$
$\quad$ End
$\quad$ End

3.5 Progressive Genetic Evolutions for Join Cost Optimization

PGE-JCO (qc_r, qc_s) BEGIN

    $qc_{rs} \leftarrow \phi$ //is a set of query chains formed from crossover of qc_r, qc_s

      For each join $\{j_k \forall j_k \in qc_r\}$ Begin

         $coj \leftarrow \phi$ //set of crossover joins

            For each join $\{j_l \forall j_l \in qc_s\}$ Begin

               If $((j_k \cong j_l))$ Begin

                   $coj \leftarrow j_k$

End

     End

      For each $\{cj \forall cj \in coj\}$ Begin

Partite qc_i in to two at cross point cj , and label the left part as $\overleftarrow{qc_r}$ and right part as $\overrightarrow{qc_r}$

Partite c_j in to two at cross point cp , and label the left part as $\overleftarrow{qc_s}$ and right part as $\overrightarrow{qc_s}$

$$if\left(\left(\left\{\overleftarrow{qc_r}\right\} \cong \left\{\overleftarrow{qc_s}\right\}\right) \| \left(\left\{\overrightarrow{qc_r}\right\} \cong \left\{\overrightarrow{qc_s}\right\}\right)\right) \text{ Begin}$$

Form query chain qc_k as

$$qc_k \leftarrow \phi; qc_k \leftarrow \overleftarrow{qc_r}; qc_k \leftarrow cp; qc_k \leftarrow \overrightarrow{qc_s}$$

//forming query chain by connecting left part of qc_r and right part of qc_s

$qc_{rs} \leftarrow qc_k$

Form query chain qc_k as

$$qc_l \leftarrow \phi; qc_l \leftarrow \overleftarrow{qc_s}; qc_l \leftarrow cp; qc_l \leftarrow \overrightarrow{qc_s}$$

//forming query chain by connecting left part of qc_s and right part of qc_r

    $qc_{rs} \leftarrow qc_l$

End

End

For each new query chain $\{qc \forall qc \in qc_{rs}\}$ Begin

     If $(c(qc) > c(qc_r) \& \& c(qc) > c(qc_s))$ Then delete qc from qc_{rs}

End

End

Return qc_{rs}

END

4 Experimental Setup and Performance Analysis

The experiments were done on the Intel i5 generation infrastructure. The implementation of the considered evolution models was done using Java. The input SPARQL query formation was done through a tailormade Java application that generates queries according to the specifications, such as the number of joins desired (join of two joins, join of two individuals of subject predicate or object, or join of an individual and other join), coverage percentage of distributed triple stores and the number of divergent query chains that target similar search output. The statistical analysis of the results obtained from these selected and proposed evolutionary models was done using expression language R [20].

4.1 The Context of Input Query Formation

To justify these constraints, the synthesized queries for experiments were formed to extract the results from the combination of benchmark RDF stores called FOAF [21], FACTBOOK [22] and LUBM [23]. The queries formed to perform the experiments may or may not be sensible toward the significance of the results, since the formation of the queries are done to fulfill the objective called "querying distributed RDF stores". The length of the query is denoted by the number of joins involved. The length is having fewer role towards the optimality. The majority of the time, though the length is high, the overall join cost can be low. Hence the length of the query is not considered as a metric to optimal query. The synthesized input queries have a number of joins between 8 and 24. The range of query chains defined for a desired search is between 12 and 30.

The experimental results obtained from PGE-JCO were compared with other benchmarking strategies called RCQ-GA [11] and RCQ-ACS [10].

4.2 Performance Analysis

The metrics used to assess the performance of the adapted evolutionary strategies are "evolutions completion time" [24], "time complexity" [24], "search space optimization" [25], "access cost optimization" [4], "root mean square distance [26] of the distributed triple store access cost" and "root mean square distance of the distributed triple store search cost".

The metric "evolution completion time" explores the scalability scale of the process. The observed results (see Fig. 1) of this metric for benchmarking models RCQ-GA, RCQ-ACS and the proposed PGE-JCO evince that the PGE-JCO is scalable compared to the other two evolution models. The evolution completion

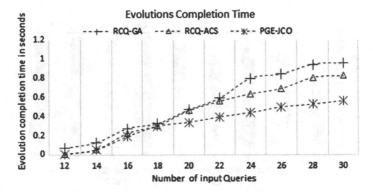

Fig. 1 Evolutions completion time observed for RCQ-GA, RCQ-ACS and PGE-JCO

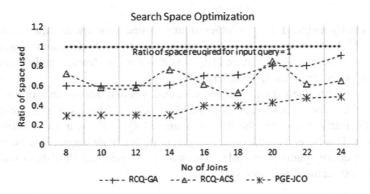

Fig. 2 The ratio of search space used against actual search space required for RCQ-GA, RCQ-ACS and PGE-JCO

time of PGE-JCO is not proportionate to the increase in the number of joins (see Fig. 1).

The other metric time complexity [24] observed for PGE-JCO is considerably low and its order of complexity is linear (O(n)) to the number of queries, whereas in the case of RCQ-GA the complexity observed is high and its order of complexity is quadratic (O(n2)) to the number of input queries. In the case of RCQ-ACS, the growth of complexity is quasi-linear (O(n*log(n))) to the number of input queries. Figure 1 indicates the growth in computational complexity, which is low in the proposed PGE-JCO compared to two other models RCQ-GA and RCQ-ACS.

The search space optimization (see Fig. 2) is another metric considered to assess the performance of the PGE-JCO. Figure 2 shows that PGE-JCO is more scalable and robust to minimize the search space compared to RCQ-GA and RCQ-ACS. The metric was often found to be optimal in the case of RCQ-ACS, which is not stable due to the search strategy of ant colony optimization. In the case of RCQ-GA, search optimization is stable but minimal compared to PGE-JCO and RCQ-ACS.

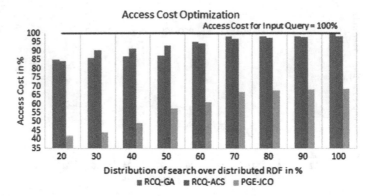

Fig. 3 The ratio of access cost against the actual cost observed for RCQ-GA, RCQ-ACS and PGE-JCO

This is clearly evinced by the other metric called root mean square distance (RMSD) [26] of search space optimization. The RMSD of search space optimization observed for RCQ-GA and RCQ-ACS is almost identical at the maximum number of joins involved in the query chain.

The results obtained for the other two metrics called access cost optimization (see Fig. 3) and root mean square distance of the access cost clearly indicate that the PGE-JCO clearly dominates the RCQ-GA and RCQ-ACS toward minimizing the access cost. This is because the other two do not differentiate the access cost if scans desired for query spanned multiple triple stores, whereas the PGE-JCO considers that while estimating the query join cost.

5   Conclusion

Optimizing the join cost of a query chain that targets the distributed triple stores is the prime objective of the proposed model. To do this, to estimate the cost of a join, the metrics called "Search of Minimal Combinations" and "Search by Minimal Access Cost" were proposed. Further, these metrics were used to define the cost function to assess the cost of each join involved in the query chain. An evolutionary approach called Progressive Genetic Evolutions-based Join Cost Optimization (PGE-JCO) was proposed, which is based on the model of a genetic algorithm. The cost function devised using the metrics "Search of Minimal Combinations" and "Search by Minimal Access Cost" was used as fitness function of the PGE-JCO. The experimental results evince the significance of the PGE-JCO toward accuracy, scalability and robustness. The performance assessment of the proposed model was done by comparing with other benchmarking models RCQ-GA [11] and RCQ-ACS [10]. The metrics defined here in this paper emerged as considerably dependable to assess the join cost of a query that aimed to search distributed triple stores. The experimental results motivated us to define the join cost optimization of query chains of distributed triple stores, using other evolutionary strategies like CUCKOO search and hybridizing one or more evolutionary strategies. Another interesting dimension of future research is to define the heuristic scale under fuzzy logic for join cost estimation.

References

1. Berners-Lee, T. H. (2001). The semantic web. Scientific american, 28–37.
2. Simmhan, Y. L. (2005). A survey of data provenance in e-science. ACM Sigmod Record, 31–36.
3. David, C. O. (2012). A survey of RDF storage approaches. ARIMA Journal, 11–35.
4. Quilitz, B. (2008). Querying distributed RDF data sources with SPARQL. Springer Berlin Heidelberg, 524–538.
5. Schmidt, M. M. (2010). Foundations of SPARQL query optimization. 13th International Conference on Database Theory (pp. 4–33). ACM.
6. Gonçalves, F. A. (2014). Query join ordering optimization with evolutionary multi-agent systems. Expert Systems with Applications, 6934–6944.
7. Whitley, D. (2014). An executable model of a simple genetic algorithm. In Foundations of genetic algorithms (pp. 45–62).
8. Jain, A. W. (2013). Learning trajectory preferences for manipulators via iterative improvement. Advances in Neural Information Processing Systems, 575–583.
9. Otten, R. H. (2012). The annealing algorithm. Springer Science & Business Media.
10. Hogenboom, A. N. (2012). RCQ-ACS: RDF Chain Query Optimization Using an Ant Colony System. International Conference on Web Intelligence and Intelligent Agent Technology (WI-IAT) (pp. 74–81). IEEE/WIC/ACM.
11. Hogenboom, A. M. (2009). RCQ-GA: RDF chain query optimization using genetic algorithm. Springer Berlin Heidelberg, 181–192.
12. H. Stuckenschmidt, R. V. (2005). Towards Distributed Processing of RDF Path Queries. International Journal of Web Engineering and Technology, 207–230.
13. Andrey Gubichev, T. N. (2014). Exploiting the query structure for efficient join ordering in SPARQL queries. 17th international conference on Extending Database Technology.
14. TanselDokeroglu, U. T. (2012). Particle Swarm Intelligence as a New Heuristic for the Optimization of Distributed Database Queries. 6th International Conference on Application of Information and Communication Technologies, (pp. 1–7).
15. Senn, J. (2010). Parallel Join Processing on Graphics Processors for the Resource Description Framework. 23rd International Conference on Architecture of Computing Systems, (pp. 1–8).
16. Sherifsakr, S. E. (2014). Hybrid query execution engine for large attributed graphs. Journal of Information Systems, 45–73.
17. C. Liu, H. (2010). Towards efficient SPARQL query processing on RDF data. Tsinghua science and technology, 613–622.
18. X. Zhang, L. C. (2012). Towards efficient join processing over large RDF graph using map reduce. Scientific and Statistical Database Management, 250–259.
19. Layzer, D. (1980). Genetic variation and progressive evolution. American Naturalist, 809–826.
20. Ihaka, R. (1996). R: a language for data analysis and graphics. Journal of computational and graphical statistics, 299–314.
21. The Friend of a Friend (FOAF) project. (2000). Retrieved from foaf-project.org: http://www.foaf-project.org/.
22. Agency, C. I. (2014). The CIA World Factbook 2015. New York: Skyhorse Publishing Inc.

23. Y. Guo, Z. P. (2005). Lubm: A benchmark for owl knowledge base systems. J. Web Sem, 158–182.
24. Wing, J. M. (2006). Computational thinking. Communications of the ACM, 33–35.
25. Chong, E. I. (2005). An efficient SQL-based RDF querying scheme. 31st international conference on Very large data bases (pp. 1216–1227). VLDB Endowment.
26. Carmines, E. G. (1979). Reliability and validity assessment. Sage publications.

Improved Apriori Algorithm Using Power Set on Hadoop

Abdullah Imran and Prabhat Ranjan

Abstract In this era of big data, a huge volume of data is produced. Storage and analysis of such data is not possible by traditional techniques. In this paper, a good method to implement the MapReduce Apriori algorithm using vertical layout of database along with power set and concept of Set Theory of Intersection have been proposed. The vertical layout has the advantage of scanning only a limited number of records for calculating the support of an item. By use of the power set concept, we are able to generate frequent item set, in just two scans of database, reducing the complexity. The concept of set intersection is used to calculate its support. The result shows good improvement over the existing MapReduce Apriori algorithm.

Keywords Apriori algorithm · Association rule · Big data · Frequent item set · Hadoop

1 Introduction

Huge amount of data are generated from diverse field of business, medical sciences, space science, etc. The attributes of such data have different degree of relation among them and are produced at high rate. Such nature of data induced Big data problem which is explained by the 3V feature [17]: Volume—the amount of data produced—is high enough to be stored efficiently by traditional techniques. Variety is a blend of structured, semi-structured and unstructured data. Velocity is the rate of generation of data which is very high. These 3V features prove to be an obstacle for traditional data processing system, as the system either fails to achieve the order of scaling required or its inability to deal with wide varieties of data [17].

A. Imran (✉) · P. Ranjan
Central University of South Bihar, Patna, Bihar, India
e-mail: abdullahimran007@gmail.com

P. Ranjan
e-mail: prabhatranjan@cub.ac.in

© Springer Science+Business Media Singapore 2017   245
S.C. Satapathy et al. (eds.), *Proceedings of the First International Conference on Computational Intelligence and Informatics*, Advances in Intelligent Systems and Computing 507, DOI 10.1007/978-981-10-2471-9_24

Data mining with acceptable degree of accuracy on such data is of great concern. In the current scene, almost every industry related to business has enormous data generated from different sources. These operational data are either stored locally or in a distributed manner. They contain hidden information which is useless until we do some processing to mine them [13]. Association rule extraction is a globally well-recognized technique to draw out frequent patterns among items. These features are very crucial for market planning and strategy making.

Apriori algorithm is one of the most widely used techniques to find frequent item set from transactional database. With the sharp increase in data size to petabytes and terabytes, the original Apriori algorithm becomes inefficient to process. This increase has pushed us toward the use of cloud. The cloud provides affordable and massive storage by allowing us to increase the capability of shared resources [9]. The Apriori algorithm based on MapReduce programming construct of Hadoop is reliable and provides safeguard against node crash during processing. The time involved in frequent item set generation is performance critical. The existing Apriori algorithm scans the database in each iteration with the number of iterations equal to the maximum size of the candidate item set. Also because of the use of horizontal data layout in most of the proposed MapReduce Apriori algorithm, the implementation takes more execution time, as to find the count of any item set the algorithm scans all the transactions in every iteration. This increases the communication cost and the performance decreases. In this paper, we have used the power set concept along with the intersection algorithm in the vertical layout of data to further improve the algorithm presented in [8]. Due to the use of power set, we were able to generate a set of all frequent item sets L using L1 (frequent size-1 item set) in a single iteration. The use of a vertical format of data layout reduced the time complexity incurred in support finding. For this purpose, the intersection algorithm is used. Thus, the communication cost is reduced and a better result is obtained.

The rest of the paper consists of the following parts. Section 2 introduces some background knowledge. Next, Sect. 3 discusses related work along with the problem statement. Then Sect. 4 explains the preliminaries, and Sect. 5 explains the proposed algorithm in detail. Section 6 discusses the experiment along with the result. Finally, Sect. 7 gives a conclusion and focuses on future work.

2 Background

Association Rules: The work of extracting the association rule was introduced in [3]. The main focus was to find a bond between items or item sets in transactional data [22]. Association rules output item sets that occur frequently in the record [4]. Its overview can be formally explained as follows. Let E = (e1, e2, ... en) be a set of elements, called items. Let T be a transaction and Dbase be a database with records of transaction with the constraint that T is a subset of E. A distinct id TID is associated with each transaction. A transaction T is said to contain L, a set of items in E, if L is a subset of T. An association rule is symbolized as L $\rightarrow$ M, where L is a subset of E, M is a subset of E and the intersection of L and M is an empty set. We have to find rules having support and confidence greater than the specified values. Support and confidence are two important measures in association rule mining. For an item e, we say its support is s if the fraction of transactions in Dbase containing L equals s. The rule L $\rightarrow$ M has a support s in the database Dbase if s of the transactions in Dbase contains L $\cup$ M, i.e., support (L $\rightarrow$ M) = P (L $\cup$ M). The rule L $\rightarrow$ M has a confidence c in the transaction set Dbase if c of the transaction containing L also contains M. Confidence is calculated as (L $\rightarrow$ M) = P(L/M).

Apriori Algorithm: Apriori is the basic and most accepted algorithm proposed by R. Agrawal and R. Srikant for finding frequent (L)-item sets based on candidate generation. The main theme of the algorithm is to generate new combination of candidate item sets from the previous combination of frequent item sets generated and then scanning the database to find their support to see if they fulfill the minimum threshold criteria [5]. The algorithm consists of two parts. The first part aims to find frequent (L1) item set by calculating the total frequency of each item present in the database. The second part, an iterative process, is iterated until the next generated frequent (K − 1) item set is empty set. It consists of two subparts: joining and pruning [13]. In the first subpart items of size (K − 1) from frequent (K − 1) set is joined with itself to produce candidate items of size K, CK. The second subpart eliminates any subset of CK which is not present in the frequent (L − 1) item set. Then the algorithm scans the database to find the frequency of each items present in the candidate item set, CK.

Hadoop and MapReduce: Encouraged by benefits of parallel execution in the distributed environment, the Apache Foundation came up with open source platform, Hadoop, for faster and easier analysis and storage of different varieties of data [1]. HDFS and MapReduce programming model are two integral parts of it. Google File System gave birth to HDFS (Hadoop Distributed File system) [12], which mainly deal with storage issues. Unlike the RDBMs, it follows WORM (write-once read-many) model to split large chunk of data to smaller data blocks and allocate them on the free node available [2]. Input data blocks are stored at more than one node to achieve fault tolerance and high performance [2]. MapReduce is a linearly adaptable programming model. It is inspired by Google's MapReduce [7]. It requires two functions—a map ( ) function and a reduce ( ) function, both of which work in a synchronous manner to operate on one set of key value pairs to produce the other set of key value pairs [2]. These functions are equally valid for any size of data irrespective of the degree of the cluster. MapReduce uses the feature known as data locality to collocate the data with the compute node, so that data access is fast. It follows shared nothing architecture which eliminates the burden from the programmer of thinking about failure. The architecture itself detects failed map or reduce task and assigns it to a healthy node [21, 23].

3 Related Work and Problem Statement

3.1 Related Work

An improvement in the process of association rule extraction remains a hot area of research today. Most of the research concentrates on frequent item set generation part of association rule extraction using the Apriori algorithm. The existing Apriori algorithm suffers from the problem of scanning the entire database in each iteration for generation of frequent item sets and the number of iterations equals the size of the largest candidate item set size. Also to find the count of each candidate item set, each of the record is scanned one by one in all such iterations making it an exhausting process. So, many parallel forms of Apriori algorithm have been proposed, such as Count Distribution, Data distribution and Candidate Distribution algorithm to reduce the execution time [4]. Count Distribution algorithm reduces communication cost, as only the count is exchanged among the processors, but it suffers from underutilization of memory due to the replication of hash tree as a whole on each processor. Data Distribution efficiently utilizes system memory, but incurs high communication cost because of the need to send local database from each node to all other nodes. Candidate Distribution algorithm gives better results, as it balances the total workload among the processors [20]. These parallel algorithms improved the execution time of the original Apriori algorithm, but such improvements are still not up to the mark to deal with data set of very large size. To cope up with shortage of storage space, cloud storage proves to be very fruitful, but data security is a major concern here [19]. Several modifications of the Apriori algorithm using MapReduce as base have been proposed to increase the scalability of the Apriori algorithm. They are based on the candidate distribution concept. In such enhancing system, the transactional data set is either represented in a horizontal format [18, 16, 14, 6, 15] or vertical format [8]. The data representation is in item set against transaction-id in the vertical format, whereas it is transaction-id against item set in the horizontal format [20]. Most of the earlier MapReduce Apriori implementation use horizontal layout of the database. In all such algorithms, to count the support of any item set, n (total records present) number of transactions is scanned one after other in each iteration which is a time-consuming task. The paper [8] implemented the MapReduce-based Apriori algorithm in the vertical format. Here, generation of candidate item set involve extra iteration. The algorithm also generates some unnecessary candidate item sets. Counts of these item sets are collected at the reducer which increases the overall communication cost involved. The proposed algorithm in this paper uses power set along with the intersection algorithm [11] in the vertical format data representation. Use of power set allows us to generate relevant items in minimum scanning of database. The intersection algorithm reduces the time requirement for calculation of the frequency of each item. To find the frequency (support), only the common transaction in each candidate item is taken into account, thus limiting our algorithm to scan relevant records only.

3.2 Problem Statement

Given a large set of transactional database, Dbase in the horizontal data layout has two threshold values minsup and minconf. The database consists of n records. We calculate a set of association rule SASR = {R1, R2,..., Ri,...} based on the support calculation of candidate item set SCS = {C1, C2,..., Ci,...}, where each Ri is a rule of form A → B and Ci is a candidate item set of size i. The time complexity for support calculation of an item of Ci requires scanning of each record of database, which is in the order of O(n). For a very large data set, this order of complexity is not acceptable.

4 Preliminaries

Theory of Set Intersection: In mathematics, the intersection of two sets A and B is denoted as $A \cap B$ and defined as the set which gives a common element set, CS, present in the given sets. For example, from Table 1 if I(S) = {1, 2}, then CS = {T1, T2, T3}.

Table Layout: Table layout is of vital importance while calculating support in the Apriori algorithm. Essentially, two types of table layout can be defined: 1. horizontal layout, 2. vertical layout. However, the vertical layout is comparatively better as compared to the horizontal layout. Table 2 and 3 represent the horizontal layout and vertical layout, respectively, for the sample transactional data set. To calculate the support of any chosen item set, we need to scan the entire data set every time in case of the horizontal layout, while for a vertical layout the number of record scanned depends on the number of literals in the chosen item. To understand better , see Table 4. From the table we can easily visualize that for the horizontal layout, the overall complexity is O(n), whereas for the vertical layout it is O(m) where m is the number of literals in the chosen item set.

Table 1 Example

Items	Transaction set
1	T_1, T_2, T_3, T_4, T_5
2	T_1, T_2, T_3
...	T_1, T_2
N	T_2, T_3

Table 2 Horizontal layout

Transaction no.	Item set
T_1	{a, b, c, d}
T_2	{b, c, d}
...	...
T_n	{a, b}

Table 3 Vertical layout

Items	Transaction set
a	$T_1,..., T_n$
b	$T_1, T_2,...$
c	$T_1, T_2,...$
d	$T_1, T_2,..., T_n$

Table 4 Comparison of scans

Support	Horizontal layout	Vertical layout
S(a)	n	1
S(a, b)	n	2
....	....	...
S(a, b, c, d)	n	4

Power Set: In set theory, the power set of any set S, denoted as P(s) or 2^S, is a set of all subsets of S, including the null set and the set itself [10]. For example, let S = {x, y, z, m}, then P(S) = {{ }, {x}, {y}, {z}, {m}, {x, y}, {x, z}, {x, m}, {y, z}, {y, m}, {z, m}, {x, y, z}, {x, y, m}, {x, z, m}, {y, z, m}, {x, y, z, m}}

5   The Proposed Algorithm

It consists of a sequence of three steps. In the first step, conversion of the horizontal data set to the vertical data set is carried out. Then in the second step, frequent item sets are generated using the intersection algorithm in the vertical format. Finally in the third step based on these frequent item sets, association rules are generated. The available data set is in horizontal format (transaction-id list against items) which is to be transformed into the vertical format (item against transaction-id). Algorithm 1 CHV is used for this transformation. Figure 1 shows this transformation.

ALGORITHM 1: Chv

Input: Horizontal data-set
Output: Vertical data-set
1. Each Mapper operates on its local database to generate local tid-list for all items.
2. Creation of global tid-list:
A. Every mapper exchanges tid-list with all other to get total number of transaction associated with every item.
B. Each Reducers collect together the tid-no associated with a particular item from all mapper output to form global tid-list for each item.
Return: Vertical data-set.

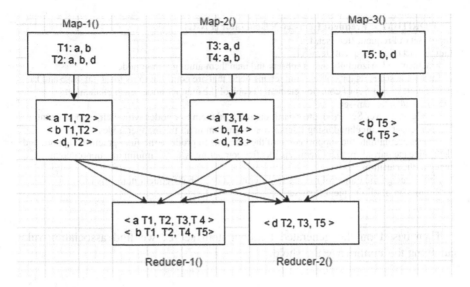

Fig. 1 Horizontal to vertical transformation

Generation of frequent item set: It consists of two MapReduce phase MR1 and MR2.

(MR1) Generation of frequent item set of size-1 L1 is carried using Algorithm 2 Frequent Item set-1 Generator (F1-G).

ALGORITHM 2: Frequent Item-set-1Generator (F1-G)

Input: Vertical data-set
Output: Set of Frequent Item-set, L1
1. First the vertical database is subdivided horizontally into p data subsets which are mapped to q nodes.
2. Then each node scan its own data subsets and generates candidate item-set of size-1 at each of the Map Node i.e. ($C_{1,1}$, $C_{2,1}$, $C_{3,1}$,..., $C_{q,1}$).
3. Collect candidate item-set of size-1 ($C_{1,1}$, $C_{2,1}$, $C_{3,1}$..., $C_{q,1}$) at the reducer. Item-set whose count is less than minimum support, minsup is removed to get Frequent Item-set of size-1, L1.
4. Return: L1

(MR2) Generation of frequent item set L from L1 using power set and intersection algorithm is done using Algorithm 3 Frequent Item set-N Generator (FN-G).

ALGORITHM 3: Frequent Item-set-N Generator (FN-G)

Input: Set of Frequent Item-set, L1.

Output: Set of all Frequent Item-set

1. Fragment L1 horizontally into u subsets and map them among v map node.
2. Each node scans its data blocks and retains items that are common to its local database and L1.
3. Generate Power Set of common element in each of the mapper node. Each element of P(S) is candidate item-set.
4. Divide the Power Set, P(S) into r partition and send them to r nodes with their support count. Support count is calculated using Intersection Algorithm in all the mapper nodes.
5. r nodes accumulate the support count of the elements to produce the final practical support, and determine the partial frequent item-sets Lk after comparing with the minimum support count msup in the partition.
6. Finally merge the output of r nodes to generate set of global frequent item-sets Lk.
7. Return: Set of all Frequent Item-set L.

Then based on the generated frequent item set L, we find association rules satisfying the confidence threshold.

6 Experiment and Results

In our approach, we have used power set, vertical database layout and Set Theory of Intersection [11]. Power set simplified the steps, as frequent item sets L2 to Lk are generated in a single step from L1 using algorithm (FN-G). The main advantage of the above system is that it reduces the overall time spent in finding support of the candidate item set. The intersection algorithm allows us to find the support by simply counting the common transactions in each element of candidate sets. The time elapsed in the scanning of the database is reduced significantly.

A sequence of MapReduce phase one for conversion from the horizontal to the vertical format for next generation of frequent item set of size-1 and for the last generation of the remaining frequent item set is used in a chain manner with the output of the first as input for the next. With the view of testing the performance of

Fig. 2 Results

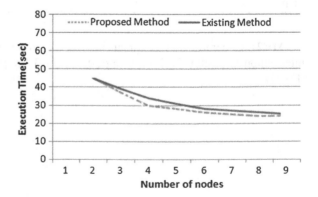

the above strategy, the experiment was done on Hadoop cluster of variable length with different capabilities. The experiment was performed on 88000 numbers of records. The data set may be obtained from http://fumi.ua.ac.be.data. The proposed algorithm is compared with the algorithm given in paper [8], which also uses the vertical format. Implementation shows desirable result with reduction in the time involved in the process . The execution time further reduced with increase in the number of nodes which is evident from Fig. 2.

7 Conclusion and Future Work

We have successfully implemented the proposed algorithm. Our algorithm for frequent item set generation gave better result than the existing one [8]. Incorporating power set, set theory concept of intersection in the vertical format of the database increased the degree of performance. Power set reduced the overall complexity with the number of steps used in the frequent item set generation minimized, and the Intersection Algorithm reduced the time involved in counting the support. Future work shall focus on implementing the algorithm in a controlled environment, where mapper ( ) and reducer ( ) work asynchronously rather than in a synchronous normal manner. This would make the system faster.

References

1. Apache Hadoop, http://hadoop.apache.org.
2. Yahoo! hadoop tutorial, http://developer.yahoo.com/hadoop/tutorial/index.html.
3. R. Agrawal, T. Imielinski, and A. Swami. Mining association rules between sets of items in large databases. In ACM SIGMOD Record, volume 22, pages 207–216. ACM, 1993.
4. R. Agrawal and J. C. Shafer. Parallel mining of association rules. IEEE Transactions on Knowledge & Data Engineering, (6):962 969, 1996.
5. R. Agrawal, R. Srikant, et al. Fast algorithms for mining association rules. In Proc. 20th int. conf. very large databases, VLDB, volume 1215, pages 487–499, 1994.
6. S.-Y. Chen, J.-H. Li, K.-C. Lin, H.-M. Chen, and T.-S. Chen. Using map reduce framework for mining association rules. In Information Technology Convergence, pages 723–731. Springer, 2013.
7. J. Dean and S. Ghemawat. MapReduce: simplified data processing on large clusters. Communications of the ACM, 51(1):107–113, 2008.
8. S. Dhanya, M. Vysaakan, and A. Mahesh. An enhancement of the MapReduce Apriori algorithm using vertical data layout and set theory concept of intersection. In Intelligent Systems Technologies and Applications, pages 225–233. Springer, 2016.
9. I. Foster, C. Kesselman, J. M. Nick, and S. Tuecke. The physiology of the grid. Grid computing: making the global infrastructure a reality, pages 217–249, 2003.
10. B. Ganter and R. Wille. Formal concept analysis: mathematical foundations. Springer Science & Business Media, 2012.

11. K. Geetha and S. Mohiddin. An efficient data mining technique for generating frequent itemsets. International Journal of Advanced Research in Computer Science and Software Engineering, 3(4), 2013.
12. S. Ghemawat, H. Gobioff, and S.-T. Leung. The Google file system. In ACM SIGOPS operating systems review, volume 37, pages 29–43. ACM, 2003.
13. J. Han, M. Kamber, and J. Pei. Data mining: concepts and techniques: concepts and techniques. Elsevier, 2011.
14. L. Li and M. Zhang. The strategy of mining association rule based on cloud computing. In Business Computing and Global Informatization (BCGIN), 2011 International Conference on, pages 175–178. IEEE, 2011.
15. M.-Y. Lin, P.-Y. Lee, and S.-C. Hsueh. Apriori-based frequent itemset mining algorithms on mapreduce. In Proceedings of the 6th international conference on ubiquitous information management and communication, page 76. ACM, 2012.
16. X. Lin. Mr-apriori: Association rules algorithm based on mapreduce. In Software Engineering and Service Science (ICSESS), 2014 5th IEEE International Conference on, pages 141–144. IEEE, 2014.
17. J. Manyika, M. Chui, B. Brown, J. Bughin, R. Dobbs, C. Roxburgh, and A. H. Byers. Big data: The next frontier for innovation, competition, and productivity. 2011.
18. W. Mao and W. Guo. An improved association rules mining algorithm based on power set and Hadoop. In Information Science and Cloud Computing Companion (ISCC-C), 2013 International Conference on, pages 236–241. IEEE, 2013.
19. Z. Qureshi, J. Bansal, and S. Bansal. A survey on association rule mining in cloud computing. International Journal of Emerging Technology and Advanced Engineering, 3(4):318–321, 2013.
20. A. M. Sakhapara and H. Bharathi. Comparative study of Apriori algorithms for parallel mining of frequent itemsets. International Journal of Computer Applications, 90(8), 2014.
21. R. SARITHA and M. U. RANI. Mining frequent item sets using map reduce paradigm.
22. O. R. Zaiane, M. El-Hajj, and P. Lu. Fast parallel association rule mining without candidacy generation. In Data Mining, 2001. ICDM 2001, Proceedings IEEE International Conference on, pages 665–668. IEEE, 2001.
23. M. J. Zaki. Parallel and distributed association mining: A survey. IEEE concurrency, (4):14–25, 1999.

Distributed Cooperative Algorithm to Mitigate Hello Flood Attack in Cognitive Radio Ad hoc Networks (CRAHNs)

Anjali Gupta and Muzzammil Hussain

Abstract Cognitive radio technology was first introduced by J. Mitola in 1999 to solve the problem of spectrum scarcity for wireless communication. A Cognitive Radio Ad hoc Network (CRAHN) is one of the cognitive radio-based architectures in which wireless unlicensed nodes communicate in infrastructureless environment. Each node in CRAHN has the cognitive capability of sensing the surrounding radio environment for accessing the underutilized licensed spectrum in an opportunistic manner. But those unlicensed users (secondary users) should not make any interference to the communication of licensed users (primary users). In CRAHNs, each node operates as an end system and also as a router to forward packets for multi-hop communication. Due to the dynamic topology, time and space-varying spectrum availability and lack of centralized system, CRAHNs are vulnerable to various security attacks. Hello flood attack is a network layer attack in CRAHNs which can drain off the limited resources of CRAHN nodes by making excessive flooding of hello messages in the network. The proposed distributed cooperative algorithm mitigates the hello flood attack in CRAHNs.

Keywords Cognitive radio ad hoc networks · Attacks · Hello flood attack

1 Introduction

Today, with the increasing growth in wireless device, also the demand for accessing the spectrum is increasing drastically. Due to the shortage problem of spectrum (i.e., 900 MHz and 2.4 GHz) freely available for unlicensed wireless communication, FCC has proposed a new communication paradigm, i.e., Dynamic Spectrum Access

A. Gupta (✉) · M. Hussain
Department of Computer Science and Engineering, Central University of Rajasthan,
Kishangarh 305817, Rajasthan, India
e-mail: 2014mtcse004@curaj.ac.in

M. Hussain
e-mail: mhussain@curaj.ac.in

© Springer Science+Business Media Singapore 2017
S.C. Satapathy et al. (eds.), *Proceedings of the First International Conference on Computational Intelligence and Informatics*, Advances in Intelligent Systems and Computing 507, DOI 10.1007/978-981-10-2471-9_25

255

and also has made available the underutilized licensed spectrum (i.e., 400–700 MHz) for the unlicensed users to access. J. Mitola in 1999 introduced the "cognitive radio (CR)" concept. Spectrum bands which are not currently being used by licensed users (primary users) and are freely available, known as spectrum holes or white spaces.

Cognitive radio is an intelligent device which can sense its radio environment for the spectrum utilization in an opportunistic manner and can adapt its transmission parameters (i.e., transmission power, frequency, modulation tech., etc.) according to the sensed available spectrum. The basic idea of cognitive radio is that secondary users (SUs) can access the licensed spectrum band in an opportunistic manner and without interfering with the communication of primary users (PUs). In CRN, the secondary user (SU) senses the existing spectrum and, if available, occupy it and vacate it when a primary user (PU) of the spectrum returns.

Cognitive radio has two primary objectives, highly reliable communication and efficient utilization of the radio spectrum. Cognitive radios are classified basically in two architectures which are infrastructure-based CR networks and infrastructureless Cognitive Radio Ad hoc Networks (CRAHNs). In the infrastructure-based CR networks, the local observation about spectrum holes made by each node is sent to the base station. The base station analyzes the free spectrum and allocates them to the nodes in the network according to their requirements. On the contrary, in CRAHNs all the nodes share their local observations with every other node in the network and cooperate in the spectrum allocation strategy. In Cognitive Radio Ad hoc network, CR users are mobile and can communicate with each other in a multi-hop manner on both licensed and unlicensed spectrum bands.

1.1 Attacks in CRAHNs

Various attacks are possible in CRAHNs at different layers [1, 2, 3, 4].

1.1.1 Physical Layer Attacks

Primary user emulation (PUE) attack: The attacker pretends to be the primary user (PU), tricks the legitimate secondary user (SU) and prevents legitimate CRs to access the "spectrum holes".

Objective function (OF) attack: The learning algorithm of the CR engine is disrupted in CR nodes and affect the objective function parameters for the CR engine.

Jamming attack: A jamming attacker may transmit continuous packets to force a legitimate SU to never sense an idle channel.

1.1.2 Link Layer Attacks

Spectrum-sensing data falsification (SSDF) attack: It is also known as Byzantine attack. It takes place when an attacker sends false local spectrum-sensing results to its neighbors or to the fusion center, causing the receiver to make a wrong spectrum-sensing decision.

Control Channel Saturation DoS (CCSD) attack: In multi-hop CRAHN, an attacker can generate forged MAC control frames for the purpose of saturating the control channel and thus decreasing the network performance due to link layer collisions.

Selfish channel negotiation (SCN) attack: In a multi-hop CRN, an untrusted CR host can refuse to forward any data for other hosts to conserve its energy and increase its own throughput.

1.1.3 Network Layer Attacks

Sink-hole attack: An attacker advertises itself as the best route so that the neighbor node sends the packet through it.

HELLO flood attack: The attacker broadcasts the HELLO message to other nodes in the network to tell that it is a neighbor node and drain off the resources of nodes.

1.1.4 Transport Layer Attacks

Lion attack: It uses the PUE attack to disrupt the TCP connection and degrades the TCP performance. An attacker emulates licensed transmission and makes all SUs to switch (frequency hand-off) to another channel. It is known as a cross layer attack.

1.2 Hello Flood Attack and Its Impacts in CRAHNs

In CRAHNs, reactive or on-demand routing protocols (i.e., AODV, DSR, etc.) require nodes to broadcast the hello messages to all the nodes within the radio range to announce their presence.

Nodes receiving such a packet may assume that the sender is within its radio range and it is one of the neighbors. Reception of a hello message in CRAHN environment indicates a viable communication channel with the source of the hello message. The hello message is sent periodically to know about the presence of a neighbor and also the link connectivity. In CRAHNs, the AODV protocol uses the hello message to know about the link breakage because of spectrum switching due to the comeback of primary users or high mobility of the nodes [5]. In mobile ad hoc networks, two variables control the determination of connectivity using hello messages:

HELLO INTERVAL: The maximum time interval between transmissions of hello messages.

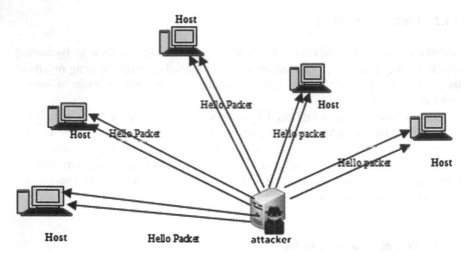

Fig. 1 Hello flood attack

ALLOWED HELLO LOSS: The maximum number of periods of HELLO INTERVAL to wait without receiving a hello message before detecting a loss of connectivity to a neighbor [5]. The recommended value for HELLO INTERVAL is 1 s and for ALLOWED HELLO LOSS 2 s in [6].

In hello flood attack, a laptop-class malicious node while being out of range can broadcast hello messages with high energy to all other nodes in the network to ensure themselves as their neighbor. So the other nodes will route their packets from the malicious node in the network. The malicious node can either simply drop all the packets or selectively forward the packet. The attacker has the purpose of hogging off the limited resources [7]. The hello message is a control packet.

An adversary can flood the control channel with hello messages and thus make the network down (Fig. 1).

2   Literature Survey

Various mechanisms have been proposed for hello flood attack prevention in mobile ad hoc networks (MANETs), but in Cognitive Radio Ad hoc Networks (CRAHNs) research on the prevention of attacks is still ongoing.

The existing mechanisms in CRAHNs for the prevention of hello flood attack is proposed by Dubey et al. [8]. They have used symmetric key sharing with a trusted base station for the prevention of hello flood attack. The base station will establish the session keys between parties in the network. But this is not possible in CRAHNs, as there is no centralized system available for key distribution.

Wanjari et al. [9] have given an approach of flooding attack prevention using node authentication and packet forwarding methods. It uses CRP with node authentication phase when a node enters into the network. Before transmitting a packet to a node in the network, malicious node table will be checked for the malicious node in the network. This method requires extra overhead for node authentication.

The first flooding attack prevention (FAP) was proposed by Royer et al. [10]. In this, they have defined neighbor suppression method for prevention of RREQ flooding where it gives more priority to those nodes which send less number of RREQ packets. For prevention of data flooding, the path cutoff method is defined in which when a node identifies data flooding, then it simply cuts off that path and uses another reliable path for communication. The disadvantage of this method is that flooding packets still remain in the network and attack is not completely prevented.

Balakrishnan et al. [11] used three threshold values: transmission threshold, blacklist threshold and white listing threshold for flooding prevention. It aids the elimination of a centralized authority for managing the thresholds. This technique is a promising method for the prevention of flooding attack in MANETs.

To deal with the high mobility situation in MANET, Shandiliya et al. [12] have proposed a method based on the trust estimation function and delay queue in basic DSR routing protocol. The proposed method works in a distributed cooperative way.

Devaraj [13] has proposed a simpler way to mitigate flooding attack using a priority-based scheme. In this, when a malicious node broadcasts RREQs more than the limit specified by RFC, the priority of those packets will be reduced and other high-priority RREQ packets will be processed. The problem with this method is that genuine RREQ packets can be at stake.

Madhavi et al. [14] used signal strength and client puzzle-based method to detect and prevent flooding attack. In this, the presence of more than one kind of attacker may affect the performance of the network.

Hu et al. [15] have presented the rushing attack defense mechanism using secure neighbor detection, secure route delegation and randomize route request forwarding. The node overhead is more in this mechanism due to the multiple nodes sending route requests at the same time.

3 Proposed Work

3.1 Pre-assumptions

The proposed algorithm will run for a unit of time and it is assumed that all the nodes behave cooperatively and no node will play selfishly. A single node through its local observation will not take decision on defining an attacker node. In one unit time, high mobility of a node will implicitly be considered as malicious node behavior. An attacker can use different identities and can flood the network with excessive hello packets using Sybil attack. In this case, preexisting localization mechanisms in CRAHNs can be used to identify a node's position.

3.2   Pre-requisites

In our proposed algorithm, hello flood prevention is based on maintaining two threshold values at each node as well as a counter for the received hello packets.
n: Threshold value for hello packets allowable in one unit of time
λ: Threshold value for count on the attack scenario of the neighboring nodes
h_i: Counter at node i for receiving the hello packet

3.3   Proposed Algorithm

Distributed cooperative algorithm to mitigate hello flood attack in Cognitive Wireless Ad hoc Network

```
begin
while(t > 0) do
        if(Node A receives a broad-casted hello message from other node) then
            if(ha < n) then
                        increment ha by 1
                        goto step 1
            else
                        Sender could be a malicious node
                        Declare it as an "Attack Scenario"
                        for k=1 to N do
                                Send request to every neighbor node to get their
                                status
                                Node A counts no. of received "Attack Scenario"
                                from      other      node     in      the      network
                        end for
                        if(No. of Attack Scenario ≥ λ) then
                                Sender is a malicious node
                                Node A will discard all the hello message received
                                in unit time t
                                Inform other N nodes in the network about the
                                attack
                                Initialize h to 0
                        else
                                Node A accepts all the received hello messages in
                                unit time t
                        end if
            end if
        else
            goto step 1.
        end if
end while
Process the recived hello packets at Node A and Perform the neccessary function.
end
```

According to our proposed algorithm, for one unit of time, a node will receive broadcasted hello packets from its neighbors and possible attacker in the network. When a node A receives a hello packet, then the number of received hello message counter, i.e., h_a will be compared with the threshold value n at node A. If h_a is lesser, then node A will increment counter h_a by 1 and will go to step 1 again. If it is more than and equal to n, then node A will assume that the sender could be a malicious node. Node A will identify it as an "Attack Scenario". Node A will then enquire from its neighbor nodes in the network to get their recorded status. If node A gets "attack scenarios" from neighbor nodes more than and equal to the threshold value λ, it will declare the sender as a malicious node and will share this information with neighbor nodes in the network. Node A will discard all the packets received in unit time, initialize the parameters and start the whole procedure again. In case the number of "attack scenarios" is not more than the threshold value, then it will go to step 1. After the timer expires, the node A will perform the necessary function on the accepted hello packets in unit time.

3.4 Evaluation

Chakeres et al. [5] have shown that two variables control the determination of connectivity using the hello message:

HELLO INTERVAL: The maximum time interval between transmission of the hello messages.

ALLOWED HELLO LOSS: The maximum number of HELLO INTERVAL to wait without receiving a hello message before detecting a loss of connectivity to a neighbor [5].

The recommended value for HELLO INTERVAL is 1 s and for ALLOWED HELLO LOSS 2 s in [6]. This means that in every 1 s, a hello packet will be received by a node and for a maximum 2 s hello packet loss is allowed. So, in our algorithm we can take the threshold value n approximately 3–5 hello packets in one unit time, but λ is yet to be defined. As nodes will only communicate to share the information about the attack scenarios that too will be one bit information only. So, the communication overhead will be negligible. Here, n computes the counts of hello packets received from other nodes in the network and very little computation overhead is involved.

Every node is required to maintain a list < **Node ID, Counter of Hello packets** > where it will maintain the received hello packet counter at each node and that too only for one unit of time. After expiration, it will discard it. Storage is required for counting the number of "attack scenarios". Both the counters at each node will sum up to a small storage overhead.

4 Conclusion

There are no concrete mechanisms to handle the hello flood attack in Cognitive Radio Ad hoc Networks, especially distributively. Here, we propose a distributed cooperative mechanism to handle hello flood attack in CRAHN and it is anticipated that the proposed mechanism will be effective in handling hello flood attack in CRAHNs at no extra cost. We are able to prevent hello flood attack with very little communication, computation and storage overheads in our algorithm.

References

1. León, O., Hernández-Serrano, J., & Soriano, M.: Securing cognitive radio networks. international journal of communication systems, 23(5), 633–652 (2010).
2. Clancy, T. C., & Goergen, N.: Security in cognitive radio networks: Threats and mitigation. In Cognitive Radio Oriented Wireless Networks and Communications, 2008. CrownCom 2008. 3rd International Conference on (pp. 1–8). IEEE (2008, May).
3. El-Hajj, W., Safa, H., & Guizani, M.: Survey of security issues in cognitive radio networks. Journal of Internet Technology, 12(2), 181–198 (2011).
4. León, O., Román, R., & Hernández-Serrano, J.: Towards a cooperative intrusion detection system for cognitive radio networks. In NETWORKING 2011 Workshops (pp. 231–242). Springer Berlin Heidelberg (2011, January).
5. Chakeres, I. D., & Belding-Royer, E. M.: The utility of hello messages for determining link connectivity. In Wireless Personal Multimedia Communications, 2002. The 5th International Symposium on (Vol. 2, pp. 504–508). IEEE (2002, October).
6. Perkins, C., Belding-Royer, E., & Das, S.: Ad hoc on-demand distance vector (AODV) routing (No. RFC 3561) (2003).
7. Singh, V. P., Ukey, A. S. A., & Jain, S.: Signal strength based hello flood attack detection and prevention in wireless sensor networks. International Journal of Computer Applications (0975–8887) Volume (2013).
8. Dubey, P., & Chaudhary, S.: A Survey-Cognitive Radio Network Attacks & Preventions (2014).
9. Wanjari, A., & Dhamdhere, M. V. Node Verification to Prohibit Flooding Attack in Mobile Ad-hoc Network.
10. Royer, E. M., & Toh, C. K.: A review of current routing protocols for ad hoc mobile wireless networks. Personal Communications, IEEE, 6(2), 46–55 (1999).
11. Balakrishnan, V., Varadharajan, V., Tupakula, U., & Moe, M. E. G.: Mitigating flooding attacks in mobile ad-hoc networks supporting anonymous communications. In Wireless Broadband and Ultra Wideband Communications, 2007. AusWireless 2007. The 2nd International Conference on (pp. 29–29). IEEE (2007, August).
12. Shandilya, S. K., & Sahu, S.: A trust based security scheme for RREQ flooding attack in MANET. International journal of computer applications, 5(12), 0975–8887 (2010).
13. Devaraj, A. F. S.: Examination of Impact of Flooding attack on MANET and to accentuate on Performance Degradation. Int. J. Advanced Networking and Applications, 4(04), 1695–1699 (2013).

14. Madhavi, S., & Duraiswamy, K.: Flooding attack aware secure AODV. Journal of Computer Science, 9(1), 105 (2013).
15. Hu, Y. C., Perrig, A., & Johnson, D. B.: Rushing attacks and defense in wireless ad hoc network routing protocols. In Proceedings of the 2nd ACM workshop on Wireless security (pp. 30–40). ACM (2003, September).

Performance Comparison of Intentional Caching Schemes in Disruption-Tolerant Networks (DTN)

S. Manju, S.J.K. Jagadeesh Kumar and V.R. Azhaguramyaa

Abstract DTN abbreviated as disruption-tolerant networks is the opportunistic network, characterized by irregular network connectivity, long and variable delays, asymmetric data rates and low node density. Data access is the big research issue in DTNs because of its distinguishing characteristics. To improve the data access in DTN, schemes like caching and replication were introduced, to provide a distributed storage of data in the network thereby increasing the data availability. This paper differentiates some of the intentional caching techniques such as cooperative caching, duration aware caching, adaptive caching and distributed caching based on various parameters like contact duration, caching cost, cache node election process and forwarding schemes. Upon simulation with ONE, these schemes were found to increase delivery probability and reduce data access delay to a greater extent, thereby improving the performance of the network.

Keywords DTN · Caching · Central node · Contact duration · Learning automata

1 Introduction

Disruption-tolerant networks (DTN) [1] are featured by long latency, variable delays measured in days, intermittent connectivity, lack of end to end connectivity and unstable network topology. The reason for these characteristics is because the DTN deployed extreme environment, where nodes get disconnected frequently. DTN is widely applied in challenging networks such as disaster recovery, terrestrial

S. Manju (✉) · S.J.K. Jagadeesh Kumar · V.R. Azhaguramyaa
Sri Krishna College of Engineering and Technology, Coimbatore, India
e-mail: 14mg011@skcet.ac.in

S.J.K. Jagadeesh Kumar
e-mail: sjkjk@skcet.ac.in

V.R. Azhaguramyaa
e-mail: azhaguramyaa@skcet.ac.in

© Springer Science+Business Media Singapore 2017
S.C. Satapathy et al. (eds.), *Proceedings of the First International Conference on Computational Intelligence and Informatics*, Advances in Intelligent Systems and Computing 507, DOI 10.1007/978-981-10-2471-9_26

mobile networks, satellite communications, military adhoc networks and underwater communication where achieving data connectivity is of major concern resulting in higher data access delay [2]. The mobile nodes in DTN contacts each other opportunistically. As there is a lack of end to end path in DTNs, the sender has to store the data in its local buffer before sending it to the destination. Hence it is mandatory to use "store, carry and forward" techniques for data transmission. In these DTN environments, accessing data is highly challenging. More research has been done in data forwarding and efficient data access methods originating from epidemic routing [3]. Later on, some studies also concentrated on analyzing the node contact patterns, mobility patterns [4] and social properties such as community, to improve data access [5, 6]. This paper deals with various caching techniques applied in a DTN environment such as cooperative caching, adaptive caching, and duration aware caching and distributed caching. All these methods were found to increase the data access efficiency by redundantly storing data at the central nodes of the network, i.e., caching nodes.

The rest of the paper is organized as follows. Section 2 focuses on the different types of data access methods. Section 3 deals with the detailed survey of four different caching techniques. Section 4 differentiates all four schemes based on the aforementioned parameters stating the advantages of all caching schemes and Sect. 5 deals with performance evaluation based on two QoS parameters (delivery probability and delay). Finally, Sect. 6 concludes the paper.

2 Data Access Methods in DTN

The data access in DTN can be improved by a technique called caching. Caching the data redundantly to improve data availability can be done either incidentally or intentionally, hence resulting in two types of caching, namely 1. incidental caching and 2. intentional caching. Intentional caching is a scheme where data are cached only at some specific set of nodes chosen exclusively, whereas every node in the incidental caching scheme caches the pass by data. The set of specific nodes, i.e., caching nodes (CN) are chosen based on various parameters like path weight, number of neighbors, energy, contact pattern and contact duration [7–9]. When opportunistic contacts and intermittent connectivity are considered, incidental caching would fail because of over simplistic consideration of query history. Hence, intentional caching schemes are employed in DTN [7].

3 Survey on Various Caching Techniques

This section deals with various caching schemes employed in DTN that improves data access. This study also includes various techniques and measures used for electing the caching nodes.

3.1 Cooperative Caching (CC)

This is a new intentional caching technique by which the data are cached only at some specific nodes called network central locations (NCLs) representing caching nodes that are easily accessed by other nodes in the network [7, 10, 16]. The caching nodes are chosen based on a probabilistic selection metric which follows hypoexponential distribution [11]. In this scheme, the data being generated at the data source are prioritized and stored at NCL based on the query history and popularity. Queries raised by requestors are initially forwarded to NCLs to serve the request at a faster rate. If the data are not found in NCL, then the request is broadcasted to the entire network until the data source is found. For every request given by the requestor, more number of copies will be sent to the requestor since the data are distributed and stored across multiple locations. Though this kind of data retrieval ensures prompt access, network resources are wasted in repeatedly transmitting the same data. Hence, the number of responses is minimized by generating probabilistic response based on a sigmoid function, thereby coordinating all the caching nodes [7].

The significance of the scheme lies in the selection of network central location (NCL). A metric C_i called selection metric (1) is calculated for each and every node in the network based on adjacent nodes. The value C_i obtained as the degree of centrality is obtained by summation of all intercontact time between nodes that fall between the source and destination [7]:

$$C_i = \frac{1}{|V|} \cdot \sum_{j \in V} p_{ij}(T). \tag{1}$$

Finally, the 'K' nodes having the highest C_i metric are assumed to be nodes having the highest potential in the network and are involved for caching. When the cache capacity of central nodes get full, then nodes nearest to the caching nodes called as relays are also involved for caching.

3.2 Duration-Aware Caching (DAC)

This is the only existing scheme that addresses the issue of contact duration among nodes. This novel scheme proposes community-based cache node election to deal with the instability of DTN [8]. The DAC scheme starts the work by detecting community using k-clique algorithm proposed by Hui et al. [12]. Since nodes that belong to the same community have a higher chance of meeting each other, it will be much easier for any node to fetch data from nodes inside the same community than the nodes outside the community. The set of nodes having the highest potential to contribute their cached data to other nodes are identified for caching based on the centrality metric, which resembles a realistic toy scenario [8]. The centrality metric represents the marginal caching benefit. The novelty about this scheme is that it restricts the amount of data to be cached at the specific node by considering the contact duration (CD) and hence more number of nodes are involved in the caching process. Since it is DTN, the nodes will be in contact for less time resulting in limited contact duration. Hence, a large data item may not be completely transmitted between two nodes within their contact. As a result, the data access will take a longer time than the usual data access time. In these cases, data are fragmented and stored in different nodes which could fall in the requestor's transmission range in the near future. Further, the requestor will collect its data packet from more number of caching nodes resembling the coupon collector's problem [13]. To solve this problem, the random linear network coding technique [14] was introduced to encode the data.

The amount of data to be stored at each node is identified by the adaptive caching bound. Each node maintains a table called community information table (CIT) that holds information like node ID, centrality value, caching bound and timestamp. Based on the information provided by CIT, caching is performed. This scheme improves the performance by reducing the data access delay which is because of considering the contact duration. On comparing this scheme with Gao et al.'s method, this scheme has proven its caching efficiency by considering a new parameter, contact duration.

3.3 Adaptive Caching (AC)

This is a new adaptive technique [9] used to select caching nodes in an efficient manner. This is a learning automata-based scheme which works based on environment rather than rules. The ultimate aim of the scheme is to choose a set of nodes for caching which could contribute more to the entire network. Hence, the scheme starts by choosing a dominating set (DS) of nodes from the entire network which have at least one neighbor. From DS, a connected dominating set (CDS) is constructed which is a connected subgraph of network. From CDS, an initial caching set (ICS), a yet another minimized set of nodes is selected based on the count of the neighbors. These ICS are nodes that could provide services to all the other nodes in the network. Then, each node of ICS is given as input to the learning automata (LA) which is a self-operating machine [9] that responds according to the environment in which it operates.

For each node in ICS, the LA will output either 0 (reward) or 1 (penalty). Depending on the output, the selection probabilities of the final caching set (FCS) are calculated based on the forwarding ratio. This function is given by the linear reward inaction scheme,

$$p_i(n+1) = p_i(n) + \lambda_i(1 - p_i(n)), \tag{2}$$

$$p_j(n+1) = (1 - \lambda_j)p_j(n) \text{ For all } j \neq i. \tag{3}$$

The process of updating the selection probability happens until the selection probability reaches one. Finally, the 'K' number of nodes having the highest selection probability is chosen as FCS [9].

3.4 Redundancy and Distributed Caching (DC)

Mikko et al. proposed a scheme of redundancy-based storage and content retrieval system for improving the data access in DTN based on the infrastructure network [15]. This is yet another cooperative scheme where the intermediate nodes are involved for redundant storage and content retrieval. Cooperation between nodes is achieved by defining a new architecture to support the message-based content storage technique. The nodes in the network are grouped as community and some nodes are identified for redundant storage, which are mostly access points or gateways. Whenever a request arises, the respective community performs a cache lookup resulting either as a hit or miss. Initially, the request will be multicasted only among their community and later on broadcasted to the entire network when data are not found within the community.

This infrastructure-based redundancy scheme works by exchanging data units at the DTN layer such as routing, forwarding and queuing. The data are stored and forwarded as fragmented bundles. Bundles are fragmented using erasure code-based scheme at the application layer using Reed Solomon codes [15]. This queuing type enables caching ADUs for a long time. The proposed architecture has a storage/retrieval module that is responsible for both store and retrieve mechanisms. Based on the passing request, the module does one of the two things:

1. Perform cache lookup for the corresponding request and retrieve the bundles stored.
2. Store the data in queue for efficient data access based on the probabilistic selection.

This scheme results in reduced latency and improved retrieval performance. Because of redundant storage and flooding, the network suffers from congestion. Because of congestion, certain bundles are also dropped.

4 Survey Table

All the aforementioned schemes are compared based on various distinguishing parameters in the following comparison Table 1.

Table 1 Comparison of all schemes

Caching technique	CN election Alg.	Fragmentation technique	Forwarding scheme	Nodes involved in caching	Pros	Cons
CC	Hypoexponential distribution	No	Multicast and broadcast	Varies	Fault tolerant	Inconsistent CN election
DAC	Community based	Contact duration (CD)	Broadcast	Fixed	Optimal election of CN	No response optimization
AC	Learning automata (LA) based	No	Broadcast	Fixed	Efficient election of CN	High traffic
DC	Community based	Erasure coding	Multicast and broadcast	Fixed	Improved cache hit ratio	Results in congestion

5 Performance Evaluation

This survey had evaluated the performance of the above mentioned schemes by extending the use of Java-based ONE Simulator [16–18]. The number of DTN nodes considered for simulation is 50 which includes a varied scale of 1–6 caching nodes. Having random walk as the movement model, we recorded traces after a simulation of DTN for 12 hours in an area of 400 m × 400 m. The buffer size is considered to be 10 Mb for the normal node and 100 Mb for the caching node. The message size is fixed as 10 Kb. Various parameters such as number of messages delivered, dropped, created or relayed are analyzed to differentiate among the above-mentioned caching schemes. The following metrics are used for evaluating the performance of four schemes.

1. **Delivery Probability**—the probability value of the number of packets successfully delivered to the destination to the number of messages generated.
2. **Delay**—the time taken to deliver a message from the source node to the destination.
3. **Caching cost**—the number of replicas of a single message.
4. **Number of caching nodes**—the number of nodes involved in caching in DTN.

From Fig. 1a, it is observed that an increase in the number of caching nodes increases the deliver probability. When the number of caching nodes is varied from scale 1–5, an increase in delivery probability is observed. However, increasing caching nodes beyond 5 contributes only a minimal change toward increase in delivery probability. It is observed that 5 caching nodes for 50 DTN nodes is more than enough, while those beyond 5 will result in under utilization of the node's buffer. Of all four schemes, cooperative caching (CC) and adaptive caching (AC) provide higher delivery probability because of its flexibility toward environmental change. Since messages are fragmented and cached around all DTN nodes, duration-aware caching (DAC) does not contribute much toward delivery probability. This scheme (DAC) would increase the delivery probability if the contacts between nodes in the network are high. In distributed caching (DC) the change in slope is high and delivery probability reaches maximum when the requestor is nearer to the gateway nodes (since caching is done on infrastructure-based nodes) and is minimal when the requestor is far away from the gateway nodes. From Fig. 1b, it is observed that an increase in caching cost reduces data access delay. The DC scheme transfers data with less delay when compared with the other three schemes because of the gateway node's involvement in caching. The queries generated from nodes nearer to the gateway nodes are handled faster accounting for a minimal delay, since the distance between them is also minimal. The DAC scheme stands behind all the schemes since it takes a longer time to reassemble all fragmented packets stored in multiple DTN nodes. Of all the four schemes, CC and AC schemes results in an acceptable delay.

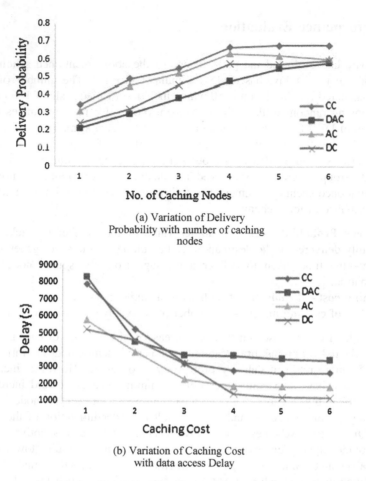

(a) Variation of Delivery
Probability with number of caching
nodes

(b) Variation of Caching Cost
with data access Delay

Fig. 1 **a** Variation of delivery probability with number of caching nodes, **b** Variation of caching cost with data access delay

6 Conclusion

Data access and its efficiency are of predominant concern in a DTN environment. In this paper, we had surveyed four different intentional caching schemes, which differ by various parameters such as caching node election and fragmentation process. Each and every scheme was found to be better in its own way of delivering data by considering different facts. Besides its uniqueness, learning automata-based scheme (AC) and cooperative caching (CC) were found to be best, resulting in higher delivery probability and less delay. In future, we would consider various social aware parameters such as energy and centrality to optimize the cache node election, thereby improving data access.

References

1. Long Xiang Gao, Shui Yu, Tom H. Luan, Wanlei Zhou: Delay Tolerant Networks, Springer (2015).
2. K. Fall: A Delay-Tolerant Network Architecture for Challenged Internets, Proc. ACM SIGCOMM Conf. Applications, Technologies, Architectures, and Protocols for Computer Comm., pp. 27–34 (2003).
3. A. Vahdat and D. Becker: Epidemic Routing for Partially Connected Ad Hoc Networks, Technical Report CS-200006, Duke Univ. (2000).
4. J. Burgess, B. Gallagher, D. Jensen, and B. Levine: MaxProp: Routing for Vehicle-Based Disruption-Tolerant Networks, Proc. IEEE INFOCOM (2006).
5. W. Gao, Q. Li, B. Zhao, and G. Cao: Multicasting in Delay Tolerant Networks: A Social Network Perspective, Proc. ACM MobiHoc, pp. 299–308 (2009).
6. P. Hui, J. Crowcroft, and E. Yoneki: Bubble Rap: Social-Based Forwarding in Delay Tolerant Networks, Proc. ACM MobiHoc (2008).
7. Wei Gao, Guohong Cao, Arun Iyengar and Mudhakar Srivatsa: Cooperative Caching For Efficient Data Access in Disruption Tolerant Networks, IEEE Vol. 13, No. 3 (2014).
8. XuejunZhuo, Quinghua Li, Guohong Cao, Yigi Dai, Boleslaw Szymanski, Tom La Porta: Social-Based Cooperative Caching in DTNs: A Contact duration Aware Approach, IEEE International Conference on Mobile Ad-Hoc and Sensor Systems, (2011).
9. Reisha Ali and Rashmi Rout: An Adaptive caching Technique using Learning Automata in Disruption Tolerant Networks, International Conference on Next Generation Mobile Apps, Services and Technologies (2014).
10. W. Gao, G. Cao, A. Iyengar, and M. Srivatsa: Supporting Cooperative Caching in Disruption Tolerant Networks, Proc. Int'l Conf. Distributed Computing Systems (ICDCS) (2011).
11. S.M. Ross, Introduction to Probability Models, Academic (2006).
12. P. Hui, E. Yoneki, S. Y. Chan, and J. Crowcroft: Distributed Community Detection in Delay Tolerant Networks, Proc. of ACM MOBIARCH (2007).
13. M. Mitzenmacher and E. Upfal: Probability and Computing: Randomized Algorithms and Probabilistic Analysis, Cambridge (2008).
14. P.A. Chou, Y. Wu, and K. Jain: Practical Network Coding, Proc. of Annual Allerton Conf. on Comm., Control, and Comput. (2003).
15. M.J. Pitkanen, Jorg Ott: Redundancy and distributed Caching in Mobile DTNs, MobiArch, ACM International Workshop on Mobility in evolving internet architecture (2007).
16. A. Keranen, Opportunistic Network Environment Simulator, Special Assignment report, Helsinki University of Technology, Department of Communications and Networking (2008).
17. A. Keranen and J. Ott, Increasing Reality for DTN Protocol Simulations, Tech report, Helsinki University of Technology (2007).
18. Barun Saha: The ONE – dtn simulator tutorial, http://delay-tolerant-networks.blogspot.ca/p/one-tutorial.html.
19. A. Balasubramanian, B. Levine, and A. Venkataramani: DTN Routing as a Resource Allocation Problem, Proc. ACM SIGCOMM Conf. Applications, Technologies, Architectures, and Protocols for Computer Comm., pp. 373–383 (2007).
20. Narottam Chand, R.C. Joshi and Manoj Misra: Efficient Cooperative Caching in Ad Hoc Networks, Proc. Int'l Conf. on Communication System Software and Middleware (2006).

A Cascaded Method to Reduce the Computational Burden of Nearest Neighbor Classifier

R. Raja Kumar, P. Viswanath and C. Shobha Bindu

Abstract Nearest neighbor classifiers demand high computational resources, i.e., time and memory. Two distinct approaches are followed by researchers in pattern recognition to reduce this computational burden. The first approach is reducing the reference set (training set) and the second approach is dimensionality reduction which is referred to as prototype selection and feature reduction (a.k.a feature extraction or feature selection), respectively. In this paper, we cascaded the two methods to achieve the reduction in both directions. The experiments are done on the bench mark datasets, and the results obtained are satisfactory.

Keywords Nearest neighbor classifiers · Prototype selection · Pattern recognition · Feature selection

1 Introduction

Nearest neighbor classification (NNC) is a popularly known classification method in pattern recognition. NNC is simple to use and easy to understand. For a test pattern 'q', NNC or 1-NNC finds the nearest neighbor in the reference set and assigns the class label of this nearest neighbor to the test pattern [1].

A simple variant of NNC is to consider the k-nearest neighbors of the test pattern and the class label infers by majority voting. NNC suffers from severe drawbacks.

R. Raja Kumar (✉)
Rajeev Gandhi College of Engineering
and Technology, Nandyal, India
e-mail: rajsri1229@yahoo.co.in

P. Viswanath
IIIT Chittoor, Chittoor, India
e-mail: viswanath.p@iiits.in

C. Shobha Bindu
JNTUA College of Engineering, Anantapur, India
e-mail: shobabindhu@gmail.com

© Springer Science+Business Media Singapore 2017
S.C. Satapathy et al. (eds.), *Proceedings of the First International Conference on Computational Intelligence and Informatics*, Advances in Intelligent Systems and Computing 507, DOI 10.1007/978-981-10-2471-9_27

275

(1) NNC requires memory space to store all the training patterns. To classify the given query pattern 'q', it computes the distances between the query pattern 'q' and every training pattern from the training set to find the nearest neighbor of 'q'. Hence, the time complexity is $O(n)$, where 'n' is the training set size. To classify the test set of size 'm', the time complexity becomes $O(mn)$. This is a severe problem if the size of training set 'n' is large. One can reduce the memory and computational needs of NNC, if the size of training set is reduced considerably.

One method of reducing the training set can be achieved by retaining the patterns which will play crucial role in the classification process and by discarding the others which are not playing vital role in the classification process. This approach is called prototype selection method. Considerable number of papers have been published on reducing the size of training set and, hence, reducing the memory and computational demands of NNC.

Similarly, the dimensionality is also a problem if the data sets are represented by large number of features as it leads to the curse of dimensionality [2].

The rest of the paper is organized as follows: Sect. 2 discusses about a review of related works. Section 3 discusses about algorithms and notations. In Sect. 4, data sets are discussed briefly. Experimental results have been presented in Sect. 5. Conclusion is presented in Sect. 6.

2 Related Work

The early method presented in this context of reducing the training set size was the condensed nearest neighbor (CNN) proposed by Hart [3]. Let $\hat{D}$ be the condensed set found by applying CNN method over the original training set D. $\hat{D}$ is called condensed iff it classifies all the training patterns from D. Then, $\hat{D}$ is used instead of D to classify the test set. CNN has two disadvantages: (1) Retention of unnecessary samples (2) Occasional retention of internal rather than boundary patterns [4]. In 1976, Ivan Tomek proposed two modifications to CNN which overcome the above disadvantages of CNN [4]. In 2002, Devi and Murthy proposed modifiedcondensed nearest neighbor (MCNN) method which is a prototype selection method [5]. In this method, prototype set was built incrementally. In 2005, Viswanath, Murthy and Bhatnagar proposed an efficient NN classifier, called overlap pattern nearest neighbor (OLP-NNC) which is based on overlap pattern graph (OLP-graph) [6]. In this paper, they built OLP-graph incrementally by scanning the training set only once and OLP-NNC works with OLP-graph. In 2007, Babu and Viswanath [7] proposed the generalized weighted k-nearest Leader Classifier method which finds the relative importance of prototypes called weighting of prototypes, and this weight is used in classification process. In 2011, Sarma and Viswanath [8] proposed an improvement to the k-nearest neighbor mean classifier (k-NNMC), finds k-nearest neighbors class-wise. Classification process is carried out with these mean patterns. In 2014, Raja Kumar et al. [9] proposed a new prototype selection method for nearest neighbor classification called other class nearest neighbors (OCNN), which is based on the retaining samples that are near to decision boundary and which are crucial in the classification task.

The reduction of dataset can also be done in other direction, i.e., reducing the dimensionality. Feature extraction and feature selection are the two distinct approaches to reduce the features of data set. Linear discriminant analysis is an example of feature extraction [10, 11]. Feature selection is another method to reduce dimensionality. Feature selection methods are well discussed in [12, 13, 14].

Considerable number of papers were published on reducing the cardinality of training set, but less number of papers have been published on reducing the training set size in both directions, i.e., cardinality and dimensionality. Kuncheva et al. proposed methods based on genetic algorithms which reduce the training sets in both directions [15, 16].

3 Algorithms and Notations

Let TS be the given training set. A pattern in TS is represented as (Xi, Yi) for $i = 1$, $2, \ldots n$ (n is the cardinality of TS), where Xi represents the d-dimensional feature vector and Yi represents the class label for ith pattern. So, $Xi = (xi1, xi2, \ldots xid)$ and $Yi = \{y1, y2, \ldots yc\}$, where c is the number of classes present.

3.1 Algorithm: CNN

Condensed nearest neighbor begins with the pattern selected from the training set which forms the initial condensed set. Then, each pattern in the training set is classified using the condensed set. If a pattern in the training set is misclassified, it is included in the condensed set. After one pass through the training set, another iteration is carried out. This iterative process is carried out till there are no misclassified patterns [5]. The condensed set has two properties:

It is a subset of original training set
It ensures 100 % accuracy over the training set which is the source for deriving the condensed set.

Let TS be the given training set and TS (CNN) is the new training set formed by applying condensed nearest neighbor rule. The algorithm can be obtained from [3].

3.2 Algorithm: OCNN

This method is based on the intuitive notion of retaining patterns that are near to decision boundary. It is obvious that the patterns which are near the decision boundary play a very important role in the classification of a process. The boundary patterns are computed using other class nearest neighbors (OCNN) method [9].

Let TS be the given training set. TS(OCNN) is the set formed after applying the OCNN method. OCNN(Xi) is the set of nearest neighbors for pattern Xi from other classes which can be computed using k-NN rule. The method is outlined in Algorithm 1.

Consider the example training set shown in Fig. 1. Two classes are present in the Fig. 1. The positive class tuples are represented by '+' and negative class tuples are represented by '−'. For example, if a query pattern '*' lies on the left side of the boundary decision, then it is classified as '+' class, otherwise '−' class. That means the patterns that are near to decision boundary are enough for classification. The other samples can be removed from the training set. This is the central idea of our method [9]. In the example present in Fig. 1, two patterns from '−' class and two patterns from '+' class which are near to decision boundary are typical patterns when compared to other training patterns. This can be found using the other class nearest neighbors algorithm (OCNN) discussed in Algorithm 1.

Algorithm 1 Other Class Nearest Neighbor algorithm

1. Initially Set S=∅.

2. **for** each X_i present in TS

3. **do**

4. Find $S_i = \text{OCNN}(X_i)$ using k-nn rule.

5. $S = S \cup S_i$.

6. **end for**

7. TS(OCNN)=S;

8. **output** TS(OCNN).

3.3 Sequential Forward Selection (SFS)

The data sets may contain large number of features (dimensions or attributes), some of which are irrelevant to the classification process [17]. Some features might be redundant features. For example, if the classification task is to classify the person with obesity or not based on his personal data which contains the information about his name, age, contact number, blood pressure, waist size, and cholesterol level, the features containing the person's blood pressure and contact number are irrelevant.

Fig. 1 Nearest neighbor classifier

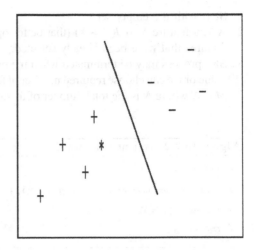

A domain expert may pick the useful features [17]. But this process is inopportune in case of inadequacy of expertise and when the data behavior is unknown. This is also time consuming in case of large number of features present in the data sets. Moreover, we need manual intervention in this process.

Hence, in machine learning, pattern recognition and allied fields dimensionality reduction is the topic of research over years. Dimensionality reduction can be done in two ways [18].

1 Feature selection.
2 Feature extraction.

Feature selection or feature subset selection is the process of selecting a subset of existing features without using any transformation. Feature extraction projects the existing features onto a lower dimensional space with the help of transformations [18]. Linear discriminant analysis (LDA) and its variants are good examples of feature extracion. LDA process is described well in [10, 11].

The formal definition of feature subset selection is given below. Feature subset selection: Let F be the feature set, i.e., $F = \{Fi \mid i = 1, 2, ..., N\}$; find a subset $Z = \{Fi1, Fi2, ..., FiM\}$ with $M < N$, which optimizes an objective function $J(Y)$.

Choosing objective function is categorized into two groups. Filter approach: the objective function evaluates feature subsets by their information content, typically interclass distance, statistical dependence or information-theoretic measures [18]. Wrapper approach: the objective function is a pattern classifier, which evaluates feature subsets by their predictive accuracy (recognition rate on test data) by statistical resampling or cross-validation [18].

Sequential Forward selection (SFS) is a feature selection algorithm which selects the subset of features based on objective function $J(.)$. It aims to find the best subset of features from the given original set of features. The process of SFS is carried as follows.

1 Start with the empty set.
2 Add a feature X to $J(Y + X)$ (that better optimizes $J(.)$), where Y is a subset of features that have been already selected.
3 SFS process may be terminated when the objective function reaches to threshold value or we can choose required number of features in the subset of features M, i.e., $M < N$ where N is the total number of features in the data set by maximizing $J(.)$

Algorithm 2 Sequential Forward Selection algorithm

1. Initially Set $Y_0 = \emptyset$.

2. Select the next best feature $x^+ = \text{argmax}[J(Y_k + x)]$

3. Update $Y_{k+1} = Y_k + x^+$; k=k+1

4. Go to step 2.

The method is outlined in [18] and reproduced in Algorithm 2 for reference. We used the wrapper approach and the optimization criterion is accuracy.

3.4 An Example

We illustrate the algorithms with the example data set shown in Table 1. It has two attributes X and Y along with class label present in the last column. Two class labels Yes and No are present. Let Xi refers to the pattern i in the data set, i.e., $X1$ is the first pattern $(1, 1)$ in the data set. The Euclidean distances between the patterns are present in Table 2.

(1) Now, we build prototype set S based on the method OCNN shown in Algorithm 1.

The $SX_i = OCNN(Xi)$ where $i = 1, 2, ..., 8$ are obtained as follows; We considered $k = 2$.

Table 1 Example data set

No.	A	B	Class
X1	1	1	Yes
X2	1	2	Yes
X3	2	1	Yes
X4	2	2	Yes
X5	5	1	No
X6	5	2	No
X7	6	1	No
X8	6	2	No

Table 2 Distances

No.	X1	X2	X3	X4	X5	X6	X7	X8
X1	0	1	1	1.414	4	4.12	5	5.09
X2	1	0	1.414	1	4.12	4	5.09	5
X3	1	1.414	0	1	3	3.16	4	4.12
X4	1.414	1	1	0	3.16	3	4.12	4
X5	4	4.12	3	3.16	0	1	1	1.414
X6	4.12	4	3.16	3	1	0	1.414	1
X7	5	5.09	4	4.12	1	1.414	0	1
X8	5.09	5	4.12	4	1.414	1	1	0

Table 3 Prototype set chosen by OCNN on Table 1

No.	A	B	Class
X3	2	1	Yes
X4	2	2	Yes
X5	5	1	No
X6	5	2	No

$$SX_1 = \text{OCNN}(X_1) = \{X_5, X_6\} \quad SX_2 = \text{OCNN}(X_2) = \{X_5, X_6\}$$
$$SX_3 = \text{OCNN}(X_3) = \{X_5, X_6\} \quad SX_4 = \text{OCNN}(X_4) = \{X_5, X_6\}$$
$$SX_5 = \text{OCNN}(X_5) = \{X_4, X_3\} \quad SX_6 = \text{OCNN}(X_6) = \{X_4, X_3\}$$
$$SX_7 = \text{OCNN}(X_7) = \{X_4, X_3\} \quad SX_8 = \text{OCNN}(X_8) = \{X_4, X_3\}$$

The final prototype set S is computed by the union of all SX_i. For this example, $S = SX_1 \cup SX_2 \cup SX_3 \cup SX_4 \cup SX_5 \cup SX_6 \cup SX_7 \cup SX_8$. Hence, we get $S = \{X5, X6, X4, X3\}$ shown in Table 3.

(2) We build the prototype set S based on the method CNN. The first pattern $X1$ forms the initial condensed set S. So $S = \{X1\}$. Using S, the dataset in Table 1 is classified. During the first iteration, the patterns $X1$, $X2$, $X3$, $X4$ are ignored as they classify using S. The pattern $X5$ is added to S because it is miscalssified by S. The patterns $X6$, $X7$, $X8$ are ignored as they classify using S. After the end of the first iteration, $S = \{X1, X5\}$. Then, the second iteration is carried out. During the second iteration, new patterns are not added to S, because all the patterns in the data set are classified. At this stage, the algorithm is terminated with set $S = \{X1, X5\}$, the final condensed prototype set.

4 Data Sets

In this section, the data sets used for the experiments are discussed. We applied the methods on three data sets. All the data sets are taken from Murphy and Aha [19]. The information about the data sets is shown in Table 4. It shows about the name of

Table 4 Description of data sets

Data set	Number of training patterns	Number of test patterns	Number of features	Number of classes
Thyroid	144	71	5	3
Iris	99	51	4	3
Pima	600	168	8	2

the data sets, number of training and testing patterns, number of features and also the number of classes.

5 Experimental Results

We experimented the method with the popular data sets. The accuracies obtained on the data sets are shown in Table 5. In this table, the accuracies obtained by OCNN and CNN are given. We got satisfactory results. We compared our method OCNN with CNN.

These results are evident enough to say that OCNN method is competing with CNN method and it is showing good accuracy. The methods are implemented on 'DELL OPTIPLEX 740 n' model having Intel Core 2 DUO 2.2 GHz processor with 1 GB DDR2 RAM capacity. The environmental setup was, all the methods were written in C language under Linux operating system. We compared the time taken to compute the prototype set by OCNN method and by CNN method. The results are shown in Table 6. From the table, it is clear that the OCNN method is showing good improvement in execution time with respect to CNN over all the data sets. OCNN method can build the prototype set faster than CNN. This is because OCNN method does not require multiple scans as CNN.

Table 5 Accuracy obtained over data sets

Data set	OCNN		CNN	
	Number of prototypes	Accuracy (%)	Number of prototypes	Accuracy (%)
Thyroid	38	92.10	26	91.55
Iris	22	98.33	16	96.08
Pima	289	76.78	266	65.23

Table 6 Comparison of time

Data set	OCNN (s)	CNN (s)
Thyroid	0.005	0.009
Iris	0.004	0.008
Pima	0.019	0.031

Table 7 Accuracy obtained over thyroid

k	SFS+CNN			SFS+OCNN			CNN+SFS			OCNN+SFS		
	# of prototypes	# of features reduced	Accuracy (%)	# of prototypes	# of features reduced	Accuracy (%)	# of prototypes	# of features reduced	Accuracy (%)	# of prototypes	# of features reduced	Accuracy (%)
3	24	1	82.10	43	1	87.27	28	2	81.57	46	1	89.47
5	24	1	84.44	45	1	86.45	28	2	83.34	48	1	88.74
7	24	1	86.22	48	1	87.22	28	2	84.42	45	1	89.32
9	24	1	85.54	46	1	86.88	28	2	85.79	47	1	88.52
			84.57 ± 1.80			86.95 ± 0.37			83.78 ± 1.78			89.01 ± 0.45

Using OCNN and CNN prototype selection methods, we can reduce the data set size, but the dimensionality remains the same. To reduce the dimensionality of the data set, we used SFS discussed in Sect. 3.3. As a result, the data set was reduced both in cardinality and dimensionality. Now, there are two possible ways of reducing the data set in both directions:

1 Applying prototype selection method (CNN, OCNN followed by dimensionality reduction method (SFS).
2 Applying dimensionality reduction method (SFS) followed by prototype selection method (CNN, OCNN).

We experimented with both posssibilities. The results are tabulated datasetwise. OCNN+SFS means first OCNN is applied and, then, SFS is applied. We compared the performance with the existing CNN method. We conducted the experiment using different values of k(3, 5, 7, 9), and the last row represents average of the results along with standard deviation.

The results over thyroid data set are given in Table 7.

The graphical results obtained over thyroid data set are present in Fig. 2a.

The results for the iris data set are present in Table 8.

The graphical results obtained over iris data set are present in Fig. 2b.

The results obtained over Pima data set are present in Table 9.

The graphical results obtained over Pima data set are present in Fig. 3.

We have considered accuracy as primary measure to estimate classifiers' performance. From the results, it is clear that our method is well competing with the existing method CNN. We are able to reduce the data sets in both directions without sacrificing accuracy.

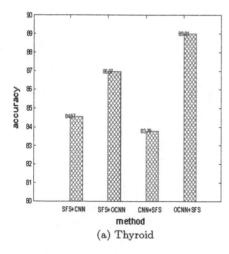
(a) Thyroid

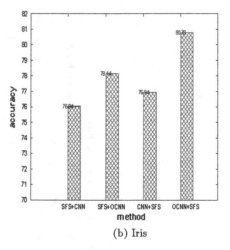
(b) Iris

Fig. 2 Results over thyroid and iris data sets

Table 8 Results over iris data set

k	SFS+CNN			SFS+OCNN			CNN+SFS			OCNN+SFS		
	# of prototypes	# of features reduced	Accuracy (%)	# of prototypes	# of features reduced	Accuracy (%)	# of prototypes	# of features reduced	Accuracy (%)	# of prototypes	# of features reduced	Accuracy (%)
3	23	1	76.33	31	2	78.83	21	1	76.84	33	1	80.48
5	23	1	75.45	35	2	76.43	21	1	77.53	35	1	81.22
7	23	1	76.67	33	2	77.88	21	1	74.55	36	1	79.67
9	23	1	75.65	38	2	79.55	21	1	78.87	39	1	81.77
			76.04 ± 0.60			78.14 ± 1.33			76.94 ± 1.80			80.78 ± 0.91

Table 9 Accuracy obtained over Pima

k	SFS+CNN			SFS+OCNN			CNN+SFS			OCNN+SFS		
	# of prototypes	# of features reduced	Accuracy (%)	# of prototypes	# of features reduced	Accuracy (%)	# of prototypes	# of features reduced	Accuracy (%)	# of prototypes	# of features reduced	Accuracy (%)
3	301	2	57.52	350	2	58.92	266	2	61.11	365	3	63.09
5	301	2	58.32	358	2	59.45	266	2	60.21	371	3	62.67
7	301	2	57.22	360	2	60.42	266	2	62.23	368	3	63.32
9	301	2	59.32	368	2	61.88	266	2	62.88	378	3	65.52
			58.09 ± 0.60			60.16 ± 1.29			61.60 ± 1.18			63.65 ± 1.27

Fig. 3 Results over Pima

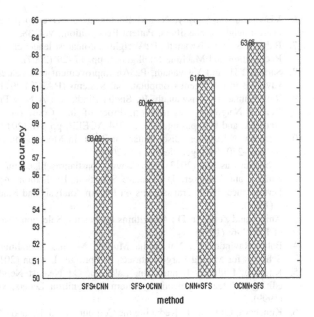

6   Conclusions and Future Enhancement

In this paper, we proposed an approach which can reduce the computational burden of NNC. This is achieved by reducing the data set both in dimensionality and cardinality. The methods are implemented and compared with the existing methods over the bench mark data sets and the results are found satisfactory.

The future enhancement of the paper is to make SFS work well with the large numbered featured data sets. One can investigate the ways of integrating SFS with the OCNN and CNN methods, such as whether it is possible without sacrificing accuracy etc.

References

1. Cover T. and Hart P.: Nearest neighbor pattern classification, Information Theory, IEEE Transactions on 13.1 pp. 21–27 (1967).
2. Verleysen, Michel, and Damien Franois.: The curse of dimensionality in data mining and time series prediction. Computational Intelligence and Bioinspired Systems, Springer Berlin Heidelberg, pp 758–770 (2005).
3. Hart P.: The Condensed Nearest neighbor rule, IEEE Trans on Information Theory, vol. IT-14 (3), pp 515–516 (1968).
4. Tomek I.: Two modifications of cnn, IEEE Trans on Syst. Man. Cybern., vol. SMC-6 no 11, pp 769–772 (1976).
5. M.N.M and Susheela Dev, V.: An incremental prototype set building technique, Pattern Recognition., vol. 35, pp 505–513 (2002).

6. Viswanath, P., Murthy, N. and Bhatnagar, S.: Overlap pattern synthesis with an efficient nearest neighbor classifiers, Pattern Recognition., vol. 38, no. 8, pp 11871195 (2005).
7. Babu, V.S. and Viswanath, P.: Weighted k-nearest leader classifier for large data sets, Pattern Recognition and Machine Intelligence, pp. 17–24 (2007).
8. Sarma, T.H. and Viswanath, P.: An improvement to k-nearest neighbor classifiers, Recent advances in Intelligent Computational Systems (RAICS) 2011 IEEE, pp 227231 (2011).
9. Raja Kumar, R., Viswanath, P., Shoba Bindu, C.: A New Prototype Selection Method for Nearest Neighbor Classification, Proc. of Int. Conf. on Advances in Communication, Network, and Computing, CNC, 2014 ACEEE, pp 1–8 (2014).
10. Izenman, A.J.: Linear discriminant analysis. In Modern Multivariate Statistical Techniques, pp. 237–280, Springer New York (2008).
11. Sebastian Raschka (2013), http://www.sebastianraschka.com/.
12. Anil, J. and Zongker, D.: Feature Selection: Evaluation, Application, and Small Sample Performance, IEEE Transactions on Pattern Analysis and Machine Intelligence, vol. 19, no. 2 (1997).
13. Anil, J. and Zongker, D.: Algorithms for Feature Selection: An Evaluation, IEEE proceedings of ICPR'96, (1996).
14. Babu, Ravindra, T., Narasimha Murthy, M. and Subrahmanya S.V.: Data Compression Schemes for Mining Large Datasets, In Springer London (2013).
15. Kucheva, Ludmila, I. and Jain, Lakshmi, C.: Nearest Neighbor Classifier: Simultaneous editing and feature selection, Pattern Recognition Letters, vol. 2, no. 11, pp 1149–1156 (1999).
16. Kuncheva, Ludmila, I.: Reducing the Computational demand of the nearest neighbor classifier (2001).
17. Han, J., Kamber M., and Pei J.: Data mining: concepts and techniques: concepts and techniques. Elsevier (2011).
18. Sequential Forward Selection, http://research.cs.tamu.edu/prism/lectures/pr/prl11.pdf.
19. Murphy D.A.P.M.: UCI repository of machine learning databases, [http://www.ics.uci.edu/mlearn/MLRepository.html], Department of Information and Computer Science, University of California, Irvine, CA. (1994).

An Innovative Method to Mitigate DDoS Attacks for Cloud Environment Using Bagging and Stacking

B. Kiranmai and A. Damodaram

Abstract This paper describes an approach for mitigating distributed denial-of-services attacks in Cloud. DDoS attacks are huge pitfall for Cloud, and still, this is not very well handled. We presented a survey of existing work to defend DDoS attacks and mechanisms. In Cloud, intruder detection systems can be deployed at various positions like front-end, back-end or at Virtual Machine. Most of the existing IDS have been deployed at Virtual Machine in cloud. We proposed a new frame work using ensemble classifiers, namely, Bagging and Stacking to detect intrusions of both insiders and outsiders, and our proposed frame work will deploy at back-end. We focused on defending DDoS attacks in cloud environment which are the bottle neck of a Cloud environment compared to other type of attacks.

Keywords Cloud computing · Intrusion detection system · Data mining techniques · Weka · WireShark

1 Introduction to Cloud Computing [1]

Cloud computing is all over the place. Distributed computing gets its name as a representation for the Internet. Cloud processing guarantees to cut operational and capital expenses. Distributed computing is a build that permits you to get to applications that really live at an area other than your PC or other Internet joined gadget. Security and protection is one of the biggest ranges of sympathy toward anybody assembling a cloud system.

B. Kiranmai (✉)
Keshav Memorial Institute of Technology, Narayanguda, Hyderabad, India
e-mail: kiranmaimtech@gmail.com

A. Damodaram
Sri Venkateswara University, Tirupathi, Andhra Pradesh, India
e-mail: damodarama@rediffmail.com

© Springer Science+Business Media Singapore 2017
S.C. Satapathy et al. (eds.), *Proceedings of the First International Conference on Computational Intelligence and Informatics*, Advances in Intelligent Systems and Computing 507, DOI 10.1007/978-981-10-2471-9_28

The main issue concern is security [2]. Cloud service providers do not bother whether the customer is a legitimate user or not. Due to this problem arises like (i) Data Breach, (ii) Insiders, (iii) Cryptographic attacks, and iv) DDoS attacks. All of these kinds of attacks we call them as Intruders [3].

People from everywhere now have passage to different people from wherever else. Globalization of handling assets might be the best duty the cloud has made to date. Dispersed registering is still in its soonest organizes. Conveyed registering might be seen as a benefit available as an organization for virtual data focuses yet disseminated figuring and virtual server ranches are not the same. People from everywhere now have passage to different people from wherever else. Globalization of processing assets might be the best duty the cloud has made to date. Distributed computing is still in its most punctual stages. Circulated figuring might be seen as a benefit available as an organization for virtual data focuses; however, appropriated processing and virtual server ranches are not the same [4, 5].

Lattice figuring is frequently mistaken for distributed computing. Network processing is a type of dispersed registering that executes a virtual super PC made up of a gathering of orchestrated or internetworked PCs going about as one to perform tremendous errands. Various dispersed registering associations are filled by structure figuring utilization and are charged like utilities; however, conveyed processing can and should be seen as a created next stride a long way from the framework utility mode 1 [6, 7, 8, 9, 10, 11].

"Distributed computing is a model for engaging supportive, on-interest framework access to a commonplace pool of configurable enrolling assets (e.g., systems, servers, stockpiling, applications, and associations) that can be promptly provisioned and discharged with unimportant association exertion or association supplier in one of the known classes to each test occasion. This is possible in light of the fact that in the midst of setting up the system takes in components from each one of the classes cooperation" [7, 8, 9, 10, 11, 12, 13, 14, 15].

Cloud computing is implemented by virtualization [7, 8, 9, 10, 11].

Virtualization is a technique for running different free virtual working frameworks on a solitary physical PC. This methodology augments the arrival on speculation for the PC [6].

Cloud registering is a processing model which deals with a pool of configurable figuring assets. Distributed computing can be ordered as open cloud, private cloud, and half-breed cloud regarding arrangement. While open cloud and private cloud are utilized by open and a solitary association, separately, half-breed cloud is an organization of open and private cloud bases. Subsequently, mixture cloud share the properties of both open cloud and private cloud. Half and half cloud permits organizations keeping their basic applications and information in private while outsourcing others to open [4, 16, 17] (Fig. 1).

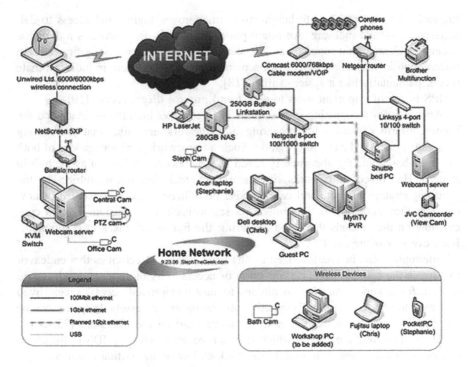

Fig. 1 Cloud computing architecture [45]

2 Introduction to Intruder Detection System

Intrusion discovery is the procedure of watching the occasions happening in a PC framework or organize and investigating them for occurrences which damages related security arrangements or practices. Interruption location methods can be named abuse identification and abnormality recognition. An intrusion discovery framework is a product apparatus used to identify unapproved access to a system. An intrusion detection system is capable of detecting all types of malicious network traffic and computer usage. This incorporates system assaults against defenseless administrations, information driven assaults on applications, host based assaults, for example, benefit acceleration, unapproved logins and access to touchy documents —and malware.

An interruption identification framework is an element detecting intrusions is to consider both the ordinary and the known strange examples for preparing a framework and after that performing grouping on the test information. Such a framework consolidates the upsides of both the mark based and the peculiarity based frameworks and is known as the hybrid system.

Cross breed frameworks can be extremely proficient, subject to the order strategy utilized, and can likewise be utilized to name inconspicuous or new cases as they allocate checking substance that supplements the static watching limits of a firewall—for example, benefit heightening, unapproved logins, and access to delicate records—and malware. An interruption recognition framework is a dynamic observing element that supplements the static checking capacities of a firewall. An interruption identification framework screens activity in a system in indiscriminate mode, particularly like a system sniffer [18].

IPS is one of important tools to detect and prevent illegal access [19].

Approach for recognizing interruptions is to consider both the ordinary and the known bizarre examples for preparing a framework and after that performing characterization on the test information. Such a framework joins the upsides of both the mark-based and the abnormality-based frameworks and is known as the hybrid system. Half-breed frameworks can be exceptionally proficient, subject to the grouping strategy utilized, and can likewise be utilized to mark concealed or new cases as they allot one of the known classes to each test occasion. This is conceivable on the grounds that amid preparing the framework takes in components from every possible one [20, 21].

Interruption can be characterized as any arrangement of activities that endeavor to bargain the respectability, classification or accessibility of an asset. With regard to data frameworks, interruption alludes to any unapproved access, unapproved endeavor to get to or harm, or vindictive utilization of data assets. Intrusion can be categorized into two classes, anomaly intrusions and misuse intrusions [22, 23].

There are different places at which IDS can be deployed [24]. IDS in the cloud can be deployed at the front end, at the back-end or on the virtual machine.

A. Actualizing IDS at the front-end of the cloud will identify assaults on the end client system.
B. Actualizing IDS at the backend of the cloud environment will identify all inner assaults on the cloud and all outside systems which originate from end client system.
C. Actualizing IDS on virtual machine (VM) inside the cloud environment will distinguish assaults on those machines as it were.

3 Previous Work

Some of the existing systems developed IDS in Cloud [25].

Sanchika Gupta et al. [11, 26, 27] recognized vulnerabilities accountable for most likely comprehended framework build strikes in light of cloud and did an essential examination on the endeavors to set up wellbeing available in cloud environment. This paper realized securing attacks from insiders and untouchables using framework profiling. A profile is made for each virtual machine in cloud that portrays framework behavior of each cloud customer. In this, the total framework development at uncommon zone is filtered in light of VM's IP addresses. Intrusion distinguishing proof is performed on the packs starting from a particular machine in light of its profile. VM profile delineates the ambushes that are possible on it. After profile creation, the data got from that virtual machine are hunt down ambushes whose imprints are accessible in VM profile database and match with strike signature database, and if a match happens, it sends information to recognizable proof and notice fragment. This procedure must be actualized totally. It is under Implementation.

Mr. Kumar [28] proposed multi-stage appropriated helplessness acknowledgment estimation, and counter measure decision part which depends on graph-based informative models. He proposed a get-together testing philosophy sent on back end servers. Assailants are expected to dispatch the application administration ask for either at high interim rate or high work stack or even both. By occasionally, checking the normal reaction time to administration demands and contrasting them

Title of the paper	Authors	IDS implemented at	Techniques employed	Merits	Drawbacks
Integrating Signature based Apriori based Network Intrusion in Cloud Computing[28]	Chirag N. Modi	Centralized or Distributed	SNORT to identify known attacks Apriori algorithm to identify unknown attacks	Method is not complex Easy to implement	May not able to identify all types of unknown attacks
A Profile based Network Intrusion Detection and Prevention system for securing Cloud Environment[28]	Sanchika Gupta etal	Virtual Machine	VM Profiling	Profiling which reduces load of server and low cost	Not Implemented
Securing Cloud Network Environment against Intrusion using Sequential algorithm[28]	Mr R.Kumar	Network Switch	Constraint based group testing model	Low detection latency and false detection	Theoretical analysis has been done
DDoS attack protection in the era of cloud computing and SoftwareDefined Networking[5]	B.Wang .et.al	Network Switch	DaMask	Data Shift	Degradation of performance
A Cooperative Intrusion Detection Model for Cloud Computing Networks[11]	Shaaohua Teng et.al.	Front End	E-CARGO model	Detected SYN attacks	Unable to trace ip address of the user

Fig. 2 Summaries of the above techniques

and the particular edge values brought from a true blue profile each virtual server is connected with a negative or positive result by this distinguishing assailant from a pool of authentic clients. Successive calculation has been proposed to hinder the aggressors. It will seek consecutively and obstruct the aggressor hub in transmission way. Hypothetical examination has been done and needs to actualize on DDOS assaults.

Shaaohua Teng et al. [29] proposed a collective interruption location design and E-CARGO model is utilized to model this framework. Collective interruption identification model comprises of five parts to be specific occasion generators, highlight indicator, measurable finder, combination focus, and reaction unit.

The occasion generators gather information from the systems and create suspicious interruption occasions. They present the suspicious interruption occasions to the component and measurable location specialists. As indicated by the system convention, these suspicious interruption occasions are partitioned into TCP occasions, UDP occasions, and ICMP occasions (Fig. 2).

4 Proposed System

While the cloud serves various genuine customers, it may in like manner have threatening customers and organizations, for instance, spam frameworks, Botnets, and malware transport channels. Cloud suppliers must think about those issues and execute the indispensable countermeasures.

Other than that, distributed denial-of-service (DDoS) ambushes can have an a great deal more broad impact on the cloud, since now various organizations may be encouraged in the same machine. Right when an attacker focuses on one organization it may impact various others that have no association with the key target. DDoS is an issue that is still not incredibly all around dealt with. On the other hand, following the cloud gives more foremost flexibility and may administer resources rapidly it ends up being more grounded to foreswearing of organization, yet it goes with a cost to the customers [30, 31].

We proposed an innovative method to implement IDS (DDoS attacks) in Cloud environment. Our idea is to implement IDS at back-end of the cloud which will be able to detect both inside attacks and outside attacks.

DDoS attack involves three parties [32].

Offender: Offender party is the person who plots the assault.
Helpers: Helpers are the machines that are traded off by the guilty party to dispatch assault against a casualty.
Victim: Victims are the ones on who assault has occurred.

Our aim is to minimize the cost overhead in terms of computation and increase detection rate. Our procedure involves ensemble classification techniques. Our frame work is modeled into three components, namely, Capturing and Recording Data, Detection Component, and Output Component.

A. *Capturing and Recording Component*: In this module, we capture network traffic and record traffic. To capture network traffic, we have so many network traffic sniffer tools available. We are using Wire Shark a network analyzer tool to capture packets. Wire shark is a GUI-based tool which runs on Windows, Linux, Solaris, Mac. "Live information can be perused from Ethernet, IEEE 802.11, PPP/HDLC, ATM, Bluetooth, USB, Token Ring, Frame Relay, FDDI, and others (contingent upon your platform)" [33, 34, 35].

Decoding support for some conventions, including IPsec, ISAKMP, Kerberos, SNMPv3, SSL/TLS, WEP, and WPA/WPA2. Shading tenets can be connected to the parcel list for snappy, natural investigation. Output can be sent out to XML, PostScript®, CSV, or plain tex [27, 33, 35, 36].

B. *Detection Component*: After capturing data, we check for any malicious activity in the traffic. There are various parameters to check whether an attack has occurred or not. Some of the filtering mechanisms we used are:

 (i)   Check if too many packets are coming from the same source: This can be known by checking the data rate of a particular user in stipulated period. tshark –G field|grep "datarate"
 (ii) Check the destination traffic by setting a filter. ipaddress = <destination>
 (iii) Check the behavioral of user profile: Like the user may log in regularly at same time from the same time zone. If a user logged in different time zone, then we say abnormal user.
       If found malicious we alert the system, otherwise perform pattern matching using Ensemble Classifiers.
 (iv) We train data using multiple learning algorithms, "Ensemble strategies utilize numerous learning calculations to acquire preferable prescient execution over could be gotten from any of the constituent learning calculations" [37, 38, 39, 40].

Ensemble technique that makes people for its gathering via preparing every classifier on an arbitrary redistribution of the preparation set. Every classifier's preparation set is created by haphazardly drawing, with substitution, N illustrations, where N is the extent of the first preparing set; a considerable lot of the first samples might be rehashed in the subsequent preparing set while others might be forgotten. Every individual classifier in the ensemble is created with an alternate random sampling of the preparation set [39, 41] (Fig. 3).

Algorithm of Bagging Classifier [42].

(1) for m = 1 to M //M ... number of emphases

   (a) Draw (with substitution) a bootstrap test Sm of the information b) Learn a classifier Cm from Sm

(2) for every test case

   (a) Try all classifiers Cm
   (b) Predict the class that gets the most elevated number of votes Training the data by taking different samples and whenever a matched packet has come it will be identified different kind of data.

Fig. 3 Bagging classifier
[46]

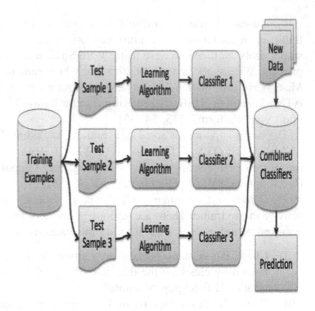

Stacking [42, 43]: In stacked speculation, the combiner f() is another learner and is not confined to being a direct mix as in voting.

The combiner framework ought to figure out how the base learners make mistakes. Stacking is a method for assessing and rectifying for the predispositions of the base learners. In this way, the combiner ought to be prepared on information unused in preparing the base learners. In typical stacking implementation, a number of initially level singular learners are created from the preparation information set by utilizing distinctive learning calculations. The individual learners are then consolidated by a second-level learner which is called as meta learner.

Algorithm:
Input t: Data set $D = f(x(1); y(1)); : : : ; (x(N); y(N))g$
In the first place level learning calculations L1; : ;LT Second-level learning calculation L

Process:

For $t = 1; : ; T$
$ht = Lt (D)$ %Train first-level individual learner ht
End
$do = ;$ %Generate another information set
For $i = 1; : ;N$ For $t = 1; : ; T$ $zit = ht (x(i))$
End
$do = d0 [f((zi1; : ; ziT); y(i))g$
End
$h0 = L(D0)$ %Train the second-level learner h0
Yield: $H(x) = h0(h1(x); : ; hT (x))$

C. Output Component: Using Ensemble Classifiers like Stacking and Bagging for training the data, and tested with new samples of data. Resulting is a classification of classes named normal and abnormal.

5 Results

Initially, we developed a prototype of the above said model using WEKA [44]. We have chosen knowledge flow WEKA Classifier. For Training data, several samples have been considered and cross validation of ten folds were done to check accuracy of training. Network traffic has been captured for a particular period of time and converted data to CSV format that has to fed as input to Classifier. Bagging and stacking algorithms are applied for the data captured and results are analyzed (Figs. 4 and 5).

	Time	Source	Destination	Protocol	Length	Info
1	0	192.168.0.2	198.41.30.198	TCP	54	52087 > 80 [ACK] Seq=1 Ack=1 Win=65340 Len=0
2	0.000098	198.41.30.198	192.168.0.2	TCP	1506	80 > 52087 [ACK] Seq=1 Ack=1 Win=15544 Len=1452
3	0.001481	198.41.30.198	192.168.0.2	TCP	1506	80 > 52087 [ACK] Seq=1453 Ack=1 Win=15544 Len=1452
4	0.001544	192.168.0.2	198.41.30.198	TCP	54	52087 > 80 [ACK] Seq=1 Ack=2905 Win=65340 Len=0
5	0.001706	198.41.30.198	192.168.0.2	TCP	1506	80 > 52087 [ACK] Seq=2905 Ack=1 Win=15544 Len=1452
6	0.001836	198.41.30.198	192.168.0.2	TCP	1506	80 > 52087 [ACK] Seq=4357 Ack=1 Win=15544 Len=1452
7	0.001873	192.168.0.2	198.41.30.198	TCP	54	52087 > 80 [ACK] Seq=1 Ack=5809 Win=65340 Len=0
8	0.001993	198.41.30.198	192.168.0.2	TCP	1506	80 > 52087 [ACK] Seq=5809 Ack=1 Win=15544 Len=1452
9	0.002039	198.41.30.198	192.168.0.2	TCP	1506	80 > 52087 [ACK] Seq=7261 Ack=1 Win=15544 Len=1452
10	0.002096	192.168.0.2	198.41.30.198	TCP	54	52087 > 80 [ACK] Seq=1 Ack=8713 Win=65340 Len=0

Fig. 4 Sample of captured data using wire shark

Name of the classifier	TP Rate	FP Rate	Precision	Recall	F-Measure
Bagging	0.591	0.485	0.362	0.591	0.362
Stacking	0.548	0.548	0.301	0.548	0.485

Fig. 5 After deploying in WEKA tool result obtained [47, 48]

6 Conclusion and Future Work

Blue print has done for the method we have discussed to mitigate DDoS attacks in cloud environment. We have taken classifiers to do pattern matching are bagging and stacking. As per our results, bagging has done well compared to stacking. Detection rate is higher for bagging than stacking. Meta Classifiers we have chosen for stacking are Id3 and J48. In the future, we are going to implement in a cloud simulator cloudsim for further improvement and analysis. This is in progress.

References

1. http://dic.academic.ru/dic.nsf/ruwiki/1070341.
2. http://www.south.cattelecom.com/rtso/.../CloudComputing/0071626948_chap01.pdf.
3. Buyya, R., Broberg, J., Goscinski, A.M.: Cloud computing: Principles and paradigms, vol. 87. John Wiley & Sons (2010).
4. DistributedComputing, https://en.wikipedia.org/wiki/Distributed_computing.
5. George Sadowsky James X. Dempsey Alan Greenberg Barbara J. Mack Alan Schwartz.: Information Technology Security HandBook Infodev (2003).
6. John W. Rittinghouse., James F. Ransome: Cloud computing Implementation, Management, and Security. CRC press.
7. Cloud Computing: http://csrc.nist.gov/groups/SNS/cloud-computing/cloud-def-v14.doc.
8. Farzad Sabahi 978-1-61284-486-2/111$26.00 IEEE (2011).
9. National Institute of Standards and Technology, http://www.nist.gov/itl/cloud.
10. CloudComputing, www.ibm.com/developerworks/community/blogs/cloudcomputing/entry/nist_s_definition_of_cloud_computing_what_is_cloud_computing.
11. www.slideshare.net/avazhamiva/cloudcomputing.
12. Cloud: https://en.wikipedia.org/wiki/private_cloud.
13. Cloud Computing.: http://adsmediat.com/overview-of-cloud-computing/.
14. Cloud Computing: http://newalboyonads.xyz/2016/01/13/cloud-computing/.
15. Cloud Computing: https://en.wikipedia.org/wiki/Distributed_computing.
16. Anthony T. Velete., Toby J. Velte., Robert Elsenpeter: cloud computing A practical approach by. Tata McGraw-Hill Edition.
17. Bing Wang, Yao Zheng, Wenjing Lou, Y. Thomas Hou.: DDoS attack protection in the era of cloud computing and SoftwareDefinedNetworking: IEEE 22nd International Conference on Network Protocols (2014).
18. A. Patcha., J-M. Park: An Overview of Anomaly Detection Techniques: Existing Solutions and Latest Technological Trends, Computer Networks, doi:10.1016/j.comnet.2007.02.001 (2007).
19. Hai Jin et al.: A VMM-based intrusion prevention system in cloud computing environment © Springer Science + Business Media, LLC J Supercomput DOI 10.1007/s11227-011-0608-2 (2011).
20. Kapil Kumar Gupta.: L "Layered Approach Using Conditional Random Fields for Intrusion Detection" IEEE Transactions on Dependable and Secure Computing Vol 7, No 1, Jan–March (2010).
21. http://www.slideshare.net/ieeexploreprojects/4-layered-approach-using-conditional-random-fields-for-intrusion-detection.
22. Srinivas Mukkamala.: Intrusion Detection using Nueral Networks and Support Vector Machines 0-7803-7278-6/02/$10.00 IEEE(2002).

23. Asmaa shaker., Ashroor: International conference on Future Information Technology IPCSIT vol.13 (2011) IACSIT Press, Singapore (2011).
24. Baraka, Hifaa Bait., Huaglory Tianfield": Intrusion Detection system for cloud environment" Proceedings of the 7th International Conference on Security of Information and Networks. SIN 14 (2014).
25. B. Kiranmai., A. Damodaram: Defending DDOS Attacks in Cloud Computing Environment A SurveyInternational Journal of Application or Innovation in Engineering & Management (IJAIEM) Volume 4, Issue 11, November 2015 ISSN 2319-4847 (2015).
26. Sanchika Gupta et al.:Copyright published by Elsevier. This is an open access article distributed under the Creative Commons Attribution License (2011).
27. http://www.hindawi.com/journals/ijdsn/2013/364575/t.
28. Mr R. Kumar: International journal of Engineering and General Science Volume 3, Issue 1, January- February,2015.
29. Shaohua Teng et al.: International Journal of Security and its applications vol. 8 N0.3, pp 107–118 http://dx.doi.org/10.14257/ijsia.2014.8.3.12(2014).
30. http://www.researchgate.net/profileCarlos_Westphall/publication/275463281_ SECURWARE_2014 The_Eighth_International_Conference_on_Emerging_Security_ Information_Systems_and_Technologies/links/553d04b70cf245bdd7696e3d.pdf#page = 48.
31. Botnets.: https://en.wikipedia.org/wiki/Botnet.
32. www.cisco.com/us/products/ps1634/products_Configuration_guide/ chapter0918690080849898.html.
33. WireShark: www.wireshark.org.
34. Wireshark: http://www.filehippo.com down load wireshark_32/11401/.
35. Wireshark: https://openrepos.net/content/nieldk/wireshark/.
36. Wireshark: http://www.darknet.org.uk/2008/04/wireshark-v100-released-cross-platform-graphical-packet-sniffer/.
37. Ensemble Learning: https://en.wikipedia.org/wiki/Ensemble_learning.
38. Ensemble Learning: http://stackoverflow.com/tags/ensemble-learning/info/.
39. GulshanKumar et al.: "The Use of Artificial-Intelligence-Based Ensembles for Intrusion Detection: A Review" Applied Computational Intelligence and Soft Computing Volume 2012 (2012), ArticleID 850160, 20 pages http://dx.doi.org/10.1155/2012/850160.
40. Ensemble Methods: www.ensemble.com/.
41. Ensemble Learning: https://www.cs.cmu.edu/afs/cs/project/jair/pub/volume11/opitz99a-html/node3.html.
42. Lecture 6: Ensemble Methods Marina Santini, Uppsala University Department of Linguistics and Philology.
43. Hong Chang.: Ensemble learning Institute of Computing technology Chineese Academy of sciences Machine Learning Methods (Fall 2012).
44. Weka tutorial.: www.cs.waikato.ac.nz/ml/weka.
45. Security in Cloud Computing, http://www.meanboyfriend.com/overdue_ideas/2008/10/send-in-the-clouds/.
46. Bagging: https://www.google.co.in/imgres?imgurl=http://csewiki.unl.edu/wiki/images/7/72/ Screen_shot_2010-1203_at_5.46.21_PM.png&imgrefurl=http://csewiki.unl.edu/wiki/index. php/Bagging_and_Boosting&h=389&w=719&tbnid=AR8WMF2_O4B1M:&docid=i0gVO8ofVs4 DM&ei=HdRiVqPCH5KeugS88oiIBA&tbm=isch&ved=0ahUKEwjjxP2I2MTJAhUSj44KHTw5 AkEQMwgiKAYwBg.
47. Jiawei Han., Micheline Kamber: Data Mining Concepts and Techniques Second Edition Morgan Kauffman Publishers (2006).
48. Evaluation Measures: http://iasri.res.in/ebook/win_school_aa/notes/Evaluation_Measures.pdf.
49. R. Jeena: International Journal of Innovative Research in Computer and Communication Engineering Vol.3,Issue 2, February 2015.

A Novel Method for QoS-Based Spectrum Allocation in Cognitive Radio Systems

Aswathi R. Nair and E. Govindaraj

Abstract In this paper, the spectrum holes in licensed frequency bands are utilized for satisfying quality-of-service (QoS) demands of cognitive users. The available spectrum holes are classified as white, gray, and black. The bands are then fairly allocated to the cognitive users based on their QoS requirements. The data traffic from cognitive users are classified as high, medium, and low QoS required flows. Then, the white, gray, and black spaces are, respectively, assigned to them. Thus, spectrum utilization is improved, and fairness is achieved, since more no of secondary users are accommodated, but QoS is not compromised.

Keywords Spectrum holes · Cognitive users · White space · Gray space · Black space · Quality of service · Spectrum utilization

1 Introduction

1.1 Background

Cognitive radio systems are a hot topic among researchers nowadays. The cognitive radio addresses the problem of spectrum scarcity, by utilizing the licensed spectrum bands whenever and wherever they are unoccupied by licensed users. The licensed users are called primary users, the spectrum opportunity is called a spectrum hole, and the users utilizing the spectrum other than primary users are called secondary or cognitive users. The cognitive users are intelligent devices that can coexist with the primary users in a network system and can utilize the spectrum without causing any harmful interference to the primary users. Thus, cognitive radios efficiently utilize the precious radio spectrum.

A.R. Nair (✉) · E. Govindaraj
Department of Computer Science and Engineering, MES College of Engineering,
Kuttipuram, Malappuram, Kerala, India
e-mail: aswathinair10@gmail.com

© Springer Science+Business Media Singapore 2017
S.C. Satapathy et al. (eds.), *Proceedings of the First International Conference on Computational Intelligence and Informatics*, Advances in Intelligent Systems and Computing 507, DOI 10.1007/978-981-10-2471-9_29

301

Spectrum holes are the part spectrum currently unoccupied by primary users that can be used by cognitive users to transmit data. The cognitive radios first find the spectrum holes to transmit in, and this step is called spectrum sensing. Spectrum sensing is an important step in the working of cognitive radio, since inaccurate spectrum sensing can destroy whole working principle of cognitive radio networks itself. There are many spectrum sensing method namely [1] energy detection, feature detection, matched filter detection, etc. to name a few. The sensed spectrum holes are then used by the cognitive radios to transmit data. Studies [2] have shown that certain portions of spectrum are largely occupied, some are moderately occupied, and remaining are largely unoccupied, they are called [3] black, gray, and white spaces respectively.

1.2 Related Works

Many attempts have been carried out to efficiently allocate available spectrum opportunities to the secondary users. QoS provisioning for heterogeneous services has been proposed in [4], and real time and non-real time secondary users in distributed cooperative cognitive radios are considered as potential users. In [5], an automatic channel selection scheme which coexists with Wi-Fi network is proposed. The primary users are considered to operate in ISM band, and on the basis of QoS requirement, the channels are selected and allocated available frequency in ISM band. The authors in [6] "QoS-based spectrum decision framework for cognitive radio networks" provide the QoS to the data packets by allocating the channel which can satisfy the QoS requirement of the user. In all related works, either it is assumed that the secondary users' traffic class is marked in packets or it is explicitly given. Some works also proposed traffic classification by complex learning methods. However, in this paper, simple classification based on power of secondary users is considered.

1.3 Purpose of This Paper

The purpose of this method is to allocate channels to secondary users exactly according to their QoS requirements. The QoS requirement is measured in SINR (signal to interference and noise ratio). Since the channel is allocated exactly to requirement, the channel utilization is improved and more number of secondary users get to transmit data.

1.4 Organization of This Paper

This paper is organized as follows. Section 2 discusses the interference temperature model, how the interference temperature is calculated and how it is used by the secondary users to access spectrum. Section 3 gives the working of the proposed system in detail. Section 4 describes the experimental setup in Matlab and the results are studied. Section 5 concludes the paper gives possible future directions.

2 Interference Temperature Model

Interference temperature [7] is a metric to measure and manage interference in an RF environment. It can also be defined as the measure of RF power at the receiver. That means measure of power from other sources and noise. Let us assume as in [7] a signal $s(t)$ is transmitted by a primary user and is corrupted by additive white Gaussian noise $n(t)$, the received signal is represented by $y(t)$. Here, we are considering discrete signals.

$$y(\mathbf{t}) = \mathbf{s}(\mathbf{t}) + \mathbf{n}(\mathbf{t}) \tag{1}$$

Suppose in a sampling period T, N number of samples is taken then, the power of received signal P_R is calculated by:

$$P_R = \frac{1}{N} \sum_{i=1}^{N} |y_i^2|. \tag{2}$$

The corresponding interference temperature T_I is given by:

$$T_I = \frac{P_R}{K \cdot B}, \tag{3}$$

where K is Boltzmann constant, $K = 1.38 \times 10^{-23}$ J/°K, and B is bandwidth of the channel.

For the secondary user to transmit without causing interference to the primary user, secondary user should adjust the power according to the formula as follows:

$$T_I + \frac{P_S}{K \cdot B} \leq T_L, \tag{4}$$

where T_L is the interference temperature limit, for a band in a given geographical area. P_s is the power at which the secondary user should adjust transmission inorder to avoid interference with primary users.

3   Proposed System

The proposed cognitive radio system works in a WLAN and operated in ISM band (902–928 MHz) as in [4]. The system assumes N users are subscribed to the WLAN access point, and they are the primary users. Channel_id uniquely identify each primary user channel. The base station senses and reallocates the spectrum every T_s seconds. Therefore, if channel state changes in this time, the secondary user has to switch the channels. There are M cognitive radios (secondary users) managed by the cognitive base station. Cognitive users are having varying QoS demands. Using interference temperature model, the channels are classified. White spaces are assumed to be unoccupied by any primary users, gray spaces are assumed to be occupied but still can satisfy needs of a moderate user, and black spaces are the rest. The traffic from cognitive users is also classified as high, medium, and low QoS services. Then, the white spaces are allocated to high QoS users, gray to medium, and black to low QoS users. The QoS is directly proportional to the power of secondary users. The basic working of cognitive base station is summarized in Fig. 1.

Fig. 1 Basic working of QoS-based channel allocation system

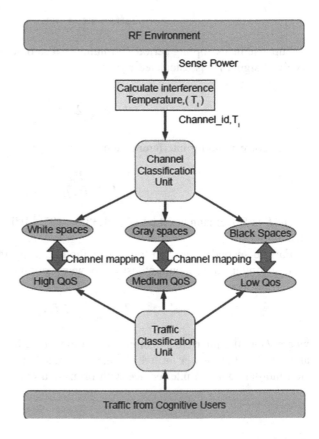

3.1 Interference Temperature-Based Channel Classification

As stated earlier, the interference temperature can measure the power over a signal due to other source. From the view of a secondary user, a primary user signal is an interference, and this can be quantified using interference temperature metric. Therefore, a larger interference temperature value means that the channel is much occupied by some other source, in our case a primary user and of-course some noise. Let us assume the interference temperature due to noise be T_N. And if the interference temperature is equal to interference temperature limit, then there is no more opportunity to transmit in that spectrum.

The classification unit (Fig. 1) works according an algorithm. The algorithm checks whether the current signal received from a primary user has interference temperature between zero and T_N, if so the channel is currently not occupied by any primary user. If the interference temperature lies between T_N and T_M, then the channel is moderately occupied T_M that will vary according to the QoS of the moderate users.

Algorithm 1: Channel Classification Algorithm

```
Channel Classification (Channel_id,T_I)
begin

    For each channel i,
    repeat
        Calculate interference temperature T_I
        If(0 <= T_I <= T_N)
            Then, channel is a White space.
        Else If(T_N < T_I <= T_M)
            Then, channel is a Gray space.
        Else (T_M < T_I <= T_L)
            Then, channel is a Black space.
End.
```

3.2 Channel Allocation

The traffic classifier unit (Fig. 1) measures the power from each secondary user. The base station will have information about the range of power of the secondary users. Then, the base station classifies the traffic according to their QoS requirement as high, medium, and low priority flows. The classified channels are then allocated to the classes of QoS.

The power at which a secondary user is defined by Eq. 4. Thus, at a given time, the power of transmission by a secondary user must lie between interference temperature of the channel I_T and interference temperature limit T_L.

The spectrum utilization can be measured by measuring occupied spectrum in a given time based on the interference temperature, if any spectrum has interference temperature close to interference temperature limit, then the channel can be said to be utilized.

4   Experiments and Results

The experiment is set up in matlab2011. There are 5 primary user channels are there and 10 secondary users. The primary users and secondary will be present or absent in poisson distributed manner, and their powers are uniformly distributed between 25 and 3 dBm. The QoS requirements and the power range of each class are summarized in Table 1. The power values are referred from [8].

The allocations are made, such that the QoS requirements are satisfied and the interference temperature is not exceeded. The white spaces can be allocated to a high and a low QoS user and still can achieve the QoS.

In general, interference temperature model the spectrum allocation is done if primary users' interference temperature is such that a secondary user can still share the channel and limit is not exceeded. The QoS of the secondary is not considered.

When the proposed method is compared with the general interference method, the spectrum utilization and number of secondary users accommodated is found improved. The channel utilization ratio is the measure of channel utilized by all secondary users to the available channel after being used by primary users. The number of secondary users is counted for different number of available primary user channels.

Table 1   SINR and Power range for different QoS classes in the experiment

QoS class	Power range (dBm)	SINR (dB)
High QoS	$18 < Ps \leq 25$	40
Medium QoS	$8 < Ps \leq 18$	25
Low QoS	$2 < Ps \leq 8$	10

5 Conclusion and Future Work

The proposed system addresses the QoS requirement of secondary users. The different QoS secondary users are assigned to different channels that can satisfy their requirement. The channel utilization and total secondary users who can share spectrum are also improved by this method. Even the heavily occupied bands are utilized whenever possible. The users can experience quality in the services they are using, since they are allocated channels according to the service requirements.

In the future, a ranking method can be used to rank the channels according to their potential and channel switching to be reduced to reduce the delay experienced by secondary users Figs. 2 and 3.

Fig. 2 Channel utilization for general interference model and QoS-based model

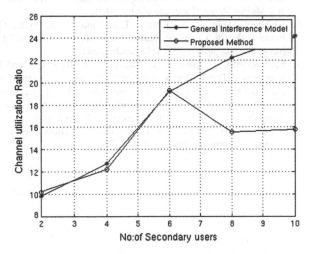

Fig. 3 Number of secondary users accommodated in general interference model and proposed model

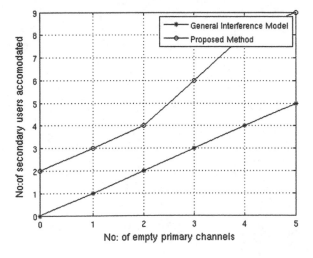

References

1. Weifang, Wang.: Spectrum Sensing for Cognitive Radio. In: Third International Symposium on Intelligent Information Technology Application Workshops, IEEE, Nanchang, 21–22 November, 2009.
2. S, Haykin.: Cognitive Radio: Brain-Empowered Wireless Communications. In: IEEE Journal on Selected Areas in Communications, Vol. 23, NO. 2, pp. 201–220, Febraury, 2005.
3. Mohammed, M. Saleh. and Hemalatha, Rallapall.: Quick Detection and Assignment of Spectrum Hole in Cognitive Radio. In: International conference on Intelligent Systems, Data Mining and Information Technology, Bangkok (Thailand), 21–22 April, 2014.
4. Ammar, Alshamrani., Xuemin, Shen. And Liang-Liang, Xie.: QoS Provisioning for Heterogenous Services in Cooperative Cognitive Radio Networks. In: IEEE Journal on selected Areas in Communications, Vol. 29, No. 4, April, 2011.
5. Mustapha, Bennai., John, Sydor., and Mahmudur, Rahman.: Automatic Channel Selection for Cognitive Radio Systems. In: 21st Annual IEEE International Symposium on Personal, Indoor and Mobile Radio Communications, 2010.
6. Vishram, Mishra., Lau, Chiew. Tong. and Chan, Syin.: QoS based Spectrum Decision Framework for Cognitive Radio Networks. In: 18th IEEE International Conference on Networks, Singapore, 12–14, December, 2012.
7. F, Benedetto., G, Giunta1., E, Guzzon1., M, Renfors2. and M, Arcangeli1.: Improving the Interference Temperature Estimation for Dynamic Spectrum Access in Cognitive Radios. In: IEEE Global Conference on Signal and Information Processing, Austin, 3–5, December, 2013.
8. E, J, Rivera-Lara., R, Herrerías-Hernández., J, A, Pérez-Díaz and C, F, García-Hernández.: Analysis of the relationship between QoS and SNR for an 802.11g WLAN. In: IEEE International Conference on Communication Theory, Reliability, and Quality of Service, Bucharest, Romania, June 29–July 5, 2008.

A Review on fMRI Signal Analysis and Brain Mapping Methodologies

S.V. Raut and D.M. Yadav

Abstract Functional magnetic resonance imaging (fMRI) is a safe non-invasive technique used for understanding the brain functions against various stimuli and hence to predict the brain disorders. The fMRI signal patterns and brain mapping have been found promising the medical science in the recent days. Adequate contributions have been made in the literature on fMRI signal analysis. This paper identifies notable research works that have contributed on fMRI signal analysis and predicting the brain functions and performs systematic review on them. The review provides the strengths and weaknesses, and the research gaps exist in the works based on 11 useful parameters, such as sparsity, non-linearity, robustness, etc.

Keywords *fMRI* · Brain region · Voxel · Brain mapping · Classification · Precision · Sparsity

1 Introduction

Functional magnetic resonance imaging (fMRI) refers to a non-invasive imaging technique [1, 2], which has evolved as a significant tool for detecting the brain regions that are engaged in the cognitive processes [3]. A typical fMRI data set is constituted of blood oxygenation level-dependent (BOLD) signals of tens of thousands voxels that are time series in nature [4]. It has been successfully applied for more than 15 years in neuroscientific and cognitive studies [5], because of its ability to offer sufficient spatial and temporal resolutions to measure the location, amplitude, and timing of brain activity [3]. The fMRI signal analysis intends to determine the brain regions that are activated during the course of a cognitive task

S.V. Raut (✉)
Rajarshi Shahu College of Engineering, Pune, Maharashtra, India
e-mail: boratesavita2006@gmail.com

D.M. Yadav
JSPM Narhe Technical Campus, Pune, Maharashtra, India
e-mail: dineshyadav_8@yahoo.com

© Springer Science+Business Media Singapore 2017
S.C. Satapathy et al. (eds.), *Proceedings of the First International Conference on Computational Intelligence and Informatics*, Advances in Intelligent Systems and Computing 507, DOI 10.1007/978-981-10-2471-9_30

[3]. It can be differentiated into two ways, namely, signal detection and charac-terization of hemodynamic response function. The former deals with locating brain region that produces high neural activity against a stimulus [6]. The latter deals with interpreting the changes in the blood flow upon stimulus [7].

The research attention on fMRI signal analysis falls on acquisition methodolo-gies [8], preprocessing techniques [9], enhancement methods [10, 11], denoising algorithms [12, 13], hemodynamics estimation [14, 15]. and post-processing methodologies [16, 17]. All such methodologies can be broadly categorized into approaches, namely, hypothesis driven approaches and data-driven approaches [3].

The traditional hypothesis driven approaches include Markov random field (MRF) models [18, 19] and hidden Markov models (HMM) [20]. However, these models are complex and require more information about the stimulus [3]. In con-trast, the statistical parametric mapping (SPM) is a simple hypothesis-driven approach that operates based on the general linear model (GLM) [21]. The refer-ence functions are generated by considering both the hemodynamic response function (HRF) and the deterministic stimulus timing function. These assumptions and the adopted deterministic timing function do not correlate with both the spatial and the temporal patterns of brain activation function [22]. Moreover, the analysis is highly univariate [22] rather than the desired multivariate analysis [2].

On the other hand, the data-driven approaches entertain multivariate analysis for fMRI signals. They include both the supervised [23, 24] and the unsupervised methods [25–28]. In contrast to the hypothesis-driven approaches, data-driven approaches make few or no assumptions about the HRF shape and do not require apriori knowledge about the stimulus timings [3]. However, the unsupervised methods find a few drawbacks over the supervised methods. A few of them include poor characterization of noisy fMRI data set [29] and biased decomposition [30]. This necessitates the need for supervised data-driven approaches to carry out the fMRI signal analysis.

A. *Background*

The fMRI is a safe non-invasive technique to measure and map the activity of the brain. It differs from MRI in measuring the brain activity rather than structuring the brain. The fMRIs are signals observed from the brain, when there is an external stimulus. It measures the variation of the brain electrical pulses, when the stimulus is observed. For instance, an fMRI produces the variation between the brain activity under normal and excited conditions. In other words, the electrical pulse variations are observed when a person goes to excited stated from normal conditions. Brain mapping is the primary application where fMRI signals play crucial role.

2 Notable Research Works

The fMRI signal-based brain mapping is vital for understanding the characteristics of brain and for diagnosis purpose. A handful of research works has been collected, and the survey has been done. The concise context behind the survey is discussed below.

Yu anqing Li et al. [4] have detected voxels from fMRI data with task relevance using their iterative sparse representation based algorithm. At every iteration of the algorithm, a sparse weight is generated using the context of linear programming. The weight vectors that have been obtained throughout the iterations are averaged to produce the final weight vector. The obtained final weight vector has peculiar characteristics, such as sparsity, stability, and the ability to represent the significance of the feature through its magnitude. The experimental results have shown that the tasks decoded by their algorithm from fMRI data are more relevant than univariate general-linear-model-based statistical parametric mapping.

Vincent Michel et al. [5] have also insisted the necessity of extracting relevant information for predictive diagnosis. The relevant information includes the organizing principles that are being followed by the brain. However, the conventional mass-univariate procedure has not considered this information. Despite the fact that the multivariate pattern analysis considers this information, if the learning method is not regularized, curse of dimensionality often occurs. The conventional regularization methods have not considered spatial structure of images too. This may result in poor representation of the extracted features. Total variation (TV), which represents the l_1—norm of the image gradient, has been identified by the authors as a prominent solution to interpret the extracted information in a more efficient manner. They have applied it for brain mapping and accomplished good compatibility with it.

Jean Honorio et al. [31] have attempted to predict diverse disorders (cocaine use, schizophrenia and Alzheimer's disease) in unseen subjects from fMRI data. They have shown that the voxels have been scattered and are unstable for group classification. Hence, they have opted for using single region per experimental condition and a majority vote classifier. The experimental results have demonstrated the performance of their method over the state-of-the-art methods. They have also asserted that their method has provided meaningful representation for voxels to retain discriminability.

Since SOMs transform high-dimensional data into low-dimensional map using the unsupervised learning method, they require a post-processing method to describe the similarity between the feature vectors as well as the clusters and the interested features. Hence, Katwal et al. [3] have proposed graph-based visualization of self-organizing maps (SOMs) to acquire fMRI features based on density-based connectivity, correlations and correlation-based connectivity. They have applied the features to determine the brain regions associated with the tasks and experimentally proved the classifying performance of visualization of SOMs over independent component analysis and voxel wise univariate linear regression analysis.

Kurt Barb et al. [1] have addressed the challenges of extracting hemodynamic response that deals with the quantification and the interpretation of the physiological processes of stimulus—activated brain regions. They have introduced fractional models to understand and quantify hemodynamics from the fMRI measurements. The models have been experimentally proven for its computational efficiency, flexibility, and robustness against noise.

Lotfi Chaari et al. [32] have attempted to perform both the two steps of within—subject analyses, namely, brain activity detection and estimation of the hemodynamic response. They have introduced the region-based joint detection–estimation (JDE) framework to handle these steps using multivariate inference for detection and estimation. The JDE has been built using regional bilinear generative model of the BOLD response and by introducing spatial, temporal and physiological priors as constraints in Markovian model. The constrained model has been used for parameter estimation.

The JDE has been reformulated in a missing data framework to obtain variational expectation-maximization (VEM) algorithm. The Markovian model has been approximated using a variational approximation in unsupervised spatially adaptive JDE inference. This has allowed automatic fine-tuning of spatial regularization parameters. The derived methodology has been experimentally proved for its computational efficiency and precise estimation.

Fan Deng et al. [33] have developed a data-driven multiscale signal decomposition framework, called as, empirical mean curve decomposition (EMCD). EMCD has been intended to optimize the mean envelopes from fMRI signals and it has iteratively extracted coarser-to-finer scale signal components. The EMCD framework has been experimentally proved on resting-state fMRI, task-based fMRI and natural stimulus fMRI to infer meaningful low-frequency information from blood oxygenation level-dependent (BOLD) signals.

Babak Afshin-Pour et al. [34] have introduced a flexible model to optimize the reliability of spatial pattern by maintaining a tradeoff between prediction accuracy and reproducibility of linear predictive models. This has further improved the detection of task—positive brain network as well as temporal variable and spatially reproducible networks. The hybrid model has been constructed based on the weighted sum of optimization functions of linear discriminant analysis (LDA) model and a generalized canonical correlation (gCCA) model. The linear discriminant term has preserved the capability of the model to differentiate the fMRI scans containing multiple brain states. On the other hand, gCCA has discovered a linear combination for the scans of all the subjects in a way that the estimated boundary map can be reproducible across subjects. Experimental investigation has been carried out in both the real and the simulated fMRI data and comparison has been made against LDA and Gaussian Naive Bayes (GNB) techniques. Their hybrid model has outperformed the other two techniques.

Chun-An Chou et al. [35] have proposed a feature selection framework based on mutual information (MI) and partial least square regression (PLS) to derive an informativeness index. The informativeness index has prioritized the voxel selection process on the basis of degree of relevance with the practical scenario. The features that have been extracted through their framework were used in standard classification algorithms and experimented on benchmark fMRI data set. The computational results have revealed that their feature selection framework has substantially improved the classification performance than the conventional features.

3 Review Outcomes

A. Summary

Numerous parameters define the quality of the research works. This paper considers 11 parameters that play key role to define the nature of the literary works. Though few works have been considered for review, we have illustrated the deviation among the works based on these parameters. The percentage of contribution of the works that are considered for review is shown in Fig. 1, which is derived from the review summary given in Table 1.

B. Parametric Study

(1) *Impact of sparsity*: The sparse representation leads to produce a combination of voxels, which is able to represent stimulus function at high efficiency. Hence, a positive impact of sparse representation on precise combination of pixels is ensured [4]. The sparse representation given in [4] has believed to be reason for high informativeness. In [5], the data representation has also been considered with sparsity, and hence, the regularization has been done. The EMCD representation given for fMRI signal has also become sparsely distributed [33]. However, most of the works do not consider the sparse representation to define or extract voxels [1, 3, 31, 32, 34, 35].

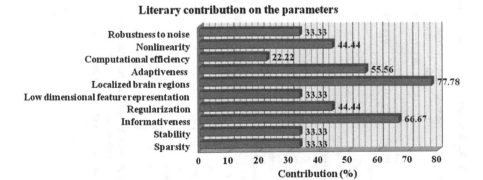

Fig. 1 Literary contribution based on the selected parameters

Table 1 Review summary based on parameters

Methodology	Sparse representation for voxel selection	Total variation regularization	Predicting disorder using single brain region	Graph-based visualization of SOMs	Fractional-order time series models	JDE-VEM framework	EMCD fMRI signal analysis	Hybrd LDA + gCCA model	MI and PLS based voxel selection framework
Year [Citation]	2009 [4]	2011 [5]	2012 [31]	2013 [3]	2012 [1]	2013 [32]	2013 [33]	2014 [34]	2014 [35]
Sparsity	✓	✓	–	–	–	–	✓	–	–
Stability	✓	–	✓	–	–	–	–	–	✓
Informativeness	High	High	–	High	–	–	High	High	High
Regularization	–	✓	✓	–	✓	✓	–	–	–
Low dimensional feature representation	–	–	✓	–	✓	–	–	✓	–
Localized brain regions	Yes	–	Yes	Yes	Yes	Yes	Yes	Yes	–
Adaptiveness	–	✓	✓	–	–	✓	✓	✓	–
Precision[1]	>50 %	>90 %	>95 %	2.5[2]	–	–	> 0.7[3]	>80 %	>75 %
Computational efficiency	–	–	–	–	–	Good	–	Good	–
Nonlinearity	–	Yes	Yes	–	–	–	Yes	Yes	–
Robustness to noise	–	–	–	–	Yes	Yes	–	Yes	–

[1]Given precision has not been considered under a common platform and hence exclusive comparison is not feasible
[2]measured with t-static
[3]Measured in terms of Pearson correlation

(2) *Algorithm stability*: Algorithm stability stands for the insensitivity of the results on varying algorithm parameters or any miscellaneous selection parameters. The voxel selection framework [4] is experimentally proved that it is insensitive to multiple voxels. Certain works such as [31, 35] have ensure the stability of the algorithm even diverse datasets have been used under a single brain region. However, majority of the works [1, 3, 5, 32–34] have not given significance for the algorithm stability.

(3) *Informativeness*: Informativeness refers to the information held by any signal representation. According to [4], sparse representation leads to high informative voxel combinations. Since the feature representation is given in high-dimensional space, the informativeness behind the extracted features is naturally high [3, 5]. The EMCD representation for the fMRI has also been proved for its high informativeness [33].

Generally, low-dimensional representation can affect the richness of the information and so the uncertainty persists in the informativeness of the voxels [1, 31, 35]. Despite low dimensionality representation has been given in [34], informativeness has also been ensured in their work. A few of the works have not discussed about the feature [32], whereas a few works are assumed to have informativeness due to high dimensionality representation [35].

(4) *Impact of regularization*: Regularizing the voxel representation for predicting the fMRI patterns has not been considered in [3, 4]. Regularization has been considered as the keen interest to handle the high-dimensional data representation [1, 5]. The generalization problem has been focused well in [31] to learn the single brain region for predicting the disorder under a regularized environment. The JDE inference adopted in [32] has offered automatic fine-tuning of spatial regularization parameters. In [33–35], proper regularization has not been discussed though adequate contribution has been given on signal representation.

(5) *Representing the features in low dimensional space:* Representing in low-dimensional space is a much needed process, so that the processing time as well as the complexity can be reduced. However, research works such as [3, 4] have concentrated on considering maximum voxel combinations, but not on representing them in low-dimensional space. In contrast, works such as [1, 31] have shown adequate concentration on representing the voxels extracted from single brain region in low-dimensional space.

The review states that when one part of the researchers have not concentrated on low-dimensional representation [5, 32, 33, 35], other part of researchers show their interest on adequate dimensionality reduction processes [34].

(6) *Localizing significant brain regions*: Localizing brain regions have been considered in the literature as a significant process for brain mapping, and

hence, hemodynamic response can be recorded [32, 33]. There is a strong correlation between voxels and brain region localization.

In [1, 3, 4], voxels have been selected in such a way that appropriate brain regions have been localized. On the other hand, localizing single brain region is asserted to extract the voxels which are helpful to predict the brain disorder [31]. Despite the advantages persist, a few of the literary works have not considered the significance of localizing the brain regions [5, 35].

(7) *Adaptiveness*: Adaptiveness refers to the ability to perform even there is an impact of external parameters. The voxel selection framework [3, 4] has not considered the impact of external parameters [1, 33, 35].

Generally, a presumption exists that the predictor in [5] is adaptive to environmental effects, since regularizing the fMRI patterns has been considered as a prominent task. Hence, the probability of being adaptive for multiple scenarios is high, when there is an impact of generalization [31, 32]. In contrast, a few works have demonstrated their adaptiveness under no regularization [34].

(8) *Precision*: The works that have been considered for review have undergone diverse experimental procedures and test benches. Hence, it is not advisable to present an exclusive review among them based on the precision published in those works. However, we discuss about the accomplishments in terms of performance metrics. While predicting three cases subjected to four different stimuli, a mean precision of about 50 % (approximated) has been achieved [4].

Experimenting in a different set of real and synthetic data, an average prediction accuracy of about 85 % has been recorded in [5]. In another investigation, 95 % accuracy has been recorded on predicting the brain disorders [31], while 80 % in [34] and 75 % in [35]. A few works have reported precision in terms of t-statistics of about > 2.5 [3] and Pearson coefficient [33]. A few works have not considered any exclusive precision calculation [32], whereas signal intensity has been analyzed rather than precision [1]. The summary of review based on the precision is illustrated in Fig. 2.

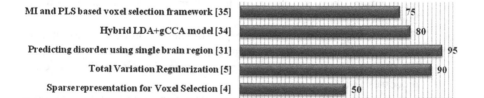

Fig. 2 Approximate precision accomplished by various methodologies

(9) *Computational efficiency*: A very few works have discussed the performance in terms of computational efficiency too [32, 34], whereas many of the methods have not been analyzed in this perspective [1, 3–5, 31, 33, 35].

(10) *Nonlinearity*: It is well known that the fMRI patterns are nonlinear. However, certain research frameworks have not shown its significance on the data nonlinearity [1, 3, 4, 32, 35]. In contrast, many of the works have considered the data nonlinearities [5]. As a solution, regularization has been adopted [31] as well as proficient signal representing methodology [33, 34].

(11) *Robustness to noise*: The fMRI patterns acquired from machines can be easily contaminated by noise. Hence, many of the researchers have considered the impact of noise in such a way that their methodologies remain robust against noise [1, 32, 34]. Though synthetic data have been used in [5], it has been simulated under a noisy environment, and hence, the methodology has been experimented. However, a few works have not discussed about the robustness of the methodologies under a noisy environment [31, 33, 35]. For instance, the voxel selection mechanism remains to be experimented under noisy environment [3, 4].

4 Research Gaps

In fMRI signal analysis, the selection of relevant voxels (or features) remains as a challenging task. The process is highly significant, because it directly influences the fMRI prediction. Moreover, the increased number of features often results in overfitting problem. This further increases the need for a voxel selection framework.

Analysis of variance (ANOVA) is the most widely used technique for voxel selection process. Alternatively, the regularization methods play crucial role for formulating the generalized learning model. The regularization intends to provide stable and sparse weight estimation for the voxels. Moreover, the selection of voxels should have high degree of relevance with the BOLD signals. Multiple voxel interactions and the extracted spatial patterns are incorporated by the multivariate classification methods.

Support vector machines (SVM), neural networks (NNs), and classification trees (CT) are few renowned classification algorithms adopted for fMRI multivariate pattern analysis. Literature supports to use these algorithms along with the dominant voxel selection framework [35]. However, non-linearities and signal compositions that exist in BOLD makes the voxel selection process challenging. This necessitates the signal decomposition methodology to be adopted in the multivariate signal analysis [33, 36, 37].

5 Conclusion

This paper has identified notable research works done on fMRI signal analysis and brain disorder prediction on which the review has been done. 11 useful parameters, such as signal sparsity, robustness against noise, data nonlinearity, precision, computational efficiency, etc., have been considered to investigate those works. The review has produced the strengths, features, and weaknesses, and the gaps persist in those methodologies. The interests of researchers have also been addressed, so that the practical issues can be understood well. The research gaps are encouraging to further carry out our research work in the field.

References

1. Barbé, K.; Van Moer, W.; Nagels, G., "Fractional-Order Time Series Models for Extracting the Haemodynamic Response From Functional Magnetic Resonance Imaging Data", IEEE Transactions on Biomedical Engineering, Vol. 59, Issue 8, pp 2264-2272, 2012.
2. K. K. Kwong and D. A. Chesler, "Functional MRI," in Medical Devices and Systems. Boca Raton: CRC Press, 2006, pp. 22–30.
3. Katwal, S.B.; Gore, J.C.; Marois, R.; Rogers, B.P., "Unsupervised Spatiotemporal Analysis of FMRI Data Using Graph-Based Visualizations of Self-Organizing Maps", IEEE Transactions on Biomedical Engineering, Volume 60, Issue 9, pp 2472–2483, 2013.
4. Yuanqing Li; Namburi, P.; Zhuliang Yu; Cuntai Guan; Jianfeng Feng; Zhenghui Gu, "Voxel Selection in fMRI Data Analysis Based on Sparse Representation", IEEE Transactions on Biomedical Engineering, Volume 56, Issue 10, pp 2439–2451, 2009.
5. Michel, V.; Gramfort, A.; Varoquaux, G.; Eger, E.; Thirion, B., "Total Variation Regularization for fMRI-Based Prediction of Behavior", IEEE Transactions on Medical Imaging, Volume 30, Issue 7, pp 1328–1340, 2011.
6. C. Genovese, N. Lazar, and T. Nichols, "Thresholding of statistical maps in functional neuroimaging using the false discovery rate," NeuroImage, Volume 15, pp. 870–878, 2002.
7. M. Lindquist, J. Loh, L. Atlas, and T. Wager, "Modeling the hemodynamic response function in fMRI: Efficiency, bias and mismodeling," NeuroImage, Volume 45, no. 1, pp. S187–S196, 2009.
8. B. Hu, G. Varma, C. Randell, S. Keevil, T. Schaeffter, and P. Glover, "A novel receive-only liquid nitrogen (LN2)-cooled RF coil for high-resolution in vivo imaging on a 3-tesla whole-body scanner," IEEE Transactions on Instrumentation and Measurement, Volume 61, Issue. 1, pp. 129–139, Jan. 2012.
9. S. Strother, "Evaluating fMRI preprocessing pipelines—Review of pre-processing steps for BOLD fMRI," IEEE Engineering in Medicine and Biology Magazine, Volume 25, Issue 2, pp. 27–41, Mar./Apr. 2006.
10. D. A. Karras and G. B. Mertzios, "New PDE-based methods for image enhancement using SOM and Bayesian inference in various discretization schemes," Measurement Science and Technology, Volume. 20, Issue 10, 2009.
11. V. Rallabandi and P. Roy, "Magnetic resonance image enhancement using stochastic resonance in Fourier domain," Magnetic Resononance Imaging, Volume. 28, Issue 9, pp. 1361–1373, 2010.
12. X. Yang and B. Baowei Fei, "A wavelet multiscale denoising algorithm for magnetic resonance (MR) images," Measurement Science and Technology, Volume 22, no. 2, 2011.

13. J. Sijbers, D. Poot, A. J. den Dekker, and W. Pintjens, "Automatic estimation of the noise variance from the histogram of a magnetic resonance image,"Physics in Medicine and Biology, Volume. 52, pp. 1335–1348, 2007.
14. C. Goutte, F. A. Nielsen, and L. K. Hansen, "Modeling the haemodynamic response in fMRI using smooth FIR filters," IEEE Trans. Med. Imag., Volume 19, no. 12, pp. 1188–1201, Dec. 2000.
15. R. Gibbons, N. Lazar, D. Bhaumik, S. Sclove, H. Chen, K. Thulborn, J. Sweeney, K. Hur, and D. Patterson, "Estimation and classification of fMRI hemodynamic response patterns," Neuroimage, vol. 22, pp. 804–814, 2004.
16. A. den Dekker, D. Poot, R. Bos, and J. Sijbers, "Likelihood-based hypothesis tests for brain activation detection from MRI data disturbed by colored noise: A simulation study," IEEE Trans. Med. Imag., vol. 28, no. 2, pp. 287–296, Feb. 2009.
17. K. Barb'e, W. Van Moer, and L. Lauwers, "Functional magnetic resonance imaging: An improved short record signal model," IEEE Trans. Instrum. Meas., vol. 60, no. 5, pp. 1724–1731, May 2011.
18. X. Descombes, F. Kruggel, and D. Y. von Cramon, "Spatio-temporal fMRI analysis using Markov random fields," IEEE Trans. Med. Imag., vol. 17, no. 6, pp. 1028–1039, Dec. 1998.
19. M. Svens´en, F. Kruggel, and D. Y. von Cramen, "Probabilistic modeling of single trial fMRI data," IEEE Trans. Med. Imag., vol. 19, no. 1, pp. 25–36, Jan. 2000.
20. S. Faisan, L. Thoraval, J. P. Armspach, J. R. Foucher, M. N. Metz-Lutz, and F. Heitz, "Hidden Markov event sequence models: Toward unspervised functional MRI brain," Acad. Radiol., vol. 12, no. 1, pp. 25–36, Jan. 2005.
21. K. J. Friston, A. P. Holmes, K. J. Worsley, J. –P. Poline, C. D. Frith, and R. S. J. Frackowiak, "Statistical parametric maps in functional imaging: A general linear approach," Human Brain Mapping, vol. 2, no. 4, pp. 189–210, 1994.
22. E. Zarahn, G. K. Aguirre, and M. D'Esposito, "Empirical analyses of BOLD fMRI statistics," Neuroimage, vol. 5, no. 3, pp. 179–197, Apr. 1997.
23. L. K. Hansen, J. Larsen, F. A. Nielsen, S. C. Strother, E. Rostrup, R. Savoy, N. Lange, J. Sidtis, C. Svarer, and O. B. Paulson, "Generalizable patterns in neuroimaging: How many principal components," Neuroimage, vol.9, no. 5, pp. 534–544, May 1999.
24. K. H. Chuang, M. H. Chiu, C.C. Lin, and J. H. Chen, "Model-free functional MRI analysis using Kohonen clustering neural network and fuzzy c-means," IEEE Trans. Med. Imag., vol. 28, no. 12, pp. 1117–1128, Dec. 1999.
25. Jingyu Liu; Lai Xu; Caprihan, A.; Calhoun, V.D., "Extracting principle components for discriminant analysis of FMRI images", IEEE International Conference on Acoustics, Speech and Signal Processing, 2008. ICASSP 2008, pp 449–452, 2008.
26. S.J. Peltier, T. A. Polk, and D. C. Noll, "Detecting low-frequency functional connectivity in fMRI using a self-organizing map (SOM) algorithm," Human Brain Mapping, vol. 20, no. 4, pp. 220–226, Aug. 2003.
27. W. Liao, H. Chen, Q. Yang, and X. Lei, "Analysis of fMRI data using improved self-organizing map and spatio-temporal metric hierarchical clustering," IEEE Trans. Med. Imag., vol. 27, no. 10, pp. 1472–1483, Oct. 2008.
28. H. Chen, H. Yuan, D. Yao, L. Chen, and W. Chen, "An integrated neighborhood correlation and hierarchical clustering approach of functional MRI," IEEE Trans. Biomed. Eng., vol. 53, no. 3, pp. 452–458, Mar. 2006.
29. Chiew, M.; Graham, S.J.; "BOLD Contrast and Noise Characteristics of Densely Sampled Multi-Echo fMRI Data, IEEE Transactions on Medical Imaging, Vol. 30, No. 9, pp. 1691–1703, 2011.
30. M. J. McKeown, S. Makeig, G. G. Brown, T. P. Jung, S. S. Kindermann, A. J. Bell, and T. J. Sejnowski, "Analysis of fMRI data by blind separation into independent spatial components," Human Brain Mapping, vol.6, no. 3, pp. 160–188, 1998.
31. Honorio, J.; Tomasi, D.; Goldstein, R.Z.; Leung, H.-C.; Samaras, D., "Can a Single Brain Region Predict a Disorder?", IEEE Transactions on Medical Imaging, Volume 31, Issue: 11, pp 2062–2072, 2012.

32. Chaari, L.; Vincent, T.; Forbes, F.; Dojat, M.; Ciuciu, P, "Fast Joint Detection-Estimation of Evoked Brain Activity in Event-Related fMRI Using a Variational Approach", IEEE Transactions on Medical Imaging, Vol. 32, Issue 5, pp 821–837, 2013.
33. Fan Deng; Dajiang Zhu; Jinglei Lv; Lei Guo; Tianming Liu, "FMRI Signal Analysis Using Empirical Mean Curve Decomposition", IEEE Transactions on Biomedical Engineering, Volume 60, Issue 1, Part: 1, pp 42–54, 2013.
34. Babak Afshin-Pour, Seyed-Mohammad Shams, and Stephen Strother, "A Hybrid LDA + gCCA Model for fMRI Data Classification and Visualisation", IEEE Transactions on Medical Imaging, Volume PP, Issue 99, pp 1–12, 2014.
35. Chun-An Chou; Kampa, K.; Mehta, S.H.; Tungaraza, R.F.; Chaovalitwongse, W.A.; Grabowski, T.J., "Voxel Selection Framework in Multi-Voxel Pattern Analysis of fMRI Data for Prediction of Neural Response to Visual Stimuli", IEEE Transactions on Medical Imaging, Volume 33, Issue 4, pp 925–934, 2014.
36. A. N. Paithane and D. S. Bormane. "Analysis of nonlinear and non-stationary signal to extract the features using Hilbert Huang transform." Computational Intelligence and Computing Research (ICCIC), 2014 IEEE International Conference on. IEEE, 2014.
37. A. N. Paithane and D. S. Bormane. "Electrocardiogram signal analysis using empirical mode composition and Hilbert spectrum." Pervasive Computing (ICPC), 2015 International Conference on. IEEE, 2015.

Static Program Behavior Tracing for Program Similarity Quantification

Dugyala Raman, Bruhadeshwar Bezawada, T.V. Rajinikanth and Sai Sathyanarayan

Abstract Characterizing program behavior using static analysis is a challenging problem. In this work, we focus on the fundamental problem of program similarity quantification, i.e., estimating the behavioral similarity of two programs. The solution to this problem is a sub-routine for many important practical problems, such as malware classification, code-cloning detection, program testing, and so on. The main difficulty is to be able to characterize the run-time program behavior without actually executing the program or performing emulation. In this work, we propose a novel behavior tracing approach to characterize program behaviors. We use the call-dependency relationship among the program API calls to generate a trace of the API calling sequence. The dependency tracking is done in a backward fashion, so as to capture the cause and effect relationship among the API calls. Our hypothesis is that this relationship can capture the program behavior to a large extent. We performed experiments by considering several "versions" of a given software, where each version was generated using the code obfuscation techniques. Our approach was found to be resilient up to 20 % obfuscation, i.e., our approach correctly detected that all obfuscated programs that are similar in behavior based on the API call sequences.

Keywords Program similarity · Static analysis · Behavior tracing · Malware

D. Raman (✉)
Vardhaman College of Engineering, Hyderabad, India
e-mail: raman.vsd@gmail.com

B. Bezawada
Koneru Lakshmaiah University, Guntur, India
e-mail: bru@kluniversity.in

T.V. Rajinikanth
Sreenidhi Institute of Science and Technology, Hyderabad, India
e-mail: rajinitv@gmail.com

S. Sathyanarayan
Citibank, Singapore
e-mail: sai.sathyanarayan@gmail.com

© Springer Science+Business Media Singapore 2017
S.C. Satapathy et al. (eds.), *Proceedings of the First International Conference on Computational Intelligence and Informatics*, Advances in Intelligent Systems and Computing 507, DOI 10.1007/978-981-10-2471-9_31

1 Introduction

Analyzing and characterizing program behavior are important and challenging problems.[1] In general, a new version of a program is created by adding additional functionality to the original vanilla version of the program. However, the core behavior of the program remains the same as the vanilla version. Such analysis has extensive applications in program classification scenarios, such as malware analysis and signature generation. In malware, the changes are made to hide the program from anti-virus signature databases. Thus, it is important to develop approaches that can characterize the core program behavior of programs, which are evolved from the original version.

Although program behavior can be learnt by executing the program or emulating it in a sand box and collecting traces, this is an expensive approach and there are no guarantees that the entire program behavior would be captured within the given sample execution. We focus on static analysis-based techniques, as they are light weight and fast when compared with run-time approaches, such as sand boxing and dynamic program tracing. In the past, system calls have been used for modeling program behavior [1–5] and for detecting malware. Other approaches like [6] used frequency distribution of critical API calls to detect the presence of malware behavior in a program. However, these approaches have high false positive rates, as frequency distributions can be common across programs that are dissimilar in behavior.

To extract behavioral aspects, API call sequences are shown to be quite useful [7–30]. For example, a particular calling sequence could be typically constant across all programs evolved from the same base program. However, extracting such a sequence could be cumbersome, as a forward analysis on the control-flow graph can have an exponential number of such sequences. Furthermore, even if the program analyst extracts all such sequences, it is non-trivial to identify which of these sequences characterizes the core program behavior.

To address this problem, we use the notion of *critical API calls*, which are the most frequently occurring API calls in the original version of the software. We focus only on the sequences generated by the critical API calls, and thereby reduce the computational effort for the program analyst. Given this insight, our approach consists of a backward tracing algorithm, which consists of a sequence of instructions that have either directly or indirectly modified the API call's arguments. For a given program, we use two versions from the same class and generate all the backward traces. Next, we use a statistical profiling technique that compares the lengths of the sequences and stores the result of the comparison. This serves as the signature for the class of programs and the statistical threshold is used to check if any other program exhibits similar behavior. We have tested our technique on two Windows executables, `notepad.exe` and `mstc.exe`, by introducing some obfuscations and adding additional call sequences.

---

[1]Malware are programs that exhibit malicious behavior that can disrupt the proper functioning of a computing system and can cause damage to sensitive data or to other resident programs.

In this paper, we present a static analysis approach which extracts the behaviors of a program as ordered sequences of critical API calls. Our contributions are as follows:

- **Capturing Program Behavior using Backward Traces.** For each critical API call in the program binary, a backward trace is the sequence of all other critical API calls that effect the arguments of the concerned API call, resulting in accurate modeling of program behavior. We describe an algorithm that can perform this tracing.
- **Program Similarity Testing.** We have experimented by obfuscating two normal Windows application programs by adding additional redundant functionality and determining their similarity to the original program. Our preliminary experiments show that a vanilla version has to be obfuscated by introducing at least 20 % of new functionality to evade detection. These results show that our approach can effectively classify programs based on their behavior.

The paper is organized as follows. In Sect. 2, we present the related work which has been done in the past in the field of program behavior analysis and malware detection. In Sect. 3, we describe our backward tracking approach for behavioral tracing. In Sect. 4, we describe a prototype implementation of our approach, present experimental results, and evaluate the effectiveness of our approach. Finally, in Sect. 5, we conclude and discuss future work.

2 Related Work

Much of the related work in this domain has come from the area of malware detection. We cover some important results. Cohen [13] and Chess and White [14] use sandboxing to detect viruses. They proved that, in general, the problem of virus detection is undecidable. Our work is closely related to the work done by Bergeron et al. [1]. They used slicing on assembly code of a program to extract code fragments that are critical from the security standpoint. In [8], the authors have used the static analysis to generate a critical API graph which they compare against a security policy to detect the presence of malicious code in the program. In [10], Christodorescu and Jha use static analysis to detect malicious code in executables. Their implementation called SAFE handles simple obfuscations used by malware writers, such as insertion of NOPs between instructions, that are used to evade detection. Bergeron et al. [8] consider critical API calls and security policies to test the presence of malicious code. Their approach does not work for obfuscated malicious executables. Bilar [15] uses statistical structures, such as opcode frequency distribution and graph structure fingerprints, to detect malicious programs.

In [16], Christodorescu et al. exploited dynamically captured semantic heuristics to detect obfuscated malware. DOME [17] uses static analysis to detect system call locations and run-time monitoring to check all system calls that are made from a location identified during static analysis. Min Sun et al. [3] use dynamic monitoring to detect worms and other exploits. Sai et al. [6] used frequency distribution of API calls in a program to differentiate between malware and benign programs.

A more closely related work is the work in [18], where the authors attempt to capture behavioral similarities across malware variants of a same family of malware. They used execution traces of a malware sample to create a malspec (malicious specification) for the sample. This malspec is able to detect other variants of the malware sample. It is a dynamic analysis technique and uses subgraph matching for comparing two malspecs, which increases detection time considerably. Similar approaches have recently been proposed in [5, 7, 12]. In [11], the authors use data flow graphs to quantify the behavior of malware, and thereby detect them effectively. However, their approach is specific to malware detection and cannot be applied to program similarity estimation.

3   Our Approach

In this section, first, we outline our approach behavioral tracing. We call the behavioral trace of a program as *signature* of the program. Next, we describe, in detail, our program behavior model used for signature generation and the statistical comparison technique. Our model generates signatures for the vanilla (the earliest version) of the program and calculates the similarity of this signature with the future versions of the program.

The behavior of a program can be specified based on the API calls that the program and its versions use. For instance, a virus trying to search for executable files will typically make use of API calls, such as FindFirstFileA, FindNext FileA and FindClose, in KERNEL32.DLL. This behavior of searching files is captured by the use of these API calls. Rather than considering all API calls, we consider only *critical API calls* [8, 18]. The critical API calls are essential calls required for the core functionality of the program and will be required across all future versions. We extract the sequences of critical API calls from the program and represent them in a compactly as signatures. Next, we give details of how we perform each of the steps of our approach.

3.1   Disassembling the Program and Converting into IR

We use IDAPro Disassembler to disassemble the program binary. Next, we convert the disassembled instructions into an intermediate representation (IR) which is in the form of C-like statements. If these steps were being carried out on a parallel processor, multiple programs could be processed in parallel and make them ready for analysis. Thus, overhead of these steps is negligible for analysis. In addition, these steps could be done offline on a cloud cluster or high-end server.

3.2 Extracting the Behavior

Each program behavior can be specified by an ordered sequence of API calls. The signature of a program is hence based on the sequences of *critical* API calls. Let the set $P = (a_1, a_2, \ldots, a_n)$ be a *profile* created from a program by extracting its critical API calls, where a_i represents the ith critical API call and n being the total number of critical API calls.

To get the sequence of statements for each critical API call, we compute the backward trace of the program with respect to set P. A trace is a subset of a program, which contains only those statements that evaluate the arguments input to the concerned tracing criterion. In our case, the tracing criterion contains the set of critical API calls P. In our approach, we need not generate the trace for all the critical API calls in P, since if the trace generated for a_i criterion contains any of the API call say a_j in P, then we need not perform tracing on a_j, since a_i already contains all the trace statements that a_j should contain. Therefore, the number of traces will be less than n, say m. The algorithm for generating the behavior traces is shown in Algorithm 1.

Algorithm 1: SIG_BEHAVIOR to compute the Behavior Traces

1 Let S be a set which contains S_i, where S_i is the trace w.r.t tracing criterion i, and API_g be the global API call monitoring.
2 Initially, API_g set contains a empty set
3 Set $API_g = \{\emptyset\}$
4 Traversing each API calls, from the set P
5 **For each** api a in P {
6 **if**$(a \neq API_g)$ {
7 S_a = Backward_Trace(a)
8 $S = S \cup S_a$
9 $API_g = API_g \cup \{\text{API in } S_a\}$
10 }
11 **Repeat till** All the API calls are traversed from set P

3.3 Computing the Longest Common Subsequence (LCS)

In this section, we compute the signature for a program by computing the LCS for each backward trace generated from the previous section. The signature is generated by computing the LCS of each trace from some versions of the program. For example, to create a signature for MyDoom family, we consider MyDoom.a and MyDoom.aa as training samples. The algorithm for generating the signature is shown in Algorithm 2. While we have considered malware samples, without loss of generality, this approach applies to any program consisting of several versions.

Algorithm 2: LCS_PROGRAM to compute the LCS

---

1 Let S be a set which contains S_i, where S_i is the behavior trace w.r.t tracing criterion i, S' be
  an another set which is similar to S, and LCS_a be the longest substring for API call a. S and
  S' are set of two variants of the same program.
2 **For each** S_a in S {
3 **For each** S'_a in S' {
4 if$(S_a = S'_a)$ {
5 $LCS_a = \mathrm{LCS}(S_a, S'_a)$
6 }
7 }
8 **Repeat till** All the trace components are traversed from set S

3.4 Classification Strategy

We use a statistical profiling technique to differentiate between programs. We measure the difference between the proportions of the critical API calls in a signature and that of a test program using the Chi-square test [19]. The chi-square test is a likelihood-ratio or maximum-likelihood statistical significance test that measures the difference between proportions in two independent samples. The signature SIG_i for a program M_i specifies the LCS of critical API calls that a version of the program which belongs to M_i is expected to have. To test the relationship of a given test file to M_i, its API calls are extracted and compared to that in the signature. The chi-square is then computed as:

$$\chi_i^2 = \frac{(O_i - E_i)^2}{E_i} \qquad ; 1 \le i \le n.$$

Here, O_i is the observed frequency/LCS of the ith critical API call in the test file and E_i is its expected frequency, i.e., frequency in the signature of the program. Now, χ^2 is compared against a threshold value ϵ from a standard chi-square distribution table with one degree of freedom. The degrees of freedom are associated with the number of parameters that can vary in a statistical model. A significance level of 0.05 was selected. This means that 95 % of the time we expect χ^2 to be less than or equal to ϵ. For one degree of freedom and significance level 0.05, $\epsilon = 3.84$. Let $U = \{API_i \mid \chi_i^2 \le \epsilon_i\}$. We define a *degree of membership* λ as

$$\lambda = \frac{|U|}{n}.$$

Degree of membership λ is a measure of similarity of a test file to the given program. The statistical profiling algorithm is shown in Algorithm 3.

Algorithm 3: STAT(LCS_i, LCS_j)

Input: API frequency/LCS set for a file, $LCS_i = \{O_1, O_2, \ldots, O_n\}$, and another API
   frequency/LCS set $LCS_j = \{E_1, E_2, \ldots, E_n\}$
Output: Degree of membership, λ
1 **for** $i = 1$ *to* n **do**
2     $\chi_i^2 = \frac{(O_i - E_i)^2}{E_i}$;
3 $U = \{API_i \mid \chi_i^2 \leq 3.84\}$;
4 $\lambda_i = \frac{|U|}{n}$;
5 **return** λ

---

4 Experimental Analysis

To test the resilience of our approach against program obfuscations, we tested with the following known obfuscation techniques. *Code Transposition*: Code is shuffled and unconditional jumps are inserted to restore the original order of execution of program. *Instruction Substitution*: Many instruction sequences can be rewritten with similar semantics. A number of such transformations were identified and replaced. *Register Reassignment*: Usage of registers is interchanged, for example, EAX is used as EBX and vice versa. *Dead Code Insertion*: Code is added to program without modifying its behavior. To challenge our technique, we added code that consisted of critical API calls that would result in no change of behavior. For each program sample, a random number of program transformations were applied using the obfuscator. A set of 1000 obfuscated binaries for the sample program were generated. The signature was generated using only the sample program binary that was used to generate the 1000 obfuscated binaries. All experiments were conducted on an Intel Celeron 1.6 GHz, 1 GB RAM machine.

4.1 Synthetic Program Similarity Test

The experiment was carried out on 245 samples of malicious programs using the synthetic obfuscations described earlier. The observations made during the experiment were:

- Our approach was resilient to control-flow transpositions. There was no effect on the critical API call sequences that were generated using backward tracing.
- Our approach was resilient to instruction substitution transposition. In our approach, one of the steps is to convert the disassembly into a C-like intermediate representation, where such substitutions are handled and replaced by a code sequence which is kept same for the whole program.

- Our approach was resilient to register reassignment transposition and dead code insertion. In both these techniques, the data dependencies were not changed. Therefore, our backward tracing analysis results were same.

4.2   Real-Life Program Similarity Test

In this experiment, semantically obfuscated program binaries of a sample were tested against the signature of the original program from which the semantically obfuscated binaries were generated. The experiment was carried out on two programs— notepad.exe and mstsc.exe. The API call sequences of a benign program were generated using backward tracing. Incrementally, some dummy noise API calls were introduced in these API call sequences. The following observations were made.

Observation 1. At some point, adding noise to API call sequences caused the program to be not detected as a variant of the original program. For the tested programs, notepad.exe and mstsc.exe, after about 20 % increase in the API call sequences length, our approach could not identify them as similar to original sequence. One of the reason for this was the threshold that had been kept for similarity detection.

Observation 2. Given, two programs, our program was able to provide an estimate percentage of similarity between those two programs based on the similarity in the API call sequences of the two programs. This metric can be used to cluster similar programs, especially, when analyzing a large number of programs.

5   Conclusion and Future Work

In this paper, we presented a novel static analysis approach to characterize program behaviors using behavior tracing We observed that the calling sequence relationship in API calls cannot be removed without changing the behavior of the program. Thus, calling sequences of API calls are able to capture program behavior to a large extent. To be able to accurately evaluate our approach, we developed an obfuscator that would apply random syntactic transformations to generate an obfuscated binary from the original binary.

Our approach is able to generate a signature, which is sufficient to characterize any number future versions of the program. Our approach is able to classify programs on the basis of behavioral similarity and also to provide an approximate measure of the quality of obfuscations. We provided the results to show the efficacy of our approach in classifying similar programs. As future work, we intend to use our approach in clustering programs on the basis of behavior.

References

1. J. Bergeron, M. Debbabi, M. M. Erhioui, and B. Ktari. Static Analysis of Binary Code to Isolate Malicious Behaviors. In the Proceedings of the IEEE 4th International Workshop on Enterprise Security, WETICE'99, Stanford University, California, USA, June 16–18, 1999, Pages 184–189, IEEE Press.
2. Sean Peisert, Matt Bishop, Sidney Karin, Keith Marzullo, Analysis of Computer Intrusions Using Sequences of Function Calls, IEEE Transactions On Dependable and Secure Computing, VOL. 4, No. 2, APRIL-JUNE 2007.
3. Hung-Min Sun, Yue-Hsun Lin, and Ming-Fung Wu. API Monitoring System for Defeating Worms and Exploits in MS-Windows System. In Proceedings of 11th Australasian Conference on Information Security and Privacy, ACISP 2006, Melbourne, Australia.
4. R. Sekar, M. Bendre, D. Dhurjati, P. Bollineni, A Fast Automaton-Based Method for Detecting Anomalous Program Behaviors. IEEE Symposium on Security and Privacy, 2001.
5. Gerardo Canfora; Francesco Mercaldo; Corrado Aaron Visaggio; Paolo Di Notte; Metamorphic Malware Detection Using Code Metrics, in Information Security Journal: A Global Perspective, Taylor & Francis, pp 1–14, 2014, DOI:10.1080/19393555.2014.931487.
6. V. Sai Sathyanarayan, Pankaj Kohli and Bezawada Bruhadeshwar. Signature Generation and Detection of Malware Families. Proceedings of 13th Australian Conference on Information Security and Privacy, ACISP 2008.
7. Ronghua Tian; Islam, R.; Batten, L.; Versteeg, S.; Differentiating malware from cleanware using behavioural analysis, 2010 5th International Conference on Malicious and Unwanted Software (MALWARE), pp 23–30, 19–20 Oct, 2010, Nancy Lorraine.
8. J. Bergeron, M. Debbabi, J. Desharnais, M. Erhioui, Y. Lavoie and N. Tawbi. Static Detection of Malicious Code in Executable Programs. In the Proceedings of the International Symposium on Requirements Engineering for Information Security SREIS'01, Pages 1–8, March 5–6, 2001, Indianapolis, Indiana, USA.
9. R. W. Lo, K. N. Levitt, and R. A. Olsson. MCF: A Malicious Code Filter. Computers and Security, 14(6):541566, 1995.
10. Mihai Christodorescu and Somesh Jha. Static Anlaysis of Executables to Detect Malicious Patterns. In proceeding of the 12th USENIX Security Symp. (Security03), pages 169–186 August 2003.
11. Tobias Wchner, Martn Ochoa, and Alexander Pretschner. 2014. Malware detection with quantitative data flow graphs. In Proceedings of the 9th ACM symposium on Information, computer and communications security (ASIA CCS '14). ACM, New York, NY, USA, 271–282. DOI=10.1145/2590296.2590319, http://doi.acm.org/10.1145/2590296.2590319.
12. Raman Dugyala; Bruhadeshwar Bezawada; Romanch Agrawal; Sai Sathyanarayan; Rajinikanth Tatiparthi; Application of Information Flow Tracking for Signature Generation and Detection of Malware Families; International Journal of Applied Engineering Research (IJAER). ISSN 0973-4562 Volume 9, Number 24 (2014), pp. 29371–29390.
13. F. Cohen. Computer Virus: Theory and experiments. Computers and Security, 6:2235, 1987.
14. D.M. Chess and S.R. White. An undetectable computer virus. In proceedings of Virus Bulletin Conference, 2000.
15. Bilar, D.: Statistical Structures: Tolerant Fingerprinting for Classification and Analysis given at BH '06 (Las Vegas, NV): Blackhat Briefings USA (August 2006).
16. Mihai Christodorescu, Somesh Jha, Sanjit A. Seshia, Dawn Song, Randal E. Bryant, Semantics-Aware Malware Detection, Proceedings of the 2005 IEEE Symposium on Security and Privacy, p. 32–46, May 08–11, 2005.
17. C.Jesse, R.Rabek. I.Khazan, M.Scott, L.Robert and K.Cunningham, Detection of Injected, Dynamically Generated and Obfuscated Malicious Code. In Proceedings of 2003 ACM workshop on Rapid Malcode October 2003.
18. Mihai Christodorescu, Somesh Jha and Christopher Krugel. Mining Specification of Malicious Behavior. In proceeding of the 6th joint meeting of the European Software Engineering Conference. ACM SIGSOFT Symp. On ESES/FSE 2007.

19. Sokal, R. R. and Rohlf, F. J.; Biometry: the principles and practice of statistics in biological research., 3rd edition. New York: Freeman (1994).
20. Mehdi, B.; Ahmed, F.; Khayyam, S. A.; Farooq, M.; Towards a Theory of Generalizing System Call Representation for In-Execution Malware Detection, 2010 IEEE International Conference on Communications (ICC), 23–27 May 2010, pp: 1–5, Cape Town, South Africa.
21. M. Pietrek, An In-Depth Look into the Win32 Portable Executable File Format, in MSDN Magazine, March 2002.
22. VX Heavens. At http://vx.netlux.org.
23. Ilfak Guilfanov. An Advanced Interactive Multi-processor Disassembler. http://www.datarescue.com, 2000.
24. Kent Griffin, Scott Schneider, Xin Hu and Tzi-cker Chiueh. Automatic Generation of String Signatures for Malware Detection. In Proceedings of the 12th Symposium on Recent Advances in Intrusion Detection (RAID), Saint-Malo, Brittany, France, September 2009.
25. N. Landi. Undecidability of static analysis. ACM Letters on Programming Language and systems (LOPLAS), 1(4):323 337, December 1992.
26. C. Willems. CWSandbox: Automatic Behaviour analysis of malware. http://www.cwsandbox.org/, 2006.
27. M. Sharif, V. Yegneswaran, H. Saidi, P.A Porras, and W. Lee. Eureka: A Framework for Enabling Static Malware Analysis. In Proceedings of the 13th European Symposium on Research in Computer Security, Malaga, Spain, October 2008.
28. Ulrich Bayer, Paolo Milani, Clemens Hlauschek, Christopher Kruegel, and Engin Kirda. Scalable, Behavior-Based Malware Clustering. 16th Annual Network and Distributed System Security Symposium (NDSS 2009), San Diego, February 2009.
29. Tony Lee, and Jigar J. Mody. Behavioral Classification. In EICAR Conference, 2006.
30. G. Mazeroff, V. De Cerqueira, J. Gregor, and M. Thomason. Probabilistic Tree and Automata for Application Behavior Modeling. Proceedings of 41st ACM Southeast Regional Conference, 2003.

Ontology-Based Automatic Annotation: An Approach for Efficient Retrieval of Semantic Results of Web Documents

R. Lakshmi Tulasi, Meda Sreenivasa Rao, K. Ankita and R. Hgoudar

Abstract The Web contains large amount of data of unstructured nature which gives the relevant as well as irrelevant results. To remove the irrelevancy in results, a methodology is defined which would retrieve the semantic information. Semantic search directly deals with the knowledge base which is domain specific. Everyone constructs ontology knowledge base in their own way, which results in heterogeneity in ontology. The problem of heterogeneity can be resolved by applying the algorithm of ontology mapping. All the documents collected by Web crawler from the Web and a document base is created. The documents are then given as an input for performing semantic annotation on the updated ontology. The results against the users query are retrieved from semantic information retrieval system after applying searching algorithm on it. The experiments conducted with this methodology show that the results thus obtained provide more accurate and precise information.

Keywords Web crawler · Ontology · Ontology mapping · Semantic annotation · SPARQL

R. Lakshmi Tulasi (✉)
School of IT, JNTU Hyderabad, Hyderabad, India
e-mail: srmeda@gmail.com

M.S. Rao
Department of CSE, QIS Institute of Technology, Ongole, India
e-mail: ganta.tulasi@gmail.com

K. Ankita
Department of CS/IT, GEU, Dehradun, India

R. Hgoudar
Centre for PG Studies, VTU, Belagavi, India
e-mail: rhgoudar@gmail.com

© Springer Science+Business Media Singapore 2017   331
S.C. Satapathy et al. (eds.), *Proceedings of the First International Conference on Computational Intelligence and Informatics*, Advances in Intelligent Systems and Computing 507, DOI 10.1007/978-981-10-2471-9_32

1 Introduction

The Web contains excessive amount of data, including large number of Web pages, colorful contents, internet applications, etc., which creates a complex and massive information resource. However, keyword-based traditional search engine minimizes the recall and precision ratio by searching low relevant information [1]. To remove the drawbacks of keyword-based traditional search, various researches has been done in this area.

Semantic annotation is performed on the data stored in kB. Annotation is simply to add up something with the existing data. Semi-automatic annotation is when some of the information is added up by the user and after that all the information is added up automatically into existing data according to the user-designed schema. In the case of automatic annotation, all the information is added up automatically without the involvement of the user.

In this paper, we have proposed an approach for semantic information retrieval. The Web crawler is used for collecting the Web pages automatically from the Web. The collected Web pages are then used as an input for further processing like performing annotation on the newly updated ontology knowledge base. Every time, the crawler crawls the Web; the knowledge base is updated. The information stored in the kB is returned to the user after applying the searching algorithm on it, for his/her query.

The rest of the paper is organized as follows: related work is described in Sect. 2; Sect. 3 constructs the proposed the overall architecture along with the proposed algorithms; experiment of the proposed system is introduced in Sect. 4; Sect. 5 concludes the paper.

2 Related Work

A number of approaches have been proposed for ontology mapping like Mematic Algorithm, SAMBO (System for Aligning and Merging Biomedical Ontologies), QOM (Quick Ontology Mapping), ASMOV (Automatic Semantic Matching of Ontologies with Verification), FALCON (Automatic Divide and Conquer Approach), Anchor Prior, and DSSim which results in either 1:1 alignment or n:m alignment. These approaches have the limitations like no Graphical User Interface, not taking input in different forms and time consuming [2–6].

Extraction of information and semantic annotation has been an active research area in recent years. Liu et al. [7] described a hybrid pattern called XPattern for the annotation of Web pages for the semantic Web. XPattern combined the structure feature of Web and lexical features of Web text. Tenier et al. [8] presented a method for the annotation of the Web pages, where O ontology is used as a resource for identifying the relationship between the concepts and a number of Web pages are given as input. Every time, a Web page is loaded, it requires concept identification, structure interpretation, and relationship between concepts. Ala'a et al. [9] discussed semantic Web annotation platforms which offer a user-friendly interface by integrating the Web pages, ontology file, and semantic annotation. Yiyao et al. [10] describe an automatic annotation approach, where the data are aligned in such a way that the same type of data is put into one group and likewise, numbers of groups are formed for annotation. An appropriate label is assigned to each group. An annotation wrapper is used to annotate the new result pages from the Web database. Garrido et al. [11] used the NASS technique to infer appropriate tags for annotation.

A SDOM (Semantic Document Object Model) is introduced by Fei et al. [12] for extracting the information from Web pages by combining the content and structural information. In this, SDOM-based information extraction system sends an inquiry request to repository and wrapper extracts the information which is then mapped into corresponding tag information. SDOM module converts this information into SDOM tree. Imdadi and Rizvi [13] described an approach to extract and integrate concept/classes across multiple ontologies and evaluated them on a well-formed ontology. They discussed the issues, where different ontologies have similar names and remove this issue by the computing prism of similarity versus dissimilarity of features. Jian et al. [14] used two approaches in semantic annotation, document preprocessing and semantic annotation algorithm. In this algorithm, the document is divided into instances by extracting ontology terms and then search for its properties and relevant value in closer location. Khan et al. [15] proposed an approach for improving the search results or maximizing the precision in multidisciplinary Web archives using ontology-based annotation. The proposed architecture includes KIM platform for supporting GATE (General Architecture of Text Engineering), Lucene, and semantic metadata. Although the previously proposed approaches for semantic annotation are effective, but needs the involvement of the user, previously proposed approaches are sometimes proved to be less efficient in case if there are millions and billions of Web documents. This limitation is being removed by our proposed system up to some extent.

3 Proposed System

Figure 1 describes the proposed architecture, where the Semantic information is retrieved for the users' query through user interface. In this system, ontology mapping is applied on the ontologies for resolving the heterogeneity in them [16]. A Web crawler is been designed that would crawl the document from the Web and creates a document base (DB). Each document from a DB is processed and updates the ontology knowledge base by performing annotation on it.

When a query is fired, the control is transferred to the searching algorithm and the required information is returned to the user through Semantic Information Retrieval System.

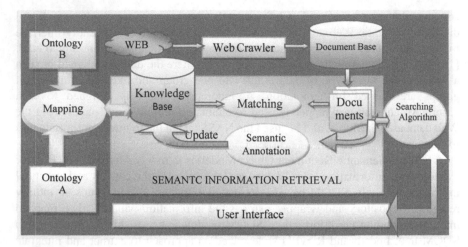

Fig. 1 Proposed system

Proposed architecture consists of four modules:

1. Ontology Mapping
2. Web Crawler
3. Automatic Semantic Annotation
4. Searching Information

3.1 Ontology Mapping

The proposed algorithm of ontology mapping results in a set of matched concepts in different ontologies. It consists of three steps: syntactic similarity; semantic similarity; and finding similarity by calculating the weight of each concept; shown in Fig. 1.

Syntactic Similarity: For calculating syntactic similarity, the Levenshtein algorithm is used.

Semantic Similarity: Word Net is used to take out the synonyms of a particular concept. After expanding the concept, Lavenshtein algorithm is used again to find the similarity (Fig. 2).

Finding similarity by calculating the weight: The weight (*W*) for each candidate is calculated and compared during mapping.

3.2 Web Crawler

A Web crawler is a program that downloads the Web page of a particular URL (given as seed) and extracts all the links contained in a particular Web page and

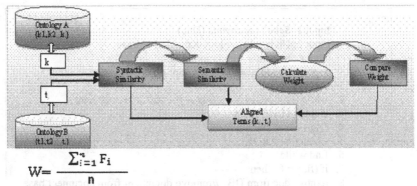

$$W = \frac{\sum_{i=1}^{n} F_i}{n}$$

Where the **F** is feature and **n** is the total number of features.

$$sim(C_1, C_2) = \begin{cases} 1 & \text{if } wc_1 \text{''} wc_2 \text{ and } n_1 \text{''} n_2 \\ 0 & \text{Otherwise} \end{cases}$$

Fig. 2 Functional structure of ontology mapping [16]

repeatedly downloads all the Web pages identified by these links either in a DFS or in a BFS manner. We have proposed an algorithm (Algorithm 1) for Web crawler. The basic need for designing the Web crawler is to retrieve the Web pages from the Web and store them in a local repository.

```
Input 1: Domain Specific Webpage URLS (Url)
Input 2: Total no of URLs (n)
Output: Set of documents along with their text file
ln: URL extracted from the document
1.    int i ← 0
2.    store the urls in string array st ; st[i] ← url ; i ← i+1
3.    if st[i].contains("http") then
4.    while (i <= n)
5.    parse.st[i]   //parse the document
6.    for each ln    st[i] ← ln   end for
7.    i ← i+1
8.    end while
9.    int k ← 0
10.   while st[i] ≠null
11.   while (tx ≠null)
12.    out[k].append(tx) //text(tx)is saved in an output file
13.   end while
14.   k ← k+1
15.   end while
```

Algorithm 1

```
Input 1: Set of documents (Doc)
Input 2: Updated Ontology Knowledge Base
Output: Semantic annotation in KB
p: property; dp: data property, op: object property
1.  for each Doc
2.  while (tx ≠ NULL)        // tx is the text in a document
3.  if (tx = TRUE) then
4.  flag ← 1
5.  Create( SPARQL)
6.  End while
7.  if (flag = 0) then
8.  remove doc from DB   //remove document from document base
9.  End for
10. for each cl              //cl  is the class in the KB
11. for each ind
12. for each p
13. if (individual ind matches the property p) then
14. if(p is data-property dp) then
15. Check for datatype //i.e. string, int, anyURI etc.
16. Create(SPARQL)
17. Otherwise
18. if(p is object-property op) then
19. Create(SPARQL)
20. End for
21. End for
```

Algorithm 2

3.3   Automatic Semantic Annotation

The previous algorithm (Algorithm 1) is designed to collect all the documents and store them in a repository or document base (DB). However, some of the documents in the DB could be non-relevant. To overcome this drawback, algorithm 2 is proposed, where the terms related to the specific domain (i.e., Sports) would be added to the ontology knowledge base. The algorithm will search for all the properties of a particular term and than a SPARQL query is fired that would annotate all related properties along with their URL of the document to the ontology KB. Every time, when a new document is processed, the ontology KB is updated.

3.4 Searching Information

When user searches for any query, the query is pre-processed, as described in Algorithm 3. Preprocessing is accomplished by: break the query, remove stop words, punctuation marks, and convert the query in lower case.

```
Input: Query (qry)
Output: Semantic Information
sw: stop words; tr: term in KB; newq: string variable; rs: results
1.   For each qry
2.   qry ← qry.remove(sw);qry ← qry.remove(regex)
3.   qry ← qry.toLowerCase()
4.   qry ← qry.split(qry, " ")
5.   int i ← 0
6.   while (i ≤ qry.length)
7.     if (qry.contain(tr))
8.       newq → append(tr)
9.     res ← create (SPARQL)
10.   return res
11.  End while
12. End for
```

Algorithm 3

The pre-processed query would be matched with the concepts in the knowledge base and the SPARQL query is generated against the pre-processed users query. The information is retrieved from the Semantic IR system for every SPARQL query.

4 Experiment and Result

A graphical user interface is created, so that user can search for any type of information for a specific domain. Searching algorithm is applied on the query and the relevant results are returned to the user through semantic information retrieval system.

In our sports ontology, we have created knowledge base for various sports categories. We tested the efficiency of the system for three subdomains, i.e., Cricket, Hockey, and Football, as shown in Table 1. Results of Precision and Recall are shown in Figs. 3 and 4.

Table 1 Documents tested for sports domain

Term	Cricket	Hockey	Football
Stored relevant data	10500	5225	8275
Retrieved data	10752	5314	8413
Relevant data	10326	5512	8176
Precision	96.03 %	96.95 %	97.18 %
Recall	98.34 %	98.36 %	98.8 %

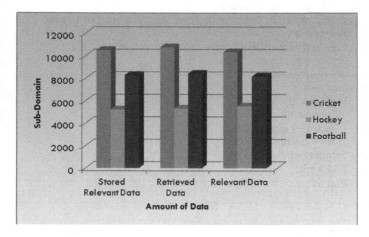

Fig. 3 Amount of data elements for various domains

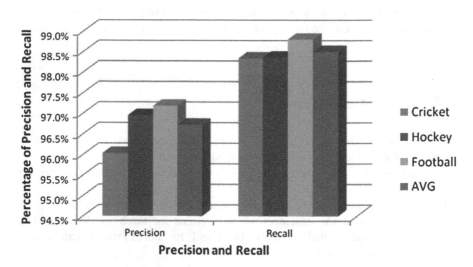

Fig. 4 Precision and recall for three sub domains

5   Conclusion

The aim of the approach presented in this paper is to retrieve the semantic information using ontology mapping, Web crawler, semantic annotation, and searching information algorithms. Unlike the existing approaches, the proposed system works in a sequential manner as: (1) ontology update, (2) collection of documents from the Web by Web crawler, (3) automatic semantic annotation, and (4) searching algorithm, that will run each time user searches for any query. The implemented system generates the information relevant to user query and provides the vision of semantic search.

References

1. Mohamed Kassim, Mahathir Rahmany, "Introduction to Semantic Search Engine," International Conference on Electrical Engineering and Informatics, pp. 380–386, Selangor, Malaysia, August 2009.
2. J.Euzenat, P. Shaviko, "Ontology Matching", IEEE Transactions On Knowledge And Data Engineering, Vol. 25, No. 1, IEEE, 2013.
3. Giovanni Acampora, Pasquale Avella, Vincenzo Loia, Saverio Salerno and Autilia Vitiello, "Improving Ontology Alignment through Memetic Algorithms" Int'l Conf. on Fuzzy Systems, IEEE,2011.
4. Patrick Lambrix and He Tan, "SAMBO - A System for Aligning and Merging Biomedical Ontologies," J. Web Semantics, pp. 196–206, 2006.
5. Junwu ZHU, "Survey on Ontology Mapping", ELSEVIER, 2011.
6. Trong Hai Duong, Geun Sik Jo, " Anchor-Prior: An Effective Algorithm for OntologyIntegration", 978-1-4577-0653-0/11, EEE,2011.
7. Yuan Liu, Li Zhanhuai, Zhang longbo, Chen Shiliang "Annotating Web Pages for Semantic Web" 2009 World Congress on Computer Science and Information Engineering, IEEE, 2008.
8. S. Tenier, Y. Toussaint, A. Napoli, X. Polanco "Instantion of relations for semantic annotation" Proceedings of the 2006 IEEE/WIC/ACM International Conference on Web Intelligence IEEE, 2006
9. Ala'a Q. Al-Namiy, Faris S. Majeed "Towards Automatic Extracted Semantic Annotation (ESA) for Web Documents" DOI 10.1109/APCIP.2009.292, IEEE, 2009.
10. Yiyao Lu, Hai He, Hongkun Zhao, Weiyi Meng, Clement Yu "Annotating Search Results From Web Databases" Digital Object Identifier 10.1109/TKDE.2011.175 IEEE, 2011 (transaction)
11. Angel L. Garrido, Oscar G´omez, Sergio Ilarri and Eduardo Mena "NASS: News Annotation Semantic System" DOI 10.1109/ICTAI.2011.149, IEEE, 2011
12. Yulian Fei, Zongwei Luo, Yun Xu, Winston Zhang "A Semantic DOM Approach for Webpage Information Extraction" 978-1-4244-4639-1/09/, IEEE, 2009
13. Nadia Imdadi and Dr. S.A.M. Rizvi "An Approach to Owl Concept Extraction and Integration across Multiple Ontologies" International Journal of Web & Semantic Technology (IJWest) Vol. 3, No. 3, July 2012 IJWesT, 2012
14. Sun Jian, Xu Jungang, Cen Zhiwang "Semantic Annotation in Academic Search Engine" Web Society(SWS), 978-1-4244-6359-6/10 IEEE, 2010
15. Arshad Khan, David Martin, Thanassis Tiropanis "Using Semantic Indexing to Improve Searching Performance in Web Archives" 978-1-61208-248-6, IARIA 2013.
16. Ankita Kandpal, R H Goudar, Rashmi Chauhan, Shalini Garg, Kajal Joshi, "Effective Ontology Alignment: Approach for Resolving the Ontology Heterogeneity Problem for Semantic Information Retrieval", Springer, 2013.

An Iterative Hadoop-Based Ensemble Data Classification Model on Distributed Medical Databases

Thulasi Bikku, Sambasiva Rao Nandam and Ananda Rao Akepogu

Abstract As the size and complexity of the online biomedical databases are growing day by day, finding an essential structure or unstructured patterns in the distributed biomedical applications has become more complex. Traditional Hadoop-based distributed decision tree models such as Probability based decision tree (PDT), Classification And Regression Tree (CART) and Multiclass Classification Decision Tree have failed to discover relational patterns, user-specific patterns and feature-based patterns, due to the large number of feature sets. These models depend on selection of relevant attributes and uniform data distribution. Data imbalance, indexing and sparsity are the three major issues in these distributed decision tree models. In this proposed model, an enhanced attributes selection ranking model and Hadoop-based decision tree model were implemented to extract the user-specific interesting patterns in online biomedical databases. Experimental results show that the proposed model has high true positive, high precision and low error rate compared to traditional distributed decision tree models.

Keywords Distributed data mining · Hadoop · Ensemble approach · Medical databases

T. Bikku (✉)
Department of CSE, VNITSW, Guntur, AP, India
e-mail: thulasi.bikku@gmail.com

S.R. Nandam
Department of CSE, SRITW, Warangal, Telangana, India
e-mail: snandam@gmail.com

A.R. Akepogu
Department of CSE, JNTUCEA, Ananthapuramu, India
e-mail: akepogu@gmail.com

© Springer Science+Business Media Singapore 2017   341
S.C. Satapathy et al. (eds.), *Proceedings of the First International Conference on Computational Intelligence and Informatics*, Advances in Intelligent Systems and Computing 507, DOI 10.1007/978-981-10-2471-9_33

1   Introduction

In the past, to identify named entities in the biomedical field, single kinds of entities such as protein and/or gene labels were used. It is more effective to use multiple kinds of named entities at the same time. For each named entity, an entity pattern recognizer could be executed, and then, after multiple runs, all kinds of named entities are annotated and, finally, the results could be merged. Gene clustering is one of the data mining models that were developed for microarray gene expression data [1]. The basic assumption with respect to training and test data is that, all distributed instances have been taken from the same feature space with the same distribution. In the biomedical field, transfer learning enables the context knowledge contained in labeled source domains to predict unlabeled data in the target domain, where the domains differ in distributions. Traditionally, the transfer learning was developed to find the correspondence between pivot features and other specific features extracted from different domains. These learning models extract persistent information that aims to reduce the difference between the domains. Domain transfer learning reduces the difference between the distributions of different domains and, thus, minimizes cross-domain prediction errors. There have been many approaches implemented for transfer learning; one of the promising ones is feature transfer learning. In this, most relevant contextual features are adopted as representative objects of both domains.

The distributed data that are generated from different sources are multiple, complex, distinct and independent. Due to the large number of instances with redundant and irrelevant attributes, an appropriate filtering model has been used to fill the noisy attribute or instances. To handle large number of attributes, an efficient feature selection model was used to select the ranked attributes for classification or clustering. Al-Khateeb and Masud [2] implemented concise set of rules using attribute selection model in rough set theory. This model has gained wide acceptance in data analysis, machine learning and statistical analysis. The important challenge in the classification algorithms is the error rate and class imbalance. Traditional models attempt to optimize the overall precision of their predictions. Therefore, we would prefer an efficient classification model which performs well on the minority class, size, complexity and arbitrary data distribution in the distributed data mining. Big data concern large volume, growing datasets that are automatic and complex to analyze. It is too complex to filter noise and recognize unstructured data for decision-making.

The main challenges to handle online medical databases are data accessing and arithmetic computing procedures, semantics and domain knowledge for a variety of big data applications and difficulties that come due to dynamic, complex, noisy and constantly changing and evolving data. The above challenges can be handled using the three-tier architecture as shown in Fig. 1. Three tiers mentioned in the

Fig. 1 Big data mining
framework

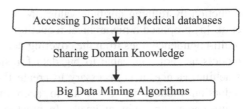

architecture are data accessing and computing, data privacy and domain knowledge
and Big data classification model.

In the first tier, data accessing and arithmetic computing procedures are performed on the distributed data. The main problem is that huge amounts of data are
shared in different locations and the data are growing in each data source enormously. Hence, to compute and analyze large-scale distributed data, we need an
effective platform like Hadoop. In the second tier, data semantics and domain
knowledge are used for a variety of applications that involve big data. In the online
biomedical applications, users collaborate with each other and share knowledge to
the groups/user communities. It is the most crucial task in both low-level and
high-level big data mining algorithms. In the third tier, problems that come from big
data size, distributed data, complexity and dynamic nature are analyzed prior to
mining algorithm. In this tier, uncertain, sparse, incomplete and multisource data
are preprocessed using filtering models and data are analyzed after preprocessing.
After the preprocessing, the local Hadoop learning models are applied to find the
hidden patterns, and the feedback is sent to preprocessing stage.

Hadoop-based classification models are one of the distributed data mining that
classifies high-dimensional unstructured data into meaningful patterns and it helps
users for decision-making and knowledge discovery. These classification techniques make the large datasets appear in a simpler form for hidden patterns.
Classification approaches on big data are broadly classified in two ways, i.e.,
supervised and unsupervised. In supervised learning, all the instances in the training
data have class labels for decision-making. In unsupervised learning, all the
instances in the training data do not possess class labels for decision-making.
Supervised classifier is used to classify the new instance using the training data.

2 Related Work

Yu and Li [1] and Zhang and Suganthan [3] implemented multi-scale decision tree
representation using granularity computation for generating mixed hierarchical
decision rules. They implemented these multi-scale decision tables under different
levels of granularity and level-wise threshold. Al-Khateeb and Masud [2] implemented Probability based decision tree (PDT) under single level granularity to
hierarchical multi-level granularity and research on attribute generalization

reduction by refining attribute values. Mendes-Moreira and Soares [4] and Mathe et al. [5] proposed a hierarchical reduction model for concept hierarchy to acquire multi-confidence rules from the covering decision systems. These attribute selection models are not applicable to big data for storage and manipulations on a single machine. Hence, it is necessary to create the most efficient hierarchical attribute reduction models for big data to accommodate a variety of user's medical requirements on different levels. In most of the big data applications, sampling techniques have been applied to find the relational features or interesting patterns. Sampling models would be practically successful only if the samples are equally distributed or satisfy the hypothetical space. Instead of using sampling techniques, the best solution to handle the large number of attributes for smaller databases is parallel computing. Traditional features selection based classifiers were applied on large databases to compute granularities separately and later combined together to find the global solution of the whole data. But there is no guarantee, as these partitioned attributes or instances could exchange relational information to each other. Thus, in the majority of the cases, these techniques fail to extract a subset of features for large datasets. Graph-based medical disease models have been gaining a lot of attention due to its uncertainty and the process they detect disease patterns or relationships between entities. The structural representation holds essential information for categorizing and visualizing entities and, hence, useful in learning, clustering and decision-making in biomedical applications. Graph-based classification models are promising and could optimize traditional keyword-based techniques [5].

Machine learning models applied on distributed clinical databases attempt to find patterns and relationships, to understand the features and progression of certain diseases. Single class models are used in noise filtering with limited instances. In many medical applications, the degree of class imbalance will alter, particularly while classifying the online medical datasets. Conventional models such as k-nearest neighbors, classify an instance set by comparing its Euclidean distances to each class without considering the features' contextual information. A multiclass imbalanced classification brings a lot of challenges in big data due to its data complexity. A typical solution is to partition the multiclass data into binary classification and then use balancing techniques to combine distributed data. Data imbalance rate in the binary classification can be defined as the ratio between the majority class instances count to the minority class instances count [6, 7]. Existing dictionary-based approach does not give optimal results for identifying protein names, because new protein names tend to build, and there are, sometimes, hundreds in terms of identical proteins are referenced. Medline is a large repository of publicly available scientific literature. Model-based clustering algorithms have been implemented on document clustering [6–8], where document clusters are represented as probabilistic methods that are conceptually separated from the data dimensions. Presently, graph-based clustering models using statistical models are also successfully applied to document clustering mechanism. These graph models are optimized using some predefined document measures on the directed graph. The hierarchical Latent Dirichlet allocation (hLDA) method was implemented in [9] as unsupervised method which is a generalization of LDA. Graphical ranking-based clustering algorithms have been implemented to construct a sentence model graph in which each node is a sentence in the overlay documents. The traditional medical clustering algorithms are not suitable because the algorithms work in batch processing, whereas iterative process merges each iteration cluster into the existing clusters which leads to duplicate attributes.

3   Proposed Model

In this proposed framework, an iterative Hadoop-based ensemble classification model was implemented to find the qualitative patterns from the associated features. Medical disease prediction was executed in three steps as shown in Fig. 2. In the first step, distributed medical data from different sources are integrated as a single dataset. In the second step, an improved distributed data filtering algorithm was applied to replace the inconsistent values in the Hadoop framework. In the third phase, the Hadoop's mapper filtered dataset was used to find the interesting patterns using a reducer's ensemble algorithm (Fig. 3).

Fig. 2 Different static pattern analyzers

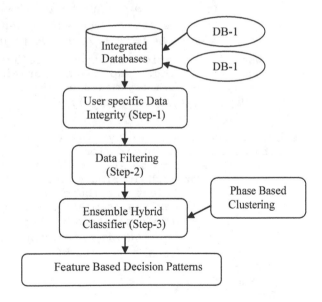

Fig. 3 Attributes similarity computation

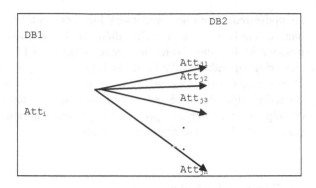

Mapper Phase:
Distributed Data Integration (Combining two databases)

```
Input: Medical datasets: MDList
Output: Single integrated data.
Procedure: Let the two databases are represented as
MDlist1 and MDlist2.
For each attribute Att_i in MDlist1 do
     For each attribute Att_j in MDlist2 do
          If (Att_i != ∅ && Att_j != ∅) then
          If (Type(Att_i)==Type(Att_j))then
```

$$Sim(Att_i, Att_j) = (P(Att_i / Att_j)*P(Att_j))*Correlation(Att_i, Att_j);$$

Where correlation is the correlation between the two attributes. Map each correlated attributes using it similarity measure as :

Map ((Att_i, Att_j) , $Sim(Att_i, Att_j)$);

```
     Done
```

Select maximum similarity values from the list of attributes for data integration.

```
     For each attribute Att_i in MDlist1 do
               For each attribute Att_j in MDlist2 do
                    Select attributes pair (Att_i, Att_j) as
                    Maximum (sim(Att_i, Att_j))
```

D'=Integrate attributes Att_i & Att_j

Map ((Att_i, Att_j) , $Sim(Att_i, Att_j)$);

```
                    Done
     Done
          End if
     Done
```

In this algorithm, medical data from different sources are integrated using relational attributes. Similarity measure was computed to find the attribute's relationship for data integration.

Mapper-Based Data Cleaning (Data filtering method)

In this Mapper preprocessing mechanism, integrated dataset is used to fill the continuous and nominal missing values. Since the distributed medical data may have numerical and nominal missing or inconsistent values, these values are replaced with the computed values. For the numerical type of attributes, the maximum possible estimators can be used to replace the null or inconsistent values. In case of nominal attributes, the conditional posterior probability estimator can be used to replace the null or inconsistent values.

```
Input: Integrated dataset D';
Output: Filtered dataset as FDList;
Procedure: For each attribute Aᵢ in D' do
For each instance I (Aᵢ) in Aᵢ do // Nominal attributes
If (Type (I(Aᵢ)==Nominal && I(Aᵢ)== φ)then
Probability estimation is computed as
```

$$P_1(I(A_i)) = \log(\Pr ob(I(A_i)/Cls_m) + e^{\log(\Prob(I(A_i))}.$$

$$P_2(I(A_j)) = \log(\Pr ob(I(A_j)/Cls_m) + e^{\log(\Prob(I(A_j))}.$$

$$A_j = Max(\{P_1(I(A_i))\}, \{P_2((A_j))\});$$

```
i.e if Aⱼ ∈ P₁(I(Aᵢ)) then
```

$$I(A_j) = val_i ;$$

```
Else
```

$$I(A_j) = val_j ;$$

```
End if // Numerical Attributes
If (Type (I(Aᵢ)==Numerical && I(Aᵢ)== φ)then
```

$$I(Aj) = (Max(num(A_j), num(A_j))/|(\mu - (e^{\log(\Prob(I(A_i))} + e^{\log(\Prob(I(A_j))})/2|);$$

```
End if
End for
End for
FData=Cleaned data;
```

To execute our hybrid ensemble model, we must distribute the filtered data among different machines. The given data are partitioned onto different parallel machines rather than replicating them to minimize memory storage. Due to the large size of data, we should take a random sampling of the training data and ensure the consistency of the ensemble model by repeating of sampling. The input data objects are distributed across the Hadoop mappers. The initial number of clusters and representative object are randomly selected as N and R. These parameters are placed in individual mappers or in a common location and accessed by all the mappers. The distance between each object to the representative object is measured, and the shortest objects are clustered together. The reducer processes accept the clustered objects and update the representative objects using the fuzzy membership value.

Hadoop-based Ensemble classifier (HBEC):

Input: Filtered dataset FData.
Output: Decision patterns for medical databases
Procedure: For each attribute FData do
  Divide the medical records FData into 'N' independent clusters.
  Select a representative point randomly R$_i$
   While i<N do
   Dist (R$_i$, x) = $\lim\limits_{p \to 0} (\sum\limits_{i=1} | x_i - R_i |^p)^{1/p}$

  Assign each data object to the cluster, which has the nearest distance.
  Update representative instance using the fuzzy membership matrix as shown below

$$Mem(\mu_X) = (1 / Dist(R_i, x))^{1/\theta - 1} / \sum_{r=1}^{N = \#\, clusters} (1 / Dist(R_r, x))^{1/\theta - 1}$$

Where $\theta = $ *fuzzy parameter*
Update representative instance= Maximum {Dist (R$_i$,x)}/ Mem(μ_X);
  End while
   Done
Done

Hybrid ensemble decision tree construction:

```
Here the clustered data can be represented as CData.
For each phase clustered data CDij in CData do

If CDij ==Null then
Return leaf node with matched medical pattern set as
empty.
Else if class (CDij) ==1 then
Return leaf node with medical patterns m.
Else
Split CDij into r disjoint partitions using random
sampling distribution where r=m-classes.
```
Let $CD_1(i1,j1), CD_2(i2,j2)...CD_r(ir,jr)$ are r disjoint partitions
with m-classes such that
$$CD_{ij} = CD_1(i1,j1) \cup CD_2(i2,j2)...\cup CD_r(ir,jr) \quad At_1(n) \quad \text{Corresponds to}$$
the attribute list of the data partition $At_1(n)$.
```
For each matched partition s do
Find the medical attribute ranking using the equation
```
$$\text{ARank}(P, At_i(n)) = \{\sum P(At_i(i)/At_j(i))\}_{m-classes} / \max\{IG(At_i(i), At_j(i))\}$$
```
Where i,j=1,2....n attributes and m=number of classes.
```
$P(A_i(n))$: Probability of the tuples satisfying
θ = Data scaling factor (0-1)
```
If ARank (P, Ati(n),m)< θ then
```
$$\text{ARank}(P, A_i(n)) = \text{AttRank}(P, A_i(n)) + \theta;$$
```
End if
End for
Select the root-node using the attribute with the
highest ARank in all the partitions.
Repeat until no more instances in the based partitions.
Display phase based patterns in the decision tree.
End for // end
```

In this algorithm, ensemble model filtered data were clustered, and patterns are extracted using hybrid decision tree construction. In each phase, cross-defect metrics and its relationships are evaluated using the pattern discovery process. These patterns are used to identify the medical documents and its dependencies in each partition. As the scaling parameter changes, different decision patterns are evaluated at each iteration.

4 Performance Analysis

In this experimental study, we have analyzed the Proposed Ensemble Classifier model with traditional algorithm model in terms of classifier accuracy, different cluster rate and run time taken to classify different medical datasets such as Medline and PubMed repositories. Separation index is measured between clusters, it tells about the compactness and measures the gap between clusters in a partition. Entropy is defined as the measure of uncertainity for partition set. Purity is a simple evaluation measure for checking quality of a cluster. The lower entropy means better clustering and greater entropy means that the clustering is not good. The quantity of disorder is found using entropy measure (Tables 1 and 2). Bad clusters have purity value close to 0, a perfect clusters has a purity of 1. To measure the efficiency of the clustering we use precision. Precision=True Positive/(True Positive+False Positive)

In Table 3, time taken to extract medical patterns is minimized using the proposed model compared to the traditional models.

Table 1 Comparing the proposed model clustering accuracy of ensemble model with other algorithms in terms of Entropy, Seperation Index and Precision

Algorithm	Avg_Cluster_Entropy	Avg_Separation_Index	Precision (%)
Hierarchical multi-class DT	0.61	0.25	78.34
P2P C4.5	0.678	0.26	74.25
CART	0.598	0.198	89.6
Neural networks	0.698	0.473	89.13
PDT	0.498	0.526	83.5
Proposed ensemble model	0.3987	0.187	94.76

Table 2 Comparing the Proposed Ensemble classifier performance with different algorithms based on different cluster rate

Algorithm	5-clusters based classifier accuracy (%)	10-clusters based classifier accuracy (%)	30-clusters based classifier accuracy (%)	40-clusters based classifier accuracy (%)
Hierarchical multi-class DT	69	78.35	67.88	79.35
P2P C4.5	74	84.5	82.34	71.45
CART	89.57	81.56	79.67	81.46
Neural networks	82.56	69.35	71.64	82.45
PDT	75.74	83.45	78.34	74.35
Proposed	91.45	88.43	89.35	92.46

Table 3 Comparing the proposed model with traditional algorithm models in terms of Runtime (ns)

Algorithm	Total patterns	Runtime (ms)
Hierarchical multi-class DT	100	2562
P2P C4.5	100	3244
CART	100	2891
Neural networks	100	2608
PDT	100	2382
Proposed ensemble model	100	1693

5 Conclusion

Pattern extraction from medical databases using a traditional rule-based approach results in more error rate for similarity identification of gene/proteins. Patterns which are subsets of medical patterns are not relevant to the biological study. Traditional models do not provide an efficient preprocessing approach for protein/gene names tokenization. Data imbalance, indexing and sparsity are the three major issues in these distributed decision tree models. In this proposed model, an enhanced attributes selection ranking model and Hadoop-based decision tree model were implemented to extract the user-specific interesting patterns in online biomedical databases. Experimental results show that the proposed model has high true positive, high precision and low error rate compared to traditional distributed decision tree models.

References

1. Zhiwen Yu, Le Li; Jiming Liu, "Hybrid Adaptive Classifier Ensemble", Cybernetics, IEEE, (2015):177–190.
2. Al-Khateeb, Masud, "Recurring and Novel Class Detection using Class-Based Ensemble for Evolving Data Stream", IEEE Transactions on Knowledge and Data Engineering (TKDE), (2015):34–45.
3. Le Zhang, Suganthan P.N, "Oblique Decision Tree Ensemble via Multisurface Proximal Support Vector Machine", IEEE Transactions on Cybernetics, (2015):2165–2176.
4. Joao Mendes-Moreira, Carlos Soares "Ensemble approaches for regression: A survey", ACM Computing Surveys, Volume 45 Issue 1, November (2012):123–136.
5. Mathe, C., Sagot, M.F., Schiex, T., Rouze, P. Current methods of gene prediction, their strengths and weaknesses. Nucl Acids Res. 2002;30:4103–4117.
6. Stein, L. The case for cloud computing in genome informatics. Rev J: Genome Biol. 2010;11:207.
7. Mason, C.E., Elemento, O. Faster sequencers, larger datasets, new challenges. Genome Biol. 2012;13:314.
8. Shanjiang Tang, Bu-Sung Lee, Bingsheng He, "DynamicMR: A Dynamic Slot Allocation Optimization Framework for MapReduce Clusters", IEEE Transactions, Vol 2 No.3 Sep 2013, pp. 333–345.
9. Wullianallur Raghupathi and Viju Raghupathi, "Big data analytics in healthcare: promise and potential", Health Information Science and Systems, pp. 1–10, 2014.

A Framework for Secure Mobile Cloud Computing

Lakshmananaik Ramavathu, Manjula Bairam
and Sadanandam Manchala

Abstract Mobile Cloud Computing (MCC) presents new kinds of offering and amenities to mobile customers to have maximum benefits and advantages of cloud computing. In spite of these advantages, protection is still the main concern and a reason of worry for cloud customers. This paper specializes in the use of biometric authentication framework for comfortable entry to confined information within the cloud making use of a mobile. Certainly, biometrics supplies a bigger measure of protection than average authentication methods. In this paper, we proposed the pre-processing steps and algorithms for extracting the features and matching the biometrics traits. By making use of the proposed algorithm, the user is not going to experience handiest benefit from local computing power and storage capacity, in addition to that they are going to get advantages and benefit of higher authentication accuracy, customized services with low hardware cost and secure entry.

Keywords MCC · Fingerprint · Biometrics · Secure · Classifiers · K-mean · Ridges · Enrollment and verification

1 Introduction

Capability to access information and application anytime anywhere at low cost is the main advantage of MMC. The most significant protection issue on MCC is protecting and securing remote application and data from illegal access. At the same

L. Ramavathu (✉) · M. Bairam · S. Manchala
Kakatiya University, Warangal, Telangana, India
e-mail: lakshman.ramavathu@gmail.com

M. Bairam
e-mail: manjula.ramavathu@gmail.com

S. Manchala
e-mail: sadanb4u@yahoo.co.in

© Springer Science+Business Media Singapore 2017 353
S.C. Satapathy et al. (eds.), *Proceedings of the First International Conference on Computational Intelligence and Informatics*, Advances in Intelligent Systems and Computing 507, DOI 10.1007/978-981-10-2471-9_34

time an authorized user easily can access information; cloud provider also can achieve this. There is also the likelihood of unauthorized access, which is access using third party such as hackers. As a consequence, the security limitation in MCC recently has become hot research for study [1, 2].

Users can employ the system of authentication to make use of cloud services via online user interface, either an internet provider application programming interface (API) or mobile software, or internet browser in cloud computing environment. Authentication on cloud is imperative to furnish simplest and secure access to cloud services through recognized users. Currently, authentication process is done in various approaches, like a simple text password [3].

MCC provides internet-based services like computing and storage facility for all kinds of users in different fields including financial, education, government, etc. MCC brings a new kind of benefits and services to mobile customers to get the whole benefits cloud computing [4]. Regardless of these benefits and advantages, security continues to be a great issue, which is considered as an anxiety source for cloud users. Provider of cloud services has accepted the complexity of cloud security; they are also working deeply to deal with it [5]. In reality, security of cloud is becoming a competitive key and edge differentiator among providers of cloud.

By making use of the approaches and practices of robust security, the security of cloud soon will be a boost far above level of IT departments obtained using their own application strategy and hardware components [6]. This paper specializes in the usage of fingerprint recognitions for securing the access to restrict the information in the cloud by utilizing a mobile phone. Certainly, biometrics supplies a better measure of protection than ordinary authentication technique, which means the resources of cloud available are most effective to authenticate users and to protect from illegal and unauthorized customers [7]. In this paper, we proposed the pre-processing steps and algorithms for extracting the features and matching biometric data to compare with template database stored in systems. In this proposed algorithm, we recommend a new and sophisticated framework, which will be used to perform the authentication of fingerprint password as web-based services within the cloud computing. Using the suggested cloud-based platform of authentication, we shall be able to apply many advanced algorithms practically for recognizing the large-scale feature.

By making use of the contribution of this proposed algorithm, the customers are not going to benefit best from nearby storage potential and computing vigor, but also, they are going to benefit from better customized service, secured access, and high authentication accuracy with low hardware expenses [8]. Individual biometrics authentication presents an effective solution to segregate portions related to the identity management. Indeed, biometrics makes use of automatic schema to be able to distinguish persons according to their own behavioral and/or physiological features [9].

2 Biometric Authentication Framework for MCC

A biometric authentication framework is a one-to-one suit that determines whether the claim of an individual is true, customarily labeled as remote server or client end authentication. There are two phases: Enrollment and Verification. The client registers his biometric data as a template to the system during the enrollment phase. During the verification phase, the client biometric information is matched with the stored template in the system and makes the decision according to the result and this can be shown as in Fig. 1.

The proposed mechanism of biometrics authentication based on recognition of fingerprint to comfortable accessing to MCC is described in detail. Not too long ago, there have been a number of research and works about making use of webcam or any digital camera (digicam) to represent the sensor. The embedding of a particular fingerprint sensor or adding external hardware to act as a reader of fingerprint will be probably highly priced, also it will have an effect on the simplicity of mobile. Using the prevailing digital camera to capture photos of fingerprint in mobile as sensor for biometric is cheap to enforce. The suggested system is utilizing the method of fingerprint recognition to acquire the tip of a finger images by means of the mobile digicam.

2.1 Fingerprint

Fingerprint is the pattern gift on a finger. It comprises intricate patterns or symbols of stripes, referred to as a ridge. In fingerprint, darkish traces of the image are known as the ridges and the white discipline in between the ridges is named valley [10, 11].

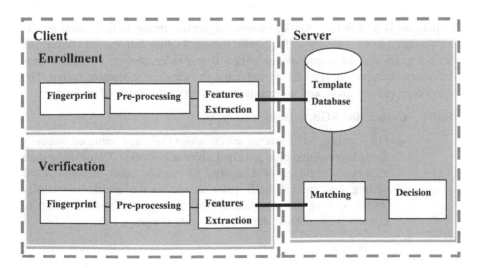

Fig. 1 Biometrics authentication framework for MCC

Fig. 2 Features of fingerprint

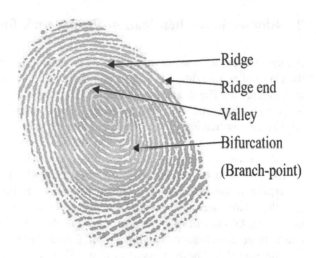

Ridge

Ridge end

Valley

Bifurcation

(Branch-point)

A ridge can unfold into two approaches, i.e., ends and fork into two ridges. If it ends, we referred as termination or ridge end. If it fork into two ridges, we referred as break-point or bifurcation. These two basic types' ridge end and bifurcation can be considered as the basic minutiae points.

Fingerprint recognition proceeds with the aid of deciding upon all of the trivia facets after reducing their lineaments and then comparing these features. Fingerprint recognition entails three important steps. These steps need to be followed in order so that correct matching of fingerprints can be performed (Fig. 2).

2.2 Pre-processing

The purpose is to transform mobile camera fingerprint image to fingerprint image, which is as equivalent as sensor fingerprint image. Mobile cam cannot convert the snapshot to be like the output picture obtained and processed via utilizing fingerprint sensor; however, at least this system aims to export an acceptable output. The proposed preprocess steps and how each steps works are explained as follows:

Step 1: Convert the RGB to greyscale images

- For instance, RGB images are composed of three unbiased channels for red, green and blue principal color accessories. Color images are as a rule developed of a number of stacked color channels. Converting the genuine RGB color image to the grayscale intensity snapshot: here, we used the *grayscale () or rgb2gray* function to convert RGB color images to grayscale images.

Step 2: Normalization

- Normalization allows standardizing the distorted levels of variation in the grayscale values among ridges and valleys by assigning pre-specified standard mean and variance values. Normalization is implemented to grayscale image to reduce the effect of illumination's differences. The equation of the grayscale normalization is:

$$K = double((A - min)*255/(max - min))$$

where A is original image value and K is opting image value (after normalization).

Step 3: Reduce blur effects

- It is possible to set a threshold in order so that only pixels which might be identical to one another are blurred collectively using selective blur filter. It is typically priceless as a device for decreasing graininess in images without blurring sharp edges. The implementation is much slower than a Gaussian blurring, so that you will have to no longer use it until you particularly want the selectivity. Here, we apply a Gaussian blur to the input snapshot.

Step 4: Segmentation

- Image segmentation is the procedure of dividing an image into a couple of elements. This is most often used to establish objects or different valuable knowledge in digital snap shots. There are lots of special methods to participate in image segmentation together with: threshold method, transform method, texture method and color-based segmentation methods. In this section, we used color-based segmentation using k-means clustering.

Step 5: Estimating the orientation

- The fingerprint image discipline orientation defines the nearby orientation of the ridges included in the fingerprint. Regional ridge orientation is estimated at each and every factor by means of finding the most important axis of version in the image gradients. In ridge frequency estimation, the frequency snapshot represents the local frequency of the ridges in a fingerprint. Window size, block measurement, min. & max. and wavelength parameters are initialized for estimation of ridge frequency. Here, we used the Gabor filter to filter the fingerprint images.

Step 6: Ridges enhancement

- After applying Gabor filter we get enhanced fingerprint which has fine quality of true trivial. Wavelet turn out is to be used to de-noising the images and increase the contrast between the ridge and background using a map operation to the wavelet coefficient set. Lastly, we take part in binarization of recent more advantageous images.

Algorithm 1: Core Ridges Extraction

```
1.   for i=1:n. do

     {

         a. Get  the  Grey-scale  image  G_SI  (VA)  /*V  is
            vertices & set of edges is A*/
2.   Minimizes the distance function

     D (G_SI) =β (G_SI) +γP (G_SI)

3.   If

     {

     Single point is observed within the extracted edges
     set

4.   Then

     It is viewed as an ending of simple ridge

5.   Else

     It's considered as an ending of a ridge bifurcation

6.   Filter  the  ridges  ending  and  ridge  bifurcation
     obtained in the extraction step

     } End if

     } End for
```

2.3 Features Extraction

A fingerprint image is viewed as a group of ridges instead of a collection of pixels. This paper adapts the well-known recognized algorithm called the Core ridges extraction to generate the skeleton. The smooth ridges are the core ridges [12, 13] if and only if they fulfill the property. It does not intersect itself, it must self-reliable and it has finite size inside any bounded subset of the discipline. Extra small print may also be found in [14]. Algorithm 1 suggests the technical steps [15].

Fingerprint images saved or stored in mobile will not be required because each time the consumer needs to access the cloud, it captures a brand new fingerprint image and it is as simple as login. After the pre-processing and the feature extraction steps are finished, data have to be transferred to the server and saved in template database. It takes all benefit from the cloud and the approach is hosted on cloud.

3 Matching Algorithm

This step involves comparing the input images with the template images. During the enrollment, template images are collected and stored in cloud as template database. The input image is compared against template images during the recognition phase. In this phase, whether the input image matches with template image to form the identical finger or not is decided. For this, a novel *matching algorithm* is proposed for score features of biometrics. This algorithm is the combination of Strong and weak Classifiers to endeavors and discovers a blend of "weak rankers" to make a precise single ranker. To score the features, the matching scores of each subsystem are combined to find multiple matching score which is then sent to the decision phase. One of the standard approaches to score the features is to attempt classifier for determining the best decision edge between imposter and genuine instances. The Strong Classifier and its pseudo-code are shown in Algorithm 2.

Matching algorithm can be referred and despite similarity of RankBoost.B and AdaBoost [16], some inconsistency exists among them and thus there is a need to improve Weak Classifier. The final output of RankBoost.B and AdaBoost algorithm is a linear combination of WeakClassifier.

In Algorithm 2, the Weak Classifier is called for each iteration and it generates a weak ranking. Matching algorithm preserves a distribution D_t over $a_0 \times a_1$ that is passed on to iteration m to the Weak Classifier. Intuitively, Matching Algorithm decides D_t to underline diverse parts of the training data. A high weight is assigned to a couple of instances, which shows an incredible significance of the Weak Classifier. Hence, the Weak Classifier matches the instances effectively. The Weak Classifier and its pseudo-code are shown in Algorithm 3. Like boosting-based algorithm, Matching Algorithm additionally incorporates training Weak Classifier subroutine with slight difference. In matching algorithm, Weak Classifier gives weak ranking rather than weak classification. Weak Classifier dependably has esteem for every instance.

Algorithm 2: Strong Classifier

---

1. Input: a_1 and a_0 are disjoint subset of A

    /* $a_1 \cup a_0 = A$ */
2. Initialize

$$w_1(a) = w \begin{cases} \dfrac{1}{|a_1|} & \textit{If } (a_1 \in A) \\[2mm] \dfrac{1}{|a_0|} & \textit{If } (a_0 \in A) \end{cases}$$

    /* total number of training sets*/
3. *for* m := 1,2, . . ., M do loop:
    {
    Call the WeakClassifier(A)

4. $D_t(a_0, a_1) = w_t(a_0).w_t(a_1)$;

5. Get weak ranking $k_t : A \in R$

6. Select $\theta_t \in R$

7. Update $w_{t+1}(a) = w_{t+1}(a_0).w_{t+1}(a_1)$ where $(a_1 \in A)$ and $(a_0 \in A)$

    } End of loop m
8. Output: Final Ranking function:
    $$K(a) = \begin{cases} 1 & \textit{If } (F(a)=1) \\ 0 & \textit{If }(F(a)=0) \end{cases}$$

The detailed explanation of matching algorithm is as follows:

The feedback algorithm operates in iterations; in each iteration, it calls Weak Classifier to find the best weak ranking. Weak Classifier is called for each iteration of m to maintain distribution D_t over a_0 and a_1. Iteratively, "matching" chooses D_t to emphasize different parts of the training data. Set ranking score or feature for finding a weak ranking k that is equal to one of the ranking features or thresholds τ_k. Assume that our model is given 'n' ranking feature or score denoted $t_1, t_2, \ldots, t_N$, C^* is a pointer variable which is used to assign higher scores that are assigned to more preferred instances. Variable r_1 is assigned with 1 if a_i is equal to 1. Variable r_2 is assigned with 0 if a_i is equal to -1. This algorithm incrementally evaluates a sorted list of candidate thresholds $[\tau_k]_{k=1}^K$ and stores the values i^* and τ^* for which $f(a_1) > f(a_0)$, means that instance a_1 is preferred to a_0 by f. For loop is used to rank all instances of one set over another set, and same as for loop, if statement is used to find the best weak ranking condition. At each iteration of the algorithm, first a ranking feature k_t and associated weight w_t are chosen and updated. Finally, ranking feature k_t can be learned efficiently and each ranking feature is valued.

Algorithm3: Weak Classifier

```
1.   WeakClassifier( A )
     {
```

2. Given: distribution over $a_1 * a_0$

3. set of instance $[t_i]_{i=1}^{N}$

4. For each t_i a set of threshold $[\tau_k]_{k=1}^{K}$ such that

   $\tau_1 \geq \tau_2 \geq,..., \geq \tau_k$

5. Initialize:

   $$C^* = 0$$

6. $r_1 = 1 \quad If(a_i = 1)$

7. $r_2 = 0 \quad If(a_i = -1)$

8. *for* i:=1 up to N do loop

   {

   8.1. *for* k:=1 up to K do loop

      {

      Q=0

      $\tau = 1$ /*threshold*/

      $$Q = Q + \sum_{a:t_i(a)=1} \beta(a) \quad \text{/* Sum of Potentials */}$$

   8.2. if $(Q > C^*)$

      {

      $i^* = i$

      $\tau^* = \tau_k$

      $r^* = w_i$

      } End if
      } End for loop k
      } End for loop i

9. Return (r^*, τ^*, i^*)
   } End WeakClassifier

---

4 Decision

In this section, first in the enrollment section, the person presents the finger to the mobile camera to receive fingerprint. These pattern fingerprints are pre-processed and features are extracted from the image and stored in a template database. After taking fingerprint image as login type in the verification phase, the login image is pre-processed and features are extracted. Then matching algorithm is performed between login image and template. If matching succeeded, then the user is accepted or rejected. The matching score (K) is the outcome of the comparison between the extracted elements and features saved in a template database.

```
If (K is less than threshold) then less matching
If (K is higher than threshold) then high matching
```

Subsequent to that, the matching rating (K) can be determined by the comparison of predefined threshold (τ) value.

```
If (K > τ) then, the user is accepted
Else if (K < τ), the user is rejected [17].
```

5 Conclusion

The mobile computing and cloud computing are combined to form MCC. At the same time MCC brings the new security threats and challenges such as unauthorized user's access. This paper focusses on how to protect the MCC resources from illegitimate access. In this regards, we used the biometric recognition system. This is used to protect resources of mobile devices and cloud. Validating cloud mobile clients utilizing the current mobile cam as a unique fingerprint sensor to get a unique fingerprint image and after that preparing and reorganization were proposed. In this paper, we have theoretically proved the authentication system for MCC. We are planning to implement this contribution and simulate the logs files in future. In light of these logs, cloud security confirmation policies will be altered and re-designed. Accessing log record will presumably be utilized to discover unapproved endeavors to get to data by outsiders, the cloud supplier or any gatecrashers.

References

1. Li X., "Cloud Computing: Introduction, Application and Security from Industry Perspectives," *International Journal of Computer Science and Network Security*, vol. 11, pp. 224–228, 2011.
2. Yu X. and Q. Wen, "Design of Security Solution to Mobile Cloud Storage," *Knowledge Discovery and Data Mining*, pp. 255–263, 2012.
3. Dinesha H. and V. Agrawal, "Multi-level authentication technique for accessing cloud services," in Computing, Communication and *Applications (ICCCA), 2012 International Conference on*, 2012, pp. 1–4.
4. Dinh R., C. Lee, D. Niyato and P. Wang "A Survey of Mobile Cloud Computing: Architecture, Applications, and Approaches," *Wireless Communications and Mobile Computing*, vol. 38, pp. 1–38, 2011).
5. Song W. and X. Su, "Review of Mobile cloud computing" *IEEE Transactions on Communication Software and Networks*, vol. 4, pp. 1–4, 2011.
6. Martinez G., D. Castano, E. A. Rua, J. A. Castro and D. R. Silva, "Secure crypto-biometric system for cloud computing" *IEEE Transactions on Securing Services on the Cloud*, vol. 1, pp. 38–45, 2008.
7. Ronghing C., R. Chumning T. Zhenghua and Z. Qingkai, "Identity Based Encryption and Biometric Authentication Scheme for Secure Data Access in Cloud Computing" *Chinese Journal o/Electronics*, vol. 21, pp. 254–259, 2012.
8. R. Lakshman Naik & S. S. V. N. Sarma, "A Framework for Mobile Cloud Computing" International Journal of Computer Networking, Wireless and Mobile Communications (IJCNWMC), Vol. 3, Issue 1, pp. 1–12, Mar 2013.
9. Ashok J., V. Shivashankkar and P. Mudiraj, "An Overview of Biometrics," *International Journal on Computer Science and Engineering*, vol. 2, pp. 2402–2408, 2010.
10. Sangram Bana and Dr. Davinder Kaur. "Fingerprint Recognition using Image Segmentation", (IJAEST) *International Journal of Advanced Engineering Sciences and Technologies*, Vol No. 5, Issue No. 1, 012 – 023
11. Graig T. Diefenderfer, "Thesis on –Fingerprint Recognition||" at *Naval Postgraduate School*, Monterey, California, June 2006
12. Milao D., Q. Tang, and W. Fu, Fingerprint minutiae extraction based on principal curves, *Pattern Recognition Letters*, vol. 28(16), pp. 2184–2189, 2007.
13. Zhang J. and J. Wang, "An overview of principal curves", *Chinese J. Comput.*, vol. 26(2), pp. 129146, 2003.
14. Ghany K.K.A., H.A. Hefny, N.I. Ghali, A.E. Hassanien,"A Hybrid approach for biometric template security", *The 2012 IEEE/ACM international conference on Advances in Social Network Analysis and Mining* (ASONAM2012), pp. 941–942, Turkey, 2012.
15. Ghany K.K.A., M.A. Moneim, N.I. Ghali, A.E. Hassanien, H.A. Hefny, "A Symmetric Bio-Hash Function Based On Fingerprint Minutiae And Principal Curves Approach", ASME press, New York, *The 3rd International Conference on Mechanical and Electrical Technology*, ICMET2011, Vol. 1, pp. 405–410 China, 2011.
16. Yoav Freund, Raj Iyer, Robert E. Schapire, Yoram Singer, "An Efficient Boosting Algorithm for Combining Preferences" *Journal of Machine Learning Research*, 4 (2003) pp. 933–969
17. Iehab AL Rassan, Hanan AlShaher, "Securing Mobile Cloud Computing using Biometric Authentication (SMCBA)" *2014 International Conference on Computational Science and Computational Intelligence*, 978-1-4799-3010-4/14 DOI 10.1109/CSCI.2014.33.

Intelligent Evaluation of Short Responses for e-Learning Systems

Udit kr. Chakraborty, Debanjan Konar, Samir Roy and Sankhayan Choudhury

Abstract Evaluation of learners' response is an important metric in determining learners' satisfaction for any learning system. E-Learning systems currently use string matching or regular expression-based approaches in evaluating short answers. While these endorse the correctness of an answer, they are limited to handling predictable errors only. The nature of errors, however, may vary and it is important to intelligently judge the nature of the error to correctly gauge the state of learning of the learner. A better learning experience requires the system to also display benevolence, which is an innately human behavior characteristic, in evaluating the response. The current paper presents a k-variable fuzzy finite state automaton-based approach to implement an evaluation system for short answers. The proposed method attempts to emulate human behavior in the context of errors committed which may be knowledge based or inadvertent in nature. The technique is explained with sample scores from test conducted on a group of learners.

Keywords e-Learning · Evaluation · Multiword · Inadvertent error · K-state fuzzy finite automaton

U.kr. Chakraborty (✉) · D. Konar
Department of Computer Science & Engineering, SMIT, Sikkim, India
e-mail: udit.kc@gmail.com

D. Konar
e-mail: debanjan.konar@gmail.com

S. Roy
Department of Computer Science & Engineering, NITTTR, Kolkata, India
e-mail: samir.cst@gmail.com

S. Choudhury
Department of Computer Science & Engineering, University of Calcutta, Kolkata, India
e-mail: sankhayan@gmail.com

© Springer Science+Business Media Singapore 2017
S.C. Satapathy et al. (eds.), *Proceedings of the First International Conference on Computational Intelligence and Informatics*, Advances in Intelligent Systems and Computing 507, DOI 10.1007/978-981-10-2471-9_35

1 Introduction

Intelligent Tutoring Systems, proposed during the early 1960s were initially modeled on the way the human tutor worked. To implement the same, artificial intelligent techniques were considered. However, in spite of few notable breakthroughs, the acceptance of such computer-aided educational tools has not been as anticipated [1]. During recent times, with the growth of Information and Communication Technologies (ICT) and the ever increasing reach of the internet, online learning systems have gained importance. According to a report by the Sloan consortium, 3.9 million (over 20 %) students in the U.S. were taking at least one online course in 2008, which shows a 12.9 % increase in online enrollment over a period of just 1 year [2]. The reason for this growth is that online courses offer "anytime," "anywhere" learning which provides flexibility and convenience for students and instructors [3].

A major issue in implementation and subsequent acceptance of e-learning systems providing online learning support is learner satisfaction. A number of factors can be listed which contribute to the success of online courses. Phipps and Merisotis (2000) provided a comprehensive list of benchmarks for measuring success in internet-based online courses. The benchmarks include Institutional support, Course development, Teaching/Learning Success, Course structure, Student support, Faculty support, Evaluation, and Assessment [4].

This paper reports work done on Intelligent Evaluation of students' responses for better learning experience in an e-learning environment. The proposed system tries to emulate the behavior of a human evaluator in benevolently accepting learner responses with inadvertent errors, which otherwise would be graded incorrect by standard string matching or regular expression-based systems.

2 Literature Survey

As reported in [5], the types of questions to be supported by e-learning systems are:

- Short answers (Short): a textual answer consisting of a few words.
- Essay: a textual answer with an unlimited or limited number of words that is not corrected automatically.
- Multiple choice questions (MCQ): choose one option out of a list of possible answers.
- Multiple response question (MRQ): choose one, more or no option out of a list of possible answers.
- Fill in the blanks (FIB): complete missing words in a sentence or paragraph.
- Match: given two lists of terms, match each term on one list with one term on the other.
- Crossword (Cross): fill out a crossword using definitions of words in horizontal and vertical positions.

Among the types listed, at least three variants exist which accept single word or short phrases as possible responses. These include short-answer questions, fill in the blanks and crosswords. While being significant pedagogically, as these aids the learner in knowledge building on a complex topic by building several small parts of a larger artifact, these types of questions also aid in developing the learners' reading and writing skills [6]. The popular acceptance of this type of questions in e-Learning systems and in systems requiring online evaluation, however, can be attributed to the ease of implementation. All major learning platforms such as IMS QTI, Moodle, and OpenMark provide support for implementation and evaluation of such questions [5].

The techniques followed in evaluating the learners' response to the types of questions requiring sentences to be of lengths varying from single word to about a phrase, however, vary from implementation to implementation. Automatic evaluation of answers to these type of questions frequently use direct string matching algorithms or pattern matching using regular expression [2] and are followed by popular tools such as IMS QTI [7], Moodle [8] and Blackboard [9]. While direct string matching supports questions with factoid answers or having only one correct answer, these are not appropriate for phrases which may have variations. While regular expressions offer a higher degree of freedom to the learner in expressing the answers, these are restricted by the evaluators imagination as the framing of the regular expressions decides the score of an answer.

Another scenario was explored in the work reported in [10], where the authors considered inadvertent mistakes committed by the learner while answering. A pertinent issue, especially for e-learning, where tests are conducted online these mistakes may occur out of typographical or phonetic errors committed by learners while typing or learning through recorded or live voice over electronic media. The string matching or regular expression-based approaches do not consider such cases and thus hinder the achievement of a fulfilling learning experience.

The limitation of the work reported in [10], however, was that it was designed for single-word responses only. The current paper proposes work done towards implementing a similarly benevolent system for multiple-word responses.

3 Problem Definition

The conventional model of response analysis as shown in Fig. 1 would classify a learners' response as correct only if it follows the mapping m-p-x. However, the modeled scenario under consideration deals with the mapping m-q, where the learner types an incorrect response in spite of being in a state of knowledge. The error committed by the learner is of a nature which would be accepted by the human evaluator who would then be grading the response as partially correct and scores it with some penalty. This intelligent response analysis, depicted through Fig. 2 is what is to be achieved for multiple-word responses.

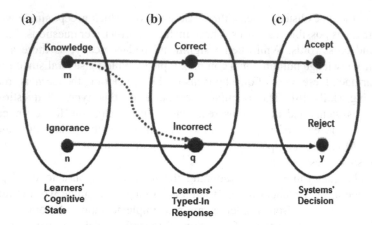

Fig. 1 Conventional model of response analysis [10]

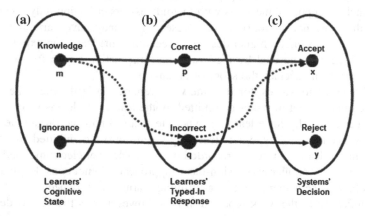

Fig. 2 Intelligent model of response analysis [10]

The problem definition given in [10] states:

Let $r = a_1 a_2 \dots a_n$ be a string of alphanumeric characters that corresponds to an expected response of a learner during an interaction between the learner and the system. There is a set $A = \{s_1, s_2 \dots, s_m\}$ of strings such that $s_i \neq r$, and $|s_i| = |r|$, for each i. Each s_i is a variation of the expected response r such that a human evaluator would interpret it as acceptable response even though endowed with some spelling mistake within a tolerable limit. The set B of unacceptable responses is given by $B = \Sigma - A - \{r\}$ where Σ is the universe of the words of length n. It may be noted that both set A and B are unknown in the sense that the entire lists of acceptable and unacceptable responses are not given at the outset. Then, given an arbitrary n character learner's response x, the issue is how to establish the mapping shown in (1):

$$f(x) = \begin{cases} \text{accept,} & \text{if } x = r, \text{ or } x \in A \\ \text{reject,} & \text{if } x \in B \end{cases} \tag{1}$$

However, the above is true only for single-word responses, which may be modeled as Σ-languages for individual responses. A Σ-language is any subset of the set Σ^* of all finite words of the alphabet Σ [11]. Extending the definition to multiple-word responses, the current paper presents a k-variable finite automaton-based model. A k-variable finite automaton on Σ is a finite automaton over the alphabet $(\Sigma \cup \{\diamond\})^k$ [11]. These typically accept k-variable languages which are subsets of $(\Sigma^*)^k$ when defined over Σ.

A k-variable language R in Σ^* is finite automaton recognizable if its convolution $c(R)$ is recognizable by a k-variable finite automaton [11].

However, the problem under consideration gains dimension due to the nature of inputs that may be typed in by the learners. Not restricted to the k-variables as under the formal definition, each answer may have multiple k-variable languages as multiword answers including phrases may be differently framed. Each of the k-variables may further be distorted due to the errors committed by the learner. The model must, therefore, be robust to accept any k'-variable although designed for k-variable language where k' is a distorted variable acceptable to the model.

The problem, therefore, can be stated as—to design a k-variable finite automata-based model to accept learner responses as a k-variable language R and its distorted representation R' consisting of k' variables and grade the same.

To incorporate the scoring of answers, the transitions had to be weighted. So each automaton had to be implemented as a k-variable fuzzy finite automaton. A Fuzzy Finite Automaton (FFA) is a six tuple, defined as:

$M = \langle \Sigma, Q, Z, q_0, \delta, \omega \rangle$ where Σ is a finite input alphabet and Q is the set of states, Z is a finite output alphabet, q_0 is an initial state, $\delta: \Sigma \times Q \times [0,1] \to Q$ is the fuzzy transition map and $\omega: Q \to Z$ is the output map [12].

4 Solution Strategy

Considering that the responses will also have to be graded, a weighted transition scheme was followed. The proposed work included the following types of transitions:

- Correct alphabet transition
- Incorrect but acceptable alphabet transition
- Incorrect alphabet transition
- Missing alphabet transition

The incorrect but acceptable alphabet transitions are based on the typographically valid substitution set (TVSS) and phonetically valid substitution set (PVSS) for each alphabet. The TVSS consists of the eight neighborhood keys for a standard

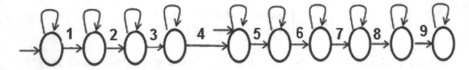

Fig. 3 Model automaton

Table 1 Transition dictionary

Transition	Correct alphabet	PVSS	TVSS
1	N	–	{B,M,H,J}
2	E	{A}	{W,R,S,D,F}
3	W	{U}	{Q,E,A,S,D}
4	Blank	–	–
5	D	–	{S,F, W,E,R, X,C,V}
6	E	{A}	{W,R,S,D,F}
7	L	–	{K,I, O, P}
8	H	–	{G, J, T,Y, U, B,N}
9	I	{E}	{U,O, J, K, L}
Loop	Corresponding	Corresponding	Corresponding

QWERTY keyboard. The PVSS consists of valid phonetic substitutions for given alphabets. These sets are prepared a priori.

As an example, for the question *"Name the capital of India"*, the following may be considered as correct responses:

- New Delhi
- Delhi

Figure 3 shows the model automaton, which is explained through Table 1. Since the automaton is designed to accept the input and its valid convolutions, it can start from either transition 1 for one input or from transition 4 for another input pattern. The transitions are defined not only for correct inputs but also for incorrect but acceptable alphabets which represent distorted signal.

The transitions numbered 1–9 are the ones which may be undertaken in case of inputs of the given three types and blank spaces. The scores for each transition are computed as:

$$T_C = \frac{1}{|S|} \tag{2}$$

$$T_T = \frac{1}{P|S-1|} \tag{3}$$

Table 2 Sample responses and their scores

Sl.No	Correct answer	Learners' response	Score
1	New Delhi	New Delhi	0.906
2	Mumbai	Bombay	0.5
3	Mumbai	Mumbai	0.84
4	Mumbai	Mumbai	0.906
5	Kerala	Kerala	1
6	Kerala	Kerala	0.875
7	Kolkata	Kolkata	1

The expression (2) is used for computing the score of the correct transition, while (3) for both TVSS and PVSS. $|S|$ represents the length of the correct answer and P represents the position of the error. Blanks and erroneous repeated entry of the same alphabet do not add any score nor penalize the user. All other incorrect transitions are allowed without scoring.

5 Results and Discussion

Experiments were conducted on a group of 150 students who were asked to answer five (05) questions through a web-based interface in a timed environment. The total time allotted was sixty (60) seconds. The incorrect entries and their scores are listed in Table 2. While some errors were caused due to the students' ignorance, most were typographical errors.

During the evaluation process, it is not possible to determine whether the error has been caused due to ignorance or otherwise unless human intelligence is used to judge it. The evaluation system does not delve into the issue of determining the type of the error, but grades it based on the response.

6 Conclusion

The methodology proposed through this paper is for intelligent evaluation of short responses of students when tested through computer-based online examinations. The proposed system, modeled on *k-variable fuzzy finite automaton* evaluates the learners responses not only for correct and incorrect answers but also judges them for partial correctness occurring out of inadvertent errors committed while answering. Effectively the automaton handles k-variables, their convolutions and distortions. The model handles multiword responses and is an augmentation over existing work in the field. The scores returned for test cases bear resemblance to human evaluators and appear acceptable.

The system being flexible to allow user-centric controls in weight adjustments and transition planning displays intelligence-based benevolence similar to the human evaluator.

References

1. Corbet, A.T., Koedinger, K.R., Anderson, R., (1997) "Intelligent Tutoring Systems", In Helander, M., Landauer, T.K., Prabhu, P., (Eds) *Handbook of Human Computer Interaction*, Second, Completely Revised Edition, Elsevier Science B.v. Chapter 37.
2. Allen, I., Seaman, J., (2008), " Staying the course. Online education in the United States", Needham, MA: The Sloan Consortium.
3. Martin, F., Parker, M., A.,(2011), "Measuring Success in a Synchronous Virtual Classroom", In Sean B. Eom, S., B., Arbaugh, J., B., (eds.), *Student Satisfaction and Learning Outcomes in E-Learning: An Introduction to Empirical Research, Information Science Reference*, pp. 249–266.
4. Phipps, R., Merisotis, J. (2000). "Quality on the line: Benchmarks for success in Internet-based distance education.", Institute for Higher Education Policy. Retrieved from http://www.eric.ed.gov/ERICDocs/data/ericdocs2sql/content_storage_01/0000019b/80/16/67/ba.pdf
5. Gutierrez, I., Kloos, C., Crespo, R. (2010). "Assessing assessment formats: The current picture", *Proceedings of IEEE Education Engineering (EDUCON), 2010*, 1233–1238. doi:10.1109/EDUCON.2010.5492384
6. Bergin, J. (2000). "Fourteen Pedagogical Patterns", Pace University. Downloaded from: https://csis.pace.edu/~bergin/PedPat1.3.html#fillintheblanks
7. Leal, J.,P., Queiros, R., (2009). "CrimsonHex: A Service Oriented Repository of Specialized Learning Objects", in Felipe, J., Cordeiro, J. (eds.), *Proceedings of ICEIS 2009*, LNBIP 24, 102–113.
8. Crisp, G. (2007). *The e-Assessment Handbook*. Continuum International Publishing Group, London.
9. Available at: http://en-us.help.blackboard.com/Learn/9.1_2014_04/Instructor/110_Tests_Surveys_Pools/100_Question_Types/Fill_in_Multiple_Blanks_Questions
10. Chakraborty, U. K., Roy, S. (2011). "Fuzzy Automata Inspired Intelligent Assessment of Learning Achievement". *Proceedings of the Fifth Indian International Conference on Artificial Intelligence*, pp. 1505–1518.
11. Khoussainov, B., Nerode, A., "Automata Theory and its Applications" (2010), First Indian Reprint, Springer International, Berlin.
12. Wen., Mo Zhi, Min., Wan, (2006) "Fuzzy Automata Induction using Construction Method", *Journal of Mathematics and Statistics*, (2):2, pp.-395–400.

Content-Based Video Retrieval Using Dominant Color and Shape Feature

Tejaswi Potluri, T. Sravani, B. Ramakrishna
and Gnaneswara Rao Nitta

Abstract Content-Based Video Retrieval (CBVR) is an approach for retrieving similar videos from the database. Need for efficient techniques of retrieval is increasing day by day. This paper used both color and shape features to retrieve the similar videos. In our system, we identified Key Frames of the video shots in the first step. We found the most dominant color of each key frame and also edge points of each key frame and stored in the feature database. The color and shape features of query video are calculated and compared with the features stored in the feature database. Videos falling within the threshold are retrieved as most similar videos. The combination of two features results in high performance of our system.

Keywords Content-based video retrieval · Edge detection · Identification of dominant color · Euclidean distance

1 Introduction

Because of the recent advancement in the technology, we require automatic retrieval of information from huge databases. Latest trend of gaining information is through videos such as you tube videos. Content-Based Video Retrieval plays a key

T. Potluri (✉)
Department of Computer Science and Engineering, VNR Vignan Jyothi
Institute of Engineering and Technology, Hyderabad, India
e-mail: Tejaswi_p@vnrvjiet.in; Tejaswi_p@yahoo.co.in

T. Potluri · T. Sravani · G.R. Nitta
Department of Computer Science and Engineering, Vignan's University, Guntur, India
e-mail: 26sravani@gmail.com

G.R. Nitta
e-mail: gnani_nitta@yahoo.com

B. Ramakrishna
Department of Computer Science and Engineering, AKRG College
of Engineering, Nallajerla, India

© Springer Science+Business Media Singapore 2017
S.C. Satapathy et al. (eds.), *Proceedings of the First International Conference
on Computational Intelligence and Informatics*, Advances in Intelligent Systems
and Computing 507, DOI 10.1007/978-981-10-2471-9_36

role in retrieving videos from database. The basic method of retrieving videos is Text-Annotation method which does not work efficiently.

Another approach for retrieving videos is to use different features of the video for matching [1–3]. The features such as color, shape and texture are used for video retrieval. Implementation of single feature alone results in retrieving dissimilar videos also along with similar videos. To overcome this, our system used combination of both color and shape features.

For our system, we considered two categories of videos such as animals and birds in our dataset. Each video in the database has to be pre-processed and obtained feature values are stored in the feature database. The pre-processing starts with segmentation of each video in the database into frames. For each frame, most dominant color is identified and stored in the feature database. Along with dominant color, the edge points of the key frame are identified and stored in the feature database. The query video is also processed in the same way and the features obtained are compared with the features stored in the database and the most similar videos are retrieved.

The rest of the paper is organized as follows: In Sect. 2, we briefly describe about the Architecture of our system. Section 3, describes the proposed system. Experimental results are highlighted in Sect. 4. Finally, our paper concludes in Sect. 5.

2 Architecture

The Architecture of the proposed system looks as shown in Fig. 1. It is organized into various blocks. Before implementing the system on the user query video, all the videos in the database have to be pre-processed and the dominant color of key frame and detected edge point values are stored in the feature database. The user query is processed in the same way and obtained dominant color feature and edge

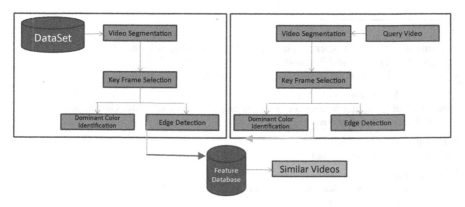

Fig. 1 Architecture diagram

point values are used for similarity matching. The videos having minimum threshold in the similarity matching are retrieved as most similar videos.

The first step of our system is video segmentation in which the video has to be divided into different frames. In the next step system selects key frames from the set of frames. By considering color look-up table all the different colors of each key frame should be mapped into desired color set in the color-mapping step. Next, the most dominant color should be identified and stored in the feature database. For each key frame, edge detection is performed; values of edges are also stored in the feature database.

The user video query also has to undergo all the processing steps mentioned above. The most dominant color of each block is compared with the dominant color of all the videos stored in the feature database. Using similarity metric, the most matching videos are retrieved. For all the retrieved videos, edge detection is performed and edge values of the query video are compared with edge values stored in the feature database. Using matching algorithm, the most similar videos are retrieved.

3   Proposed System

3.1   Video Segmentation

Any video is the sequential arrangement of the images. The video is segmented into spatio-temporal regions termed as shots which are in turn fragmented into frames using the technique [4]. The frames of a video are obtained as shown in Fig. 2.

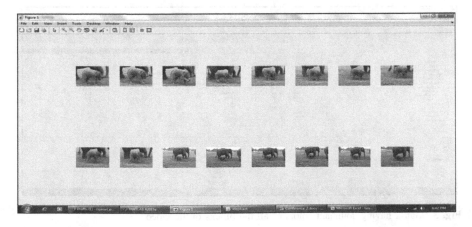

Fig. 2 Frameset of a video after video segmentation

3.2 Key Frame Selection

Key Frames should be selected in such a way that all the contents of a shot can be described using these frames. There are many techniques to select the key frame of a shot. A blind conclusion is that first and last frames of a shot can act as Key Frames.

3.3 Dominant Color Identification

The dominant color of a key frame can be calculated by making a count of all the colors in the key frame. To reduce the complexity, system calculates the count of color bins instead of all colors. The Color bin is a range of colors with their Red, Green and Blue triplets. In our system, we are considering twenty different color bins.

Still more we can reduce the complexity of system by excluding some pixels from processing. As the maximum area of background may be covered with sky or grass, we can check the RGB value of nth pixel instead of processing all the pixels. Here, in our system, we are considering 'n' value as 10. Every 10th pixel's RGB values are extracted. We have to check the bin in which the RGB triplet value falls and increment the appropriate bin by 1. After completing the entire process, the Color bin having highest value will be the most dominant color of the Key Frame. The dominant color bin values of all the frames are stored in the matrix form in the database as shown in Fig. 3.

Fig. 3 Matrix having dominant colors of all the blocks of the video

3.4 Edge Detection

As we are processing color videos, we obtain key frames of more than 5000 color combinations. System is reducing 24-bit color image to 8-bit color image using a simple color quantization to reduce the space and time complexities. This results in 256 different combinations of RGB color spaces. If RGB value of pixel is 0, it stands for 'black' and if RGB value of pixel is 255, it stands for 'white'. System is detecting the edge points by considering the differences of pixel gray values of each pixel and its neighbor's pixel gray values [1]. For a pixel (x, y) with pixel value $p(x, y)$, let $p1(x, y)$, $p2(x, y)$, and $p8(x, y)$ denote the gray values of its neighbors in 8 directions as shown in Fig. 4.

System finds the difference between the pixel value of center pixel and pixel value of its all neighbors as in Eq. 1.

$$D_i = |P(X, Y) - Pi(X, Y)| \text{where } i = 1 \text{ to } 8 \tag{1}$$

In this way, system calculates $D1$, $D2$, $D3$, $D4$, $D5$, $D6$, $D7$ and $D8$. Calculate D by summing up all the differences. System marks pixel 'P' as edge point if it satisfies inequality as in Eq. 2.

$$T1 \le D \le T2 \tag{2}$$

The threshold values $T1$, $T2$ are set based upon the dataset we are considering for our system. The edge point values of all the key frames are stored in the feature database [1]. The edges of various frames are as shown in Fig. 5.

3.5 Color Matching

The dominant color of each key frame of the query video is identified. This dominant color is compared with the dominant colors stored in the feature database using Euclidean Distance metric as in Eq. 3 and calculates Distance1.

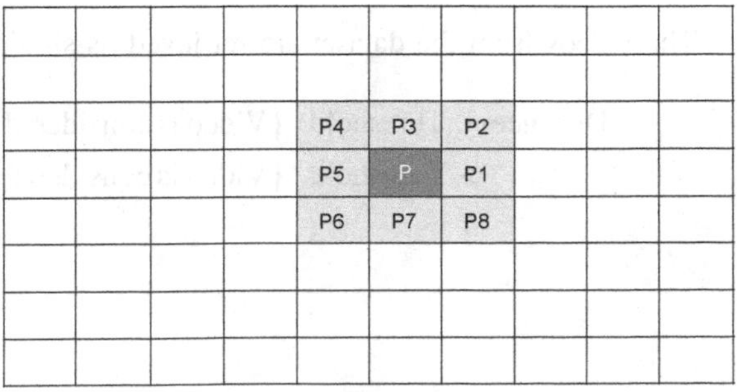

Fig. 4 Neighbor pixels of a pixel 'P'

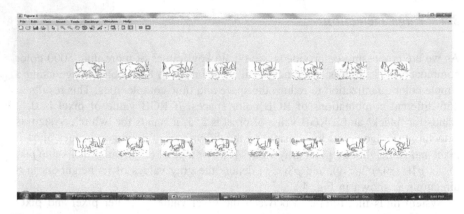

Fig. 5 Edge points of various frames

$$\text{Distance1} = \sqrt{(r_2 - r_1)^2 + (g_2 - g_1)^2 + (b_2 - b_1)^2} \tag{3}$$

3.6 Edge Matching

System calculates edge point values for all the key frames of the query video. Next, system finds the distance between the edge points of the query video and feature database which is denoted by Distance2.

3.7 Similarity Matching

As the first step of matching query video with the videos of the dataset, Distance is calculated as in the Eq. 4 using Distance1 and Distance2 from Eqs. 2 and 3.

$$\text{Distance} = \text{Distance } 1 + \text{Distance } 2 \tag{4}$$

The videos from the dataset are retrieved as similar videos using Eq. 4.

$$\text{Distance} < \text{Threshold} \quad \{\text{Video is considered as similar video}\}$$
$$> \text{Threshold} \quad \{\text{Video is considered as dissimilar video}\}$$

4 Experimental Results

Our dataset consists of two categories of images such as animals and birds. Each category consists of 50 videos. The size of each video varies from 4 MB to 50 MB. The videos of the dataset are of the mp4 format. Based upon the size, the total number of key frames varies. Our results of two queries are shown in the Fig. 6. Our system works efficiently in retrieving the videos of same item even with different colors as similar videos with less time complexity. From the results, we can observe there will be a problem when we are trying to match the video of a single item with video of group of items. For example, if the query video is of single elephant, we can find difficulty in retrieving the videos of group of elephants as similar videos.

For quantitative measurement, we have used precision–recall metric. The plots in the Fig. 7 show the superiority of this technique over CBVR using Colour Feature alone and CBVR using Shape Feature alone.

Fig. 6 Experimental results (the *first slot* is for query video)

Fig. 7 Precision–recall metric chart

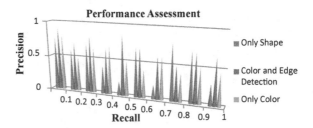

5 Conclusion

We have developed an efficient Content-Based Video Retrieval System using both
Color and Shape features. We have identified the most dominant color of each Key
Frame and edges of each key frame. The Color and shape values are stored in the
feature Database. Videos having similar Dominant Color as well as similar edges
are retrieved as most similar videos. The results show that our technique achieved
promising and effective result in video retrieval

References

1. B V Patel and B B Meshram: 'Content Based Video Retrieval Systems' In: International
   Journal of UbiComp (IJU), Vol. 3, No. 2, April 2012.
2. Vaidehi Bante, Avinash Bhute: A Text Based Video Retrieval Using Semantic and Visual
   Approach In: International Research Journal of Engineering and Technology (IRJET) e-ISSN:
   2395–0056 Volume: 02 Issue: 07 | Oct-2015.
3. M. Petkovic: Content-based Video Retrieval, In: https://doc.utwente.nl/64151/1/Petkovic.pdf.
4. Matthias Grundmann, Vivek Kwatra, Mei Han, Irfan Essa: Efficient Hierarchical
   Graph-Based Video Segmentation.

Intelligent SRTF: A New Approach to Reduce the Number of Context Switches in SRTF

C. Shoba Bindu, A. Yugandhar Reddy and P. Dileep Kumar Reddy

Abstract Throughput of the system in multiprogramming and time sharing systems mainly depends on the careful scheduling of the CPU and other I/O devices. CPU scheduling should control the waiting time, response time, turnaround time, and number of context switches. One of the most extensively used scheduling algorithms is shortest next remaining time first (SRTF), which gives the reduced amount of average waiting time. But this algorithm suffers from some drawbacks. One such is that, every upcoming process if selected for execution, causes a context switch even though it is slightly shorter than the currently running process. As the number of such situations increases, the number of context switches increases, causing the reduction in performance of the system. In this paper, we modify the traditional SRTF to intelligent SRTF, by changing the decision of the preemption, to decrease the number of context switches. The main idea of our proposed algorithm is to make a context switch only if the next process plus context switch over head is shorter than the currently running process. By this we can reduce the number of context switches and thereby the performance of the system is improved.

Keywords Throughput · Scheduling · Burst · Preemptive · Performance · Queue · Multiprogramming

C. Shoba Bindu (✉) · A. Yugandhar Reddy · P. Dileep Kumar Reddy
JNTUA College of Engineering, Anantapur, India
e-mail: shobabindhu@gmail.com

A. Yugandhar Reddy
e-mail: yugandharakkisetty@gmail.com

P. Dileep Kumar Reddy
e-mail: dileepreddy503@gmail.com

© Springer Science+Business Media Singapore 2017
S.C. Satapathy et al. (eds.), *Proceedings of the First International Conference on Computational Intelligence and Informatics*, Advances in Intelligent Systems and Computing 507, DOI 10.1007/978-981-10-2471-9_37

381

1 Introduction

Performance of the operating system mainly depends on the scheduling of CPU and other devices. Scheduler selects a process from the ready queue, which are waiting for service of the processor. Selection of the process should optimize the average waiting time, turnaround time, and first response time. Careful scheduling of the processor is required for improving the throughput of operating system.

Basically there are two types of schedulers, short-term scheduler: selects a process from ready queue and assign the same to processor. Long-term scheduler: selects a job from the disk and places the same in the main memory so that it controls the degree of multiprogramming. The scheduler is anxious primarily with:

- **CPU utilization**: reducing the idle time of CPU as much as possible.
- **Throughput**: number of processes executed per time unit.
- **Waiting time**: time spent in the ready queues
- **Response time**: time taken by the process to generate first response.

There exists different traditional CPU scheduling algorithms, each having its own merits and demerits. Some of the scheduling algorithms are as follows. **First Come First Serve (FCFS)** is simple to implement, process is scheduled according to their arrival time but this algorithm has disadvantage of **convoy effect** (When one CPU intensive process blocks the CPU, a number of I/O intensive processes can get backed up behind it, leaving the I/O devices idle) [1, 2]. **Priority scheduling**: in this, the processes are scheduled according to their priorities. Priorities can be defined either internally or externally [1, 2]. Internally defined priorities are set by operating system; external priorities are set by the criteria outside the operating system. Problem with this scheduling is **starvation** (low-priority process waits indefinitely). Solution to starvation is **aging** [3–6]. **Round Robin scheduling** is designed especially for time sharing systems. A small amount of time, called time quantum is defined. Ready queue is assumed as a circular queue. The processor scheduler traverses the queue, allocates the processor to each process for a time interval of up to 1 time slice. The processor time is shared among the processes equally [7]. Selection of time slice is the key in this algorithm, a poor selection of time slice increases the context switch overhead [8, 9].

In multiprogramming or multitasking systems, context switching means the switching from one process or thread to another in CPU. Multitasking or multiprogramming is not possible without context switches. At the same time it is a overhead to the system. The overhead of context switch may come from several aspects. The processor registers need to be saved and restored, the Operating Systems kernel code (scheduler) must execute, and processor pipeline must be flushed, the TLB entries need to be reloaded [9, 10]. This kind of modifications are involved almost for every context switch in multiprogramming or multitasking system. This kind of switching is called as internal context switching.

In Round Robin Scheduling, a technique called integer programming has been projected to resolve equations that decide a value of quantum time that is neither too large nor too small such that each process has affordable time interval and the throughput of the system is not attenuated as a result of unnecessarily context switches [8, 11]. This is a costly process as it takes time to calculate the value for time quantum. If any new process arrives in middle, again we need to start the calculation from beginning, why because, each of the process is dependent on each other for calculating the quantum time. This system may not work in distributed systems effectively. The shortest next CPU burst algorithm which is the main concern of this paper is discussed in the next section. Shortest next CPU burst algorithm always gives less average waiting time and less average turnaround time in multiprogramming or multitasking systems compared to other scheduling algorithms.

2 Shortest Next CPU Burst Algorithm

Shortest next CPU burst schedules the processes based on the length of next CPU burst. When the CPU is ready to accept the new process, the scheduler selects a process from pool of waiting processes, which has the smallest next CPU burst. FCFS is used to break the tie when the two processes have same next CPU burst. This algorithm is probably best possible; it gives minimum average waiting time for a given pool of processes. Moving the longer job after shorter job increases the waiting time of the longer process and decreases the waiting time of the shorter jobs. As a result, the average waiting time decreases.

Although this algorithm is optimal, there is no way to know the next CPU burst. For this reason, it cannot be implemented in the short-term schedulers. Generally, this algorithm is implemented in the long-term scheduler for long-term (job) scheduling in a batch system; we can use this algorithm as the length of the process time is specified by the user when he submits the job. Thus, users are motivated to estimate the process time limit accurately, since a lower value may mean faster response. (Too low a value will cause a time limit-exceeded error and require resubmission.) SJF scheduling is used frequently in long-term scheduling.

We may not know the next CPU burst but we can predict it. By predicting the next CPU burst, scheduler picks the process with the smallest predicted next CPU burst. To predict burst, we use exponential average of the measured lengths of previous CPU bursts. Let t_n be the length of the nth CPU burst and let $P_{(n+1)}$ be our predicted next CPU burst. Then, for β, $0 \leq \beta \leq 1$, characterize

$$P_{(n+1)} = \beta t_n + (1 - \beta)P_n \tag{1}$$

In the above mathematical relation, t_n contains most recent information, P_n saves the olden times. The parameter β controls the relative weight of topical and antecedents in our prediction. If $\beta = 0$, then $P_{(n+1)} = P_n$ and topical history has no effect. If $\beta = 1$, then $P_{(n+1)} = t_{(n)}$ and only the most recent CPU burst matters (the past is assumed to be old and immaterial). Generally, $\beta = \frac{1}{2}$, so the recent t and the past are equally weighted. The initial P_0 can be defined as an overall system average.

To realize the behavior of the exponential average, we can enlarge the formula for P_{n+1} by substituting for P_n as in (2).

$$P_{(n+1)} = \beta t_n + (1 - \beta)\beta t_n + \cdots + (1-)^j \beta t_{(n-j)} + \cdots + (1 - \beta)^{(n+1)} P_0. \qquad (2)$$

Since both β and $(1 - \beta)$ are less than or equal to 1, each successive term has less weight than its predecessor. Preemptive version of shortest next CPU burst is spoken as **shortest remaining time first** scheduling. Decision of preemption is made when new process arrived at read queue while a process is still running on the processor. Problem with this algorithm is even if the just arrived process is slight shorter than currently running process, it will schedule the new process by doing context switch. If the same thing repeats for more number of times, the system performance will go down because of the context switch overheads.

3   Intelligent Shortest Remaining Time First

As discussed in the previous section, performance of SRTF will drop down because of more context switches. Proposed intelligent SRTF reduces such overhead by altering the SRTF algorithm. Our proposed algorithm preempts the currently running process, only if the just arrived process burst time plus context switch overhead time is less than the remaining burst time of currently running process. By doing so, we can improve the performance of the SRTF by reducing the context switches.

3.1   Algorithm for Intelligent SRTF

```
While(ready queue <> NULL)
Begin
   If(processer is idle)
 Begin
      P := select_shortest(ready queue);
   End
  If(new process arrived)
Begin
      RT=remaining time of currently running process;
       If(new processs burst time + context switch time<RT)
then
Preempt(current process, new process);
End
End
```

Table 1 Process and their arrival and CPU burst times

Process	Arrival time	CPU Burst time
P1	0	10
P2	1	8
P3	2	6
P4	3	4

If the processor or CPU is idle then a new process P which is shortest among the available processes from the ready queue is selected and is assigned to the processor. Whenever a new process enters into the ready queue, if (new process's burst time + context switch time < RT) is met then the processor preempts the currently running process and switches to the new process, otherwise continues with the current process.

3.2 Case Studies

In this section, we compare the algorithms with the help of an example. Time taken for the context switch is highly dependent on the machine and also it is not same for all the context switches, i.e., all context switches do not take uniform time.

we take a scenario depicted in Table 1 for our study to compare average waiting time and turnaround time of the proposed algorithm with SRTF.

3.3 Case Study 1

In this case, we assume context switch time is one time unit.

Now we calculate the turnaround time and the average waiting time using SRTF and Intelligent SRTF. For SRTF, P1 has arrived at time 0, it is scheduled immediately, after completion of one unit of time P2 is ready to run now, P2 preempts P1 as the burst time of p2 is less than RT of p1, in this case processor makes a context switch which takes one unit of processor time as in the Fig. 1.

Now process P2 is to run but at time 2 process P3 is ready to run, now processor has to setup the environment to run P3, and scheduling continues as in Fig. 2.

Now process P3 is to run but at time 3 process P4 is ready to run, now processor has to setup the environment to run P4, and scheduling continues as in Fig. 3. Waiting

Fig. 1 Gantt chart after one context switch

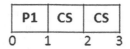

Fig. 2 Gantt chart after two context switch

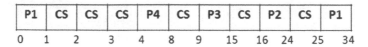

Fig. 3 Gantt chart for traditional SRTF for one unit of context switch

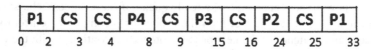

Fig. 4 Gantt chart for ISRTF for one unit of context switch time

times of P1, P2, P3, and P4 are 24, 15, 7 and 1 units, respectively, average waiting time is 11.75 units, turnaround time is 34 units.

With intelligent SRTF algorithm, Process P1 is scheduled immediately after one unit of time process P2 is ready to run but it is not scheduled because P2's burst time + context switch time is not less than P1's remaining burst time, i.e., 8 + 1 is not less than 9, so process P1 continues the execution. At time 2 process P3 is ready to execute and obeys our rule of preemption, at this movement processor has to setup the environment to run the process P3, that is process P3 preempts P1, which requires a context switch. At time 3 process P4 is ready to run, and it satisfies our rule of preemption, now processor has to setup the environment to run P4, and scheduling continues in this fashion as in Fig. 4.

Waiting times of P1, P2, P3, and P4 are 23, 15, 7, 1 units, respectively, average waiting time is 11.5 and turnaround time is 33 units. Table 2 summarizes case 1.

3.3.1 Case Study 2

In this case, we assume the context switch over head requires two time units. Figure 5 is the Gantt chart with traditional SRTF.

Waiting times of P1, P2, P3 and P4 are 28, 18, 9, and 2 units, respectively, average waiting time is 14.25 units and turnaround time is 38 units. Figure 6 is the Gantt chart using intelligent SRTF.

Table 2 SRTF versus ISRTF for context switch time of one unit

Algorithm	Context switches	Turnaround time	Average waiting time
SRTF	6	34	11.75
Intelligent SRTF	5	33	11.5

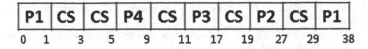

Fig. 5　Gantt chart for traditional SRTF for two time units of context switch time

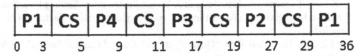

Fig. 6　Gantt chart for traditional SRTF for two time units of context switch time

Table 3　SRTF versus ISRTF for context switch time of two units

Algorithm	Number of context switches	Turnaround time	Average waiting time
SRTF	5	38	14.25
Intelligent SRTF	4	36	13.75

Waiting times for P1, P2, P3, and P4 are 26, 18, 9 and 2 units, respectively, average waiting time is 13.75 units and turnaround time is 36 units. Table 3 summarizes case 2.

4　Conclusion and Future Enhancement

The proposed intelligent shortest remaining time first scheduling algorithm offers better performance compared to the traditional SRTF by decreasing the number of context switches, the turnaround time and the waiting time. This is achieved by considering the context switch overhead while making the decision of preemption. In the future, this algorithm can be made efficient by quantifying the time required for the context switch.

References

1. Silberschatz, A.; Galvin, P.B.; Gagne, G. (2013). Operating Systems Concepts (9th ed.). Wiley. p. 161. ISBN 0-471-69466-5.
2. Tanenbaum, A. S. (2008). Modern Operating Systems (3rd ed.). Pearson Education, Inc. p. 156. ISBN 0-13-600663-9.
3. "Computer Scheduling Methods and their Countermeasures" Edward G. Coffman and Leonard Kleinrock AFIPS 68 (Spring), April 1968.
4. Andrew S. Woodhull, Andrew S. Tanenbaum. Operating Systems Design and Implementation, 2nd Edition. Prentice-Hall, 1997.

5. Con Kolivas, Linux Kernel CPU Scheduler Contributor, IRC conversations, no transcript. December 2004.
6. Stallings, William (2004). Operating Systems Internals and Design Principles (fifth international edition). Prentice Hall. ISBN 0-13-147954-7.
7. "Analysis of a Time-Shared Processor" Leonard Kleinrock Naval Research Logistics Quarterly, 11: 1, pages 5973, March 1964.
8. D Praveen Kumar, T Sreeninvasula Reddy and A Yugandhar Reddy; "Finding Best Time Quantum for Round Robin Scheduling Algorithm to avoid Frequent Context Switch" IJCSIT, Vol 5 Issue 5; Sep 2014.
9. Chuanpeng Li, Chen Ding, Kai Shen, "Quantifying The Cost of Context Switch", ACM Trans. Comput. Syst.
10. "The effect of context switches on cache performance" Jeffrey C. Mogul and Anita Borg ASPLOS, 1991.
11. S. Khanna, M. Sebree, and J. Zolnovsky. "Realtime scheduling in SunOS 5.0". Proceedings of the USENIX Winter Conference, 1992: 375390.

Anomaly Detection Using New Tracing Tricks on Program Executions and Analysis of System Data

Goverdhan Reddy Jidiga and P. Sammulal

Abstract Now the security of information and applications is getting abnormal attention in the public. Because the millions of expenditure spending to combat on continuous threats. The threats (anomalies) are widely occurred at programming scope by exploitation of coding and other side is at application scope due to bad structure of development. Today various machine learning techniques are applied over application level behavior to discriminate the anomalies, but not much work is done in coding exploits. So in this paper, we have given some rich extension work to detect wide range of anomalies at coding exploits. Here, we used some standard tracing tricks and tools available in Linux platform, which describe how to observe the behavior of program execution's outcomes and model the necessary information collected from system as part of active learning. The experimental work done on various codes of artificial programs, Linux commands and also compared their performance on artificial datasets collected while program normal runs.

Keywords Anomaly detection · Function call · System call · Tracing tricks

1 Introduction

The security of information is addressed by advanced technical concepts to satisfy the users due to different levels of attacks. Today different malicious codes are injected into programs, applications and those are run into machine in an authentic way, but are not able to find their vulnerable entries [1, 2]. Now we are using several methods and algorithms at application level, still unable to identify the narrow-level attacks due to similar signatures and profiles. So we need to go with

G.R. Jidiga (✉)
Department of Technical Education, Government of Telangana, Hyderabad, India
e-mail: jgreddymtech@gmail.com

P. Sammulal
JNTUH College of Engineering, JNTU University, Karimnager, Hyderabad, India
e-mail: sammulalporika@gmail.com

© Springer Science+Business Media Singapore 2017   389
S.C. Satapathy et al. (eds.), *Proceedings of the First International Conference on Computational Intelligence and Informatics*, Advances in Intelligent Systems and Computing 507, DOI 10.1007/978-981-10-2471-9_38

alternate techniques like use the system data level to find unknown attacks. For this, we use the basic process organization's virtual memory structure and in this, the stack is a data source for modeling attacked profiles. There are many stack smashing sources like buffer overflows; format string injections leads to injecting attacker's malicious code into primary memory and diverting system control to out of sequence.

The stack trace is a kind of lightweight technique to automate the solution process by tracing function calls and set of system calls. Here, the set of PC values treated as return addresses associated with function calls and the sequence of system calls exhibits the order of program executions. The stack frames are tracing most frequent for 'n' number of calls by caller (in main/function) and continue the repeated sequence.

In our work, we have focused on command and program-based tracing techniques. In that, the BACKTRACE is a clever tracing of function calls and also took supporting stack tracing techniques like PTRACE, STRACE and LTRACE for tracing function calls, system calls and library calls. The GUI-based tools such as KERN-SHARK are also available working based on trace-root method.

So the anomaly detection which is done at system coding levels is carried out by popular and powerful tracing techniques shown in this paper. In our work, we have followed the multiple methods and multiple combinations of data collected from stack while program running. The reason for selecting the combinations of stack data is for measuring the truthful rate of inconsistent activities in coding exploits. The early level of work on this context is considered either system calls or library calls or function calls. So, all those methods are showing merits as well as some demerits. The majority of work done previously took signature profiles, but now focused on dynamic behavior of program to focus on the zero-day attacks.

1.1 Importance of Anomaly Detection in Two Levels

The attacks (anomalies) are found in two levels, one is application level and other one is system level. So the anomaly detection criterion is different in each level, but meaning is same like finding abnormal (strange) patterns are indicating negative behavior. Hence, sure that anomaly detection is an approach of reporting abnormal and unknown patterns (behavior) caught. This kind of method helps to thwart the intrusive patterns or exploits which deviate to normal. The anomaly detection system (ADS) is the subset of intrusion detection system (IDS) primarily elaborated by Denning [3] and Anderson [4] for securing the information. He proposed IDS with two kinds of system observations on behavior of programs, but he concludes that the same diversion is happening in the execution paths.

The importance of ADS is given in both levels for maintaining the security of information available for every request made, for this make sure that no unauthorized, misuse, disclosure and modification [3]. In both the levels of ADS, many of the techniques are designed to model the anomalies for different kinds of data in real-time (critical infrastructure) applications, but still impossible to catch all latest anomalies. For application level of ADS survey is presented well investigation of both data mining and machine learning [5, 6]. The machine learning is a adorned technique in ADS by applying different combinations and giving good results, but this type can be suitable for context and continuous anomalies [7]. The machine learning at application level does not focus much on point-based anomalies. In some cases like coding exploits, the machine learning is not perfect. Anomaly detection is done at executable programs by top down and also observing in reverse by analyzing coding exploits including trace points extracted from assembly code chains generated in memory stack space [1, 2, 8]. In this paper, more topics are presented on system data level. The system data are convenient source to find any kind of attack which is not met in machine learning at application level. So data are trained by tracing necessary levels in the system and use it for next testing. This scenario is good to know more about facts of debugging, which is not at all known at application level.

Finally the structure of paper is: the previous work done in this field is given in the next section as related work, the system design for development of work is given in Sect. 3, the execution of work and results are shown and discussed in Sect. 4 and finally scope of work in future is given in conclusion.

2 Related Work

In this section, the previous work carried out and how it is helpful for the progress of work is given. The initial stage of work took data from UNIX OS and system calls are part of referring anomalies in coding exploits. In [9], they expressed their logic in specific order of taking the system calls and proposed length of system calls occurring in sequence. These data are to be used as modeling based on normal profiles and anomalies are identified in successive testing. The other side of modeling is done based on window of system calls occurring in particular time span is good to improve the detection of anomalies [7]. Later in [8], Feng et al. proposed Vt-Path (Virtual path) on set of sequence of function calls between last calling and current, which is based on stack frame information like RA values collected from the stack while training period. They used a setjmp () function andQuery longjmp () function to save the information or data while normal execution of programs and use this data later for validation to find the anomalies. But the drawback of this pair is, if any intermediate functions use any temporary resources, such as memory allocation or opening files, setjmp and longjmp will not release these resources until they are free, so we have to be very careful with using these. On the other hand, the excessive use of longjump can create very confusing execution sequences and function call hierarchies that are difficult to model and keep the track of source information. The use of model with this code will also become hard to read and use the easiest tools and technique as alternative to avoid Flattening hazards encountered previously. In paper [10], flattening hazard is discussed more and how to prevent this kind of hazard occurs in program after execution also while allocating memory portions for loop spin recursively.

There were many authors using these methods, also planning that only system call data will give solutions easily, but failed in some cases. So tracing of data considered the additional adornments such as combination of system calls and function calls into a sequence pattern and used as training set to observe the anomalous appearance of activities. In real-time applications, due to the millions of system calls creating overhead to run training and testing activities [7, 11]. For this kind always try to simplify the sample data and minimize the quantity of data collected during training period to get fast response, otherwise alternative is plan like use hashing for getting better results in less time. The other methods like N-gram technique is proposed on system calls based on data mining techniques, but this also consumes time and space creating rift in performance [9, 11]. In [12], they have given good concepts about only using function calls which got tremendous results by extracting tracing data with the help of PIN tool and STRACE. In that they have done work with PIN tool mounting on windows libraries which may be time consuming compared to Linux tracing techniques. In our previous work [13], we use a simple and clever function tracing with BACKTRACE. We hope that this method is good for any kind of function call-based ADS.

Finally, we planned for better approach by combination of different data points found in stack and with randomized sequence of control points in program execution. But ultimately the buffer and heap overflow attacks are very common in coding exploits and penetrate unauthorized sequence of data in control line of memory. Figure 1 (bottom) shows how the stack is ready for smashing due to vulnerable entries made. The buffer growing beyond the limit may be the cause of losing control sequence. For this Stack Guard [14] method was introduced by Cowan to thwart those kind of attacks. This can prevent by putting efforts of programming skills in the coding.

3 Proposed Work: System Overview

In this paper, we have used advanced tracing tricks available in Linux. All are stack-based, command-based tricks and some GUI-based tools are also available, but we have preferred command based due to swiftness in training and collecting necessary data from system. The proposed model of our work is given in Fig. 1 (top). The popular tracing tricks are BACKTRACE, PTRACE, STRACE, LTRACE, FTRACE (all are common to Linux platforms), DTRACE, TRUSS are used in Solaris OS and KTRACE, TRUSS are used in BSD.

In our paper, first we have given overview of function calls (F_n) generally represented by popular model called call-graph model shown in Fig. 1 (top left). This model give the brief view of function calls how the program execution

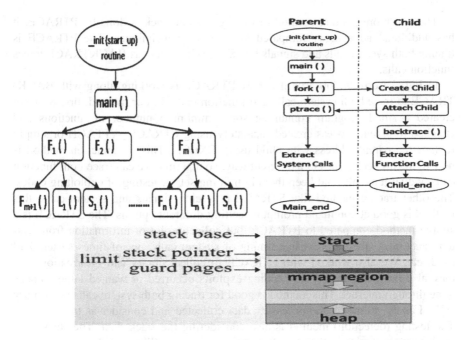

Fig. 1 The simplified call-graph model for program (*top left*), combined tracing techniques of PTRACE and BACKTRACE (*top right*), bottom figure shows the stack usage limits

behavior is shown in the form of model. This model is based on our work to initiate BACKTRACE to collect the set of return addresses, which was well explained in [13]. The backtrace (*bt*) also works well in GDB of Linux. So we have collected the system data information in multiple ways to know the correctness of source. Generally, the BACKTRACE is also used in last function to trace all return addresses back to main function. The function calling method has some additional addresses like start_up routine or _init which was calling main (). The set of system calls (S_n) generated set of libraries (L_n) called and finally some set of signals also identified during program execution are sources for finding anomalies in system level.

Let us take one snapshot of BACKTRACE from [13] shown here. Here, the RA Set is a set of return addresses creates in stack from sequence of function calls made in the program. In this paper, some set of return address paths (RAP) are generated to improve the strategy of finding anomalous execution paths.

RA Set= $\{r_1, r_2, r_3, r_4, r_5 \dots r_s\} \Rightarrow RA_{23}$=$\{8049a46, 80493da, 804938c, 8048f28, \ bbcb56 \ 8048d01\}$

$\{ \ backtrace() \rightarrow f23() \rightarrow f11() \rightarrow f5() \rightarrow main() \rightarrow init() \ \}$

The one more popular method of tracing system stack is done by PTRACE, it has additional advantages compared to all tracing techniques. The PTRACE is tracing both system calls and signals like STRACE, where as BACKTRACE traces function calls.

In Fig. 1 (top), we observe that how PTRACE is working along with BACK-TRACE to trace both system calls and function calls. In our method, first we have created a small program containing some minimum number of functions and attached to parent process created separately using PTRACE_ATTACH or simply replace the parent address with child using EXECL. This combination method is new in tracing techniques in a different way, so at a time we can trace both function calls and system calls and keep them in training set for testing of exploiting codes. The other tracing method like STRACE is tracing signals and system calls; this method is good at command prompt by setting different options. The STRACE is a simple method compared to PTRACE, but collecting lot of information from system stack and other sources like timing of system calls, no. of times system call used, etc. We also worked with LTRACE to collect library call information, this may also be useful to find any coding exploits occurred or wanted target attacks done through internet. This method is good for tracing both system calls and library calls. Finally, to model the necessary data collected and consider as training sets. The testing (detection) method is also considering the stack data. The stack generally grows downward and buffer grows upwards (like heap) in the system memory shown in Fig. 1. This is for testing phase, if any coding exploits, programmer mistakes, internet buffering and other kind of methods lead to change in the stack data. So, according to our system model overview, we have done a number of experiments using different kinds of tracing techniques and finally four base datasets are taken for testing discussed in next section.

4 Datasets: Artificial Datasets as Datasets

We have taken four basic and artificial datasets collected from normal program runs dynamically as a part of active learning of system.

Training Set: (1) Return address set, (2) System call set, (3) Library call set and (4) Return address path (RAP) set. Testing outcomes are evaluated based on the above sets and identify the anomalous or suspected points.

5 Implementation: Experiments and Results

In this section, we have shown how the work is implemented and verify the results. For this, we have selected 1.9 GHz I3 processor, Ubuntu 2.6.41 Linux (built in C) and for tracing: BACKTRACE, PTRACE, STRACE, LTRACE are considered

[15]. The results are shown in Tables 1 and 2; here we took two kinds of programs for implementing the task identified in the abstract. The first kind of programs are artificial programs written in C and applied tracing techniques, where as second type is built in command programs. Table 1 shows the results on three tracing tricks, where experiment is done on four programs P1.c, P2.c, P3.c and P4.c with increasing number of function calls 5, 10, 20, 30, respectively. Each program has been incorporated with some standard C input, output functions for generating system calls and library calls. The program behavior is trained by collecting no. of function calls, library calls and system calls. For these programs we have used PTRACE method first along with BACKTRACE to collect function calls. Generally, PTRACE trace system calls after attaching the process, trace the return addresses for each frame by taking value of EBP + 4 into EAX register. In our work, the return address is key data for identifying anomalies in coding exploits. In the same way STRACE and LTRACE are used for all programs to collect data in a simple way. The programs are exploited with injecting some shell codes or manually executing some buffer overflow attacks. The shell code is generated by taking OBJDUMP of some program and preparing small hexadecimal code into array. The other way of injecting shell into program is by calling with the help of system () function. In this case, there are three kinds of anomalies found in program behavior. The return address (RA) anomaly is found if any new RAs are present in the testing, even lost RAs some time due to stack smashing may lead to complete anomaly. In the same way, other datasets are compared between training and testing found library call anomaly and system call anomaly.

Table 2 shows the popular Linux commands tracing information collected from system using PTRACE, STRACE and LTRACE. The training sets for this case are some way different for first one. In this, the individual and combined trace data are collected for each command and stored in text files. For this, we wrote a C code to find the necessary system information while executing the command. These Linux commands are debugged using different tracing techniques available in Linux and prepared the datasets; the datasets are PC set, RA set, SC set, LC set, and signal also

Table 1 The artificial programs behavior shows the training sets and its observation by different tracing techniques

Before injecting shell code					After injecting shell code		
Sl no	Program	No. of FC	No. of SC	No. of LC	Ptrace()—o/p observation	Strace()—o/p observation	Ltrace()—o/p observation
1	P1.c	5	25	12	RA anomaly	SC Anomaly	SC, LC Anomaly
2	P2.c	10	36	15	RA anomaly	SC Anomaly	SC, LC Anomaly
3	P3.c	20	71	30	RA anomaly	SC Anomaly	SC, LC Anomaly
4	P4.c	30	98	42	RA anomaly	SC Anomaly	SC, LC Anomaly

Table 2 The Linux commands tracing information and test case outcomes after modification of command usage

S. no	File	Training original (before modification)							Testing in attacked mode (% change after modification)			
		1 PC (IP)	2 FC(R)	3 S.C	4 L.C	5 RAP	6 C1	7 C2	1 PC(IP)	6 C1	7 C2	Kind of anomaly
1	ls	5.4×10^5	1630	191	1557	210	1821	1840	5.20	3.5	40.28	A_1, A_2, A_3, A_4
2	tar	1.3×10^6	10332	151	10332	450	10483	10782	10.89	1.4	85.26	A_1, A_2, A_3, A_4
3	ps	4.4×10^5	359	1357	359	25	1716	384	4.56	0.96	43.23	A_1, A_2, A_3, A_4
4	stty	2.2×10^5	118	122	118	8	240	126	3.28	1.02	63.21	A_1, A_2, A_3, A_4
5	ping	1.4×10^5	6	41	6	2	47	8	3.56	10.25	50.00	A_1, A_2, A_3, A_4
6	dash	4.5×10^3	125	82	100	15	207	140	2.12	3.67	32.57	A_1, A_2, A_3, A_4
7	bash	3.9×10^6	451	128	450	30	579	481	12.03	1.67	59.38	A_1, A_2, A_3, A_4

*PC(IP)—program counter (instruction pointer), FC—function call (R-return address), SC—system call
LC—library call, RAP—return address path, C1—combination of FC and SC, C2—combination of FC and RAP
A_1—return address anomaly, A_2—System call anomaly, A_3—Library call anomaly, A_4—Return address anomaly

traced (but excluded in testing). In this case, we have planned two more sets called combinational sets: C1 (Function calls and System call) and C2 (Function calls and Return address path sets). Initially, the no. of function calls are depending on no. functions available in each command program. The ls command has 25 functions actually, but these are called 1630 times approximately. The time complexity of this program measurably depends on no. of times each function is called. So for the ls program maximum time is allocated for set of functions {errorno_location, ctype_get, strcoll, overflow} and other functions like memcopy, malloc, fwrite are called average. The maximum time 70 % is set for above set of functions. The same kind of scenario is available all the commands.

The anomaly detection is found based on changes in the level of directory or placing malicious program in particular directory will change the outcome of command execution. So these kinds of attacks are very common in system level and codes are exploited by malicious network programs. We have collected the total information on all commands shown in Table 2 based on the normal runs of commands. The different test cases evaluated during a complete day on all command object dumps and identified little bit change for every run. So based on this running commands we can identify any unauthorized changes to the accessing of commands and to the directory. From Table 2, we can see the percentage of exploits to be done to each program after modification of directory of command access. In that RAPs are return address paths are constructed from set of function calls randomly. But actually RAPs are special cases of impossible paths found in the program executions. Due to the huge length of the RAPs, we only took limited number of paths considered in training due to large size of tables needed to store. For this kind of scenario, sometimes it is unable to identify anomalies in small coding scripts; therefore, large set of PC values is to be considered in testing. The result shows the changes of C1 and C2 from normal running to unauthorized running of commands. The C2 changes are high compared to the C1, because single function call or system call is part of many RAPs.

5.1 Discussions

The kinds of anomalies are identified by comparison of training sets and test data collected during latest runs. In this A1, A2, A3, A4 are kind of anomalies referred through the different test cases. The tar command gives more changes in the C2 compare to all; this is change of affect from no. of PC (EIP) values. So in this case study, we have utilized all kinds of techniques to find the changes to the original programs and modified programs. The different types of anomalies found from both case studies are similar in nature, but the way of attacking or exploiting of code behavior is different. So the majority of cases in system calls may not find some anomalies, so using this kind of approach is not useful for emerging internet programs running round the clock. The combination methods of system data are useful for improving the detection of more anomalies in the system level. Therefore, we were able to design on-the-fly ADS model to observe new kind of anomalies always.

6   Conclusion and Future Work

The main goal of this paper is to show some improvements on previous work done in this field to find low-level attack changes on memory. In this paper, we mention different kinds of possibilities to thwart the security breaches including specific anomaly detection systems for future applications. Many of the algorithms are system call based and consume training time, so we need to optimize the time and then combine multiple approaches used in this to get better results. In future, we need to apply machine learning to the system level datasets to improve performance like faster detection of anomalies.

References

1. R. Sekar M. Bendre, P. Bollineni, and D. Dhurjati, A Fast Automaton-Based Method for Detecting Anomalous Program Behaviors, IEEE Sympo on Security and Privacy, Oakland, CA, 2001.
2. D. Wagner and D. Dean, Intrusion Detection via Static Analysis, IEEE Symposium on Security and Privacy, Oakland, CA, 2001.
3. Denning: An intrusion detection model, In IEEE Transactions on Software Engineering, 1987.
4. J.P. Anderson, Computer security threat monitoring and surveillance, USA, Technical Report 98-17, April 1980.
5. S. Axelsson, IDS: A Survey and Taxonomy, Chalmers Univ'y, Tech. Report 99-15, March 2000.
6. Goverdhan Reddy Jidiga and P. Sammulal, Foundations of Intrusion Detection Systems: Focus on Role of Anomaly Detection using Machine Learning, ICACM - 2013 Elsevier 2nd International Conference ISBN No: 9789351071495, Aug-2013.
7. W. Lee and S. J. Stolfo, Data mining approaches for intrusion detection, in 7th USENIX Security Symposium (SECURITY-98), Berkeley, CA, USA, 1998, pp. 79–94.
8. H. H. Feng et al., Anomaly Detection Using Call Stack Information, IEEE Sympo on Security and Privacy, Date: 11–14 May-2003 pp: 62–75, ISSN: 1081-6011 Print ISBN: 0-7695-1940-7.
9. S. Forrest, S. A. Hofmeyr, A Sense of Self for Unix Processes, in IEEE Symposium on Research in Security and Privacy, Oakland, CA, USA, 1996, pp. 120–128.
10. Yang X et. al, Eliminating the call stack to save RAM. In Proceedings of the ACM Conference LCTES, June 19–20, Dublin, Ireland, p. 1–10, ACM 978-1-60558-356-3/09/06.
11. S.A. Hofmeyr, A. Somayaji, and S. Forrest, Intrusion Detection System Using Sequences of System Calls, Journal of Computer Security, 6(3), pp. 151–180, 1998.
12. Sean Peisert, Matt Bishop, Analysis of Computer Intrusions Using Sequences of Function Calls, IEEE Trans. on Dependable Sec & Computing. 4(2): 137–150 (2007).

13. Goverdhan Reddy Jidiga and P.Sammulal, Anomaly Detection Using SmartTracing Tricks on System Stack, IEEE International Conference on Convergence of Technology I2CT-2014, April 6–8, 2014, ISBN: 978-1-4799-3759-2, pp 1-6, DOI:10.1109/I2CT.2014.7092136, Pune, India.
14. C. Cowan, A. Grier, Stack-Guard: Automatic Adaptive Detection and Prevention of Buffer-Overflow Attacks, 7th-USENIX Security Symp San Antonio, TX, 1998.
15. http://www.linuxjournal.com, http://linux.die.net and http://www.gnu.org.

Aesthetics of BharataNatyam Poses Evaluated Through Fractal Analysis

Sangeeta Jadhav and Jyoti D. Pawar

Abstract Fractals are known for their aesthetic appeal. We have calculated the Fractal Dimension (FD) with the Box-Counting method for the *Adavus*, pure dance movements in BharataNatyam. These poses were found to be Fractal with the FD in the range of 1.3–1.5. This FD range has already been proved to be naturally aesthetically appealing to the human eye. Fractals have not been used so far for Indian Classical Dance (ICD) pose analysis. In this paper we have used FD for auto classification of system generated dance poses. This experimental study also reveals that the dance poses in the FD range of 1.5–1.6 are also found to be creative and appealing by the dance experts. Considering the classification ratings of system generated dance poses by an International dancer as Gold Standard data we have found that the Accuracy, Recall and F_Score to be 46 %, 52.63 % and 48.77 % respectively. The results are promising and encouraging for further research using FD with other parameters to measure the aesthetics of dance pose.

Keywords Fractal dimension · Box-counting method · Classification · Aesthetics

1 Introduction

Indian Classical Dance poses are very unique and sculpture like. BharataNatyam (BN) is one such major and well-known ICD that has been in existence for centuries and follows the ancient text of *NatyaShastra* (NS). This theory of ICD and Dramaturgy has not only laid down the rules for various choreographic practices but also allowed innovations in it. We have successfully experimented with Genetic Algorithms by modelling the human body [1] and were able to give various choreographic options [2, 3] for the pure dance movements of BN called as *"Nritta"*. These

S. Jadhav (✉) · J.D. Pawar
Department of Computer Science and Technology, Goa University, Taleigao, Goa, India
e-mail: dcst.sangeeta@unigoa.ac.in

J.D. Pawar
e-mail: jdp@unigoa.ac.in

© Springer Science+Business Media Singapore 2017   401
S.C. Satapathy et al. (eds.), *Proceedings of the First International Conference on Computational Intelligence and Informatics*, Advances in Intelligent Systems and Computing 507, DOI 10.1007/978-981-10-2471-9_39

dance poses were rated by experts from various dance schools and the average ratings were used as a training set for the untagged poses. We also used 2D Stick figures to display the modelled 30 attribute Dance Position (DP) vectors through [4] and used Rough Set Theory in [5] for reducing the dimensions of our 30 attribute DP vectors.

Our results of Single-beat [2] and Multi-beat Choreographic sequences [3] were heavily dependent on human experts for evaluation. We noticed that these dance experts could oversee sometimes the finer aspects of creativity such as novelty, complexity, unpredictability and surprise value due to mainly lack of time and interest. While analysing the averaged ratings data, we observed discrepancies in the ratings given by the dance experts. Although NS allows various innovations in the ICD, most of the choreographers and teachers always follow the same dance routine taught by their teachers. So while we experimented a system with several constraints to generate choreography, some of the experts were skeptical to follow and see a totally new dance sequence which they had not practiced or even heard of earlier.

Dance is learnt through rote system and there exists many choreographers who simply follow their teacher or *guru* without even thinking of the new possibilities. Thus, we noticed a clear bias mostly towards traditional choreography. In such a situation, where our system was generating novel moves that were unheard and unpracticed, it was very difficult to prove their expectation. Few teachers were extremely creative and accepted all the moves as the best ones while some of them rated only on their experience. Thus, we could not have a basic agreement and felt that these experts could oversee a potential good pose to a bad pose. The diverse ratings by them prompted us to use various measures for finding the aesthetic appeal of a pose. Fractal Dimension was one such measure which has helped us to determine the aesthetic quality of a dance pose.

Beauty lies in the eye of the beholder and hence most of the times we find it very difficult to quantify the measure of beauty. We introduced the same set of poses to different experts at different locations. Some found a pose to be lowest on the Likert scale ranging from 1 to 5 while others had innovative reasons to give a higher rating [6]. We could not come to a conclusion as to why certain poses were favourable than others, nor could the experts specify explicitly the parameters on which the ratings had been done. Due to such vast discrepancies we had to tag each picture as per the expert opinion and then finally an average rating was calculated for the same. Due to the law of averages, these ratings were giving a different interpretation to the same dance pose [6].

We had a total of 224 tagged instances out of which 130 were from the existing repertoire of BN called as *Adavus*.[1] We termed these *Adavu* as "Excellent" since they were taught by the teachers. The remaining of the dance poses were tagged by the experts as per their liking. Although we had expected them to rate our poses on a scale of 1–4 where 1 was "not acceptable" and 4 was "Good", most of the experts

---

[1] Adavu(s): A basic step in BN, taught to a beginner and used for pure dance movement. They are aesthetically pleasing movements which can be used to convey no meaning in a dance sequence. They are rhythmic patterns helpful for building a beautiful sequence. They form the basic of learning dance just like alphabets are helpful in forming sentences.

were of the opinion that our poses were "Excellent" (5) and at par with innovative choreographic skills. Thus, they have rated few of the poses as 5 also. With the help of these trained data set (224 instances), we tagged unknown poses (277 instances). WEKA classified these unknown poses and gave us an accuracy of **87.42 %** for 501 instances. This proved to be better than the 66.9 % accuracy with the earlier 224 instances. To find out the genuineness of these ratings, we wanted to find a measure of beauty for a dance pose. These measures could help us in uniquely identifying the basis on which they were tagged. The art of BN choreography has a lot of aesthetic appeal and the reasons for choosing fractal geometry for verifying the same are explained in the following section in detail.

2 Related Work

Fractals for Aesthetics The numerical order in the apparent chaos of nature through self-similar repeating patterns at different scales has been found to be fractal in nature by the Polish-born mathematician Benoit Mandelbrot. He found various patterns in nature to be interesting and rated the roughness of natural objects such as the coastline of Britain, clouds, mountains, and trees, through its fractal shapes and the **Fractal Dimension (FD)** [7]. The human eye has been used to fractals everywhere around them says Richard Taylor and that is the reason why we find ourselves attracted to fractal patterns always unconsciously since we find it soothing. He also states that the eye is a natural fractal detector and has categorically proved in [8] how people have aesthetic preference for fractal images in the mid-range from **1.3 to 1.5** of its FD, irrespective of the method used to generate them by human with use of computer, maths or through nature.

Fractal Analysis Used in Dance FD through Box-Counting method was calculated through the several patterns generated on the floor by various Latin American Dances such as Salsa, Cha-cha, and Merengue by Tatlier et al. [9] and it showed that Rumba had the highest Fractal pattern with a FD of 1.36 while Merengue had the lowest FD score of 1.16. Our work presented here is completely different from the above. An inter-disciplinary work [10] involving mathematicians, scientists and dance artists used the Fibonacci Sequence and Golden Ratio along with digital images, movies and fractal-generating programs by a choreographer for a Dance show.

Some other measures of Image Aesthetics Visual Image Processing for aesthetic values have been explained through a formula by Machado et al. [11] in a different way. It states that higher the image complexity better is the aesthetics since processing complexity is also lesser. Additionally, he states that a tired state of mind chooses the images with lower processing complexity.

Dance Aesthetics The changing aesthetic preference of audience for classical Ballet was experimented with linear regression in [12]. Complex dance patterns are beautiful since they stimulate the brain says Hagendoorn [13].

3 Problem Statement

Aesthetics in the field of fine or performing arts is a very subjective decision. There are so many parameters which determine the choice of a particular piece of art that it becomes difficult to measure the importance of all these scales for a qualitative judgment. Some of the parameters could be the cognitive processing stages of the individual, emotional quotient, perceptual analysis, familiarity, personal taste (which can be a major variable), symmetry, order and so on. Thus, to automate this process we need to clearly define the aesthetic parameters for every individual. The Accuracy, Precision, Recall and F Score for the three experts are shown in Table 1.

Dance Choreography is a highly creative domain. Our ArtToSMart (System Modelled art) system generated novel choreographic suggestions for pure dance movements of BN [3]. The expert ratings of these BN poses and consequently for the multi-beat sequences [6] showed that every individual had a different taste. Thus, it was an up-hill task for us to generalize the situation. We tried to average their scores and classify them through Machine Learning techniques [5]. This resulted in a higher accuracy for the newly created instances. To verify the correctness of the classification, we had to ensure some additional measures for tagging the dance poses.

Fractal art has an aesthetic appeal of its own but for this paper our task is to find the fractal dimension for each expert and display those poses which are in their choice range. Spehar et al. [8] have already proved the universal aesthetic appeal of a fractal for the range of 1.3–1.5. Hence, we need to evaluate the FD of images to get an idea whether our poses are really having an aesthetic appeal and also to check whether they are fractal or not. So we used the Box-Counting measure for calculating the FD of the rated images.

4 Implementation Details

A study of various applications using fractals showed the methods available for the fractal analysis. Also this study revealed the use of a well-researched software like ImageJ [14] for fractal analysis. Its an open source, Java-based, public domain image processing software providing several plugins and macros for various different platforms.

Table 1 Precision, Recall, Accuracy, and F_Score for of "Good" DP images

Expert	Precision (%)	Recall (%)	Accuracy (%)	F-score (%)
1	45.45	52.63	46.0	48.77
2	33.0	45.45	37.28	38.23
3	42.00	50.00	43.63	45.65

4.1 Box Counting Method

We have used the Box Counting method for fractal analysis [15] which is widely used to determine the FD of an image. In this method, the image is covered with grids of various sizes and then the box covering the image is counted for several magnifications. This procedure is repeated for shrinking box sizes and the data are plotted on the X–Y plane. The value of Log(N) is plotted on the Y-axis while Log(r) is plotted on the X-axis. N represents the number of boxes that cover the pattern and r is the magnification scale. The FD is the slope of the line which is defined by the following equation $D = Log(N)/Log(r)$. Linear Regression is used to find the line of best fit for the data. The nature of the images can be conformed to be fractal if they are linear, i.e. the fractal patterns are data on a straight line. FD are one of the most important parts of fractal Geometry because it takes into account the changing nature of the pattern due to measuring with smaller magnification scales. Thus, with the increase in structural complexity the FD value shall also increase and higher FD could signify the image's higher artistic value. Thus, the degree of complexity can be measured by evaluating how fast the measurements change depending on the change of the magnification scale.

4.2 Fractal Analysis of "Adavus"

More than 50 pictures of *Adavus* were analysed and we have found that they were in the range of 1.3–1.5 ($\pm\delta = 0.03$ to 0.05). This already shows that they are aesthetically pleasing since they are in the range of fractals. This clearly proves that BN poses are also fractal in nature. These *Adavus* are geometrical in nature, i.e. according to the dance texts these poses form various angular movements of triangle, rectangle, square, circle, etc. [16, 17] and are always symmetrical in nature thus keeping the Euclidean Geometry [17] intact. We need to verify the FD of some new poses other than *Adavus*. These poses are deviating from Euclidean Geometry and hence will have higher FD. Our dance poses generated by the ArtToSMart software are not always self-similar unlike nature's fractals or even as seen in the *Adavu* patterns and hence Box-Counting method is the most appropriate to evaluate them. The *Adavu* patterns have lesser FD as compared to the generated images since they are symmetrical and geometric in nature.

4.3 Data Preparation for ImageJ Software

We have used the Box Counting method for fractal analysis. Two of the well-studied techniques of Box-Counting Method are for Binary and GrayScale images and hence we chose to use ImageJ for converting our coloured images to the binary mode.

4.4 Experimental Results

Aesthetic choices can be different due to variety of factors such as age, familiarity, creative skills, temperament and so on. Thus, it was not an easy task to find the exact fractal dimension which could cater to everyone's needs. This study also showed that although the preferred FD range is 1.3–1.5 for images (as seen in Sect. 2 earlier) [8], the expert liking is more in the range of 1.5–1.6 ($\pm\delta = 0.08$) for our system generated dance poses. Another important feature to be noted while calculating FD was that each image was dependent on the costume worn by the dancer. This was affecting the result of the FD. The finer gestures of BN, the leg positions and the costume problems were tackled to some extent with the help of finding the edges of the figure through the ImageJ software. While converting these images some of these information was lost to some extent. Since a novice in BN will find it difficult to understand the intricate hand and foot patterns, it has to be noted that this study can help experts or BN connoisseurs only. In addition to the factors mentioned above, we also notice that the human mind has a liking for certain objects depending on their temperament and various other factors like the time of the day, inclination towards new ideas and many more facets. This study helped us in generalizing aesthetic content in a certain way. Considering the consistency showed for majority of the images ranked by any individual, we can generalize the trend for their liking. It may not hold true for every single image but overall the results are promising.

4.5 Estimation of FD Range for BN Dance Classification

After calculating the FD range for *Adavus*, we analysed around 100 odd DP images generated by our system (the Genetic Algorithm driven ArtToSMart system can generate several DP vectors at every time and we had around 2000 such DP vectors in the database. We selected around 100 odd DP vectors and with the help of our dance expert modelled it into DP images). These were rated by 9 experts who performed at International, National and State levels on a scale of 1–5 where 1 (Not Acceptable) was the least and 5 (Excellent) was the highest. We noticed that 93 DP images had a higher rating from atleast one of the experts. Thus, we tabulated these pictures on the basis of being liked by more than 50 % of experts. This resulted in 32 such images which were either tagged as "Excellent" or "Good". Here, it was noticed that the highest FD was of 1.68 while lowest was 1.42.

With 98 % confidence level, we could prove that *Adavus* were in the pleasing range of 1.42–1.46 (Table 2) and so on for others with help of standard normal probability variable. The confidence intervals were calculated with Z statistic measures.

On the basis of this proved FD range as seen in Table 2 we created a confusion matrix for 3 experts of International, National and Local repute. Due to space constraint, we have displayed here Expert 1 and 3 only as seen in Tables 3 and 4, respectively. Due to the new statistically estimated FD, we found many instances that were

Table 2 Statistically estimated FD for the BN poses

Tag	Fractal dimension
Excellent (Adavu)	1.42–1.46
Good	1.54–1.58
OK	1.57–1.61

Table 3 Confusion matrix for Expert 1

Statistically predicted

Expert	Excellent	Good	OK	Total
Excellent	0	3	6	9
Good	0	**10**	9	19
OK	0	9	13	22
Total	0	22	28	50

Table 4 Confusion matrix for Expert 3

Statistically predicted

Expert	Excellent	Good	OK	Total
Excellent	2	3	4	9
Good	0	**11**	11	22
OK	1	12	11	24
Total	3	26	26	55

beyond the bound. We have not considered these data while calculating these scores. For, e.g. anything below 1.42 or above 1.61 was categorized as others in the confusion matrix. This resulted in 16, 23 and 24 instances not being in the confusion matrix for the three experts, respectively. Although the accuracy is not very high but we can still say that images in the statistically proved fractal range are aesthetically pleasing because they are complying with the fractal standards.

5 Conclusions

Domain experts are unable to articulately point out the measures which define the beauty of a dance pose. There were several factors mentioned such as familiarity, symmetry, creativity and so on, but we could not summarize all of them to give a common aesthetic appeal to every picture. Thus, we chose a unique methodology to identify the aesthetic appeal of a dance pose which shall also help us in the classification process. FD gave us an idea of the liking and disliking for every individual. Such findings on a bigger tagged data set can help us to refine the classification process of all unknown poses. The more the complexity of the pose, more is the FD and our experiments have revealed the same. Some experts may find these images to be highly creative while others may not like the same. Our findings show that the highest ratings by them are for the FD from 1.5 to 1.6.

The FD may differ due to the following reasons: costumes worn by the dancer, angle of photography, background, imperfection in the dancer posing in front of the camera and so on but given similar environment of the above parameters, the results shall remain consistent. Thus, we can successfully conclude that any dance pose that is fractal is pleasing to the eye (1.3–1.5) and creative if within the permissible range of 1.5–1.6. If we use additional measures along with FD for aesthetic preference of a dance pose, we can definitely come up with a full proof formula for automated classification of BN dance poses.

References

1. Jadhav, S., Joshi, M., Pawar, J.: Modelling BharataNatyam dance steps: Art to smart. In: Proceedings of the CUBE International IT conference & Exhibition Proceedings, PUNE, ACM DL (September 2012).
2. Jadhav, S., Joshi, M., Pawar, J.: Art to SMart: an evolutionary computational model for BharataNatyam choreography. In: IEEE Xplore. (December 2012) 384–389.
3. Jadhav, S., Joshi, M., Pawar, J.: Art to Smart: Automation for BharataNatyam Choreography. In: 19th International Conference on Management of Data, Ahmedabad, Gujarat, India, CSI, India (December 2013) 131–134.
4. Jadhav, S., Aras, A., Joshi, M., Pawar, J.: An Automated Stick Figure Generation for BharataNatyam Dance Visualization. In: Proceedings of the 2014 International Conference on Interdisciplinary Advances in Applied Computing, Amritapuri, Coimbatore, India, ACM DL (Oct. 2014).
5. Jadhav, S., Joshi, M., Pawar, J.: Towards automation and classification of bharatanatyam dance sequences. TechnoMathematics Research Foundation **11**(2) (2014) 93–104.
6. Jadhav, S., Joshi, M., Pawar, J.: Art to SMart: An Automated BharataNatyam Dance Choreography. Taylor & Francis **29**(2) (February 2015) 148–163.
7. Mandelbrot, B.B.: The fractal geometry of nature. Volume 173. Macmillan (1983).
8. Spehar, B., Clifford, C.W., Newell, B.R., Taylor, R.P.: Universal aesthetic of fractals. Computers & Graphics **27**(5) (2003) 813–820.
9. Tatlier, M., Şuvak, R.: How fractal is dancing? Chaos, Solitons & Fractals **36**(4) (2008) 1019–1027.
10. Burg, J.J., Miller, T.: Fractal computation in step with real-time dance. In: ISCA PDCS. (2004) 1–6.
11. Machado, P., Cardoso, A.: Computing aesthetics. In: Advances in artificial intelligence. Springer (1998) 219–228.
12. Daprati, E., Iosa, M., Haggard, P.: A dance to the music of time: aesthetically-relevant changes in body posture in performing art. PLoS One **4**(3) (2009) e5023.
13. Hagendoorn, I.: Emergent patterns in dance improvisation and choreography. In: Unifying Themes in Complex Systems IV. Springer (2008) 183–195.
14. Abràmoff, M.D., Magalhães, P.J., Ram, S.J.: Image processing with imagej. Biophotonics international **11**(7) (2004) 36–42.
15. Gouyet, J.F., Rosso, M., Sapoval, B.: Fractal surfaces and interfaces. In: Fractals and disordered systems. Springer (1996) 263–302.

16. Kalpana, I.M.: Bharatanatyam and mathematics: Teaching geometry through dance. Journal of Fine and Studio Art **5**(2) (2015) 6–17.
17. Banerjee, S.: Emerging contemporary bharatanatyam choreoscape in Britain: the city, hybridity and technoculture. (PhD. Thesis, Roehampton University, 2014).

Algorithm to Find the Dependent Variable in Large Virtualized Environment

M.B. Bharath and D.V. Ashoka

Abstract Virtualized environment generates a large amount of monitoring data; even then it's very hard to correlate such a monitoring data effectively with underlying virtualized environment, due to its dynamic nature. This paper introduces new method of mapping relationship in a virtualized data center by identifying dependent variables within monitored performance data. Dependent variables have an association relationship which can be measured and validated through statistical calculations. The new algorithm introduced here, automatically searches such relationship between various devices of the virtualized environment. Due to its dynamic nature of the virtualized environment, we have to take a measurement at multiple points of time, any relationship which holds good across these time intervals are considered as dependent variables. These dependent variables are used to characterize the complex interaction of the virtual data center device. Such relationship details can be used to build model to predict the fault occurrence. Paper explains the algorithm and experimental results obtained during our validation phase.

Keywords Virtual machines · Performance metrics · Regression logic · Data center monitoring

1 Introduction

IT Industry is on exponential growth trajectory from past two decades. Along with this, growth complexity of the underlying infrastructure is also growing, making managing/monitoring data center much more complex. In early 90's cost IT operation used to be 20 %, as per IDC study. But now dramatic growth, combined

M.B. Bharath (✉)
EMC Software & Services India Pvt. Ltd., Bangalore, India
e-mail: Bharath.Basavarajappa@emc.com

D.V. Ashoka
JSS Academy of Technical Education, Bangalore, India
e-mail: dr.ashok_research@hotmail.com

© Springer Science+Business Media Singapore 2017   411
S.C. Satapathy et al. (eds.), *Proceedings of the First International Conference on Computational Intelligence and Informatics*, Advances in Intelligent Systems and Computing 507, DOI 10.1007/978-981-10-2471-9_40

with multiple abstraction layers through virtualization made it reach ~70 % [1, 2]. Large virtualized data centers consist of wide range of elements, like networking switch/routers, storage arrays, and virtualization servers. Dynamic interaction of these elements coupled with heterogeneous nature makes it very hard to build and maintain effective monitoring model for large complex system.

It is very common to collect performance and monitoring data as a part of data center operation. This monitoring data can be treated as different internal states of the underlying system. For example, any fault/failure will have its traces in this monitoring data; if interpreted logically, these monitoring data can explain those failure (or fault occurrence), but it is not easy to correlate such data dynamically. One of the primary reasons is that—multiple abstraction layers and dynamic relationship association in virtualized setup. In a real world scenario, it's very difficult to (almost impossible) build capability to interpret such dynamic behavior. Most of these behaviors (or symptom) are situational specific, depending on the varying load on the system.

The number of read/write requests, the number of network IO packets, Logical Unit Number (LUN) throughput, and the CPU utilization % are typical examples of performance data or metrics in data center monitoring system. We use datacenter monitoring products to collect the performance metrics from different device type configured in a virtualized environment. Then an algorithm automatically searches for dependent variables in that data set. Two performance metrics are called dependent variables if one can statistically explain the changes in behavior of the other performance metrics. If such a relationship holds good for multiple performance data sampling (for reasonable time period like couple days), we regard them as dependent variables. We can build model from these discovered dependent variables, which characterize the current operational behavior status of the existing system. Once such relationship model is identified, it can be used for system management tasks such as identifying performance bottleneck, capacity provisioning, fault mapping, resource optimization, building redundancy, etc. Identification of dependent variables is the first step in building dynamic relationship detection and monitoring process. This will help us to map the ever changing relationship and correlation in virtualized production server environment, making day-to-day management of complex system more manageable.

2 Related Works

The primary motivation of this research work is to find efficient and effective monitoring system for ever changing dynamic virtual environments. As the virtual environment matures, monitoring system should be able to cope with scale. Our research is the first attempt to use the statistical modeling to explain the dynamic virtual environment. Our aim is to prove that we can successfully use statistical modeling for monitoring and fault isolation in virtualized environment. The framework used for this has been generalized, so that it can be easily adopted by the other monitoring solution also.

There is a good amount of research work done on program invariants. Ernst et al. built a system called Daikon to find the likely invariants from program source code [3, 4]. Program function entry and exits are the place, where Daikon identifies the likely invariants. Finding program invariants and their usage is completely different from our research work. Only similarity is that both use statistical model to find related elements.

Knowledge management and data mining research have frequently used these methods for analyzing Web logs. Main aim of this research is to build customer profiles using https page access patterns [5–7]. Using these access patterns, developers can easily redesign the pages to make it more effective. Similarly, system performance monitoring and analysis [8, 9] of http web servers have been done using data mining methods. Arlitt and Williamson [10] have done research on varying user work load and its effects on the Web server invariants. Menasce et al. [11] worked on identifying hierarchical structure for e-business workloads. The concept of invariants in these papers is thoroughly different from data center-dependent variables identification. At last, Jiang et al. [12] introduced ADMiRe, used for performance analysis of interconnected networking system. They identify the related elements through regression by analyzing the system performance data. End result of this process is a set of regression rules, which identifies the system relational characteristics.

3 Finding Dependent Variables

Let's consider minimalistic view of the virtualized environment in large datacenter as shown in the Fig. 1. A virtual machine has been deployed on a hypervisor (like VmWare ESX servers) which in turn connected the Storage server through two duplicate Storage Attached Network (SAN) fabric path (it's very common to have duplicate paths in SAN for fault tolerance purpose). In large data centers, we can have thousands of virtual machine (VM's) like this, interconnected by hundreds of SAN switch and storage devices, thus forming complex interconnected topology structure.

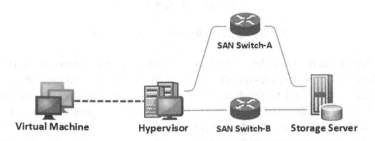

Fig. 1 Simplistic view of virtualized data center server deployment

Please note this topology is not static, as work load varies individual VM's can be migrated to different hypervisor on fly. Thus making this topology dynamic in nature, this keeps changing over time as system load varies. Even though we don't have any constant structure at any given point of time, each of these individual components will interact with each other. This interaction can be visualized as traffic flowing through interconnected roads in a city. Every junction in a road can be considered as functioning element node in virtualized world. These functional nodes/elements generate good amount of operational data and fault events on daily basis. This can be used to find the relationship between these components. For example, virtual machine IO Read/Write will be directly related/proportional to the logical unit number (LUN) read throughput attached to that hypervisor. This will have an impact on the connected FC port CFInFrames and CFOutFrames, number of FC frame received and transmitted by the port connecting storage array LUN to server. Please note that there exists a relationship between these logically connected elements and their performance metrics in virtualized data center, because these elements are responding in sync with virtual machine load. Hence, we can easily extract such a correlated element links from exiting data center monitoring data. Most of the time, these logical topology will also exhibit the underlay system characteristics and the limitation imposed by current design.

4 Model of Dependent Variables

Regression is a tool for finding existence of an association relationship between a dependent variable Y and one or more independent variables X_1, X_2, ..., X_n. This relationship can be linear or non-linear. Mathematically, this relationship can be represented as (which is an exact relationship):

$$Y = \beta_0 + \beta_1 X. \tag{1}$$

Statistical relationship is represented as Eq. 2, which is not an exact relationship. Equation 2 represents the simple linear regression and Eq. 3 represents the multiple linear regressions.

$$Y = \beta_0 + \beta_1 X_i + \varepsilon \tag{2}$$

$$Y = \beta_0 + \beta_1 X_1 + \beta_2 X_2 + \cdots + \beta_k X_k + \varepsilon \tag{3}$$

An independent variable(s) explains changes in the response variable or dependent variables. Regression often tries to find the set of explanatory variables (independent variables) to see how it affects response variables (dependent variables). In other words it tries to predict the values of the dependent variables based on the input of given set of independent variables. Purpose of the regression is to predict the value of dependent variables given the value(s) of independent variable (s). Regression helps to model the existence of an association between two vari-ables, but not their causation. All regression models use the following four assumptions.

- The error term, ε_i, follows a normal distribution.
- For different values of X, the variance of ε_i is constant (means homoscedasticity).
- There is no multi-collinearity (no perfect linear relationship among explanatory variables).
- There is no autocorrelation between two ε_i values.

We use the least square estimation (LSE) to find the betting fitting model. Use the LSE because ordinary least square beta estimates are best linear unbiased estimates (BLUE), provided the error terms are uncorrelated and have equal vari-ance. Also, normally, estimations tend to have minimum variance and they have consistency as the sample size increases; estimate β_i' converges to the true popu-lation parameter value β_i. Least square function is given by Eq. 4.

$$SSE = \sum_{i=1}^{n} \varepsilon_i^2 = \sum_{i=1}^{n} (y_i - \beta_0 - \beta_i x_{1i})^2. \tag{4}$$

5 Algorithm to Find the Dependent Variables

This section will discuss how to automatically build dependent variable relationship model from data center monitoring metrics. It is quite normal to collect large number of performance metrics for monitoring virtualized environment, but not all pair of metric will have correlation or linear relationship. Most of the production system has dynamic nature, where their association keeps on changing based on the load and other factors within the environment (some of the VM may be migrated to other server, if the load goes above certain threshold values). Now the challenging question is how do we find these dependent variables at what interval. One of the key advantages of virtualized environment is that basic model will remain the same irrespective of its elements' association. For example, even though VM are moving between the servers, their logical path remains the same. That is from Virtual Machine ≫ hypervisor ≫ data-store ≫ host hba (host bus adapter) ≫ portwwn (port worldwide name) ≫ array initiator which is mapped to a LUN. Based on this logical structure, we can always start from the VM and reach the associated LUN mapped to that virtual machine.

Figure 2 describes the different stages within the algorithm, which automatically find the dependent variables and their association in virtualized data center

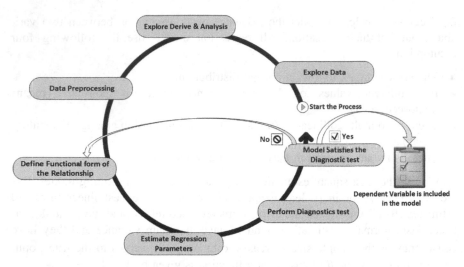

Fig. 2 Process of selecting the dependent variables

monitoring system. We always start from the VM and trace the association down the line to its mapped LUN. It starts from exploring the monitoring data set and then does the preliminary analysis (based on the domain knowledge) and selects few metrics for further processing. Third stage in the algorithm does the preprocessing and normalization. It is very important before comparison that metric time needs to be aligned and their units of measurement need to be normalized. Then, actual algorithm kicks start by defining the formal function and its relationship. Next stage, we calculate the regression parameters, which are validated through diagnostic tests. If the model satisfies diagnostic test, it will be included in model. If not, process again starts by evaluating and defining the formal function with new metric.

Let's say that we have m number of performance measurement data. We randomly start from a VM and find the fitness score $F_i(\theta)$ along the path. In each iteration, we validate our model with new measurement made at ith time window. Using the following function (Eq. 5), we select the discovered dependent variable having fitness score more than τ to be part of the discovery data set or not.

$$f(F_i(\theta)) = \begin{cases} 1 & \text{if } F_i(\theta) > \tau, \\ 0 & \text{if } F_i(\theta) \leq \tau. \end{cases} \tag{5}$$

Once we have k number of monitoring windows of data, confidence score will be calculated using $M_k = \{\theta \mid P_k(\theta) > P\}$, where P is the confidence threshold used by the model decided selection condition by the algorithm. Based on this, it selects the metric to be part of the dependent variable discovery pool.

Algorithm 5.1
Input: $I_i(t)$, $1 \leq i \leq m$
Output: M_k and $P_k(\theta)$ for each time window K
at time $t = l$ (i.e., $k = 1$),
    for each I_i and I_j, $1 \leq i, j \leq m$, $i \neq j$
        learn a model θ_{ij} using equation (5)
        compute $F_1(\theta_{ij})$
        set $M_1 = M_1 \cup \{ \theta_{ij} \}$ and $P_1(\theta_{ij}) = f(F_1(\theta_{ij}))$.
for each time $t = k * l$, $k > 1$,
    for each $\in M_k$
        compute the $F_k(\theta_{ij})$ with equation (5) using
            $I_i(t)$ and $I_j(t)$, $(k-1) * l + 1 \leq t \leq k * l$;
        update $P_k(\theta_{ij})$
        if $P_k(\theta_{ij}) \leq P$ and $k \geq K$,
            then remove θ_{ij} from the M_k
        output M_k and $P_k(\theta)$.
        $k = k + 1$

Algorithm 5.1 starts to build a model from virtual machine performance metrics and travels down the path till the attached LUN. After multiple time period sampling measurements (like couple of days of measurement data), if the regression confidence value is more than our static threshold value of 65 % (value of P), then those performance elements will be included in dependent variable discovery pool.

6 A Case Study

In this section we tried using this model in real world example. The experimental setup and results are explained in the following section

Experimental system: We have selected typical production environment where VM BAPH001WSAP01 is deployed on the hypervisor baph001a.servers. chrysler.com as shown in the Fig. 3, which is clustered with other two ESX servers. They are connected to EMC storage server BAPEM1616 (EMC VNX5600 model) on FA port 4 and 5, respectively, for both SP A and SP B controller (for duplicate path). Server is connected through two Cisco SAN switches BAPSANSWA1 and BAPSANSWB1.

Since this is a production configuration, VMs are allowed to move dynamically between three ESX servers and its associate elements in SAN and storage are mapped, respectively, to allow this movement. This dynamic nature makes it challenging to find the association and correlation between these elements which we are trying to address in this paper. This is an example of one VM; normally on a production setup, you will have thousands of such VMs, each with its own association. The production setup that we selected for our case study was medium size data center with 450 VMs deployed on the 15 ESX server connected via four cisco

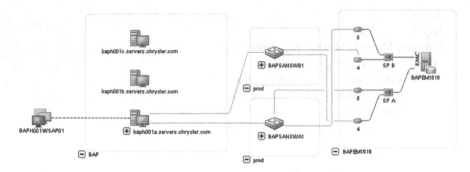

Fig. 3 Typical data center server deployment

switches to four storage arrays (two EMC VNX 5600 and two IBM DS 8000 service array).

Measurements: Monitoring data are collected using EMC SRM (EMC Storage Resource Monitor) an enterprise data center monitoring product, which discovers storage arrays, virtualization infrastructure, and SAN network. SRM has different collectors which can interact directly with device interface to extract the performance and topology details from the data center assets as shown in the Fig. 4. Default polling cycle for most of the assets is 5 min. This gets aggregated as hourly data, which in turn will be rolled up as daily and weekly average data. We have used last 1-year-data for our experiments.

Here are the details of the performance metrics (monitoring data) collected for three logical components of virtualized datacenter, used for result validation (Table 1).

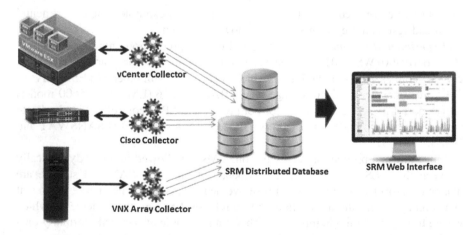

Fig. 4 Monitoring data collection by EMC SRM

Table 1 Performance metric used of modeling from different device type

Device type	Components	No. metrics	Device type	Components	No. metrics	Device type	Components	No. metrics
Virtual machine	Disk	6	Cisco switch (UCS/nexus)	FCS	9	VNX storage array	Controller	30
	FileSystem	7		Power supply	1		Disk	34
	Processor	3		VSAN	1		FAST cache	7
	Virtual disk	6		Processor	1		LUN	87
	Interface	6		Memory	1		Meta member	18
	File	1		Port	19		Port	16
	Data store	1		Logical port	3			
	Memory	7		Interface	15			

7 Experiments and Results

This section describes the framework developed for testing and validating our regression model. It's a java-based code running on Open SuSe 11 Linux machine. The program queries the EMC SRM Database using web-service API, which has last 1 year data. One of the key challenges when modeling the performance of virtual machines is that, we need to normalize the collected performance data, before it gets used in regression model. For example, to run a regression model between VM CPU utilization and storage server read/write throughput, first, one needs to make sure that time range is same for both performance metrics. Then, make sure that granularity of the data is same. Our framework takes care of these conversions automatically and gives the fitness score value for each month. Dependent variables with high fitness score are used for building final model. Due to space constrains, we have shown only two set of results. One selected by our model as shown in Fig. 5 and other one is rejected by the model as in Fig. 6. Figure 7 shows the probability plot selected (right side image) and rejected (left side image) metric. It is very evident that there is no linear relationship in discarded metric.

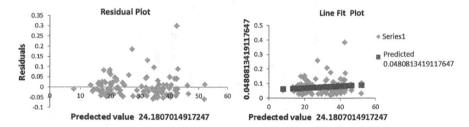

Fig. 5 Model selected dependent variable residual plot and line fit plot

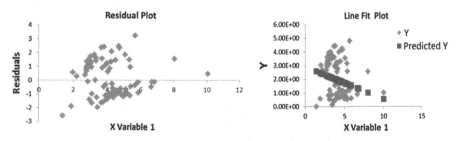

Fig. 6 Model rejected dependent variable residual plot and line fit plot

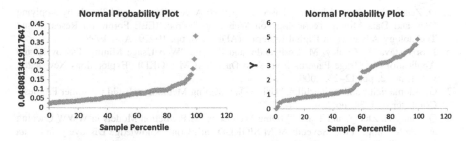

Fig. 7 Normal probability plot for selected and reject metric by model

8 Conclusion

In a large highly virtualized data center environment, tracking logical relationship between Virtual machines and its underlying infrastructure is a daunting task. But these details are very much essential for day-to-day administrative works. This virtual to logical mapping problem gets aggravated due to dynamic nature of the virtual environment, where these VM's get migrated to different hypervisor and SAN fabric based in the operational load. Because of this, there is a need for a way to automatically identify the Virtual machine and its associated elements (like switch port, LUN, etc.) in a SAN environment. We tried addressing this problem by identifying related elements from data center monitoring data. Using regression, we identify the metrics which have high correlations and which hold good for multiple sample of measurements. Such metrics are called as dependent variables, which are used for the construction of the virtual to logical path. This dynamic mapping helps in debugging operational and performance issues in field. We were able to find around 65–70 % of related metrics. In next stage of our research, we will try to map the faults/events with dependent variables.

References

1. D. Patterson and A. Brown et al., "Recovery-Oriented Computing (ROC): Motivation, Definition, Techniques, and Case Studies," Technical Report UCB//CSD-02-1175, UC Berkeley, Dept. of Computer Science, http://www.roc.cs.berkley.edu, 2002.
2. D. Oppenheimer, A. Ganapathi, and D. Patterson, "Why Do Internet Services Fail, and What Can Be Done about It," Proc. Fourth Usenix Symp. Internet Technologies and Systems (USITS '03), pp. 1–16, 2003.
3. M. Ernst, J. Cockrell, W. Griswold, and D. Notkin, "Dynamically Discovering Likely Program Invariants to Support Program Evolution," IEEE Trans. Software Eng., vol. 27, no. 2, pp. 99–123, Feb. 2001.
4. J. Perkins and M. Ernst, "Efficient Incremental Algorithms for Dynamic Detection of Likely Invariants," Proc. ACM 12th Symp. Foundations of Software Eng. (FSE '04), pp. 23–32, Nov. 2004.

5. O. Zaiane, M. Xin, and J. Han, "Discovering Web Access Patterns and Trends by Applying Olap and Data Mining Technology on Web Logs," Proc. IEEE Forum on Research and Technology Advances in Digital Libraries (ADL '98), pp. 19–29, Apr. 1998.
6. J. Srivastava, R. Cooley, M. Deshpande, and P. Tan, "Web Usage Mining: Discovery and Applications of Usage Patterns from Web Data," ACM SIGKDD Explorations Newsletter, vol. 1, no. 2, pp. 12–23, 2000.
7. G. Adomavicius and A. Tuzhilin, "Using Data Mining Methods to Build Customer Profiles," Computer, vol. 34, no. 2, pp. 74–82, 2.
8. Q. Yang, H. Zhang, and T. Li, "Mining Web Logs for Prediction Models in WWW Caching and Prefetching," Proc. Seventh ACM SIGKDD Int'l Conf. Knowledge Discovery and Data Mining (KDD '01), pp. 473–478, 2001.
9. M. Spiliopoulou, C. Pohle, and L. Faulstich, "Improving the Effectiveness of a Web Site with Web Usage Mining," Proc. Int'l Workshop Web Usage Analysis and User Profiling (WEBKDD '99), pp. 142–162, 2000.
10. M.F. Arlitt and C.L. Williamson, "Web Server Workload Characterization: The Search for Invariants," ACM SIGMETRICS Performance Evaluation Rev., vol. 24, no. 1, pp. 126–137, 1996.
11. D. Menasce, V. Almeida, R. Riedi, F. Ribeiro, R. Fonseca, and W. Meira, "In Search of Invariants for E-Business Workloads," Proc. Second ACM Conf. Electronic Commerce (EC '00), pp. 56–65, 2000.
12. N. Jiang, R. Villafane, K. Hua, A. Sawant, and K. Prabkakara, "ADMiRe: An Algebraic Data Mining Approach to System Performance Analysis," IEEE Trans. Knowledge and Data Eng., vol. 17, no. 7, pp. 888–901, Aug. 2005.

Challenges of Modern Query Processing

Archana Kumari and Vikram Singh

Abstract Efficient query processing plays a critical role in numerous settings especially in case of data centric applications. Starting from the architecture to the final stage of result compilation a query processing system has to undertake several challenges. In this paper, we compare different architectures and existing approaches of query processing. In modern days, database systems have to deal with data distribution. To make challenges more complex, the participating databases might be heterogeneous in nature. In this paper, we discuss various challenges of query processing systems in centralized, distributed, and multidatabase backgrounds. We also analyze various parameters and metrics that directly impact query processing.

Keywords Distributed database systems · Multidatabase systems · Query optimization · Query processing architecture · Query processing challenges

1 Introduction

One of the key motivations behind the use of database systems is the aim for integration of functioning data of an enterprise and to deliver unified, thus controlled access on the data. Data is retrieved by posing a query to the database system. A query is a language construct that helps to retrieve a part of data from database [1]. When a user frames a query, a query processor frees the user from specifying the exact procedure to get the required answer via various transparency mechanisms. It takes a query, parses it, optimizes it, and finally retrieves the intended result. Query processing performs transformation on a high-level query and converts it into low-level query. This transformation must be complete and

A. Kumari (✉) · V. Singh
Computer Engineering Department, National Institute of Technology,
Kurukshetra 136119, Haryana, India
e-mail: archana19dhankar@gmail.com

V. Singh
e-mail: viks@nitkkr.ac.in

© Springer Science+Business Media Singapore 2017   423
S.C. Satapathy et al. (eds.), *Proceedings of the First International Conference on Computational Intelligence and Informatics*, Advances in Intelligent Systems and Computing 507, DOI 10.1007/978-981-10-2471-9_41

correct [2]. The Main aim of query processing is to find transformations that are most efficient for generating query result. In centralized databases, query processor finds the most efficient equivalent relational algebra query. While in distributed systems, the query processor takes into account the cost of data transfer.

The main contribution of the paper is to discuss various aspects of query processing and related modern-day challenges in it. The paper has illustrated existing architectures of query processing in Sect. 1 and subsequently highlight query processing in the centralized database systems and distributed systems in Sects. 2 and 3, respectively. The inherent challenges in modern-day query processing are discussed in conclusion, in Sect. 6.

1.1 Query Processing Architecture

Architecture of a query processing system determines the internal representation of query, enumeration of query execution plans, and the phase at which system has the

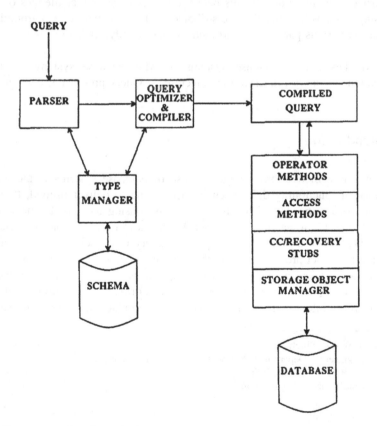

Fig. 1 Query processing phases in Carey et al. architecture

Table 1 Comparison of Haas architecture and Carey architecture for query processing

Parameter	Haas et al. architecture	Carey et al. architecture
Extending DBMS	Customizable System—The relations of a query language can be modified and can be made more comprehensive using new operators. For instance, In STARBURST [19], the query rewriting part can be modified by adding new rewrite rules	Toolkit Systems—They have libraries and modules that offer various alternate methods to perform a task. For Example, In EXODUS [20], architecture type independent access methods, like B-tree, linear hashing, can be extended
Phases	Query rewriting and query optimization are two different phases	Query rewriting and query optimization are merged in one phase
Cost availability	The query rewrite phase does not have the cost information available	All transformations are algebraic and cost-based
Internal Representation	The building block in this case is an enhanced query graph	Query trees and operator trees are used for all phases of optimization

knowledge of the cost of transformations. These factors are crucial in evaluating the complexity of the system. There had been many proposed architectures, two of them are Haas et al. [3] and Carey et al. architecture [4].

Haas Architecture: The main component is parser that translates the input query into an internal representation like query graph. Query rewrite transforms a query to carry out optimizations. Query Optimizer enumerates several plans that specifies how given queries should be executed. Plan Refinement and Code Generation converts the plan made by query optimizer into executable code. Catalog stores the necessary information for parsing, rewriting, and optimizing a query. Query Execution Engine gives implementation details for the operators used.

Carey Architecture: In addition to the Haas Architecture, the main component is storage object manager for providing the capabilities for modifying storage objects. Access methods give associative access to all files of objects and support for versioning. Schema-related information in the databases is provided by type manager to support a wide variety of applications. The various methods or operations on objects are supported by Operator Methods. These components provide the dedicated functions related to query processing. The schematic architecture of Carey for query processing is shown in Fig. 1 (Table 1).

2 Centralized Database Systems

Query processing has three steps: decomposition, optimization, and execution. Query decomposition is done by query compiler. Query compiler goes through several phases like parsing, checking for semantics, query rewrite [5]. Compiler

Table 2 Comparison of bottom-up and top-down query optimization process

Parameter	Bottom up	Top down
Query representation	Generally a query graph is used for internal representation of the queries	Relational calculus expressions are normally used for internal representation
Find upper and lower bounds	The upper/lower bounds are not available to bottom-up optimizers, since such optimizers generate the plans before they considered the larger containing plans	Top-down optimizers are superior allowing to this degree because they can use upper and lower bounds to avoid generating groups of plans
First step in optimization	First, an internal query representation is made so that user queries can easily be mapped	In the first step, the algorithm constructs an access plan for all tables involved in the query
Advantage	The inferior plans are discarded very early	Every stage has a larger and richer set of plans to choose from
Disadvantage	This algorithm has exponential running time and space complexity, therefore, it is not viable for complex queries	Top-down optimizers normally retain all multi expressions. Due to this there is memory wastage
Example	IBM's System R project	INGRES relational DBMS

validates the syntax of the input query and returns appropriate error, if it finds any discrepancies. In query graph model of compilers, an internal representation of the query is formed. A compiler also checks for any inconsistencies in the sub parts of the input query. A compiler utilizes the global semantics to transform the query into a format that is easier to optimize [6]. Recent advancements in programming languages have led to the development of highly efficient query compilers [7].

Query optimization is the most critical phase in query processing and in centralized database systems; it is a two-step process. First is the logical optimization that involves the classic transformation rules on the algebraic trees which helps to reduce the manipulated data volume. Second is the physical optimization that includes the process of determining an appropriate join method as well as their ordering. The decisions are made by taking into account the size of the relations, the physical organization of the data, and access paths. Another approach is based on response time of a query. Response time can be defined as the time elapsed between the submissions of a user query till the time first result is generated [2]. The response time is calculated as the sum of three separate costs, first communication cost, second input/output cost, and final CPU cost.

There are generally two approaches for query optimization. In top-down approach, a query expressed in relational calculus is broken down into smaller sub queries. While, in case of bottom-up approach, many plans are built from simpler sub-plans (Table 2).

3 Distributed Database Systems

A distributed DBMS is distributed over a computer network such that its existence is transparent to the user. The participating databases must be logically related. The only means of communication is with the help of message passing [3]. In case of distributed databases, the communication cost plays the most critical role.

3.1 Challenges in Distributed Database Query Processing

This data distribution makes the query processing a multifaceted procedure [2, 3, 8, 9].

Fragmentation and Replication: The distributed DBMS has to take into account all the fragments and the replicated copies relevant to the query.

Cost Model: The cost model selection is application specific. A database system can choose to follow the path of minimum resource consumption or minimizing the response time.

Search space Exploration and Exploitation: The search space consists of all the equivalent query execution plans that are made after using various transformation rules. The search space could grow exponentially larger once the number of relation involved in the query increases. The search strategy has to be really efficient since there could be a very large search space available.

Control Site Selection: The appropriate control site for final query processing should be selected carefully.

Where and When to Optimize: The timing and placing of the optimization process is deciding factor of a distributed DBMS performance. The alternate execution plans can be made at compile time and the optimizer can select from the available options at run time. The phases like parsing and rewriting of the query can be done at client side while QEP (Query Execution Plans) refinements can be done at server site. In crowd sourcing systems the complexity is relatively high [10]. Hence, the system must free the user from over burden and is solely responsible for compilation, generation of execution plans, and evaluation of the marketplace.

3.2 Challenges in Distributed Database Query Execution Process

Once all the execution plans are generated, the most optimal plans are selected and the rest are pruned at successive stages. The real challenge now is to execute these plans efficiently [2, 3, 8, 9].

Selecting the data transfer mode: Typically row blocking technique is used. In this, a whole block of data is transmitted rather than a tuples.

Use of cache results: Whether to use the data caches that temporarily store copies of data to utilize the benefits of temporal locality.

Organization of sites on the network: The sites are typically arranged in a hierarchical way so sending and receiving costs can vary significantly [11–13].

Optimization of Multicasts: Multicast communication can be done to increase the degree of parallelism.

Use of semi joins: Semi joins reduce the size of intermediate results, since it is based on exit clause. However, this method is suitable where join selectivity factor is good.

Execution Control: The control site could be one single site that is centralized control or a group of sites can act as control sites [13].

The final challenge in query processing is to combine the various building blocks and build a full-fledged efficient query execution plan [2].

4 Multidatabase System

Multidatabase systems (MDBS) are a collection of individual DBMSs that are fully autonomous and cooperate at their own will [2, 3]. In case of multi database, the global conceptual schema is only a sub-part of the logical integrated schema formed by the individual schemas of component databases.

4.1 Challenges in Multidatabase Query Processing

Query processing in a multidatabase system is highly complex and requires more efforts than in a distributed DBMS [2, 3].

Computing Capability: The participating DBMSs may have varying computing capabilities. This hinders the uniform distribution of the query distribution process.

Cost Function: The participating DBMSs may have different cost functions. Hence, the local optimization is different for the same query posed.

Language Models: The participating DBMSs may have different language models for example relational, object oriented, and XML. This creates a problem in translating the query to component DBMS and in integrating heterogeneous results.

Autonomy: The participating DBMS exhibit autonomy. The autonomy can be defined across three dimensions: Communication, design, and execution autonomy.

Semi–Join-based optimizations: The semi-join-based optimizations are extremely difficult, since the source and the target tables are present on different component DBMS. The retrieval of the join attributes and shipping them from target site to host site, performing the join operation at the source site and modifying the target relation DBMS adds complexity.

Architectural Difficulties: The architecture of the multidatabase system is quite complex. The design of wrappers is difficult, since it has to take into account the translation art of the queries. With any change in local schemas, the wrapper and mediator have to be modified accordingly.

Data distribution: The query needs to reach every possible source to obtain the maximally contained query. For systems that involve continuous processing of data streams query optimization becomes much more complex, since the data never ceases to come to the databases. In such cases to efficiently overcome the heterogeneity event processing middlewares [14] are used.

Integration of results: After obtaining the results for an input query, it is sent back to the mediator to carry out necessary translations. The integration of the heterogeneous results thus obtained is a daunting task.

5   Query Processing: Work Done So Far

Query processing has become a major area of research and study in various computing applications. Earlier query processing was confined to single databases stored at a single geographical location. The increase in hardware and software systems has given modern query processing immense power over traditional query processing systems [15]. These days, databases must adapt to this variety of options available. Modern query processors have to make decisions and must deliver advance utilities and services. Due to availability of huge amount of data, modern query processors must provide scalability [16]. Traditional query processor did not exploit parallelism. They neither have the expertise nor potential to perform operations any faster. Modern query processor must provide quick and effective results to meet the never ending demands of the user. Modern query processors are capable of processing time-continuous data [17, 18]. Traditional query processor used to select a single query execution plan out many possible, while modern query processors have the capability to change plans to deal with non-uniform distributions of data. Modern query processing has certainly come a long way in efficiency, reliability, and performance. Table 3 summarizes the available surveys on query processing.

6   Analysis and Challenges

The recent ground breaking developments in database systems led to spectacular results in data management. Organizations, these days find it difficult to find approaches to manage their databases efficiently, conveniently, and in a more controlled fashion. The performance and efficiency of a database system can be analyzed across various metrics.

Table 3 Surveys on query processing

Title and authors	Aim
Donald Kossmann: The State of the art in distributed query processing [3] (ACM Computing Survey, 2000)	This paper presents a textbook architecture for distributed databases. It also discusses challenges faced in query optimization and query evaluation. A brief introduction is also present about multi database
Yu and Chang: Distributed Query Processing [21] (Computing Surv., Vol. 16, 1984)	Various heuristic based, semi-join-based both enumerative and non-enumerative query optimization techniques are used. Types of query, like cyclic and tree queries, are also discussed
Matthias Jarke and Jurgen Koch: Query Optimization in Database Systems [1] (Computing Surv., Vol. 16, 1984)	A variety of methods that are logic and heuristic based are discussed. The focus of the paper is on query optimization in centralized databases
Liu and Yu: Performance Issues in Distributed Query Processing [9] (IEEE transactions on parallel and distributed systems, 1993)	Various performance issues of query processing are discussed. It is shown that these algorithms can be divided into algorithm-based issues and implantation-based issues
Tamer Özsu and Patrick Valduriez: Principles of Distributed Database Systems [2] (Computing handbook Springer ISBN-978-1-4419-8833-1, Springer, 2014)	This handbook is about everything related to distributed database from database design to query processing in detail. Every aspect of the distributed database both in homogeneous and heterogeneous environment is presented in minute details

6.1 Analysis of Impact Parameters

Response Time: While analyzing the performance parameters in case of distributed system all three cost namely communication, I/O, and CPU time plays equal role.

Cost: Since various methodologies like parallel and distributed processing has become a possibility. The communication cost is much larger than I/O and CPU cost in case of distributed databases.

Heterogeneous databases: The query processing optimization is done into two phases. One is the global optimization that is understood by component databases and the second phase is localized optimization implemented individually.

Modern-day database software: The database software available these days are much more stable and highly user friendly.

Advanced Hardware technology: Also, in this modern era, which we can rightly call as digital age, the cost of hardware has gone down and the efficiency has gone up

Network: Due to availability of fast network the query processing approaches have to give equal weightage to all the three costs—communication, I/O, and CPU costs.

Bandwidth: Availability of larger bandwidth provides faster and more reliable connections between the distributed sites.

6.2 Inherent Challenges of Data

Talking about the challenges faced by databases, we have to admit that we are now living in the "ocean" of data. We have to consider various aspects of data analysis.

Data Dimensions: The size of data is too large to be efficiently handled by conventional databases. Hence, the need of the hour is to enhance the features of the present database systems which can incorporate as much data as possible.

Data Stride: The data is huge not only in dimensions but the pace is high too. Every day, millions and millions of data feeds are coming to the system, making it impossible for the systems to adapt at the same rate.

Data Diversity: One query processing approach might be considered for a given type of data set while the other for other dataset.

Data Changeability: In this modern age, we need database systems that can handle variability of data efficiently. The inherent interpretations of this massive amount of raw data depend on the underlying context. For example, in case of natural language processing, a single word may have different meaning.

Data Accuracy: The user will get inaccurate results, if the principal database gathers the data from diverse sources which might not be complete. In such cases, query containment concept is used. The user will get a subset of the complete answer.

Data Envisioning: One of the core tasks of a database system is to present the actual result of the query under processing in a readable and understandable format.

All of these factors demand careful examination, in particular for enterprises not already on the efficient query processing systems bandwagon.

References

1. Jarke, M., Koch, J.: Query optimization in database systems. ACM Comput. Surv, vol. 16 issue 2, pp. 111–152 (1984)
2. Özsu, M.T., Valduriez, P.: Principles of Distributed Database Systems: Third Edition Springer Science (2011)
3. Donald Kossmann: The State of the art in distributed query processing. ACM Comput. Surv., vol. 32, issue 4, Dec. pp. 422–469 (2000)
4. Carey, M.J, DeWitt, D.J., Frank, D., Graefe, G., Richardson, J.E.: The Architecture of the EXODUS Extensible DBMS: A Preliminary Report, International Workshop on Object-oriented Database Systems, pp. 52–65 (1986)
5. Aho, A.V., Lam, M.S., Sethi, R., and Ullman, J.D.: Compilers: Principles, Techniques and Tools, Pearson Education, Second Edition (2006)
6. IBM Knowledge Center, https://www.ibm.com/support/knowledgecenter

7. Rompf, T., Amin, N.: Functional pearl: a SQL to C compiler in 500 lines of code, ACM SIGPLAN, vol. 50 Issue 9, pp. 2–9 (2015)
8. Ceri, Pelagatti: Distributed Databases: Principles and Systems, McGraw-Hill (1984)
9. Liu, C., Yu, C.: Performance Issues in Distributed Query Processing, IEEE transactions on parallel and distributed systems, vol. 4, issue 8, pp. 889–905 (1993)
10. Fan, J., Zhang, M., Kok, S., Lu, M., Ooi, B.C.: CrowdOp: Query Optimization for Declarative Crowdsourcing Systems, IEEE Transactions on Knowledge and Data Engineering, vol. 27, issue 8, pp. 2078–2092 (2015)
11. Kumar, T.V., Singh, V., Verma, A.K.: Generating distributed query processing plans using genetic algorithm. In: Proceedings of the 2010 International Conference on Data Storage and Data Engineering, pp. 173–177. IEEE, USA (2010)
12. Singh, V., Mishra, V.: Distributed query plan generation using aggregation based multi-objective genetic algorithm. In: Proceedings of International Conference on TCS, pp. 20–29. ACM, USA (2014)
13. Mishra, V., Singh, V.: Generating optimal query plans for distributed query processing using teacher-learner based optimization. In: Proceedings of 11th International Conference on Data Mining and Warehouse, vol. 54, pp. 281–290. Elsevier, India (2015)
14. Pinnecke, M., Hobach B.: Query Optimization in Heterogenous Event Processing Federations, Datenbank-Spektrum, vol. 15, issue 3, pp. 193–202 (2015)
15. Tomas, K. T., Hille, M., Ludwig, M., Habich, D., Lehner, W., Heimel, M., Markl, V.: Demonstrating efficient query processing in heterogeneous environments, ACM SIGMOD International Conference on Management of Data, pp. 693–696 (2014)
16. Doulkeridis, C., Norvag, K.: A survey of large-scale analytical query processing in MapReduce, The VLDB Journal, Volume 23, Issue 3, pp. 355–380 (2014)
17. Nehmea, R., Worksb K., Leib, C.: Multi-route query processing and optimization, Journal of Computer and System Sciences, vol. 79, issue 3, pp. 312–329 (2014)
18. Agarwal, S., Milner, H., Kleiner, A., Talwalkar, A.: Knowing when you're wrong: building fast and reliable approximate query processing systems, ACM SIGMOD International Conference on Management of Data, pp. 481–492 (2014)
19. Pirahesh, H., Hellerstein, J., Hasan, W.: Extensible/rule based query rewrite optimization in starburst, ACM SIGMOD Conf. on Management of Data, pp. 39–48 (1992)
20. Graefe, G., Dewitt, D.: The EXODUS optimizer generator. In: ACM SIGMOD Conference on Management of Data, pp. 160–172 (1987)
21. Yu, C.T., Chang, C.C.: Distributed Query Processing, ACM Computing Surveys (CSUR) Surveys, vol. 16, issue 4, Dec., pp. 399–433 (1984)

IOT-Based Traffic Signal Control Technique for Helping Emergency Vehicles

Sneha Tammishetty, T. Ragunathan, Sudheer Kumar Battula,
B. Varsha Rani, P. RaviBabu, RaghuRamReddy Nagireddy,
Vedika Jorika and V. Maheshwar Reddy

Abstract Increase in traffic in cities makes emergency vehicles, like ambulance, to take more time to reach the destination from the source due to which the life of human beings are in danger. So, emergency vehicles, like ambulance and fire engines, require better traffic management for safe and fast travel to safeguard the lives of human beings. In this paper, a new method is proposed for the better management of the traffic of emergency vehicles through the use of internet of things (IOT). The proposed method enables the emergency vehicles to signal the traffic signal controller placed in the traffic junction regarding their arrival so that the traffic will be regulated. This system requires the users traveling in the emergency vehicle to signal the traffic controller hardware through the android application deployed in their mobile phones. We have also proposed the idea for an advanced system which controls the traffic automatically.

Keywords IOT · Smart things · Embedded systems

S. Tammishetty (✉) · T. Ragunathan · S.K. Battula · B. Varsha Rani · P. RaviBabu ·
R. Nagireddy · V. Jorika · V. Maheshwar Reddy
ACE Engineering College, Hyderabad, India
e-mail: tammishetty.sneha27@gmail.com

T. Ragunathan
e-mail: deanresearch@aceec.ac.in

S.K. Battula
e-mail: sudheer.itdict@gmail.com

B. Varsha Rani
e-mail: bedrevarsha03@gmail.com

P. RaviBabu
e-mail: raviram08@gmail.com

R. Nagireddy
e-mail: raghuram18113@gmail.com

V. Jorika
e-mail: v.jorika@gmail.com

V. Maheshwar Reddy
e-mail: v.maheshwarreddy@aceec.ac.in

© Springer Science+Business Media Singapore 2017   433
S.C. Satapathy et al. (eds.), *Proceedings of the First International Conference
on Computational Intelligence and Informatics*, Advances in Intelligent Systems
and Computing 507, DOI 10.1007/978-981-10-2471-9_42

1   Introduction

Internet of things (IOT) define today's generation briefly where things are connected to each other for exchange of some data. The physical devices may be anything around the world which consist of sensors, adaptable networks or anything which is used for data exchange [1–3]. IOT is the collection of smart things which communicate or exchange data in a much secured manner.

The vehicle traffic in cities has been exponentially increased due to a large number of vehicles plying on the road. Due to this heavy traffic, often traffic jams occur on roads because of which the emergency vehicles like ambulance and fire engines have to get stuck in traffic which maybe the cause for losing human lives. In [4], the authors have suggested a manual technique by which it is possible to allow the emergency vehicles to pass through the traffic signals without waiting and hence the human lives can be saved. In this paper, we have proposed a technique which uses android-based application deployed in the mobile phone and the GSM module to be deployed in traffic controller hardware. In the proposed method, the android application deployed in the mobile phones of the users (who are traveling in the emergency vehicles) generates messages to the traffic controller hardware placed in the traffic junction for controlling the traffic so that the emergency vehicle can pass through the junction without any hindrance. The proposed method will also be useful for the vehicles of the important government officials to pass through the traffic junctions without waiting so that they can reach their destination quickly.

We have proved our concept by building a prototype with the required hardware and software. The proposed software is an android application which can be deployed in mobile phones which supports android operating system. We have also used global system for mobile communication (GSM) module in the hardware for the development of the prototype.

We have also discussed regarding the enhancements to be made to this system so that the resultant system will be more efficient and control the traffic automatically. Note that, the users of the mobile phones have to give *starting place* and *destination place* as input in the enhanced system so that the android application deployed in their mobiles will control all the traffic junctions located in the route, to make the emergency vehicles to ply on the road without any hindrance.

The rest of the paper is organized as follows. In the next section, we discuss the related work. In the third and fourth sections, we cover the details of the implementation of the prototype and results of our experiments with the prototype. The fifth section discusses regarding the enhancements to the proposed system to make it more efficient. The final section covers the conclusion.

2 Related Work

In this section, we review the works carried out in the literature for the effective traffic management purpose.

The authors in [5, 6] described that sensors can be used for measurement of traffic density. Traffic lights will be managed at a particular junction, based on the density of the traffic which is measured using (radio-frequency identification) RFID technique. However, using this technique it is not possible to detect the density of the traffic beyond a particular distance. Note that, the devices using RFID are too expensive and the radio-frequency signals can easily be disrupted from their path. Sensing becomes difficult when the weather conditions are unfavorable.

In [7], traffic management, where vehicular adhoc networks *(vanets)* or road side units *(rsus)* are used to sense and communication takes place between two vehicles and later information regarding the situation are sent to trusted traffic authority, was discussed. Through these vanets, emergency requirements are recognized and further actions can be taken. But, in cases of emergency even the trusted traffic authorities may not be able to react as quickly as possible. Note that, it would take a lot of effort and cost to install *rsus* on road.

The authors in paper [8] describe traffic management for emergency signals by capturing physical queues, fixed and variable cycle length etc., using all these values, some computations are applied for managing the signals. But, this computation may not be able to respond as quickly as required for an emergency vehicle and also it is not necessary that the formation of traffic or its related problems are same or similar at all the traffic junctions. Therefore, this computation creates complexity.

In [9], authors gave optimal algorithms and ways to integrate and collect the emergency medical services data generated by IOTs. The data can be collected, stored integrated, and further retrieved for any reasons. Brief explanation of managing the data which is generated by the devices was presented. Authors give better storage and collection of data for emergency medical services.

Authors in [10] gave a detailed structure of how messages are exchanged between IOTs. This papers describes that how messages are scheduled while sending or receiving through any of the IOT devices. The authors gave the optimal way of scheduling the messages and sending them using shortest processing time algorithm.

3 The Proposed System

In this section, we discuss first, the proposed hardware which has to be placed at the traffic controller. Next, we discuss the algorithm followed in the android-based application which is used for controlling the traffic controller hardware.

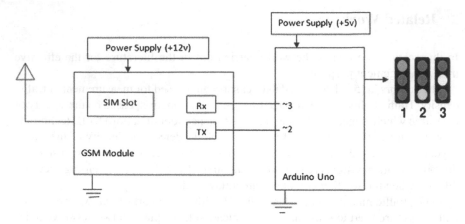

Fig. 1 Block diagram of the proposed hardware system

3.1 Proposed Hardware

We require a special hardware to be available with the traffic signal controller system which is shown in Fig. 1. The component "Arduino Uno" shown in the figure is a PCB (Printed Circuit Board) which contains a micro-controller, namely ATMEGA 328. This microcontroller is used to control the GSM module to send and receive messages from the android application deployed in mobile phones of the users traveling in the emergency vehicles. The messages include (i) the direction from which the emergency vehicle is coming and the direction to which it has to travel after passing the traffic junction (ii) "Green On" message (iii) approximate time that the vehicle requires to reach the traffic junction.

The hardware does the following on receiving these two messages.

1. If the signal is red in the direction to which the emergency vehicle to travel, the signal is changed to green and remain in the same status till ambulance passes the junction.
2. If the signal is green in the direction to which the emergency vehicle to travel, the same status should continue until the ambulance passes the junction.

The normal operation of the traffic controller system is restored once the emergency vehicle passed the traffic junction.

3.2 Resolving Priority of the Emergency Vehicles

If two or more emergency vehicles approach same junction, approximately, at same time, then the following procedure is followed.

Each emergency vehicle will be given some priority value which is predetermined in the system. For example, an ambulance carrying patients in a serious health condition will be given the highest priority and the vehicles of government officials may be given next level of priority. The vehicles are allowed to pass the traffic junction as per the following procedure.

1. The messages from vehicles are saved in the traffic controller system and these messages are sorted according to priority values of the vehicles. Note that, if two vehicles have same priority value, then based on first come first serve method, the messages of the vehicle are stored.
2. Following steps are executed until all the vehicles in the list are served.

   (a) The vehicle which has got the highest priority is selected and it is allowed to pass through the junction.
   (b) The entry related to the above vehicle is deleted from the list.

3.3 The Proposed Android Application Software

The proposed android application software is deployed in the user mobile phones kept in the emergency vehicles.

Figure 2 depicts the components of the proposed android application software deployed in the user mobile phones. One-time password (OTP) system is used in the software for user authentication purpose. After the user authentication process is completed, the current location (source) of the vehicle is tracked using GPS (geographical positioning system). The destination location should be entered manually

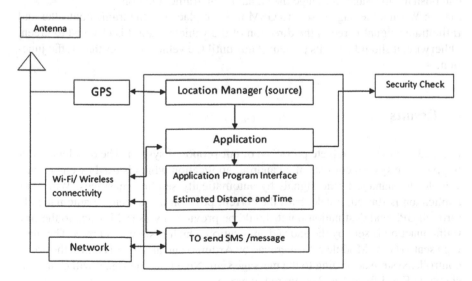

Fig. 2 The components of the proposed android application software

by the user. Distance and time of travel are estimated by considering current location of the vehicle and destination location that the vehicle to reach. Note that, speed sensors are used to calculate approximate speed of the vehicle. Time that vehicle would reach next traffic junction is calculated using speed sensor and also the distance which is calculated using latitudes and longitudes.

Formula for time calculation:Time = distance/speed.

Whenever the emergency vehicle is some n (has to be fixed for the system) kilometers away from the traffic junction (T), the android application sends messages to traffic controller system placed in T. These messages include (i) the direction from which the emergency vehicle is coming and the direction to which it has to travel after passing the traffic junction (ii) Green On message (iii) approximate time that the vehicle requires to reach the traffic junction. The traffic controller system controls the signals according to this message which is described in the Sect. 3.1.

3.4 Algorithm

Step 1: The location (Source) of emergency vehicle will be obtained using GPS which has two co-ordinates (latitude, longitude).
Step 2: The destination will be selected by the user.
Step 3: The messages will be sent to the GSM module at the traffic junction using short message service (SMS) facility available in the mobile phone of the user.
Step 4: In case of any network error or the message could not reach to GSM module, the user can connect to wireless connectivity facility available for that junction and can push the message to change the signal in the traffic junction.
Step 5: When a message is sent to GSM module placed in the traffic controller and if the traffic signal is red in the direction of the vehicle, then it is changed to green. Otherwise, it should be in its present status until the vehicle crosses that traffic junction.

4 Results

Figure 3 shows the complete prototype of our proposed system. The developed prototype of the system consists of an android application which is used by emergency vehicles to manage traffic signals by automatically sending messages. When the application is started, it calculates the current position of the vehicle automatically through GPS and destination point should be provided by user. Message to the first traffic junction is sent by the android application which is shown in Fig. 3. The message sent to GSM Module is further read by Arduino uno and it will control the traffic controller system according to the message sent. Note that, the signal will change as shown in Fig. 3 from 1 to 2 (from red to green).

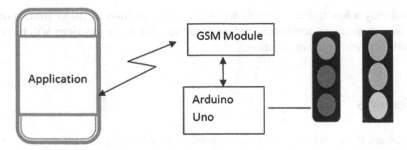

Fig. 3 Prototype of the proposed system

5 Enhanced System

In the system discussed in Sect. 3, the communication is done between the android application deployed in the user mobile phone and the traffic controller hardware placed in a traffic junction and switching of signal is done only for that junction. In the enhanced system, the communication happens between android application deployed in the user mobile phone and traffic controller hardware placed in all the traffic control junctions located in the route from the source to the destination of user travel.

In the enhanced system, the user travels in the emergency vehicle as the first step is to provide destination place to the android application software. Note that, the source place is automatically calculated by the system through the use GPS facility. Next, the software identifies the traffic junctions (TJ) located in the route (R) from the source to destination. After this, the software generates traffic control messages to the traffic controller hardware placed in TJ to see that there is no vehicle traffic in the route R and hence the emergency vehicles can travel on the road without any hindrance.

We wish to work further on priority resolving for serving emergency vehicles effectively. We also plan to develop the system to control only a few traffic junctions within the time duration of travel from source to intermediate destination and on need basis few more junctions will be controlled for reaching the destination.

6 Conclusion

In this paper, we have discussed the smart way of carrying out traffic management of emergency vehicles for rapidly growing traffic in smart cities. We have built the prototype and have shown that the proposed system can manage the emergency vehicles to reach the destination by avoiding much of the traffic. We have also discussed the idea of an enhanced system. In future, we wish to work further on priority resolving for serving emergency vehicles effectively. We also plan to develop the system to control only a few traffic junctions within the time duration of travel from source to some intermediate destination and on need basis few more junctions will be controlled for reaching the destination.

References

1. Kantarci, B.; Mouftah, H.T., "Trustworthy Sensing for Public Safety in Cloud-Centric Internet of Things," in Internet of Things Journal, IEEE, vol.1, no.4, pp.360–368, Aug. 2014.
2. Shanzhi Chen; Hui Xu; Dake Liu; Bo Hu; Hucheng Wang, "A Vision of IoT: Applications, Challenges, and Opportunities With China Perspective," in Internet of Things Journal, IEEE, vol.1, no.4, pp.349–359, Aug. 2014.
3. Bontu, C.S.; Periyalwar, S.; Pecen, M., "Wireless Wide-Area Networks for Internet of Things: An Air Interface Protocol for IoT and a Simultaneous Access Channel for Uplink IoT Communication," inVehicular Technology Magazine, IEEE, vol.9, no.1, pp.54–63, March 2014.
4. Sundar, R.; Hebbar, S.; Golla, V., "Implementing Intelligent Traffic Control System for Congestion Control, Ambulance Clearance, and Stolen Vehicle Detection," in Sensors Journal, IEEE, vol.15, no.2, pp.1109–1113, Feb. 2015.
5. Salama, A.S.; Saleh, B.K.; Eassa, M.M., "Intelligent cross road traffic management system (ICRTMS)," in Computer Technology and Development (ICCTD), 2010 2nd International Conference on, vol., no., pp.27–31, 2-4 Nov. 2010.
6. Eslim, L.M.; Hassanein, H.S.; Ibrahim, W.M.; Alma'aitah, A., "A cooperative localization scheme using RFID crowdsourcing and time-shifted multilateration," in Local Computer Networks (LCN), 2014 IEEE 39th Conference on, vol., no., pp.185–192, 8–11 Sept. 2014.
7. Lo-Yao Yeh; Yen-Cheng Chen; Jiun-Long Huang, "ABACS: An Attribute-Based Access Control System for Emergency Services over Vehicular Ad Hoc Networks," in Selected Areas in Communications, IEEE Journal on, vol.29, no.3, pp.630–643, March 2011.
8. Wei-Hua Lin; Chenghong Wang, "An enhanced 0-1 mixed-integer LP formulation for traffic signal control," in Intelligent Transportation Systems, IEEE Transactions on, vol.5, no.4, pp.238–245, Dec. 2004.
9. Zanella, A.; Bui, N.; Castellani, A.; Vangelista, L.; Zorzi, M., "Internet of Things for Smart Cities," inInternet of Things Journal, IEEE, vol.1, no.1, pp.22–32, Feb. 2014.
10. Jenq-Shiou Leu; Chi-Feng Chen; Kun-Che Hsu, "Improving Heterogeneous SOA-Based IoT Message Stability by Shortest Processing Time Scheduling," in Services Computing, IEEE Transactions on, vol.7, no.4, pp.575–585, Oct.–Dec. 2014.

Wireless Sensor Sequence Data Model for Smart Home and IoT Data Analytics

Nagender Kumar Suryadevara

Abstract In this paper, Wireless sensor sequence data mining model is demonstrated for the smart home and Internet of Things data analytics. Exploration of the sensor data patterns by correlating with the multi stream sensor data that are fused from the wireless sensor network is presented. The effective realization of the sensor data patterns from heterogeneous sensing systems for various applications of IoT can be known from the proposed conceptual data model. The conceptual data model includes the discovering of frequent pattern item sets using various computational archetypal. Results of the explicit patterns augmented for data analytics are encouraging as the prototype was tested through real-time data rather than test bed scenario data or synthetic data.

Keywords Wireless sensor networks · Activities of daily living · Wellness · Smart home · Internet of Things · Data analytics

1 Introduction

The processes of determining frequent patterns in the area associated with data mining are a great deal of results for a list of crucial problems. The conditions associated with discovering item set patterns from the frequent patterns through significant sources have to be examined methodically that can led to the generation of a number of algorithms. The wireless sensor sequence data model of the Smart Home or Internet of Things (IoT) consists of a couple of objects (or items) that can exist to be a pattern. By generalizing the actual item set type, the actual series modeling can be introduced, along with various algorithms [1, 2].

The sequence model brings out two types of organizing the item sets: (i) placing the order amongst a few item sets that are required for specific purpose; and

N.K. Suryadevara (✉)
Geethanjali College of Engineering and Technology,
Cheeryal(V), Keesara(M), R.R. Dist, Hyderabad, Telangana, India
e-mail: suryadevara99@gmail.com

© Springer Science+Business Media Singapore 2017

441

S.C. Satapathy et al. (eds.), *Proceedings of the First International Conference on Computational Intelligence and Informatics*, Advances in Intelligent Systems and Computing 507, DOI 10.1007/978-981-10-2471-9_43

(ii) those requested item sets like a pattern. Both the patterns could be effectively utilized to generate association rules or perhaps acquire prevailing arrangements by item set or perhaps serial sources in several realms along with applications. The significant feature practically in several cases of these procedures is that they command the inherently exponential difficulty in the issue by simply acquiring merely the patterns of which occur in a huge variety of input item sets or maybe sequences, called the support. Any limitation with this model intended for making typical patterns is the fact that the item set works on the frequent support importance, irrespective of the size of the identified patterns [2].

Unfortunately, when constant-support-based repeated arrangement procedures are employed to discover occasional patterns, they are going to land up creating a good greatly large number of limited patterns. If at all possible, typical patterns whose support decreases being to operate of the length and never have to come across a lot of uninspiring occasional quick patterns can be discovered. Useful algorithms for locating haunt item sets or perhaps sequences throughout large item set or perhaps succession directories are on the list of key success stories associated with Frequent Pattern Sequence Sensor data mining analysis [3, 4].

One of the initial computationally effective process for discovering frequent pattern item sets was Apriori [5], which discovers repeated item sets of size l based on previously generated size 1 repeated item sets. The Generalized Sequential Pattern (GSP) procedure updated the Apriori-such as amount-prudent mining strategy to discovery frequent patterns inside sequential databases. The fundamental levels-clever formula has been lengthy in a number of other ways resulting in extremely effective procedures such as Direct Hashing Pruning (DHP). An entirely distinct approach for obtaining typical item sets along with sequences would be the comparability class-based procedures that will breakout the larger lookup area associated with repeated patterns into small and impartial pieces and employ a top to bottom data source formatting that allows them to determine how often aside computing arranged crossing points are obtained. There are also, a set of database projection-based approaches has become established that appreciably slow up the complexness associated with acquiring repeated patterns [1–5]. The key strategy driving them is to use the patterns simply by developing all of them one particular item each period, and also concurrently division (i.e. projecting) the initial source into pattern-specific sub-databases. The procedure of pattern-growth as well as database-projection is definitely repetitive for all recurrent patterns to be generally observed. The original algorithmic program was established for locating non-sequential patterns, nevertheless it continues to be lengthy to find sequential patterns [4, 5].

Another identical algorithm would be the Frequent Pattern (FP)-growth algorithm [5] that associations projection if you routine the actual FP-tree structure to within computer memory to generate/form item sets from the authentic data source. The usual suggestions in this particular algorithm have been not long ago accustomed to establish a related procedure for locating consecutive patterns [5]. The problem of discovering common patterns was extended to that of discovering recurrent maximal patterns [5] and finding frequent closed patterns [5]. Those two problem supplements can be used to decrease the quantity of arrangements that gets ascertained along with assistance in discovering extended patterns specific to the item sets data. However, both these difficulty formulations can still beget a very numerous quick irregular item sets in the event that these kinds of item sets are maximum or un-open.

Liu et al. introduced a calculation in which each pattern of the item has its own Minimum Item Support (or MIS). The actual minimal support associated with an item set would be the most competitive MIS one of those valuables in the particular item set. Through working components of ascending acquisitions with their MIS values, the lower limit support from the item set never diminishes because the amount of item set grows, doing the assist connected with item sets descending sealed. Thus an Apriori-based algorithm can be useful.

Wang et al. introduced the general support constraints for the recurrent pattern growth. Specifically, they associate any support concern on the item sets. Simply by introducing a new function known as P-minsup containing a "Apriori-like" residence, they recommended the Apriori-based algorithm for finding the actual regular item sets. However, the actual "pushed" minimum support of every item set is forced to become adequate to the actual support importance related towards the lengthiest item set. Thus, it wouldn't prune the search space. Cohen et al. follow an alternative approaching; in this they do not utilize any support restriction. Alternatively, they search for identical item sets utilizing probabilistic algorithms that do not promise all regular item sets are available. The Top K-Frequent Pattern does not consider a predetermined minimum support beginning from zero the particular minimum support is definitely increased depending on Top-k recurrent arrangements establish. The LPMiner and Top-k Frequent Patterns are effective for discovering long recurrent item sets pattterns. In addition, TFP allows end users to help input the volume of patterns to be identified as opposed to fewer non-rational minimum supports. One issue using Top-k FP procedure is that any kind of patterns shorter when compared to a presumption lower limit size can never be found, which cannot be right for many applications.

Finally, a great discussion on the distinct type of constraints which were found in your context connected with pattern breakthrough discovery is definitely displayed throughout. Especially, the actual authors class the many constraints directly into four classifications, i.e., anti-monotone, monotone, concise, and convertible. Pertaining to these types of limitations it is revealed which successful mining can be carried out by means of forcing the particular difficulties into mining practice. Nevertheless, the length-decreasing assistance restraint will not fall under all of these several types.

The remaining part of the paper is organized as follows: Sect. 2 provides system description and detailed implementation of the developed system; Sect. 3 presented the experimental results and Sect. 4 concluded and discussed about the future work.

2   System Description and Implementation Details

The developed framework is clever, powerful and does not utilize any camera or vision sensors as it involves privacy protection and is not adequate by numerous tenants of the smart home [6–11]. The wise programming, alongside the electronic framework, can screen the use of various family unit electronic apparatuses and perceive the sensor action design in close ongoing. Additionally, the framework deciphers the entire crucial tenant every day activities, for example, get ready breakfast/lunch/supper, rest-room use, dinning, sleeping and self-prepping. Essentially, the framework functions on the usage of electrical and non-electrical apparatuses inside a home. Framework is more adaptable in its operation for a current home instead of a recently developed house."

The framework comprises of two essential modules. At the low level module, Wireless sensor system of mesh topology exists catching the sensor information taking into account the use of house hold devices and stores the information in the PC framework for further information handling [6–11]. Gathered sensor information is of low level data containing just status of the sensor as dynamic or inert and personality of the sensor. To sense the movements of tenant continuously, the following level programming module will break down the gathered information by taking after a shrewd system at different level of information deliberation taking into account time and arrangement conduct of sensor use.

The low level module comprises of number of keen sensors interconnected to recognize utilization of electrical gadgets, bed use and other non-electrical gadgets [6–11]. The designed and developed sensing systems convey through radio recurrence conventions and give sensor data that can be utilized to screen the day by day activities of smart home occupant. A savvy sensor facilitator gathers information from the gadget end shrewd sensors and forward to the PC framework for information preparing. The caught information is progressively changing and requesting quick continuous reaction time for anticipating the conduct of the tenant; for breaking down the information legitimately an effective procedure of capacity component of sensor information onto the PC framework is produced and executed [6–11]. In the present framework, programs for sensor data collection and information generation are coded using "C#". The Web interface is created utilizing PHP Script and Java Scripts. The constant sensor information investigation is finished with the assistance of R studio [12].

3   Experimental Results

Developed system is tested by installing the Smart sensing units and setting up a Zigbee based WSN at an elderly house [6–11]. Integrated system [6–11] was continuously used and generated real-time graphical representation of the sensing information. Figure 1 depicts the screen shot of the Apriori running program of the

R-Studio on the sensor data on a particular day (Tuesday's) from 21 May 2013 to 06 Aug 2013.

Figure 2 shows the frequent patterns sequences generated from the 476 transactions on the sensor data (21-May-2013 to 06-Aug-2013) using R Studio. Apriori algorithm was implemented with a support value of 0.1 and max and min size of the pattern to be of value 10. The labels of the sensor are indicated as [FR]: front door movement; [KT]: Kitchen door movement; [CR]: Centre of the house; [BK]: Back door movement.

Figure 2 demonstrates the Graph-based methods (Klemettinen, Mannila, Ronkainen, Toi-vonen, and Verkamo 1994; Rainsford and Roddick 2000; Buono and Costabile 2005; Ertek and Demiriz 2006) envision affiliation rules utilizing vertices and edges where vertices regularly speak to things or thing sets and edges show relationship in tenets. Interest measures are ordinarily added to the plot as names on the edges or by shading or width of the bolts showing the edges. Chart based perception offers an unmistakable representation of tenets however they tend to effortlessly get to be jumbled and in this way are reasonable for little arrangements of guidelines [12].

Figure 3 depicts the Parallel coordinate plot (reordered) for the generated 71 rules of the cspade Algorithm on the sensor data using R Studio. Parallel directions plots are intended to imagine multidimensional information where every measurement is shown independently on the x-hub and the y-hub is shared. Every information point is spoken to by a line interfacing the qualities for every measurement. Parallel directions plots were utilized beforehand to picture found arrangement rules (Han, An, and Cercone 2000) and affiliation rules (Yang 2003). Yang (2003) shows the things on the y-pivot as ostensible qualities and the x-hub speaks to the positions in a standard, i.e., first thing, second thing, and so forth.

Fig. 1 Implementation of Apriori algorithm on the sensor data using R studio

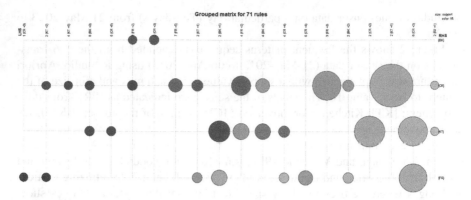

Fig. 2 Rules (patterns) generated from the 476 transactions on the sensor data (21-May-2013 −06-Aug-2013) using R studio

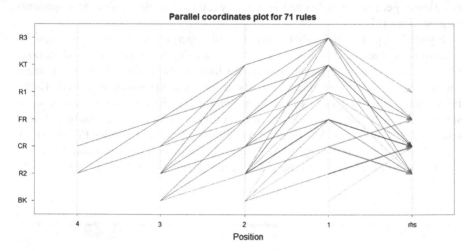

Fig. 3 Parallel coordinate plot (reordered) for the generated 71 rules of the cspade algorithm on the sensor data using R studio

Rather than a straightforward line a bolt is utilized where the head focuses to the resulting thing. Bolts just traverse enough positions on the x-hub to speak to every one of the things in the tenet, i.e., rules with less things are shorter arrows [12].

4 Discussion and Future Work

The probable sensor sequences are generated from the sensor stream data by applying frequent pattern sequence process. The temporal reasoning on the activities of the inhabitant is considered in the frequent pattern sequence process. Based on the real-time sensor stream and the likelihood patterns of the sensor data, the inhabitant behaviour is determined. In turn, the same pattern generation process can be applied to the IoT domain. Several pruning mechanisms can be applied or existing techniques can be modified to determine the optimal support value for a pattern to occur based on the temporal reasoning. The pruning mechanism is applied in order to reduce the search space in the depth first or breadth first traversals of the pattern sequences of the sensor stream. Thus, the frequent pattern sensor sequence models can be applied of the IoT and Big Data application for effective data analytics.

References

1. Agrawal R., Srikant R., "Fast Algorithm for Mining Association Rules", VLDB. Sep 12–15 1994, Chile, 487–499, pdf, ISBN 1-55860-153-8.
2. Huiping Peng, "Discovery of Interesting Association Rules Based on Web Usage Mining" 2010 International Conference.
3. Han J., Pei J., Yin Y., Mao R., "Mining Frequent Patterns without Candidate Generation: A Frequent-Pattern Tree Approach" Data Mining and Knowledge Discovery, 2004, 53–87.
4. K. R. Suneetha and Dr. R. Krishnamoorthi "Identifying User Behavior by Analyzing Web Server Access Log File", International Journal of Computer Science and Network Security, vol. 9, no. 4, 2009.
5. Goswami D.N., Chaturvedi Anshu., Raghuvanshi C.S., "An Algorithm for Frequent Pattern Mining Based on Apriori", IJCSE, Vol. 02, No. 04, 2010, 942–947.
6. N.K Suryadevara, S.C. Mukhopadhyay, "Wireless Sensor Network based Home Monitoring System for Wellness Determination of Elderly", IEEE Sensors Journal-2012, Vol: 12 Issue: 6, Page(s): 1965–1972.
7. N.K. Suryadevara, A. Gaddam, R.K. Rayudu, S.C. Mukhopadhyay, "Wireless Sensors Network Based Safe Home to Care Elderly People: Behaviour Detection", Elsevier Sensors and Actuators: A Physical (2012), http://dx.doi.org/10.1016/j.sna.2012.03.020, Vol 25, Pages 96–99.
8. N.K. Suryadevara, S.C. Mukhopadhyay, "Determining Wellness through an Ambient Assisted Living Environment," IEEE Intelligent Systems-May 2014, Page(s): 30–37, ISSN:1541-1672.
9. N.K Suryadevara, S.C. Mukhopadhyay, R. Wang, R.K Rayudu, "Forecasting the behavior of an elderly using wireless sensors data in a smart home", Elsevier: Engineering Applications of Artificial Intelligence, Vol: 26, Issue: 10, Page(s): 2641–2652. ISSN:0952-1976.
10. N.K Suryadevara, S.C. Mukhopadhyay, S.D.T Kelly, S.P.S. Gill, "WSN-Based Smart Sensors and Actuator for Power Management in Intelligent Buildings", IEEE Transactions on Mechatronics, doi:10.1109/TMECH. 2014. 2301716. Page(s): 564–571, Vol: 20, Issue: 2, ISSN:1083-4435.
11. S.D.T Kelly, N.K. Suryadevara, S.C. Mukhopadhyay, "Towards the Implementation of IoT for Environmental Condition Monitoring in Homes", IEEE Sensors Journal, Vol: 13, Issue: 10, Page(s): 3846–3853. ISSN:0952-1976.
12. https://cran.r-project.org/web/packages/arulesViz/vignettes/arulesViz.pdf.

Automatic Detection of Violent Scenes in Tollywood Movies—A Rough Set Approach

T. Prathima, A. Govardhan and Y. Ramadevi

Abstract Exposure to violent content in movies will have an effect on both behavioural and psychological aspects of children. Efficient algorithms are needed to automatically detect the violent content in movies. Existing work in identifying violent content in Hollywood movies is studied along with the datasets on which the evaluation is carried out. We have also examined the different set of challenges posed by Tollywood movies and how the existing work may not be able to handle the challenges is presented. An approach based on rough sets is proposed to handle the highly dynamic nature of violent scenes.

Keywords Violence · Tollywood · Modalities · Rough sets · Rough classifier

1 Introduction

Exposure to uncensored content on motion picture will have an impact on children's behavior and their psychology [1]. The violence content that can induce aggression in children behaviour ranges from high adrenaline chasing sequences, dead bodies, atrocities, gore scenes, struggle or high action sequences, screams, etc. Automatic detection and scissoring out the above mentioned shots from motion pictures or movies is important to protect children from the ill effects of watching them which in turn could change their attitude and behavior. This paper is

T. Prathima (✉)
Department of IT, CBIT, Hyderabad, India
e-mail: prathima@cbit.ac.in

A. Govardhan
Department of CSE, JNTUCEH, Hyderabad, India
e-mail: govardhan_cse@jntuh.ac.in

Y. Ramadevi
Department of CSE, CBIT, Hyderabad, India
e-mail: yrdcse.cbit@gmail.com

© Springer Science+Business Media Singapore 2017   449
S.C. Satapathy et al. (eds.), *Proceedings of the First International Conference on Computational Intelligence and Informatics*, Advances in Intelligent Systems and Computing 507, DOI 10.1007/978-981-10-2471-9_44

organized as follows: Sect. 2 discusses the existing literature on violence detection in Hollywood movies and the datasets on which experiments are conducted, Sect. 2.1 deals with datasets, Sect. 3 presents the challenges posed by Tollywood movies, Sect. 4 introduces the proposed plan of action to address the challenges in Tollywood movies and Sect. 5 concludes the paper.

2 Existing Work

In this section review of the existing work on violence detection in Hollywood movies is presented. Yu Gong et al. [2] work considered low level visual and auditory features. Initially probable candidate shots were identified using movie making grammar and the shot boundaries were detected using color histogram. Nineteen features are extracted to denote motion activity of the candidate shots. Audio features like spectrum flux, spectrum power, pitch, high ZCR, onset, harmonicity prominence, brightness and rhythm are computed on audio frames. A semi-supervised classifier was built on top of Support Vector Machine and Radial Basis Function kernel to perform the task. Training and testing data was taken from four movies, and they have achieved 83 % precision, 96 % recall and F1-measure of 89 %.

Theodoros Giannakopoulos et al. [3] have identified three violent and four non-violent classes based on audio data by examining the movies. Violent classes are screams, fights and shots and non violent classes are basically music, speech and environmental sounds observed from the movies. 12 features are computed for each non-overlapping frame of the audio segments, audio features considered are MFCC, energy entropy, chroma, ZCR, Zero pitch ratio, spectrogram, spectral roll off, etc. To classify the incoming sample into one of the seven classes mentioned above a 1-Vs-all approach is used. K-binary classifiers are constructed using Bayesian network, the probabilities given by the Bayesian Network are inputted to the k-NN classifiers and the winner class is identified. Three classes are considered for depicting the type of video segments, they are no activity, normal activity and high activity based on the people in the shot and the activity they are performing which ranges from no activity to high activity like fighting, fast movement etc. The authors observed that if a shot is classified as high activity shot then there is likelihood that it could be a violent shot. To obtain the classification of video shots motion features and object recognition are used. The classifications done using audio and video modality are given as inputs to k-NN classifier and a binary decision is made to decide whether the shot is violent or non violent. Ten movies were considered for the task and they have achieved 83 % recall, 45.2 % precision and 58.5 % F1-measure.

Perperis et al. [4] explored both auditory and visual modalities to derive mid level semantics. They built a k-NN binary classifier on the extracted features using late fusion. Evaluation of the performance when single and multimodalities are used in discriminating violence is presented. They have also developed ontological framework/inference engine which includes audio ontology, video ontology and violence ontology in order to distinguish violence and non violence content. They have achieved a detection recall of 91.2 %, precision of 34.2 %, and F1-measure of 50 %.

Jeho Nam et al. [5] proposed a technique based on both audio and video tracks. The dynamic activity of a video shot is detecting considering the spatial and temporal relationship. Using motion sequence and spatio-temporal changes action shots are identified, from these shots, violent shots are separated by exploiting the other visual and audio features. Audio track information which is temporally correlated with the high action video shot is combined to label the shot as either violent or non-violent. For this purpose mean vector and co-variance matrix are computed for the audio track along with energy entropy. The authors have used Guassian modeling technique to classify video shots which has beating, gun fire and explosion as violent with classification accuracy of 88 %.

Esra Acar et al. [6] work exploited both audio and visual modalities to perform violence detection. They have constructed audio dictionary of size 1000 words using Bag of Audio Words (BoAW) concept extracted from the low level Mel-Frequency Cepstral Coefficients (MFCC) of uniform audio frames. Video shots are represented using motion vectors computed across successive frames. 3 two-class Support Vector Machines (SVM) are used respectively on low-level audio features, mid-level audio features and visual features to classify the input as either violent or non-violent. Results achieved from the SVM's are fused to predict the final class. To address the issue of class imbalance the authors have adopted random under-sampling strategy with SVM. The proposed method has achieved an average precision@20 of 54.5 %.

Jian Lin et al. [7] considered both audio and video modalities for their work. Video is split into shots and from each shot audio track and video frames are separately processed. Each audio track is further divided into one second frames and low level features like MFCC, Zero cross rate (ZCR), spectrum power, harmonicity prominence, bandwidth, brightness, spectrum flux and pitch, are extracted. And the computed feature vectors of the audio tracks are clustered using K-means technique. To classify the audio tracks and the cluster centres as either violent or non violent Expectation-Maximisation (EM) method is used. Motion complexity and intensity of the blocks of frame of the video shot are computed to identify high tempo scenes. To detect shots with flame and explosion frame is divided into yellow tone and non-yellow tone regions, if yellow tone region exceed a certain threshold then the shot is termed as shot with flame and explosion. To detect scenes with blood the red tone and non-red tone regions of the frame are identified, assuming that blood regions always move fast, if the red tone regions and the average motion intensity exceed respective thresholds then the shot is considered as a shot with blood in it. Computed audio and video features on the shots are co-trained on two classifiers and the best results achieved are 85.07 % precision, 95.85 % recall and 90.58 % F1-measure.

Liang-Hua Chen et al. [8] designed a violence detector by means of visual cues and temporal characteristics of a video. Features like average motion intensity, camera motion ratio, average shot length base and shot cut frequency are extracted and an SVM classifier is used to detect action scenes and frames with blood. Instead of shot level analysis proposed method considers semantically meaningful scenes and in the process they have achieved recall of 82.35 % and 100 % precision.

Bogdan Ionescu et al. [9] trained 10 multilayer perceptrons to classify 10 different violence types. Features extracted include 196 audio features, 11-color dimensions, 81-histogram of gradient and a feature for temporal structure. They have achieved overall precision of 46.14 %, recall of 54.40 % and F1-score of 49.94 % when evaluated on dataset used for multimedia evaluation.

2.1 Data Sets

Most of the existing work is trained and tested on MediaEval benchmark datasets [10], currently it consists of 31 movies out of which 24 movies are provided for training and the rest 7 are provided for testing. Following is the list of movies: The Bourne Identity, Léon, Armageddon, The Wizard of Oz, Pulp Fiction, I am Legend, Eragon, Reservoir Dogs, Fight Club, Harry Potter and the order of the Phoenix, Pirates of the Caribbean—the curse of the black pearl, The Sixth Sense, Billy Elliot, Saving Private Ryan, The Wicker Man, Midnight Express, Independence Day, Dead Poets Society, Fantastic Four 1, Fargo, Forrest Gump, The Pianist, Legally Blond, The God Father 1. Annotations are provided for the violent content in training set. Every year mediaeval conducts affect task in which the participants present their work. The concepts of violence that needs to be detected are fights, presence of fire, presence of fire arms, presence of cold arms, gory scenes, car chases, gun shots, explosions, screams [11].

3 Challenges

From Table 1 of Sect. 2 we can confirm that any work covers only subset of the following scenes classified as gore, sad, one on one fighting, sword fighting, mob fighting, ammunition usage, vehicle chases, murder, etc. Even if most of the violent concepts are covered then the recall, precision values are low, and moreover the evaluation is conducted on limited movies. The other aspects specific to tollywood movies which can pose challenges are (a) most of the movies revolves around the hero character rather than the story, (b) common movie genres are action, comedy, romcom, thriller, etc., percentage of telugu movies which fall under more than one genre is more as most of the movies are formula based, (c) bright colors are used especially variants of red and orange as they are considered more auspicious as per the regions tradition this results in failing of most of the existing techniques, (d) in some cases scenes with violence are repeated in slow motion this again put forth before us a lot of challenge in identifying that particular scene as repetitive and

Table 1 Comparison of existing work in violence detection in hollywood movies

S. no	Author (s)	Considered scenes of interest	Technology used	Size of dataset*	Evaluation criteria (%)
1.	Yu Gong et al.	Gun-shot, explosion, smash, sword	Semi-supervised cross feature learning, SVM, RBF Kernel	4	R—96 P—83 F—89
2.	Theodoros Giannakopoulos et al.	Shots, fights, screams	Bayesian Network kNN classifier	10	R—83 P—45.2 F—58.5
3.	T. Perperis et al.	Fight, torture, rape, war, riot, battle, mental violence	Bayesian Network kNN classifier Ontology framework	10	R—91.2 P—34.2 F—50
4.	Jeho nam et al.	Blood, Gun fire, Beating, Explosion	Gaussian modeling	5+	–
5.	Esra Acar	General violence	Bag of Audio Words, SVM	18	Ap@20 54.5 Ap@100 42
6.	Jian Lin et al.	Flame, explosion blood	Co-training algorithm on two classifiers	4	R—92.02 P—83.82 F—87.78
7.	Liang-Hua Chen et al.	Action scenes frames with blood	SVM	4	R—85.03 P—100
8.	Bogdan Ionescu et al.	Blood, cold arms, fire arms, gore, gunshots, screams, car chase, explosions fights, fire	Feed forward neural network	18	R—67.69 P—32.81 F—44.58

R—Recall, P—Precision, F-F1—Measure, Ap—Average Precision
*No. of Movies

moreover violent (e) the presence of song sequences, creates lot of ambiguity between bloody scenes and fight scenes when detection is based on film making rules and color features, this can happen as most of the songs and the dance sequences in the songs are choreographed where the actors are performing fast paced dance, when we go with motion intensity this results in false negatives again. And the other challenges that can crop up are group dancers in the background will add confusion to the classifier as they will fall under high activity scenes. And even some violence scenes are associated with a sad song in the background, even this will affect the recall and precision values when existing techniques are used. Theme song or title song of a movie may be repeated as a background score for crucial parts of the movie (f) in some comedy scenes/situational comedy scenes of the tollywood movies the body language of the actors involved may be false detected as violent which should be addressed.

4 Proposed System

From Fig. 1 we can see that instead of probable shots, we consider entire movies as input, from this three modalities of features are extracted a. audio features, b. video features and c. features from the rules of movie making. Audio features are extracted from the training data sets, these features are continuous in nature and by using rough clustering continuous data is discretised. From the discretised data, homogeneous segments are detected and a rough classifier [12] is trained to classify the detected segments into different audio classes like melody, speech, screams, fight, silence, etc. Total feature set is analysed for data dependencies and minimal set of attributes are identified i.e., reduct is generated using rough sets. From the frames of the video low level features are extracted and using these features, regions of blood in a frame, presence of weapons, presence of more no. of persons, etc. are detected. The issue of detecting objects of interest with cluttered back ground can be addressed using granular computing with the help of lower and upper approximations of rough sets. Image granules can be created by grouping indiscernible pixels and computation is done on these image granules. Along with these image granules using the rough sets approximation theory, rough entropy can be devised to measure the uncertainty caused by the background. Maximising the rough entropy in turn minimizes the impreciseness of the boundary of region of the object from its background [13]. From the third modality we try to extract features which help us in detecting fast camera movement, slow motion shots, etc. In building the rough classifiers on top of both audio features and video features, shots detected from the third modality are also used in training to improve the accuracy. The results from both the classifiers are fused and a decision is made whether the current frame is violent or non violent. Keeping in mind the highly volatile nature of violent scenes we propose to use rough set theory as it is a popular tool to handle imprecise data.

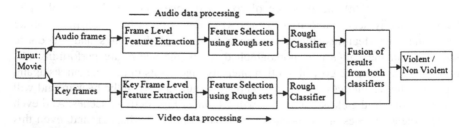

Fig. 1 Block diagram of the proposed system

5   Conclusions

In this paper we have reviewed the existing literature on violence detection in Hollywood movies, along with the datasets on which their results are presented. We have also observed the challenges that Tollywood movies pose when compared to Hollywood movies. We proposed a system which considers features from three modalities and using rough set theory which handles imprecise data, classifiers are built and decision is made to classify whether the scene is violent or non violent and this work can further be enhanced to find the degree of violence in a scene.

References

1. B. Bushman, L. Huesmann: Short-term and long-term effects of violent media on aggression in children and adults. Archives of Pediatrics & Adolescent Medicine, 160(4):348, 2006
2. Gong, Y., Wang, W., Jiang, S., Huang A., and Gao, W. 2008. Detecting Violent Scenes in Movies by Auditory and Visual Cues, Advances in Multimedia Information Processing - PCM 2008. Lecture Notes in Computer Science, 2008, Volume 5353/2008, pp. 317–326
3. Giannakopoulos, T., Makris, A., Kosmopoulos, D., Perantonis S. and Theodoridis, S.: Audio-visual fusion for detecting violent scenes in videos, Artificial Intelligence: Theories, Models and Applications. Lecture Notes in Computer Science, Volume 6040/2010, pp. 91–100
4. Thanassis Perperis, Theodoros Giannakopoulos, Alexandros Makris, Dimitrios I. Kosmopoulos, Sofia Tsekeridou, Stavros J. Perantonis, Sergios Theodoridis: Multimodal and ontology-based fusion approaches of audio and visual processing for violence detection in movies, Expert Systems with Applications 38 (2011) 14102–14116
5. JeHo Nam; Alghoniemy, M.; Tewfik, A.H.: Audio-visual content-based violent scene characterization, in Image Processing, 1998. ICIP 98. Proceedings. 1998 International Conference on Image Processing, vol.1, no., pp. 353–357 vol. 1, 4-7 Oct 1998
6. Esra Acar, Frank Hopfgartner, Sahin Albayrak.: Violence Detection in Hollywood Movies by the Fusion of Visual and Mid-level Audio Cues, Proceedings of the 21st ACM international conference on Multimedia, Pages 717–720, ACM New York, NY, USA ©2013
7. Jian Lin and Weiqiang Wang.: Weakly-Supervised Violence Detection in Movies with Audio and Video Based Co-training, Advances in Multimedia Information Processing - PCM 2009, Volume 5879 of the series Lecture Notes in Computer Science pp 930–935
8. Liang-Hua Chen; Hsi-Wen Hsu; Li-Yun Wang; Chih-Wen Su: Violence Detection in Movies, in Computer Graphics, Imaging and Visualization (CGIV), 2011 Eighth International Conference on, vol., no., pp. 119–124, 17-19 Aug. 2011
9. Bogdan Ionescu, Jan Schlüter, Ionuţt Mironicˇa, Markus Schedl.: A Naïve Mid-level Concept-based Fusion Approachto Violence Detection in Hollywood Movies, Proceedings of the 3rd ACM conference on International conference on multimedia retrieval pp 215–222 ACM New York, NY, USA ©2013
10. Technicolor Violent Scenes Dataset, http://www.technicolor.com/en/innovation/research-innovation/scientific-data-sharing/violent-scenes-dataset
11. C.-H. Demarty, C. Penet, G. Gravier, and M. Soleymani. A benchmarking campaign for the multimodal detection of violent scenes in movies. In ECCV, 2012

12. Andrzej Lenarcik, Zdzisław Piasta.: Rough Classifiers, Rough Sets, Fuzzy Sets and Knowledge Discovery, Proceedings of the International Workshop on Rough Sets and Knowledge Discovery (RSKD'93), Banff, Alberta, Canada, 12–15 October 1993, pp 298–316
13. Pal, Sankar K., B. Uma Shankar, and Pabitra Mitra.: Granular computing, rough entropy and object extraction, Pattern Recognition Letters 26.16 (2005): 2509–2517

Prism Tree Shape Representation Based Recognition of Offline Tamil Handwritten Characters

M. Antony Robert Raj, S. Abirami, S. Murugappan and R. Baskaran

Abstract Optical Character Recognition (OCR) is a unique and challengeable filed in pattern recognition. Identically demand is still present in OCR, where various works has been coming up to provide acceptable solutions. In this field, Tamil hand written recognition is getting popular due to the desire of Tamil lovers in computerizing the Tamil language which includes handwritten documents also. This task is not an easiest one, due to its curvy shape and variation in structure of shape when people are writing. Here the treatment must be happened on the structure level to address entire shapes. This paper mainly focuses on identifying the shape of the structure, where the shape of the structure is derived from the formation of the triangle based hierarchical representation. Prism Tree algorithm is utilized to complete this task, where the shape is located by the tree representation. Finally vector values are extracted from the shape representation of tree. Hierarchical based Support Vector Machine (SVM) is used for predicting the character from those vector values. Good results are achieved when the shape of the character structure is well suited for real character nature.

Keywords Prism tree · Chain code · Hierarchical SVM

M.A.R. Raj (✉) · S. Abirami · R. Baskaran
Department of Information Science and Technology, Anna University,
Chennai 600025, India
e-mail: antorobert@gmail.com

S. Abirami
e-mail: abirami_mr@yahoo.com

S. Murugappan
School of Computer Sciences, Tamilnadu Open University, Chennai, India

© Springer Science+Business Media Singapore 2017
S.C. Satapathy et al. (eds.), *Proceedings of the First International Conference on Computational Intelligence and Informatics*, Advances in Intelligent Systems and Computing 507, DOI 10.1007/978-981-10-2471-9_45

1 Introduction

In this competitive world, computerizing each and every documents is necessary. Increasing the affection of the Tamil language, Tamil lovers are in the process to convert every Tamil document in a digital form. Increasing the demand of the old documents in Tamil and palm script documents, for the first step, researchers are converting Tamil letters into computerized form. This can be done in two ways, online and offline [1]. In online mode, algorithm is directly applied on the pen tip moment to get the character structure and to recognize the same. This can be achievable in the present scenario but the offline mode is complex, wherein, in the offline mode the recognitions can be done for printed and handwritten documents. Here less complication exists in the recognition of the characters from printed documents because, the shapes of the printed character is natural for all Tamil language clauses. The Shape of the Tamil handwritten documents always depends on the writers, where the complexity level is very high. Matching the general shape of the characters and the characters which are written by various writers are not an easy task because the shape will vary from writer to writer. Not only these factors, additionally, the general structure of Tamil character are complex.

The recognition of Tamil characters was achieved by using both statistical [1, 2] and structural way in all the previous works [3–6]. Pixel counts, density values and various pixel based locations were applied on the character image in order to test whether the statistical way was successful and considerable result has been achieved by the statistical way [3, 4, 6]. Directions of character structure and area of the structure were analyzed and tested in the structural mode of verification [3, 5, 6]. Everywhere the work is depended on the shape of the character whereas the shape variation always depends on the writers. The sample shape variations of the character is 'ஜ' shown in the Fig. 1.

Few of the statistical concepts like transition and profile were discussed in the work of Sigappi et al. [7], where the Hidden Markov Model was utilized to achieve the results. The zone and the pixel variation that was based on density were implemented successfully in many works of the Rajashekararadhya et al. [8–11].

Raj Kumar and Subbiah Bharathi [12] implemented few statistical concepts of the pattern recognition in their works. The ancient style of character was considered by them, but still this has been open for researchers due to its shape variation.

Addressing all the shapes in character is another complex issue, due to the character complexity and writer complexity. So this work uniquely focuses on the shape of the character. The tree representation is chosen for representing the curvy shapes where the Prism tree is used to achieve this task.

Fig. 1 Shape variation of the character 'ஜ'

1.1 Prism Tree

Prism Tree [13] is used to construct the polyhedral approximation of objects in the two dimensional plane usually. This type of tree representation is utilized to address the boundary of any object. The shape is divided into the triangular form and the adjacent portion of the final triangle addresses the shape. The Triangular formation takes place in the curvy part of the shape and the same continues until the end points of the triangular formation (adjacent) cannot be extended further. These final end points denote the boundary of the object. This concept is highly matched with this work to represent the shape of the character and is highly suitable for treating the curvy structure of Tamil character.

2 System Implementation

This work has been experimented by implementing the general phase of the image processing. Some important phases are picked up from the general phase of image processing and the same has been implemented in this work. But majorly the feature pre-extraction and feature selection plays a crucial role here. This task is achieved by implementing a chain code algorithm for pre-extracting the features and Prism tree for extracting the features. The phases of this work are described in the Fig. 2.

The essential prepressing steps such as Binarization, Skeletonization, Noise Removal and Normalization is experimented in this work in order to get clear image

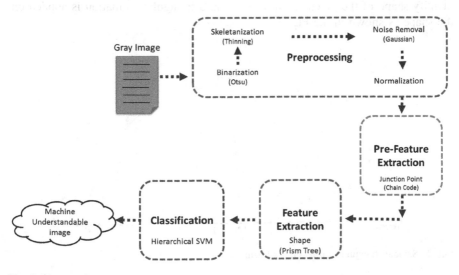

Fig. 2 Phases of system implementation

for further process. The general preprocessing steps are implemented as like the previous works [3–6], the Otsu's procedure is procured for taking out the two tone image (binary image) from the gray image, the thinning algorithm that is utilized to get a skeletonized image, The Gaussian process that is used to get noise less image and finally the normalization procedure that is applied to convert the image into a standard size image.

2.1 Feature Pre-extraction

The pre-extraction is an essential one when considering the shape as features. The character shape is segregated from the entire character structure in order to avoid the whole complexity of the character at a time. The chain code algorithm is applied on the entire character image to separate the character structure as denoted earlier [3–6]. The chain code separates the character portion based on the junction point's availability. The chain code algorithm breaks its travel when it meets any junction point (three pixels grouped together) or end points (pixel with one neighborhood pixel).

2.2 Feature Extraction—Prism Tree

The tree which is used for the polyhedral approximation of objects [13] is implemented here to extract the features from the character image. The feature separated from the pre-extraction procedure has been taken into consideration here. To identify shape of the given character portion, a triangular formation is applied on the shape as shown in the Fig. 3.

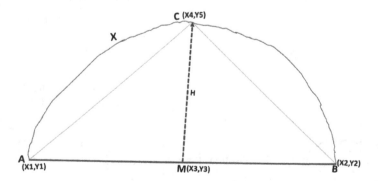

Fig. 3 Sample triangular formation for Prism tree

Triangular formation.

1. Two end points of the character portion is found and taken as A (x_1, y_1) and B (x_2, y_2) (black pixel with one neighbor black pixel).
2. A straight line has been drawn between the points A, B using the equation

$$\frac{y - y_1}{y_2 - y_1} = \frac{x - x_1}{x_2 - x_1} \tag{1}$$

The point values of A and B (x_1, y_1) and (x_2, y_2) has been substituted in the Eq. 1, which will provide the equation of the straight line (Eq. 2) lies between the points A and B.

$$ax + by + c = 0 \tag{2}$$

3. The midpoint (M) of the line $ax + by + c = 0$ lies between of the A and B is calculated using the equation Eq. 3

$$M = \left(\frac{x_1 + x_2}{2}, \frac{y_1 + y_2}{2} \right) \tag{3}$$

which gives the midpoints M (x_3, y_3).
4. Then the line which perpendicularly passing through the midpoint M (x_3, y_3) has been computed, where the coefficient of the x, y has been interchanged and the sine of those is changed which leads to the equation Eq. 4.

$$bx_3 - ay_3 + c_1 = 0 \tag{4}$$

Where, when substituting the values (x_3, y_3), the value c_1 is found. This equation provides the line which start the point at M and traversing towards the point C perpendicularly. To find the exact point of C, various points has been chosen from where the curve is lying and the same has been applied on the equation (Eq. 4). The points (x_4, y_4) which gives zero when substituting on x_3 *and* y_3 in equation (Eq. 4) is taken as C
5. The equation (Eq. 5) is used for finding the value of the perpendicular line, lies between M and C is

$$H = \left| \frac{ax_4 + by_4 + c}{\sqrt{a^2 + b^2}} \right| \tag{5}$$

6. If the value of H is greater than one then the steps continued, or else it is stopped. If it is stopped, then the triangle formation is not possible for this character shape.
7. Further, the adjacent lines (AC, BC) is drawn towards the point C from the end points A and B using the equation (Eq. 1).

8. The shape approximation of the character portion AC and BC, provides the triangle AMC and BMC.
9. The steps 2 to 8 are continued for AC and BC to check if any triangle is possible further.
10. The final adjacent shape points represent the shape of the structure.

The following procedure takes place for representing the shape in the hierarchical tree formation

Procedure 1: To represent a 'dot'

1. One or two pixels (A) is found in the character portion, which is marked as R.
2. No end points is found in the character portion,
3. No triangle can be possible and H = null;

Tree and feature formation

Input: A, R, H : Output: Feature values;
   Begin
       Initiate P, FEA, i=1, Fea1;
   If (found R)
       R ←P; R← Root; Fea1=A;
       If (H==null)   *Description*
       For each i of 1   P   : Pointer
          FEA=Fea1; Return FEA   FEA : Variable for store features
   Exit   i, Fea1 : Variables

The following figure (Fig. 4) shows the tree and feature formation process from the character portion 'dot'

Fig. 4 Feature extraction process from 'dot'

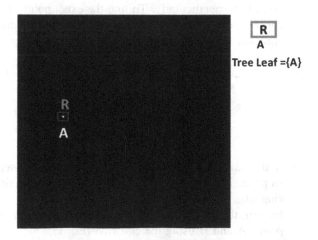

Tree Leaf ={A}

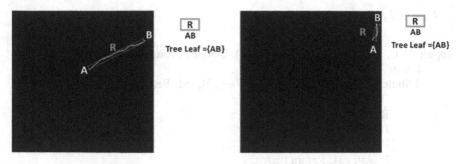

Fig. 5 Feature extraction process from 'short line or long line'

Procedure 2: To represent 'short line or long line or short shape'

1. Few pixels are found in the character portion, and is marked as R.
2. The end points (A, B) are noted from the character portion.
3. The triangle is tried to form as discussed in "Triangle formation".
4. No triangle is possible if H < 1;

Tree and feature formation

Input: AB, R, H: Output: Feature values;
Begin
      Initiate P, FEA, i=1, Fea1;
If (found R)
     R ←P; R← Root; Fea1=AB;
     If(H<1)
     For each i of 1
        FEA =Fea1; Return FEA
Exit

	Description
	P : Pointer
	FEA : Variable for store features
	i, Fea1 : Variables

The above figure (Fig. 5) represent the tree formation for given structure
Procedure 3: To represent 'closed curve'

1. Pixels are found in the character portion, and is marked as R.
2. Any two pixel points (A, B) are set on the character portion.
3. The triangle is formed as discussed in the triangle formation.
4. Triangle can be possible if H > 1;
5. Three base line (AB, BC and CA) is chosen instead of choosing two adjacent points and the triangles are formed (R1, R2, R3 and son on) as discussed earlier.

Tree and feature formation

Input: ABC, R, H, Adjacent Values (AV): Output: Feature values;
    Begin
      Initiate P, FEA, FEA_R, FEA_L, FEA_B, i=1, Fea1;
         If (R found)
         R ←P; R← Root; Fea1=ABC;
         If (H>1)
         For each i of l
         FEA=Fea1; Return FEA
             If (AV found)
                If (AV→left_adjacent)
                    Left_Leaf_of_root ← left_adjacent
                    Left_Leaf_of_root ← P
                    Left_Leaf_of_root ← root
                    FEA_L= Left_Leaf_of_root
                Return FEA_L
                If (AV→right_adjacent)
                    Right_Leaf_of_root ← right_adjacent
                    Right_Leaf_of_root ← P
                    Right_Leaf_of_root ← root
                    FEA_R= Right _Leaf_of_root
                Return FEA_R
                If (AV→bottom_adjacent)
                    Bottom_Leaf_of_root← bottom_adjacent
                    Bottom_Leaf_of_root ← P
                    Bottom_Leaf_of_root ← root
                    FEA_B= Bottom_Leaf_of_root
                Return FEA_B
                If (Found Adjacent)
                    AV← Adjacent
          Repeat until no adjacent has been found
      Exit
Description
      P   : Pointer
      FEA, FEA_R, FEA_L and FEA_B : Array declaration for store features
      i, Fea1   : Variables

The following figure (Fig. 6) shows the sample features that are formed from closed curve.

Procedure 4: To represent 'any shape'

1. Pixels are found in the character portions, and is marked as R.
2. The end points (A, B) are noted from the character portion.

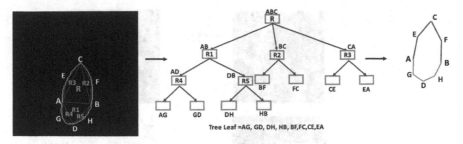

Fig. 6 Feature extraction process from 'closed curve'

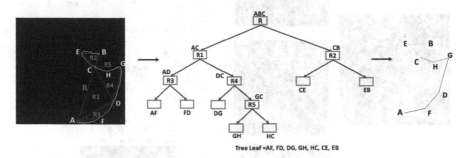

Fig. 7 Feature extraction process from 'any shape'

3. The triangle is formed as discussed in "Triangle formation"
4. Triangle can be possible. H > 1;

Tree and feature formation

First two phases of tree formation (left and right adjacent) have been taken out from the tree formation of procedure-3 and the same is implemented here for extracting the shape features. If the shape is complex, then the triangle formation has extended till gets the real shape.

The Fig. 7 describes the feature formations of the different shape of the character portion. If the complexity of the character structure is increased then the feature count will be increased.

Extraction process is applied on 28 Tamil characters in order to recognize them. A number of samples were collected from HP India Tamil Data set [14] and various hand written materials to achieve this task. The sample features extracted from the character '*ழ*' and the comparison are shown in the Fig. 8.

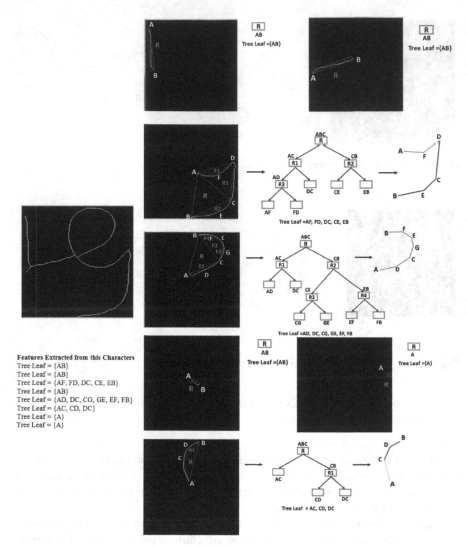

Fig. 8 Feature extracted from the character '*ψ*'

2.3 Classification

Support vector machine algorithm [15] works through the learning process, which is giving competitive success in pattern recognition. SVM is used to classify two classes using the hyper plane. The hyper plane $W^T \cdot X - b = 0$ is set by the SVM which is based on the support vectors, where W is the weight factor and b is the bios. These values are generated from the training samples. The X values are collected from the testing samples which are used to get classified character. The hyper plane makes a margin $W^T \cdot X - b = 0$ between the positive $W^T \cdot X - b > 0$ and negative values $W^T \cdot X - b < 0$. With the help of the Lagrangian theory the two classes are classified. In order to achieve this for multiclass, implementation of the hierarchical SVM [15] is useful, and the comparison takes place through One-vs-One (OVO) or One-vs-Rest (OVR) procedure. Predicting the result one class against all classes (OVR) is practically a time consuming process than OVO. This paper uses OVO procedure to predict the results. The comparison is done on first two classes, and who gets the best vote is the winner. The next class is compared with the winner of the first class. This process is continued until all classes are compared. The final best voting class will be the resultant class of this work.

3 Results and Discussion

3.1 Data Collection

100 samples were collected from each 28 Tamil characters, where 40 samples are testing samples and others are training samples. The samples are collected from HP India–Tamil handwritten character data sets [14], where each character is taken from handwritten of different individuals. With 1120 testing samples and 1680 training samples, a total of 2800 characters are taken into consideration.

3.2 Experimental Results

The experiment is done using the Matlab software. The features were extracted and store in a excel file and this file is taken as an input for the hierarchical SVM. Average of 90.08 % result is achieved for limited number of unique character. The following Table (Table 1) and figure (Fig. 9) shows the result of this experiment in detail.

This work seems to be highly advantageous as this is attached with the shape character nature. The failure is due to the high-end variation which affects the shape. The pros and cons of this work is discussed in the Table 2.

Table 1 Achieved results

Vowels and consonants	அ	ஆ	இ	ஈ	உ	ஊ	எ	ஏ	ஐ	ஒ	ஓ	ஔ	ஃ	ங
Accuracy achieved	92.5	90	95	92.5	87.5	85	90	92.5	90	87.5	92.5	95	85	90
Vowels and consonants	ல்	ண	ந	ன	ப	ம	ய	ர	ற	ல	ள	ழ	வ	ஜ
Accuracy achieved	95	95	87.5	92.5	97.5	95	85	90	82.5	92.5	82.5	87.5	90	90

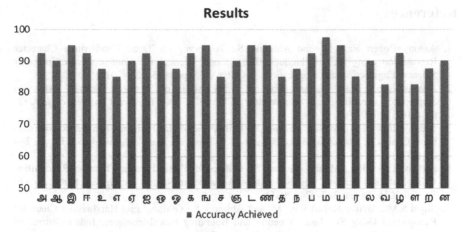

Fig. 9 Accuracy graph analysis

Table 2 Pros and cons of the shape analysis

S. no	Pros	Cons
1	Highly fit for shape analysis	When considering high variations, the shape was not enough to address all character
2	Returns good result for limited shapes	Abnormal variation is to be considered
3	Prism tree is highly coupled with nature of the character	Feature pre-extraction has been improved
4	Considering similar shapes	Feature pre-extraction has been improved
5	Writers independent	

4 Conclusion

Prism tree based hierarchical representation comprising of numerous features and values returns good results with an average of 90 %. This structural representation of the prism tree can be solved with a set of writer's issues. This tree returns the exact structure of the shape. This work fulfills the aim of this paper successfully in order to face the challenge in high variation. Addition of feature representation is to be essential for the further improvement. It can solve the issues of entire Tamil characters. Future work proceeds through finding the next level of algorithm which is combined with this process to get a successful frame work in Tamil handwritten character recognition.

References

1. Antony Robert Raj., M. and Abirami, S.: A Survey on Tamil Handwritten Character Recognition using OCR techniques, The Second International Conference on Computer Science, Engineering and Applications (CCSEA), 05, pp. 115–127 (2012).
2. Antony Robert Raj., M. and Abirami, S.: Analysis of Statistical Feature Extraction Approaches used in Tamil Handwritten OCR, *13th Tamil Internet Conference- INFITT,* pp. 144–150 (2013).
3. Antony Robert Raj., M. and Abirami, S.: Offline Tamil Handwritten Character Recognition using Chain Code and Zone Based Features, *13th Tamil Internet Conference- INFITT,* pp. 28–34 (2014).
4. Antony Robert Raj., M. and Abirami, S.: Offline Tamil Handwritten Character Recognition Using Statistical Features. AENSI Journals, Advances in Natural and Applied Sciences, 9(6) Special 2015, Pages: 367–374 (2015).
5. Shyni S.M., Antony Robert Raj., M. and Abirami, S.: Offline Tamil Handwritten Character Recognition Using Sub Line Direction and Bounding Box Techniques. Indian Journal of Science and Technology, Vol 8(S7), 110–116, (2015).
6. Antony Robert Raj., M. and Abirami, S.: Hybrid Features based Offline Tamil Handwritten Character Recognition" 14th International Tamil Internet Conference, ISSN: 2313-4887 Pages: 360–370 (2015).
7. Sigappi, A. N. Palanivel, S. and Ramalingam, V.: Handwritten Document Retrieval System for Tamil Language, International Journal of Computer Application, ISSN: 0975–8887, Vol.31, No.4, Page(s) 42-47 (2011).
8. Rajashekararadhya, S.V. VanajaRanjan, P and ManhunathAradhya. V. N.: Isolated Handwritten Kannada and Tamil Numeral Recognition: A Novel Approach. First IEEE International Conference on Emerging Trends in engineering and Technology, Page(s): 1192–1195 (2008).
9. Rajashekararadhya, S.V. and VanajaRanjan, P.: Zone-Based Hybrid Feature Extraction Algorithm for Handwritten Numeral Recognition of two popular Indian Script. World Congress on Nature & Biologically Inspired Computing, page(s): 526–530 (2009).
10. Rajashekararadhya, S.V. and VanajaRanjan, P.: Neural Network Based Handwritten Numeral Recognition of Kannada and Telugu Script. IEEE TENCON Conference, pp 1–5 (2008).
11. Rajashekararadhya, S.V. and VanajaRanjan, P.: Efficient Zone based Feature Extraction Algorithm for Handwritten Numeral Recognition of Four Popular south Indian Scripts. International journal of Theoretical and Applied Information Technology, pages: 1171–1181 (2008).
12. Raja Kumar. S, Subbiah Bharathi V.: Ancient Tamil Script Recognition from Stone Inscription Using Slant Removal Method. International Conference on Electrical and Electronics and Biomedical Engineering, Page(s) 80–84, (2012).
13. Hanan, Samet.: Foundation of Multidimensional and Metric Data Structures. Book Published by Morgan Kaufmann, PP. 386–388 (2006).
14. http://lipitk.sourceforge.net/hpl-datasets.htm.
15. Hichem Sahbi, Donald Geman,: A Hierarchy of Support Vector Machines for Pattern Detection. Journal of Machine Learning Research, Vol: 7, pp: 2087–2123, (2006).

The Emergence of Internet Protocol Television as Next Generation Broadcast Network

P.L. Srinivasa Murthy and T. Venu Gopal

Abstract Due to the emergence of innovative technologies over Internet, the traditional broadcasting of TV is changed to use IPTV for next generation networks. IPTV is the delivery of multimedia content to multiple subscribers through multicasting over well known IP. The IPTV depends on the Internet and telecommunications. Therefore, the inherent issues are to be identified and resolved for the success of IPTV. In this paper, we review the IPTV concept and its related aspects. The paper throws light into the issues of IPTV, integration of different communication media and approaches like P2P (unicast model) and IPTV (multicast model), and the need for QoS requirements and evaluation of QoS in the context of IPTV networks. Moreover, the paper provides user behaviour that can have impact on the content distribution infrastructure and the underlying strategies for content delivery and reduction of zapping time. We also present a hybrid approach that combines the features of delay insensitive and delay sensitive approaches to strike balance between highly popular and least popular IPTV channels. The insights of the paper are useful in further research in the area of IPTV which is going to be the next generation network for high quality and complex content delivery.

Keywords IPTV · Multicasting · Interactive video streaming · User behaviour modelling · QoE

P.L. Srinivasa Murthy (✉)
Gokaraju Rangaraju Institute of Engineering & Technology, Hyderabad, India
e-mail: plsrinivasamurthy@gmail.com

T. Venu Gopal
JNTUH College of Engineering Sultanpur, Hyderabad, India
e-mail: t_vgopal@rediffmail.com

© Springer Science+Business Media Singapore 2017   471
S.C. Satapathy et al. (eds.), *Proceedings of the First International Conference on Computational Intelligence and Informatics*, Advances in Intelligent Systems and Computing 507, DOI 10.1007/978-981-10-2471-9_46

1 Introduction

In the wake of technological advancements Internet Protocol (IP) is being used for delivering multimedia content over Internet. In fact, leveraging IP for global communications became a cheaper alternative for reliable and instant communication over global communication medium such as information super highway. It is a wise idea to utilize already established public network such as the Internet for public communications. With IP v6 and security considerations, IP became a viable alternative for cost effective global communications. Many technologies such as Voice over IP (VoIP) are already in place to exploit delivery of voice communications and multimedia content over Internet Protocol. This has motivated broadcasting networks like Television to adapt IP based content delivery as its next generation broadcasting method. The phenomenon which facilitates deliverance of multimedia services of Television via Internet Protocol is known as Internet Protocol Television (IPTV). Thus, the TV network can leverage the advantages of well established public network (Internet) and its underlying protocols for broadcasting content. IPTV when realized fully can bestow plethora of advantages like interactive TV, personalization, low bandwidth requirement, and accessibility on multiple devices besides a low-cost service.

Traditionally, TV is a telecommunications medium for delivering broadcasting content like text and moving images. Right from the inception TV had undergone many changes such as mechanical television, electronic television, colour television, digital television, smart television, and 3D television. With respect to broadcasting medium TV is categorized into terrestrial television, cable television, satellite television, and Internet television [24]. With the emergence of IPTV, there will be revolutionary changes in the way TV works as next generation mass communication medium with many benefits as said before. However, IPTV can throw many challenges as it is based on the public network Internet and over IP which is considered an untrusted public network. The challenges are pertaining to security, service quality, privacy and content quality. Reducing risk of unauthorized access to digital services rendered via IPTV is the major concern that needs to be addressed for taking the TV services to the next generation network IPTV successfully.

This paper throws light on the IPTV and its present state-of-the-art. Our contribution in this paper is to review IPTV as next generation broadcasting network and related aspects with valuable insights. The remainder of the paper is structured as follows. Section 2 provides integration of WiMAX and GEPON for seamless integration of underling wireless and wired counterparts. Section 3 throws light on the need for QoS and its evaluation in IPTV. Section 4 reviews interactive usage dynamics in IPTV. Section 5 focuses user behaviour modelling for reducing zapping. The Sect. 6 hybrid approaches for IPTV. Section 7 provides conclusions and recommendations for future work.

2 Integration Of WiMAX and GEPON

According to Obele et al. [1] bandwidth–hungry applications like IPTV and high-definition television (HDTV) have been around for some time and they keep on driving even for higher bandwidth in the networks. To overcome this problem it is possible to integrate both wired and wireless networks Gigabit Ethernet Passive Optical Network (GEPON) and Worldwide Interoperability for Microwave Access (WiMAX) respectively for improving Quality of Service (QoS). Obele et al. reviewed a convergence architecture that makes use of QoS and queuing concept to provide highly reliable treatment to traffic based on different classes of traffic such as long-range-dependent and self-similar. The performance was evaluated under realistic load conditions.

3 Need for QoS and Its Evaluation in IPTV Networks

Habib [2] proposed a method to evaluate QoS evaluation in IPTV networks. The research was carried out on the QoS evaluation of bandwidth scheduling. Traffic policing and call admission control (CAC) were applied to evaluate the QoS performance of IPTV networks. As Markov chain based models are widely used for QoS analysis, Habib used a model known as Markov Fluid Flow,model (MFFM) to evaluate QoS pertaining to bandwidth scheduling. With the proposed model the waiting time and probability of loss were reduced in video content delivery. This model was found good for evaluation of IPTV performance with experiments on five movie traces. QoS differentiation is explored in [3] for having diversified traffic handling mechanism to accommodate service classes that need different quality. There must be different approaches for traffic classes like delay-sensitive and delay-insensitive traffic. The economic analysis pertaining to QoS differentiation was performed in presence of different traffic classes expecting varying quality requirements. The real time applications like IPTV and VoIP need to have well defined strategies to ensure that the quality is not lost. Caching strategies as explored in [4] can help commercial IPTV to have high quality content delivery. For improving quality of services for IPTV, Ethernet networks have built in broadcasting capabilities [5].

IPTV is one of the transport layer protocols that needs high bandwidth besides quality of service networks [6]. Real Time Transport Protocol (RTP) is the underlying protocol for various bandwidth intensive applications like IPTV. Link functioning under such applications were explored in [7]. For high QoS delay-oriented analysis of networks that demand high bandwidth is essential. This is explored in [8] with respect to the design and implementation of optimal scheduling algorithms. This will help in improving quality of applications like IPTV for broadcasting multimedia content. Therefore, delay is an essential metric to measure the performance of IPTV. Not only IPTV there are many delay-sensitive applications in the real world. They include YouTube, UStream, Video on Demand, Goggle Hangouts, Google Voice, FaceTime, Line, Skype, and so on. Distortion and Error Propagation modelling are explored by Chakareski [9] in the presence of predictive video coding. Loss of packets leads to distortion of video generally. However, the research revealed that distorting is caused not only by data loss but also patio-temporal distribution of multimedia content. These are all capable of affecting quality communications in IPTV. Similar kind of research was carried out in [10].

Adaptive video streaming is one of the approaches to increase quality of video content delivery for IPTV. Video streaming is characterized by delivery of video to multiple users at a time with initial buffering. Adaptive streaming will help IPTV to enhance quality of content deliverance. Markov chains are generally used to represents different states in the content delivery process of IPTV. The underlying Hidden Markov Process can take into account bandwidth variations at run time and the application layer encoding constraints and strikes balance between then for improving quality of services [11].

4 Interactive Usage Dynamics of IPTV in Large Scale

Video on demand (VoD) is one of the means in which consumers like to have access to video content. The consumers wanted to have interactivity to watch videos selectively. Gopalakrishnan et al. [12] explored the behaviour of consumers with respect to interactive viewing of videos in the context of IPTV. They investigated how users make use of advanced streaming features and controls and how the user behaviour causes overhead on the content distribution systems. Thus, their research provides useful insights pertaining to user behaviour and the impact of the same on the distribution infrastructure. They also proposed a comprehensive model in which a large number of subscribers watch videos on demand interactively. Towards this end, they used two components known as arrival process model and stream control usage models, respectively. They are meant for determination of arrival process and interactive usage of videos. the model includes many control features to end users such as stop, no action, pause, replay, skip, rewind, fast forward, play, exit and start. The two models hey presented have their strengths based on the application.

IP network traffic management is essential for VoD purposes. Mirtchev et al. [13] reviews different models for managing IP traffic in the context of IPTV. They focused on experimenting on the traffic measures on different traffic types like VoIP flows, TV, Point-to-Point (P2P), and HTTP transfer. The parameters considered for measuring traffic include priority of packets, distribution of packets, and distribution of session duration, mean inter-arrival time, mean packet size, and packet intensity for the four kinds of network traffic. Thus, they explored the possible optimization of different traffic flows based on traffic analysis. There were many video traffic models that can be explored for IPTV. According to Tawnier and

Perros [14], the VBR video traffic models [15] can be classified into autoregressive models, and models that are based on Markov process. Auto regressive models were explored by them in terms of video coding, level, scene changes, residual error, sources, and publication date. They also investigated markovian models in terms of video coding, level, scene changes, sources and publication date. There are other models such as Wavelet based models that decompose signals for traffic analysis. Their experiments reveal that wavelet-based model was found good for having SRD and LRD co-existed.

Hossfeld [16] focused on how Internet applications and users behave in future. The investigation was made on emerging user behaviour in the context of next generation Internet applications. Quality of Experience (QoE) was used as a measure to know how users are satisfied with the content delivery. QoE largely depends on the jitter, packet reordering, packet loss and degradation of QoS. The quality of content distribution networks are determined by certain characteristics such as ability to handle flash crowds, handle churn, mobility and true participation in the network. There is a relation between selfishness of users and the robustness exhibited by the system.

5 User Behavior Based Model for Minimizing Channel Zapping Time in IPTV

Channel zapping is one of the concerns in IPTV. The content delivery quality gets reduced and this leads to the reduction in QoE of end users. Lee et al. [17] proposed a predictive tuning method that can reduce channel zapping time. This model is based on user behaviour. They used semi Markov process to model the channel zapping time based on user behavioural dynamics. User behaviour patterns are used as inputs to have tuning with pre-defined knowledge that can cater to the improved quality of services. This will lead to high QoE due to the fact that the model reduces channel zapping. This kind of model is very useful even when bandwidth availability is less and there needs to be high quality output in the content delivery. When channel preferences are considered for tuning, it will be useful in achieving the reduction in channel zapping time. In the process both channel preference and button preference are exploited.

6 P2P Approach for IPTV Streaming Over Internet

P2P approach for IPTV has been a very interesting research area. It is possible to combine IPTV and P2P where multicast and unicast are utilized, respectively. Both multicast and unicast channels work together for IPTV. The hybrid approach is shown in Fig. 1.

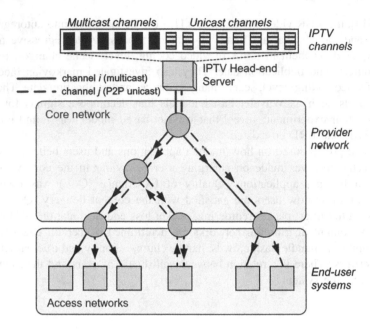

Fig. 1 Hybrid approach for IPTV [18]

As can be seen in Fig. 1, it is evident that there are unicast and multicast channels combined to form IPTV channels. For unpopular TV channels unicast approach is sufficient while the popular channels need to utilize multicast. Therefore, it is possible to achieve both the benefits using the hybrid approach. When the TV channels are not much popular, the content delivery is made using P2P networks while the TV channels with high popularity are given multicasting capabilities. Thus, it is possible to optimize bandwidth besides improving QoE [18]. Other contributions towards IPTV include modelling multiplexed traffic [19], exploring challenges in future Internet applications [20], P2P streaming in distributed fashion [21], Push-to-Peer VoD systems [22], VoIP [23] and quality of service management approaches in IPTV content delivery.

User Behaviour Analysis of IPTV Usage of VOD

As explored in [12] the IP based video distribution traces provided valuable insights on interactive usage of the IPTV. The experiments were done on both weekends and week days. The user activities considered for analysis include play, fast forward, rewind, pause, skip, and replay. Average count and average duration were considered in both time periods. The difference between the weekend and week day behaviour can provide useful information to estimate general user behaviour with respect to video distribution over IPTV.

As can be shown in Figs. 2 and 3, it is evident that there is clear difference between the user behaviour with respect to the viewing of videos of VOD fashion in terms of the difference count of different events aforementioned and the duration of

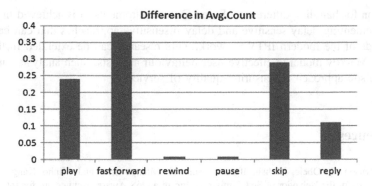

Fig. 2 Shows the difference in average count of events between weekend and week day

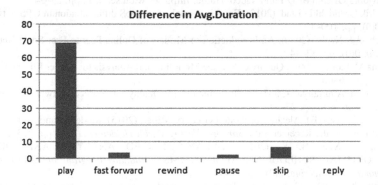

Fig. 3 Shows the difference in average duration of events between weekend and week day

each event. The usage patterns reveal that in weekend the count of events and the duration are more when compared to week days. This behaviour paves way for formalizing policies and planning for VOD content distribution through IPTV.

7 Conclusions and Future Work

In this paper we studied the concept of IPTV and its present state-of-the-art in the real world. It has been understood that many bandwidth-hunger applications like VoIP and IPTV demand high bandwidth. Therefore, there should be cost-effective strategies to ensure that the IPTV is a reality and successful phenomenon. Towards this end, this paper focuses on the IPTV and its related issues and solutions. This paper also throws light into the combination of P2P and IPTV, integration of wireless and wired communication protocols, need for QoS and evaluation of IPTV networks as next generation content delivery networks. User behaviour with respect to interactive behaviour, reduction of zapping and modelling of IPTV with provision for handling different QoE and QoS requirements. It is achieved in terms of implementing delay sensitive and delay insensitive approaches that can cater to the needs of the modern IPTV network. This research can be extended further to explore Markov model to effective user behaviour analysis. Such analysis can lead to a holistic approach for ensuring quality of services.

References

1. Brownson O. Obele, Mohsin Iftikhar, Suparek Manipornsut, and Minho Kang. (2009). Analysis of the Behavior of Self-Similar Traffic in a QoS-Aware Architecture for Integrating WiMAX and GEPON. *Optical Society of America*. 0(0), pp. 12–17.
2. Mohammad Ahasan Habib. (2009). QoS evaluation of Bandwidth Schedulers in IPTV Networks Offered SRD Fluid Video Traffic. http://www.du.se. 0(0), pp. 25–34.
3. Fredrik Grøndahl Leistad. (2012). Economic analysis of QoS differentiation in OPS networks. *o*. 0(0), pp. 100–112.
4. Michael J. Neely. (2010). An Introduction to Models of Online Peer-to-Peer SocialNetworking. *o*. 0(0), pp. 32–44.
5. Richa Malhotra. (2008). Quality of Service Modeling and Analysis for Carrier Ethernet. *CTIT*. 0(0), pp. 56–60.
6. Sanjeewa Athuraliya. (2007). TCP Behavior in Quality of Service Networks. *Sanjeewa Athuraliya*. 0(0), pp. 25–34.
7. Andrey Borisov(B), Alexey Bosov, and Gregory Miller. (2015). Modeling and Monitoring of RTP Link on the Receiver Side. *Springer-Verlag Berlin Heidelberg*. 0(0), pp. 213–313.
8. Dongyue Xue, B. S. (2013). DELAY-ORIENTED ANALYSIS AND DESIGN OF OPTIMAL SCHEDULING ALGORITHMS. *Graduate Program in Electrical and Computer Engineering*. 0(0), pp. 32–44.
9. Jacob Chakareski. (2011). Error Propagation and Distortion Modeling in Loss-Affected Predictive Video Coding. *Springer-verlag Berlin Heidelberg*. 0(0), pp. 25–34.
10. Laith Al-Jobouri 1, Ismail A. Ali, Martin Fleury, and Mohammed Ghanbari. (2015). Error and Congestion Resilient Video Streaming over Broadband Wireless. www.mdpi.com/journal/computers. 0(0), pp. 501–522.
11. Stefania Colonnese a, n, Pascal Frossard b, Stefano Rinauro a, Lorenzo Rossi a, Gaetano Scarano. (2013). Joint sourceandsendingratemodelinginadaptivevideostreaming. *ELsevier*. 0(0), pp. 23–33.
12. Vijay Gopalakrishnan, Rittwik Jana, K. K. Ramakrishnan, Deborah F. Swayne, Vinay A. Vaishampayan. (2001). Understanding Couch Potatoes: Measurement and Modeling of Interactive Usage of IPTV at large scale. *0*. 0(0), pp. 213–313.
13. Seferin Mirtchev, Constandinos X. Mavromoustakis, Rossitza Goleva, Kiril Kassev, and George Mastorakis. (2014). Generic IP Network Traffic Management from Measurement through Analyses to Simulation. *Springer International Publishing Switzerland 2014*. 0(0), p. 32–44.
14. Savera Tanwir and Harry Perros. (2013). A Survey of VBR Video Traffic Models. *IEEE*. 0(0), p. 56–60.
15. Savera Tanwir and Harry Perros. (2013). A Survey of VBR Video Traffic Models. *IEEE*. 0(0), pp. 213–313.
16. Tobias Hossfeld. (2009). Performance Evaluation of Future Internet Applications and Emerging User. *0*. 0(0), pp. 23–33.
17. Chae Young Lee, Chang Ki Hong and Kang Yong Lee. (2009). Reducing Channel Zapping Time in IPTV Based on User's Channel Selection Behaviors. *Springer-Verlag Berlin Heidelberg*. 0(0), pp. 56–60.

18. Alex Bikfalvi, Jaime García-Reinoso, Iván Vidal, Francisco Valera, Arturo Azcorra. (2010). Comparing approaches to IPTV streaming based on TV channel popularity. *ACM.* 0(0), pp. 755–768.
19. Aggelos Lazaris, Polychronis Koutsakis. (2010). Modeling multiplexed traffic from H.264/AVC videoconference streams. *ELsevier.* 0(0), pp. 100–112.
20. Marco Conti, Song Chong, Serge Fdida, Weijia Jia, Holger Karl, Ying-Dar Lin, Petri Mähönen, Martin Maier, Refik Molva, Steve Uhlig, Moshe Zukerman. (2011). Research challenges towards the Future Internet. *ELsevier.* 0 (0), pp. 32–44.
21. Shaoquan Zhang, Ziyu Shao, and Minghua Chen. (2004). Optimal Distributed P2P Streaming under Node Degree Bounds. *CENS.* 0(0), pp. 56–60.
22. Kyoungwon Suhy, Christophe Diot, Jim Kurosey, Laurent Massouli´e, Christoph Neumann, Don Towsleyy, Matteo Varvello. (2005). Push-to-Peer Video-on-Demand system: design and evaluation. *0.* 0(0), pp. 213–313.
23. Angelos D. Keromytis. (2008). A Comprehensive Survey of Voice over IP Security Research. *CENS.* 0(0), pp. 32–44.
24. Television. (2015). Television. Available at https://en.wikipedia.org/wiki/Television. Accessed on 10 August 2015.

Cognition-as-a-Service: Arbitrage Opportunity of Cognition Algorithms in Cloud Computing

V.S. Reddy Tripuram, Subash Tondamanati and N. Veeranjaneyulu

Abstract Present era of web services are evolving and extending its services with Cloud computing based remote mobile devices rapidly. The contact of the information world moving from programmable system era to cognitive systems era (Kelly III John E (2015) IBM Research Whitepaper on Cognitive computing, [1]). So the web users have high levels of expectations for the quality of interactions, accuracy of the results and the availability of the services. In the world of cognition and reasoning based web systems to answer the selected problems, to recommend a product information result in right 'context', we need more diverse reasoning strategies and on the fly strategies based intelligent Cognition Algorithms to supply Cognition-as-a-service to the cloud based web users. Here we summarize recent works and early findings such as: (1) How natural extension takes place in cognitive knowledge of dynamic web services. (2) How information processing time differs in between HDES and AES. (3) How arbitrage opportunity of cognition algorithm improves the frequency of exact findings in information search.

Keywords AI—artificial intelligence · HDES—human decision eco system · AIDSS—artificial intelligence decision support system · AES—artificial eco system

V.S. Reddy Tripuram (✉) · S. Tondamanati · N. Veeranjaneyulu
Department of Information Technology, Vignan's University, Guntur, India
e-mail: tripuram@gmail.com

S. Tondamanati
e-mail: subash.edu@gmail.com

N. Veeranjaneyulu
e-mail: veeru2006n@gmail.com

© Springer Science+Business Media Singapore 2017
S.C. Satapathy et al. (eds.), *Proceedings of the First International Conference on Computational Intelligence and Informatics*, Advances in Intelligent Systems and Computing 507, DOI 10.1007/978-981-10-2471-9_47

481

1 Introduction

In present ongoing web systems cognitive technology that have begun to use the way into product and services that we are using every day with internets automated Web Service discovery [2]. Cognitive behavior [3] deeply involved and self-sustaining gradually in e-service [4] like email, social media, e-commerce sites, online shopping, advanced web based news, health systems where we communicate and use them for knowledge.

The social web systems or personal mail program were able to recognize date, time, contacts, names, locations, attachments and email messages because of its advancements in natural language process [5]. Whereas in online shopping system dynamically enhanced machine learning—AI [6, 7] information—recommends a product to customer based on frequency of parameters like history of locations, seasons, product visits, cart information and confirmed purchases information.

2 What Is Cognition-as-a-Service

Cognition-as-a-Service (CaaS) [8] is an extension of augmenting human intelligence and Artificial Intelligence to existing cloud services like Software-as-a-Service (SaaS) [9], Platform-as-a-Service (PaaS) [10, 11] and infrastructure-as-a-Service (IaaS) [12], etc., where we reason about the simple problem to complex problem with supporting evidences to answer unsupervised deep learning and self-reinforcing adaptive training models [13] and new channels of sensory communication we need.

It can be explained by deconstruction of the terms:

A. Implicit Cognitive Service
This cognitive service already began to use their way into products and services that we use every day web services like email, social media, e-commerce, etc. So, there is a beginning to see the benefits of the advances in the machine learning and natural language processing techniques providing cognition behavior along with web services to suggest, recommend a particular content or product. This kind of traces of past history and experiences of web user information based reasoning strategy is called Implicit Cognitive service [14].

B. Interactive Cognitive Service
Technology has evolved to the point for enabling completely new problems could be solved, completely new paradigms, new systems that consumer explicitly walk up to use in and out active manner that actually possess a significant challenge for cognitive algorithms. These systems tend to be more probabilistic, deterministic but consumer wants them to work in user friendly interface. Whereas in websites there is a conversational agent providing answers to the raised questions and guiding to solve the consumer problem. Here consumer go directly to the cognitive system

with high levels of expectations for the quality of those interactions and accuracy of these results and the availability of services. This type of cognitive system is called as Interactive Cognitive Service [15].

C. Active Cognitive Service
Active Cognitive Service [16] systems are going to help us to manage our time and infrastructure resources by learning to act on our behalf with appropriate controls. So imagine the system that can learn about and ask about may be your diet, may be your health situations and any restrictions you might have. You might have the better activity you have to do today, you might be able to suggest to have a specific type of exercise where we are allowed to eat and not to eat. This is about multiple passions of lifestyle and it can make proactive recommendations for us. These systems are fabric of our technologically connected www world, and really it is a transition from interactivity with cognitive system into place where we really learn and live with it.

3　The Limits of Strategic Thinking

Human mind is really a very fascinating object. Mind is behaving like infinity with its capacity. How fast human brain can process an information to take an action.

The Mathematical core of best developed behavioral game theory [17] is what players think other players will do. In most theories this reasoning is iterated.

A guesses what B will do by guessing what B will guess A will do, ad infinitum. Until mutually consistent responses—an equilibrium is reached (Fig. 1).

An Expert will solve the puzzle in 650 ms rather than the Non Expert who solves the puzzle in 900 ms. This tiny spice will open up a huge world for Cognitive algorithm [18]. In many areas of web based services like Facebook, Twitter, Google, Wikipedia including financial markets, etc., now governed by high frequency algorithms. In this case the financial trading is just beyond the human forte.

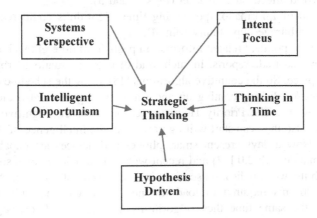

Fig. 1 The elements of strategic thinking

4 Information Processing Time in Between HDES and AES

Human Decision Eco System (HDES)

HDES combines different human decision making models to identify factors such as preferences, social norms and incentives that are most relevant for explaining human decision making in different conditions, assess their consequences are largely determined by the influence of human decisions.

Algorithmic Decision Ecosystem (AES)

Algorithmic Decision eco system considers HDES models as available database and also it includes different entities factors providing and requesting very similar services, which results in an ambitious situation. Consequently, automatic entity has to choose one out of multiple co-operation factors that suits targeted entity's needs better.

Artificial Intelligence Decision Support System (AIDSS)

Artificial intelligence knowledge-based techniques are used in Servers centric web based cloud services to integrate and enhance different tasks such as monitoring, data analysis, communicate, information storage and retrieval. In this process several techniques like Data Interpretation, Data mining, problem Diagnosis techniques and Decision support techniques are automatically selected with the support of AI for prediction, planning a better decision system model to execute and provide results to service request.

Present scenario of share markets and trading behaviors are influenced by many factors. Out of all those factors majority of the people depend on web related article and updates to trade. Sometimes stock markets of particular stock prices rise and fall, fluctuation takes place based on the reason of world events such as war, civil unrest, sudden political decisions, natural disasters and terrorism or any other insecure information news that is posted or published via web media. When chaos around the world affects stock market (Figs. 2 and 3).

So this kind of Information processing time, algorithms could read, it would open up huge arbitrage opportunity (Fig. 4).

They can read political related information posts from different web sources and also they can read trade reports. In such kind of market situations arbitrage algorithms takes place. So the cognitive algorithms [18] know the schedule of the news announcements/publishing coming up and starts reading news also they logically extracts the events. The Priority Heuristic reasoning based cognitive algorithms read the news also they can start writing news with the intelligence of AIDSS [19–21] as well. Now a days present smart phones and devices are capable to communicate semantic web 2.0 [22] and recent web technologies allows such kinds of news through the web application services in a short span of time, i.e., nearly about less than a 800 ms required to load the news through smart phone app and browsers. In the same time these algorithms upgrades the knowledge of trade systems and brings the variations of stock trade facts to the users devices as news.

Fig. 2 A web post in CNBC Web site related to biggest earthquake in Japan sent stock markets across the globe sharply lower

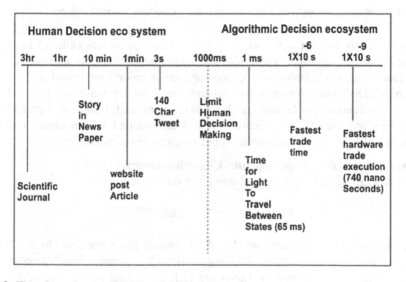

Fig. 3 Time factor between HDES and AES

5 Cognition Algorithms Reasoning in Cognition-as-a Service

These Cognition Algorithms [23] provides Decision making on targeted or selected problem (Reading, Analyzing, and Writing) by utilizing all available information source from web exhaustively to make perfectly accurate judgements.

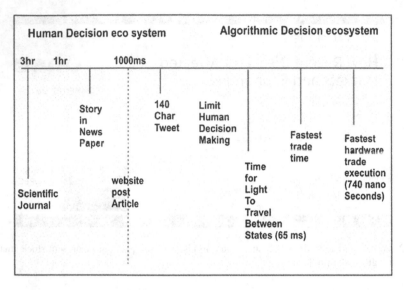

Fig. 4 Arbitrage opportunity of cognitive algorithms time factor to read web post

When users interact with web they use different reasoning skills to find confirmed results, Accordingly Cognitive decision algorithms do understand user information access patterns with reasons and those models are stored as database. Cognitive Decision algorithms also depends on Inductive reasoning [24, 25] and Deductive Reasoning solutions also depends on heuristics[c]. Heuristics (or) Mental short-cuts are used to make sense of the world around information. Cognitive Simple heuristics makes algorithms smart to guess result accurately.

A. How Conceptual Language and Cognition Interact

Each concept **m** has linguistic and cognitive dual model.

$$Mm = \{Mm^{\text{cognitive}}, Mm^{\text{language}}\}; \tag{1}$$

Language and cognition are fused at vague pre-conceptual level. Initial concept-models are fuzzy blobs. Language models have empty "slots" for cognitive model (objects and situations). Language is learned "ready-made" from the surrounding language from the sources of computer-human interaction. Cognitive Concepts are learned to match language models (Fig. 5).

High level cognition implementation is only possible due to language, i.e., based on situational awareness, layered sensing.

When cognitive algorithms try for optimal solutions they do exhaustive search, which is slow and demanding on time and memory resources. Cognitive algorithms can be divided into characteristic principles and Aided principles. The quality of these Learning cognitive algorithms design and usage depends on selection of one or more principles. These principles are selected according to the kind of learning base.

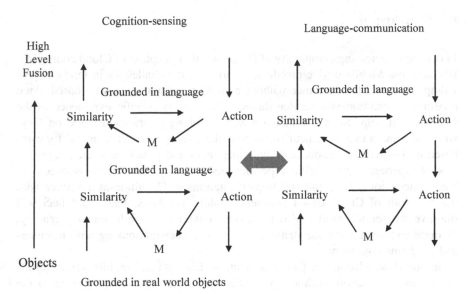

Fig. 5 High level cognition

6 CaaS Limitations

Lack of External and Internal input sources create information result mismatches in cognitive systems dynamical information processes for Cognition-as-a-service.

Cloud computing Services specific proprietary algorithms dependencies limit the intelligence of intensive processing between Natural Language Processing (NLP) and Sensed user information.

In CaaS Cognitive augmented search with the World Wide Web servers, Cloud Servers all maximum functions depends on internet connection.

7 CaaS Challenges

Developing the smarter Cognitive Operating systems, Cognitive Apps than the existing traditional operating systems, Applications to achieve higher intelligence in cloud based Cognition-as-a-Service.

To refer trillions of English sentences words, information databases for conjunction with NLP require optimal Dynamic statistical algorithms.

Present Internet of things cloud based services migrating toward Cognitive services required different new paradigms of dynamic frameworks models.

8 Conclusion

In our view, increasing complexity of IT service after adaption of Cloud computing, Big data and Mobile device proliferation are creating challenges in terms of providing service to the customer without cognitive techniques in web based cloud platforms. Cognition-as-a-Service delivers the context specific experience to the web users. It helps to next generation web searching information based on 'keyword' references to a more heuristic personalized sense of exact 'context' for users based on language, location, activity, experiences and preferences, etc., cognitive type of experiences are specific to personalized and they 'learn' from users activity based interactions. Adaption and implementations of Cognition-as-a-Service with the first path of Cloud Based Business models like SaaS, PaaS, and IaaS will improve the service results with the exact context base while sensing, reading, Humans cognition and understanding, analyzing, Decision making with machines and programs cognitions.

In our view, adaption of Cloud computing, Big data and mobile device proliferation in Information Technology services which are providing service to the customer is increasing in complexity and creating challenges without cognitive techniques in web based cloud platforms.

References

1. Dr. John E. Kelly III.: Computing, cognition and the future of knowing how humans and machines are forging a new age of understanding. In: IBM Research Whitepaper on Cognitive computing (2015).
2. Ali Shaikh Ali, Simone A. Ludwig, Omer F. Rana.: A Cognitive Trust-Based Approach for Web Service Discovery and Selection. In: Third IEEE European Conference on Web Services 14–16 Nov. 2005.
3. Paul R. Smart.: Understanding the Cognitive Impact of Emerging Web Technologies: A Research Focus Area for Embodied, Extended and Distributed Approaches to Cognition. In: 1st International Web for Wellbeing & Human Performance Workshop, Paris, France (2013).
4. Sophea Chea, Margaret Meiling Luo.: Cognition, Emotion, Satisfaction, and Post-Adoption Behaviors of E-service Customers. In: IEEE 40th Hawaii International Conference on System Sciences (2007).
5. Kanmani Sivagar.: Dynamic Referencing of Web Services via Service Discovery and Natural Language Processing. (IJCSIT) International Journal of Computer Science and Information Technologies, Vol. 5 (2), 2014, 2019–2022 www.ijcsit.com 2019. ISSN: 0975-9646.
6. Mehmet Kuzu, Nihan Kesim Cicekli.: Dynamic Planning approach to automated web service composition. Article On Applied Intelligence Volume 36, Issue 1, pp 1–28. Springer, January 2012,
7. Manoranjan Parhi, Binod Kumar Pattanayak and Manas Ranjan Patra.: A Multi-Agent-Based QOS-Driven Web Service Discovery and Composition Framework. In: ARPN Journal of Engineering and Applied Sciences, ©2006-2014 Asian Research Publishing Network (ARPN), VOL. 9, NO. 4, APRIL 2014. ISSN 1819-6608.
8. Roberto Verdone, Editors: UniBO, IT, Antonio Manzalini.: 5G Experimental Facilities in Europe. White Paper, Telecom Italia, IT.

9. Majed Alhaisoni.: On the Advancements in Cloud Computing Technology: An Overview. In: (IJCSIT) International Journal of Computer Science and Information Technologies, Vol. 6 (3), 2772–2778, ISSN: 0975-9646 (2015).
10. A Semantic Interoperability Framework for Cloud Platform as a Service. In: IEEE Third International Conference on Cloud Computing Technology and Science (CloudCom) Athens, Greece, Nov. 29, 2011 to Dec. 1, 2011, ISBN: 978-0-7695-4622-3, pp: 280-287.
11. Eleni Kamateri, Nikolaos Loutas, Dimitris Zeginis, James Ahtes, Francesco D'Andria, Stefano Bocconi, Panagiotis Gouvas, Giannis Ledakis, Franco Ravagli, Konstantinos A. Tarabanis.: Cloud4SOA: A Semantic-Interoperability PaaS Solution for Multi-cloud Platform Management and Portability. Springer, Service-Oriented and Cloud Computing, Volume 8135, pp 64–78 (2013).
12. Aniruddha S. Rumale, D.N.Chaudhari.: Cloud Computing: Infrastructure as a Service. International Journal of Inventive Engineering and Sciences (IJIES), ISSN: 2319-9598, Volume-1, Issue-3, and February 2013.
13. Minmin Chen, Jian-Tao Sun, Xiaochuan Ni, Yixin Chen.: Improving Context-Aware Query Classification via Adaptive Self-training. In: 20th ACM Conference on Information and Knowledge Management, pp: 115–124 (2011).
14. Xiaocao Hu, Zhiyong Feng, Shizhan Chen.: Analyzing Distribution of Implicit Semantic Information in Web Services. In: IEEE 37th Annual Computer Software and Applications Conference Workshops (COMPSACW) on Cognitive Computing and Application. pp: 415–420. 22–26 July 2013.
15. Qihui Wu, Guoru Ding, Yuhua Xu, Shuo Feng, Zhiyong Du, Jinlong Wang, and Keping Long.: Cognitive Internet of Things: A New Paradigm beyond Connection. In: The IEEE Internet of Things Journal (IoT-J) On Artificial Intelligence (2014).
16. Chi-Yuan Chen, Shih-Wen Hsu, Han Chieh Chao. CogIMS: an active service-oriented cognitive networks over IP multimedia subsystems. Cluster Computing, Vol. 18, No. 1, pp. 135–145, March 2015.
17. Colin F. Camerer.: Behavioural Studies of strategic thinking in games. In: ELSEVIER Trends in Cognitive Sciences Vol.7 No.5 May 2003.
18. Kristina Machova, Jan Paralic.: Basic principles of cognitive algorithms design. In: Department of Cybernetics and Artificial Intelligence, Technical University of Kosice, Letná 9, 042 00 Kosice, Slovakia.
19. Yuri Boreisha.: Web-Based Decision Support Systems as Knowledge Repositories for Knowledge Management Systems. In Ubiquitous Computing and Communication Journal, Minnesota State University Moorhead, USA (2016).
20. Li Jun ab, Wen Juna.: Cloud Computing Based Solution to Decision Making. Elsevier Procedia Engineering Volume 15, 2011, pp: 1822–1826, CEIS (2011).
21. Stephen Russell, Victoria Yoon, Guisseppi Forgionne..: Cloud-based Decision Support Systems and Availability Context: The Probability of Successful Decision Outcomes. In: 7th Workshop on E-Business Volume 22, pp: 96–109 June 2010.
22. Amith Sheth, Krishnaprasad Thirunarayan.: Semantics Empowered Web 3.0 Managing Enterprise, Social, Sensor, and Cloud-based Data and services for Advanced Applications. Morgan & Claypool Publishers (2012).
23. Steve Oberlin.: Machine Learning, Cognition, and Big Data. In: IBM's Cognitive Computing Initiative.
24. Joshua B. Tenenbaum, Thomas L. Griffiths and Charles Kemp.: Theory-based Bayesian models of inductive learning and reasoning. In: Elsevier TRENDS in Cognitive Sciences Vol. 10 No. 7 July 2006.
25. Daniel Lassiter, Noah D.: How many kinds of reasoning? Inference, probability, and natural language semantics. In Elsevier on Cognition Volume 136, pp: 123–134. October 28, 2014.

Directional Area Based Minutiae Selection and Cryptographic Key Generation Using Biometric Fingerprint

Gaurang Panchal and Debasis Samanta

Abstract Biometric is gaining its important in security due to its unique characteristics. Number of methods in the biometric and cryptography has been proposed for secured message communication. Cryptography based security that employed cryptographic keys which are generated using various key generation algorithms. However, a user has to remember the cryptographic key or maintain it in the database in secure manner by his own risk. Once the stored key is compromised, then an attacker can access the user's data easily. This motivates us to develop a new security mechanism to protect the user's persona data. In this paper, we present a novel approach to generate a biometric-based cryptography key generation using fingerprint data of a user. With respect to the security aspects, our approach is secure and more flexible for the key generation. This key can be used to encrypt the user's personal data. The fingerprints features (Reference Points) have been generated by directional area (directional component) and probability distribution. These points are having uniform probability across all the points. The advantage of this approach includes the reduction in pre-processing time. A 1024-bits key is generated from the extracted fingerprints attributes. Experimental result has been discussed in this paper.

Keywords Fingerprints · Directional area · Reference points · Probability · Cryptography · Key scheduler

G. Panchal (✉) · D. Samanta
Department of Computer Science and Engineering, Indian Institute
of Technology Kharagpur, Kharagpur, India
e-mail: contactgaurangp@gmail.com

D. Samanta
e-mail: debasis.samanta.iitkgp@gmail.com

© Springer Science+Business Media Singapore 2017   491
S.C. Satapathy et al. (eds.), *Proceedings of the First International Conference
on Computational Intelligence and Informatics*, Advances in Intelligent Systems
and Computing 507, DOI 10.1007/978-981-10-2471-9_48

1 Introduction

Due to the message exchanges on the Internet and personal data storage on the remote location, cryptography becomes very much essential for information security. The existing security mechanisms are suffering from many limitations in terms of security [1, 2]. These limitations can be overcome by the combination of biometric with cryptography [3, 4]. The feature extraction and detection in the biometric based system is a challenging issue of fingerprint verification [5]. We have used various reference points as features which are important area of fingerprints verification. The process of detection for the position points (i.e. reference points) proficient by the search of the points which has the minimum relative entropy among the possible all pixel in the global area. The location point can be obtained using the method called relative entropy which represents the directional area (sometimes it is called as direction component) for the feature extraction. The advantage of this method is that it reduces the pre-processing time. Our approach is to generate 1024 bit cryptographic key using various reference points extracted from the fingerprints. The standard cryptographic algorithms which are used publicly are DES, AES and RSA [6].

Our proposed approach uses key generation scheduler, from which the bio-crypto key is calculated from the captured biometric fingerprint image of a user. It may be noted that we do not store the biometric-based cryptographic key in the database. Here, we present a novel approach to generate a secure biometric-based cryptographic key based fingerprint image. Primarily, we extract the biometric features from the fingerprint image. Our proposed approach follows the segmentation, Orientation, estimation and morphological operators which is used to improve the performance accuracy in the feature extraction process. Note that, the fingerprint-based cryptography system is mostly use in security system [7].

2 Proposed Approach

We proposed a novel approach to generate cryptography key from the fingerprint. Generation of cryptographic key is a challenging problem now a days. Cryptography with biometric has been seen as a strong security solution. The biometric-based security system is suffering from its intrinsic problems. Hence, same cryptographic key from one session to another session is not guaranteed. Therefore, some reliable mechanism is required to ensure the same key every time.

To address the limitations of the existing mechanism, which uses the tented arch and arch type approach to generate cryptography biometric key, we focus on the relative entropy to generate reference points. We proposed to use the directional area. This directional area is extracted from the captured fingerprint image. From this directional area, we calculate the relative entropy. We calculate the directional area using the convolution operation. To calculate the directional area, we proposed to use the different four directional area (i.e. directional filters). The different directions orientations includes $0°, 45°, 90°, 180°$ along with the captured biometric fingerprint image.

This approach essentially calculates the global area using the calculated directional areas. Now, we calculate the probability for the calculated directional area. From the global area, we find a location point. With respect to the location point, the distribution of the calculated probability for the directional area becomes flatter. We consider this estimated location point as a reference points. The estimated point can be considered as required reference point due to its reproducibility along with the consistency for location finding. Moreover, the proposed method uses the probability density; there is no need to use the pre-processing mechanism to improve the quality in the feature extraction. Then by approaching DES algorithm on more relevant reference points, we can generate 1024 bit bio-key.

2.1 Pre-processing

We have used here $m \times m$ input gray scale images to generate the 48 bit cryptography key. We detect the minutiae point, core point and delta point. The we apply the low-pass filtering using following.

$$M(i) = \frac{1}{\sqrt{2 \times \pi \times \sigma^2}} \times e^{\frac{-i^2}{2 \times \sigma^2}} \tag{1}$$

Basically, it is the product of two Gaussians. Each dimension contains one Gaussian.

$$M(i,j) = \frac{1}{\sqrt{2 \times \pi \times \sigma^2}} \times e^{\frac{-i^2 + j^2}{2 \times \sigma^2}} \tag{2}$$

Here, i and j are the distance calculated from the origin $(0, 0)$ with respect to the x-axis and y-axis, respectively. We calculate the standard deviation for Gaussian distribution (Fig. 1).

To improve the quality in the feature extraction process, we first convert gray-scale image to the binary image. Basically, it increases the contradiction between the valleys and ridges of the captured fingerprint image, and hence improve the minutiae extraction. We apply morphological operation to remove unnecessary bridges and spurs.

2.2 Directional Areas Extraction

We proposed a mechanism to extract features using the four directional areas. We use the areas $(0°, 45°, 90°$ and $180°)$ to obtain the target point in the captured fingerprint

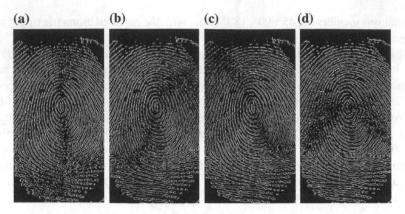

Fig. 1 Directional binary image at different angle. **a** 0° orientation (directional) image; **b** 45° degree orientation (directional) image; **c** 90° degree orientation (directional) image; **d** 180° degree orientation (directional) image

image. Sometime, it become challenging to extract the directional area which distinguish the feature like ridge and valleys from the binary fingerprint image. Therefore, we proposed to use the direction area in the feature extraction.

We proposed to use four different direction areas (0°, 45°, 90° and 180°) as shown in (Fig. 1). Here, each orientation area is defined $A0$, $A45$, $A90$ and $A180$.

Now, we have computed convolution action using the binary image and directional filter. It can be symbolized in the following.

$$D_z(l, m) = \sum_{a=0}^{x-1} \sum_{b=0}^{x-1} D(l - a + c, m - b + c) Flt(a, b) \tag{3}$$

Where $Z = 0°, 45°, 90°, 180°$ and D represents the directional areas. This directional area is calculated by the convolution operations. This convolution operation is done between the Flt and binary image. It may be noted that we keep the size of the directional filter as 5×5. The obtained result shows that there is a chance of detection of better directional area in the fingerprint image Fig. 2.

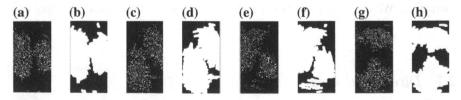

Fig. 2 **a** 0° directional binary image; **b** 0° degree directional area; **c** 90° degree directional image; **d** 90° degree directional area; **e** 45° degree directional image; **f** 45° degree directional area; **g** 180° degree directional image; **h** 180° degree directional area

2.3 Global Area Estimation for the Reference Points

At this point, the directional distributions may not be equally identifiable because the captured image includes other features also (e.g. ridge). Therefore, we first find the global area rather than the reference (i.e. target point) point in the captured fingerprint image. There is a higher chance that there can be a reference point. It may be noted that the target point may be located in fingerprint image and multiple fingerprint area. To overcome this problem, we first go through the directional area (area) using the operation called convolution between filter and binary image.

$$NF(i,j) = \frac{V(p,q)}{\sum_{l=0}^{m-1} \sum_{m=0}^{m-1} U(p-l, j-m)} \tag{4}$$

where the possible value of k is 0, 45, 90, 180. We represent the directional areas as DA0, DA45, DA90 and DA180. The figure represents that the bright part (white part) represents the directional area. Whereas the dark part (black color) does not represents the directional areas. Therefore, in the fingerprint image, for the white part the intensity value varies between 1 and 2. While in the dark part, the intensity value varies between 0 and 1 [3, 8].

Now, we use the possible combination of all directional areas (DR) in the image. The distance is calculated with respect to the different directional area of the captured fingerprint image. We, now use all the images and calculate the identical pixel using the following.

$$A(p,q) = \sum_{i} \sum_{j} K_{ij}(p,q) \tag{5}$$

Thus, we get the combined image of all four subtracted images as a Global area. To calculate the reference points the appearance probability of each directional area needs to be calculated. We calculate the probability of each area as follows.

$$P(\theta_n(p,q)|Z(p,q)) \Rightarrow \sum_{n} (\theta_n(p,q)|K(p,q)) > \frac{1}{2} \tag{6}$$

Once the probability of the entire directional area is calculated, we traverse the entire pixel one-by-one and find the most appropriate points where there is a uniform probability. The selected points, which we called as reference points (Fig. 2). We select all the minutiae points surrounding the references points. We convert the coordinate value of selected minutiae points into binary value. We generate hash value of the calculated binary value and generate 1024 bit hash key.

2.4 User Registration Information

We extract the minutiae points including core point, delta points, bifurcation and termination. We also extract the minutiae orientation of each extracted points. We store the extracted biometric features in the database. Later, the query biometric data will be compared with the stored biometric data. If there is a match, then we generate the biometric-based cryptographic key using the proposed method mentioned in the above section.

3 Experiments and Experimental Result

In this section, we discuss our experiment to substantiate the efficacy of our proposed approach and present the observed results.

3.1 Objectives of Experiments

Our first objective is to examine the effectiveness of the proposed approach to generate similar feature vectors of a user. To do this, we compare among intra-instances and inter-instance of users. Next objective is to analyze the accuracy of our proposed approach. For this, we generate feature vectors of a user and compare it with the feature vectors generated using different quality of fingerprint images of the user. To test the distinctiveness of bio-crypto keys from one user's instance to others.

3.2 Experimental Setup

In our experiments, we have used fingerprint images from publicly available fingerprint databases that is FVC 2004 [8] and NIST Special database [7]. We have done our experiments with an Intel(R) Core(TM) i5-2400 CPU @ 3.10 GHz processor. We use MATLAB 7.11 to develop our program in windows 7 OS.

3.3 Experimental Results

3.3.1 Similar Feature Vector Generation

We measure the amount of fingerprints, which generate similar feature vectors. We have done this experiment from two perspectives: First, we compare the feature vectors generated from intra-instances of users. Second, we compare the feature vectors generated from inter-instances of users. The observed results are expressed in terms

Table 1 Similarity in feature vectors

	Intra-instance		Inter-instance	
Database	True positive (%)	False negative (%)	False positive (%)	True negative (%)
DB1	98.97	1.03	0.08	99.92
DB2	99.15	0.85	0.05	99.95
DB3	98.72	1.28	0.10	99.90
DB4	99.60	0.40	0.28	99.72
NIST	99.89	0.11	0.17	99.83
Average	99.27	0.73	0.14	99.86

of confusion matrix and shown in Table 1. It is observed that on the average 99.27 % cases are true positive, 0.14 % false positive, 0.73 % false negative, and 99.86 % true negative. We could not see absolute "True Positive" results as some intra-instance images are too noisy to have a matchable score.

3.3.2 Performance with Different Size of Target Area

The performance of the different size of target area is shown in Fig. 3. It is observed that as the size of the target area increases, the TP rate decreases. This is because as the target area increases, the uncertainty of the features increases and hence the matching with the intra-instance will have higher chance.

3.3.3 Distinctiveness of Bio-Crypto Key

We measure the amount of dissimilarity of keys generated from one instance of a user to inter-instances of other users. It is observed that a user's keys disagrees with imposter's keys from 189 bits (minimum Hamming distance) to 841 bits (maximum Hamming distance) for 1024 bits keys. A similar trend has been observed in other databases used in our experiment.

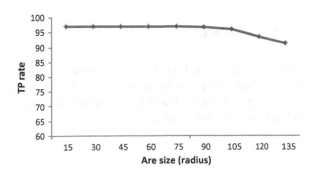

Fig. 3 Effect of target area on true positive rate

Table 2 Comparison with respect to accuracy measures

	Nandakumar et al. (2007) [11] (%)	Lee et al. (2008) [13] (%)	Martin et al. (2009) [10]	Nagar et al. (2012) [9]	Proposed approach (%)
TP Rate (GAR)	97	81.01	Not reported	75 %	99.27
FP Rate (FAR)	0.24	0.0005	0.28 %	Not reported	0.14

4 Comparison with Related Work

Related to our work, a number of researches have been reported in recent literature. In this section, we briefly survey the existing work and compare our work with them. Several mechanism have been proposed in the literature for securing user's personal information using biometric based crypto-system [9–13].

Related to our work, a number of researches [14, 15] have been reported in recent literature. It is evident that our research is comparable to the existing research in several respects. We follow a simple feature extraction procedure which is fast enough compared to the computationally expensive techniques followed in [14, 15]. Our approach is also advantageous in terms of key generation for encryption and decryption. Not only, our approach allows to generate a key from a small sized to any arbitrary large size, but also it enables to generate different keys in different sessions controlling the values of the hidden parameter in our algorithm. Last but not the least, our approach avoids deciding certain values such as threshold value, while a user in authentication check. Our proposed approach to verification is not only a novel approach to solve the problem, it is also more accurate compared to others [15]. Furthermore, we compare our approach with the existing approaches with respect to the accuracy (true positive rate (TP) also called as genuine acceptance rate (GAR), false-positive rate (FP) also called as false acceptance rate (FAR)). The results are furnished in Table 2. The data in Table 2 are self-explanatory, and indeed results are in favor of our approach both in terms of TP and FP rates.

5 Conclusion

We presented a novel approach to generate a biometric-based cryptographic key. We discussed a directional area based feature extraction approach. Our approach generate the biometric-based key using which is random enough and applicable in information security system.

References

1. H Choi, K Choi, J Kim, "Fingerprint Matching Incorporating Ridge Features With Minutiae," *IEEE Transactions on Information Forensics and Security*, vol. 6, pp. 338–345, 2011.
2. A. K. S. Li, "Fingerprint Combination for Privacy Protection," *IEEE Transactions on Information Forensics and Security*, vol. 8, pp. 350–360, 2013.
3. A Nagar, K. Nandakumar, A Jain, "Multibiometric Cryptosystems Based on Feature-Level Fusion," *IEEE Transactions on Information Forensics and Security*, vol. 7, pp. 255–268, 2012.
4. M. Upmanyu, A. Namboodiri, K. Srinathan, C. Jawahar, "Blind Authentication: A Secure Crypto-Biometric Verification Protocol," *IEEE Transactions on Information Forensics and Security*, vol. 5, pp. 255–268, 2010.
5. A. Jain, K Nandakumar, "Biometric authentication: System security and user privacy," *Computer*, vol. 45(11), pp. 87–92, Nov. 2012.
6. D. Stinson, *Crypto. Theory and Practice*, 2, Ed. CRC Press, 1995.
7. "NIST Special Database 4 (Fingerprint)," Dec. 2013. [Online]. Available: http://www.nist.gov/srd/nistsd4.cfm
8. FVC2004 Fingerprint Databases. [Online]. Available: http://bias.csr.unibo.it/fvc2004/Downloads
9. F. Hao, R. Anderson, J. Daugman, "Combining Crypto with Biometrics Effectively," *IEEE TC*, vol. 55, no. 9, pp. 1081–1088, 2006.
10. K. Martin, H. Lu, F. Bui, N. Konstantinos, D. Hatzinakos, "A Biometric Encryption System for the Self-Exclusion Scenario of Face Reco." *IEEE SJ.*, vol. 3, no. 4, pp. 440–450, Dec. 2009.
11. K. Nandakumar, A. Jain, S. Pankanti, "Fingerprint-Based Fuzzy Vault: Implementation and Performance," *IEEE Transactions on Information Forensics and Security*, vol. 2, no. 4, pp. 744–757, Dec. 2007.
12. W. Sheng, G. Howells, M. Fairhurst, and F. Deravi, "Template-Free Biometric-Key Generation by Means of Fuzzy Genetic Clustering," *IEEE Transactions on Information Forensics and Security*, vol. 3, no. 2, pp. 183–191, 2008.
13. Y. Lee, K. Park, S. Lee, K. Bae, J. Kim, "A new method for generating an invariant iris private key based on the fuzzy vault sys." *IEEE Trans. on Sys. Man and Cybernetics*, vol. 38, no. 5, pp. 1302–1313, Oct. 2008.
14. L. Eryun, Z. Heng, L. Jimin, P. Liaojun, X. Min, C. Hongtao, L. Yanhua, L. Peng, T. Jie, "A key binding system based on n-nearest minutiae structure of fingerprint," *Pattern Recognition Letters (Elsevier)*, vol. 32, pp. 666–675, 2011.
15. L. Peng, X. Yang, K. Cao, X. Tao, "An alignment-free fingerprint cryptosystem based on fuzzy vault scheme," *Journal of Network and Computer Appl. (Elsevier)*, vol. 33, pp. 207–220, 2010.

Combined Effect of Soft Computing Methods in Classification

Vijaya Sri Kompalli and Usha Rani Kuruba

Abstract Feature Selection can be done in most of the medical domains to identify the most suitable features that result in the accuracy of classification and to reduce time of computation; as it works on reduced number of features. The nature of the problem domain and the design issues of soft computing methods used determines the effectiveness of feature selection methods. The study includes the feature selection using Genetic Algorithm (GA), to generate the best feature subset of WBCD breast cancer dataset. The features with the best fitness value are selected for classification. Classification is done using a guided approach called Support Vector Machine (SVM) along with some constraints to specify the performance measures of classification.

Keywords Feature selection · Classification · Soft computing · Genetic Algorithm · Breast cancer · Support Vector Machine

1 Introduction

The aim of Soft Computing is to provide solutions that exploit the tolerance to imprecision and uncertainty to achieve tractability, robustness and low solution cost [1]. Soft Computing techniques are most applicable on human adaptability as they need to be processed soft, without showing effect on the other issues. Soft Computing methods deal with intelligent systems. These methods derive better solutions when applied collaboratively with other techniques. Soft Computing methods are

V.S. Kompalli (✉)
Devineni Venkata Ramana & Dr. Hima Sekhar MIC College of Technology,
Vijayawada, India
e-mail: vijayasri.kompalli@mictech.ac.in

U.R. Kuruba
Sri Padmavati Mahila Visvavidyalayam (Women's University), Tirupati, India
e-mail: usharani.kuruba@gmail.com

© Springer Science+Business Media Singapore 2017   501
S.C. Satapathy et al. (eds.), *Proceedings of the First International Conference on Computational Intelligence and Informatics*, Advances in Intelligent Systems and Computing 507, DOI 10.1007/978-981-10-2471-9_49

chosen to soften the possibility of attaining a better outcome to a problem. The principal components of soft computing are fuzzy logic, neural networks and probability neural networks. Later, few additions like genetic algorithms, Bayesian belief networks, etc., are made to work.

Support Vector Machine (SVM) follows guided approach of classification. SVM is widely applied on the techniques like classification, regression and clustering. SVM deals with high-dimensional data of various domains. Nearly, 54 % of the applications are solved using classification methods [2]. Multi-Layer Perceptron [16] is another widely used network that can be compared with SVM. MLP is made to work on training dataset. The decision making limits can be set based on the training set in an indirect manner. In contrast, the training data directly specifies the SVM boundaries. These boundaries can be maximized to form the clear margins of classification in the feature space. The domain parameters are adjusted to derive good results using SVM. Classification using SVM results in better solutions based on working conditions of the domain. To work with SVM some of the base requirements are essential. Some of them are to set the kernel parameters and also to select the optimal feature subspace.

Feature Selection facilitates to work with domain specific features bearing a constraint of totality to attain good classification performance [3]. Feature Selection is stated as one of the best solutions based on natural evolution to solve evolutionary problems [4]. Genetic Algorithms provide optimal solutions which are the best solution in many applications.

To diagnose the breast cancer [5] by classifying it into benign or malignant classes, soft computing methods provide many solutions based on the domain constraints and the working conditions that are considered. The effectiveness in the diagnosis can be observed by applying the feature selection and classification. As it is well known that *"prevention is better than cure"*, breast feeding and physical activity help to keep the women away from the dreadful symptoms and the risk of breast cancer. The use of clinical samples shows the incidence of breast cancer. Various classification methods and ensemble techniques [6–10] are used to diagnose the breast cancer.

2 Review of Literature

Stefan Lessmann, Robert Stahlbock, Sven F. Crone (2006) used GA to automate the model selection for SVM classifier [11]. The study includes parameter tuning and combined use of kernel methods. The fitness value is used to derive an empirical model using CV method. When unknown data is used, the selection of parameters using GA and processing classifier using SVM method results in appropriate way of generalization of domain constraints. Omar S. Soliman (2014) proposed a hybrid classification algorithm using DEs and LS SVM. Cross Validation of 80 by 20 had highest classification accuracy of 99.75 % [12]. Polynomial functions and radial basis functions are used to improve the data segregation using SVM along with integer genetic and also binary coded genetic algorithms. Binary GA is proved to be better than Integer GA [13].

3 Dataset Description

Empirical analysis of machine learning methods is carried on the datasets provided by the UCI Machine Learning Repository [14, 15]. The dataset contains 699 patients' clinical samples. Each patient data is collected for 10 attributes defining the features of breast cancer. The features are tabulated in Table 1.

Sixteen instances have the missing attribute values. Mean of the attribute values are used to replace missing values. The first attribute has no significance in its usage. So, it is eliminated to work with Genetic Algorithms. Another attribute that is eliminated is the class attribute. This attribute is used in classification, but it does not show its effect in feature selection; class is also eliminated to consider for feature selection. Finally, a total of nine features are considered to work with genetic algorithm whose domain values are distributed within the interval of 1–10. The dataset contains 65.5 % of benign data and 34.5 % of malignant data when used to work with GA and SVM.

4 Methods Worked

In this study, Genetic Algorithms are used to select the best features and then the network is trained using selected features to classify the data using Support Vector Machine. The principle of natural evolution, *"Survival of the fittest"* as stated by Darwin forms the skeleton of the principles of Genetic Algorithms. SVM suits well for high-dimensional data to determine the boundaries of classification. SVM along

Table 1 WBCD Dataset Details

Sequence.	Attribute list	Range
1.	Code number	Id number
2.	CT	1–10
3.	UC Size	1–10
4.	UC Shape	1–10
5.	MA	1–10
6.	SEC Size	1–10
7.	BN	1–10
8.	BC	1–10
9.	NN	1–10
10.	Mi	1–10
11.	Cl	(benign-2, malignant-4)

with radial basis function kernel and the polynomial kernel derived good results of classification of medical data.

Genetic Algorithms is an intelligent random search method of exploitation. GA generates the genomes that have the best fitness value at each generation. For every generation, the population evolves towards an optimal solution when operated with selection, cross-over and mutation. The fitness function is used to lead the problem to tend towards the optimal solution. Fitness function values score to evaluate each individual. An array of individuals forms the population. The distance between each of the individuals in the computation is termed as *diversity*. The best fitness value is determined using any heuristics. Initial operating strings of chromosomes are the *parents* and the parents when processed with the operators derive the next generation offsprings called *children*.

Genetic Algorithm

Step 1: *Choose the initial state of population.*
Step 2: *The current generations are parents that operate to generate the children.*
Step 3: *Scores of individuals are assigned to determine the fitness.*
Step 4: *Fitness scores are made to suit the domain specification.*
Step 5: *Choose the best fit parents using heuristics and pass to the next generation.*
Step 6: *Parents are then mutated to generate children.*
Step 7: *Set the best children as the parents of the next generation.*
Step 8: *Specify the stop condition based on the fitness of the optimal solution.*

Cross-over and mutation can be performed based on encoding methods. In this study, binary encoding is used. The termination condition needs to be specified to end with some optimal solution. It is domain relevant. In the study, the stopping condition is reached at 100 generation which is observed to be the best fitness value of individuals.

4.1 Support Vector Machine

SVM specifies the boundaries of classification efficiently to classify high-dimensional data. A hyper-plane of larger margins is set to show the margin of separation. In case of the problems of finite dimensional space, they are mapped to higher dimensional spaces. Kernel functions are used to balance the mapping of points in the input space. The points in the hyper-plane are calculated [16] as specified in Eq. (1). It uses the linearity of parameters with α_i and $p_{i,}$ which are the points in the available stored values and the p data samples in the feature sub space.

$$\sum_i \alpha_i k(p_i, p) = \text{constant.} \tag{1}$$

SVM models can be classified as C-SVM, nu-SVM, epsilon-SVM, nu-SVM. The first two are classification methods and the last two are regression methods.

Kernel functions being used for various domains are [16]

$$k(p_i, p) = p_i.p \quad \text{Linear}$$
$$(\gamma p_i.p + C) \quad \text{Polynomial}$$
$$\exp\left(-\gamma |p_i - p|^2\right) \quad RBF$$
$$\tanh(\gamma \cdot p_i.p + C) \quad \text{Sigmoid}$$

where $k(p_i, p) = \Phi(p_i) \cdot \Phi(p)$.

5 Proposed Method

The proposed method specifies the use of feature selection method along with a classifier. Feature selection is done by Genetic Algorithm. Binary genetic algorithm is used to generate the best fit chromosomes of the features that are the key to diagnose the disease. The selected subset of features is then submitted to Support Vector Machine. Genetic Algorithms mimic the nature of biological evolutions. Hence, GA is assumed to suit the best to the medical domain problems and is applied to breast cancer feature selection. SVM specifies the boundaries of classification with the provision of margins on either side of the hyper-plane. Hence, it is suitable to classify the high-dimensional data like the breast cancer data.

Proposed Algorithm

Step 1: *Load all the instances of the dataset.*
Step 2: *Apply pre-processing of data.*
Step 3: *Choose the pre-processed data to be given as input to the GA.*
Step 4: *Set the conditions required and choose the methods of selection, cross-over and mutation.*
Step 5: *Specify the fitness function.*
Step 6: *Iterate until the number of generations where optimal solution is met.*
Step 7: *Select, cross-over and mutate each of the iteration and find the best fitness score.*
Step 8: *Choose the parents of next generation until optimal values are reached.*
Step 9: *Store the best evaluated features using GA.*
Step 10: *Submit the features selected as input to the SVM network by dividing the data into train set.*
Step 11: *Note the performance measures.*
Step 12: *Test the accuracy of classification using test set and note the performance measures.*

When the results attained are observed, it is evident that the accuracy is determined based on the operational conditions of the domain knowledge. GA and SVM obtained better results.

6 Experimental Setup and Performance Analysis

Genetic Algorithm is used for feature selection. The intension in the use of GA is to choose the best fit chromosomes of features which are prone to the disease diagnosis to be classified as benign or malignant. The population type is taken as bit strings and is applied with selection, cross-over and mutation. Genome length is considered to be the nine features of the dataset. Id and class features from the WBCD dataset are not considered for feature selection process. The process of finding the best chromosome stopped after 100 generations. Number of generations at 100th iteration is giving sufficient results to choose reduced number of features when made to work with number of trials with a tournament size of 10. Hence, it is chosen to limit the iterations by 100. The selection method used is tournament selection having a limit of two tournaments. The mutation type is uniform mutation. Mutation point is a single bit mutation. Crossover chosen is arithmetic cross-over. The fitness function considered to identify the related features of disease state is k-nearest neighbour. Number of neighbours when considered as 3, by choosing the Euclidian distance, the best features derived are 3. The features selected are Clump Thickness (CT), Bare Nuclei (BN) and Normal Nucleoli (NN). These three features are then taken as input to classify using SVM. The kernel function used is radial basis function. This is better for most of the trials. And the KKT constraints are set to 0.06. The reason is that up to this level of constraint mapping, literature specifies that the working conditions of SVM are giving good results.

The performance is evaluated based on Accuracy, Sensitivity and Specificity. The confusion matrix [15] is shown in Table 2.

The predictors in the columns are compared with the actual class of the rows or vice versa. The performance measures used are: [15].

Accuracy	Total number of correct predictions of the data
True Positive Rate	Correctly classified positive data samples
False Positive Rate	Incorrectly classified positive data samples
True Negative Rate	Correctly classified negative data samples
False Negative Rate	Incorrectly classified negative data samples

Table 2 Confusion matrix

Population		True condition	
		Positive	Negative
Predicted condition	Positive	True Positive (TP)	False Positive (FP)
	Negative	False Negative (FN)	True Negative (TN)

The results of classification are determined using the following formulae [15].

$$Accuracy = (True\,Positive + False\,Positive)/(True\,Positive + False\,Positive$$
$$+ False\,Negative + True\,Negative)$$
$$Sensitivity = True\,Positive/(True\,Positive + False\,Negative)\,\%$$
$$Specificity = True\,Negative/(False\,Positive + True\,Negative)\,\%$$

Experimental Results

Data is organized separately for training and testing purpose. The data set is segregated containing 80 % of data is considered for training and 20 % of the data is considered for testing. SVM using RBF kernel is used to train and test the network. Software used is MATLAB R2013a after the study of various tools for classification [17]. Few test cases are conducted and are compared with SVM without feature selection. SVM alone is not up to the mark when compared the results with Genetic Algorithms and SVM. The figures of the experiment are given in Fig. 1. Values are tabulated and are shown in Table 3.

The analysis shows that SVM using RBF kernel has 90.72 % of accuracy. Genetic algorithms derived three attributes as best features. Using these three

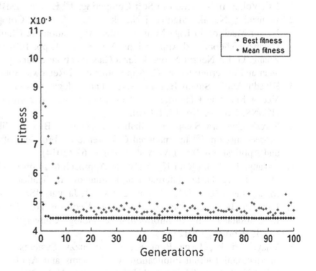

Fig. 1 Best fitness *(dotted in black)* and mean fitness *(dotted in blue)* determines the best feature values up to 100 generations using bit string Genetic Algorithm

Table 3 Experimental Results

	SVM	GA and SVM
Accuracy	0.9072	0.958
Sensitivity %	0.0857	0.0932
Specificity %	0.01	0.01
Error rate	0.0928	0.0412

attributes SVM gave good results when compared with SVM without feature selection. SVM classifier with the use of RBF kernel method shows that the classification has 95.88 % accuracy.

7 Conclusions

Classification is done using many techniques to increase the accuracy. Sometime, negligible difference in evaluation may also derive better conclusion of the diagnosis. In this study, Genetic Algorithms are applied using bit string method on WBCD dataset to select the best fit features and the selected features are submitted as input to the SVM. Later, the results of SVM alone and the results of SVM with GA methods are compared. SVM with GA is observed to be better than SVM alone. The study can be extended to carry with cluster and classify the data and compare with the outcomes of the current study.

References

1. Eva Volna: Introduction to Soft Computing. 1st Edition, ISBN 978–87-403-0573-9 (2013).
2. Janmenjoy Nayak, Bighnaraj Nayak, Behra H. S.: A Comprehensive Survey on Support Vector Machine in Data Mining Tasks: Applications & Challenges. International Journal of Database Theory and Applications. Vol. 8, No. 1, pp. 169–186 (2015).
3. Zhang, G. P.: Neural Network for a Classification: a survey. IEEE transactions on Systems, Man and cybernetics-part C: Applications and Reviews, 30(4), 451–462 (2000).
4. Riyadh AbdEl-Salam Hassan, AbdEl-Fatah Hegazy, Amr Ahmed Badr: Optimize Support Vector Machine Classifier based on Evolutionary Algorithm for Breast Cancer Diagnosis. IJCSNS, Vol.10, No. 12, (2010).
5. Suresh Chandra Satapathy, Bhabendra Narayan Biswal, Siba. Udgata K., Mandal J. K.: Proceedings on 3rd International Conference on Frontiers of Intelligent Computing: Theory and Applications (FICTA). Vol. 1, Page 367 (2014).
6. Lavanya D., Usha Rani K.: A Hybrid Approach to Improve Classification with Cascading of Data Mining Tasks. International Journal of Application or Innovation in Engineering & Management (IJAIEM). Volume 2, Issue 1, January ISSN:2319-4847 (2013).
7. Lavanya, D., Usha Rani K.: Ensemble Decision Tree Classifier for Breast Cancer Data. International Journal of Information Technology Convergence and Services (IJITCS). Vol. 2, No. 1, February (2012).
8. Vijaya Sri K., Usha Rani K.: Neuro-Fuzzy Systems and Applications – A Review. International Journal Publications of Problems and Applications in Engineering Research. CSEA2012 Vol. 04, Special Issue 01; Pg. No's 202–205, ISSN: 2230-8547; e-ISSN: 2230-8555. (2013).
9. Naga Rama Devi G., Usha Rani K.: Importance of Feature Extraction for classification of Breast Cancer Datasets – A Study. International Journal of Scientific and Innovative Mathematics Research (IJSIMR), Vol. 3, Special Issue 2, July, PP 763–768, ISSN 2347-307X (2015).

10. Naga Rama Devi G., Usha Rani K.: Evaluation of Classifier Performance using Resampling on Breast Cancer. International Journal of Science and Engineering Research, Vol. 6, Issue 2, February, ISSN 2229-5518 (2015).
11. Stefan Lessmann, Robert Stahlbock, Sven F. Crone: Genetic Algorithm for Support Vector Machine Model Selection. International Joint Conference on Neural Networks, Canada, July 16–21 (2006).
12. Omar S.Soliman, Eman AboElHamd: Classification of Breast Cancer using Differential Evolution and Least Square Support Vector Machines. International Journal of Emerging Trends and Technology in Computer Science, Vol. 3, Issue 2, April. (2014).
13. \Nithya, D., Suganya V., Saranya Irudaya Mary R.: Feature Selection using Integer and Binary Coded Genetic Algorithm to improve the performance of SVM Classifier. Journal of Computer Applications ISSN: 0974-1925, Vol. 6, Issue 3 (2013).
14. UCI Machine learning Repository, http://archive.ics.uci.edu/ml/.
15. Mangasarian O. L., Wolberg W. H.: Cancer diagnosis via linear programming. SIAM News, Volume 23, Number 5, September, pp 1 & 18 (1990).
16. www.wikipeida.com.
17. Vijaya Sri K., Usha Rani K.: Ingenious Tools of Soft Computing. International Journal of Engineering Sciences Research-IJESR http://ijesr.in/ ACICE Vol. 04, Special Issue 01, Pg. No. 1305–1312, 2013 ISSN: 2230-8504; e-ISSN-2230-8512 (2013).

Weighted Co-clustering Approach for Heart Disease Analysis

G.N. Beena Bethel, T.V. Rajinikanth and S. Viswanadha Raju

Abstract A Co-clustering approach for heart disease analysis using a weight based approach is presented. Towards the performance improvement in database mining, co-clustering approaches were used to minimize the search overhead. For the co-clustering of data, information based co-clustering (ITCC) has been used as an optimal means of clustering. However, in this co-clustering approach, elements are clustered based on Bregman divergence criterion, following the convergence of Bregman Index optimization using Euclidean distance (ED) approach. The ED approach works over the magnitude values of the elements, without consideration of the data relations. In many applications, relationship between elements played a significant role in making decision. In this paper, a relation oriented co-clustering logic following weight allocation process is presented. The proposed Weighted ITCC (W-ITCC) method/technique is applied over Cleveland data set for heart disease analysis to do performance comparisons.

Keywords Co-clustering · Weighted co-clustering · Cleveland data set · Bregman divergence criterion

G.N. Beena Bethel (✉)
Gokaraju Rangaraju Institute of Engineering and Technology, Hyderabad, India
e-mail: beenabethel@gmail.com

T.V. Rajinikanth
SNIST, Hyderabad, India
e-mail: rajinitv@gmail.com

S. Viswanadha Raju
JNTUH (Jagityal), Karimnagar, Telangana, India
e-mail: viswanadha_raju2004@yahoo.co.in

© Springer Science+Business Media Singapore 2017   511
S.C. Satapathy et al. (eds.), *Proceedings of the First International Conference on Computational Intelligence and Informatics*, Advances in Intelligent Systems and Computing 507, DOI 10.1007/978-981-10-2471-9_50

1 Introduction

In today's world, many people are being affected by various kinds of life threatening diseases. Among which, heart disease has received more attention. According to Centre for Disease Control and Prevention report, heart disease was the main cause of death all over the world. Yet, there is no perfect diagnosis system to diagnose the patient's condition who is suffering from this disease. There are many factors which affect the diagnosis of heart disease. Some of them being pathological and functional relationships of the heart and some others are clinical symptoms. All these factors affect the diagnosis accuracy and also delay the diagnosis decision. Thus, there is a need to develop an accurate heart diagnosis system for medical industry and so far many researchers have tried to establish a medical decision support system (MDSS) which will be helpful for physicians. The main aim of these systems is to enhance the diagnosis accuracy and also to reduce the diagnosis time in some complicated diagnosis decision processes [1, 2]. For an efficient diagnosis, there is a need to extract potentially useful information from large patient's data. This can be achieved through Data Mining (DM) techniques which aims to mine useful information from the available dataset.

Different DM techniques have been used to find a number of heart diseases. University of California Irvine (UCI)/Cleveland heart disease dataset [3] having 76 attributes is being used. There are both continuous and discrete attributes in this dataset. But, only 14 attributes were used so far. By considering these 14 attributes, various computational intelligence techniques were proposed in early literature. The feature selection process was made in two ways, Medical Feature Selection (MFS) and Computerized Feature Selection (CFS) [4]. Since the CFS process may select some features which are very less likely related from clinical view or may remove important features, there is a need of medical knowledge for important feature selection. In view of time complexity, CFS process gives better performance in selecting the features mathematically with a significant predictor, but did not consider the medical importance of features. Hence, there is a need to cluster the features to obtain the optimal performance between MFS and CFS. Clustering is the process of grouping the similar items and can also be applied to detect anomalies.

Various approaches were proposed earlier by considering the feature selection over the Cleveland dataset and with clustering as their prime objective. In [5], K. Rajeswari et al, proposed a novel feature selection approach by considering the correlation and association existing between the features and also proposed a classification method, which aims to reduce the patients multiple visits for test. In [6], an integer coded genetic algorithm was proposed to enhance the accuracy of diagnosis by selecting the relevant features. In [7], Subanya B et al, has proposed a Binary Artificial Bee Colony (BABC) algorithm to find the best attributes from the medical dataset based on the fitness of each attribute. [8] Proposed a novel feature selection algorithm through backward elimination criterion, and an improved feature selection procedure and accurate classification approach was suggested in [9].

Although clustering approaches give better accuracy with less time complexity, co-clustering provides still a better performance compared to clustering. Unlike regular clustering approaches like k-means, co-clustering can be considered as concurrent clustering of features by considering more than one property [10]. The co-clustering [11] approach was proposed to cluster the multiple properties of Cleveland dataset to diagnose the heart diseases. It was explained in two approaches as block co-clustering [12] and Information Theoretic Co-clustering (ITCC) [13]. ITCC proposed in [13] considered the Bregman divergence for information clustering along with its property. However, the Bregman divergence is directly related to Euclidean distance, which gives the difference of the magnitudes of attributes only. The proposed ITCC has been tested over Cleveland dataset by considering only attributes. This paper proposes a weighted ITCC co-clustering to increase the diagnosis accuracy and to reduce the time complexity. By considering the attributes having continuous property, the diagnosis accuracy will be increased, because, the attributes having continuous variations give more information about the heart condition. The attributes having continuous property are co-clustered and weight was given for each and every class (healthy, sick-1, sick-2, sick-3 and sick-4 [4]). The remaining part of the paper is well-organized in 7 sections, where Sect. 2 gives the details of Cleveland dataset. Section 3 gives a brief idea about the basic clustering approach with MFS and CFS. Section 4 illustrates the details about the ITCC. Section 5 presents the complete details about the proposed weighted ITCC, and Sect. 6 gives the simulation results for the developed approach.

2 Cleveland Dataset Representation

Cleveland Dataset is an important UCI's heart disease dataset which is widely used in heart disease diagnosis by many researchers. The most commonly used 13 attributes [14] along with their data types distinguished as Continuous (C) and Discrete (D) are Age (C), Sex (D), Cp—Chest pain (D), Resting Blood Pressure (Trestbps)—(C), Serum Cholesterol(Chol)mg/dl—(C), Fasting Blood Sugar (Fbs) —(D), Resting ECG results (Restecg)—(D), Max Heart rate (Thalach)—(C), Exercise Induced Angina (Exang)—(D), Depression Induced due to Exercise Relative to rest (Old Peak ST)—(C), Peak Exercise segment slope (Slope)—(D), No. of major vessels colored by fluoroscopy (Ca)—(D), Fixed Defect, reversible defect, normal (Thal)—(D). From the above description, a continuous data type represents continuous variations of an attribute over a period of time. For example, age varies continuously with time. The data type discrete refers to a constant value over a period of time. By considering the attributes having continuous nature, the diagnosis accuracy will increase. The dataset contains five values for class label attribute representing either healthy or one of four sick types. The five classes along with their labels are listed in Table 1.

Table 1 Class attribute level assigned

Dataset name	Class label
H-O	0 (Healthy)
Sick-1	1 (sick-1)
Sick-2	2 (sick-2)
Sick-3	3 (sick-3)
Sick-4	4 (sick-4)

3 Subset Clustering Approach

3.1 Medical Feature Selection (MFS)

Sub-Clustering is an unsupervised method of selecting and grouping the items of similar type [11]. There are various clustering approaches such as density based, Hierarchical, partitioned and sub space clustering. In [4], the clustering approach divides Cleveland datasets attributes into two clusters, a cluster having only Medical Features and a cluster having only computer features. In case of MFS, the patient will go through physical tests. When the patient is subjected to rest or exercise his/her ECG showing an abnormal reading will indicate a heart disease status. An abnormal horizontal or down sampling ST segment depression can be obtained by the positive exercise stress test [31]. From this analysis, it is clear that factors such as diabetes, resting ECG, blood sugar, cholesterol, heart rate, stress and older age, hypertension (blood pressure) and exercise induced angina are important factors for heart disease diagnosis. The selection of all these features is called medical feature selection and its clustering is called sub-clustering of MFs. Nearly, there are eight factors that are considered to have medical significance from UCI Cleveland dataset. They are: maximum heart rate, fasting blood sugar, cholesterol, resting blood pressure, resting heart rate (ventricular hypertrophy, abnormal and normal), chest pain type (asympt, notang, abnang, angina), age and exercise induced angina.

3.2 Computer Based Feature Selection (CFS)

For a given dataset, the computer selects some features by performing feature selection process. For the computer based feature selection (CFS) process, CfsSubsetEval() attribute selection (using Breadth First search strategy) provided by Weka [15] was used. The complete process can be carried out in two phases, attribute selection and clustering. CfsSubsetEval performs the attribute selection by weighing the subset of attributes, considering the degree of redundancy among all the features and individual predictive ability of each feature. Those features having low inter-correlation in the subset and high correlation with class will be preferred. In the second phase, the Breadth First search strategy groups the obtained computer features into a cluster.

3.3 MFS-CFS

CFS Approach selects the features through a computational model which was completely based on a significant predictor. Due to this, CFS may not consider the medically important factors during feature selection. From the above analysis, it can be clear that medically significant attributes such as resting ECG, resting blood pressure, fasting blood sugar, cholesterol and age will be discarded by CFS. This section is an extension to MFS. The features obtained by CFS have been combined with the features obtained by MFS and the performance was analyzed. The performance analysis is shown in the result section. Though clustering approach achieves better diagnosis accuracy and less time complexity, still there is a need to increase the diagnosis accuracy and reduce time complexity. The approach of clustering was performed for the diagnosis considered only attributes. If the attributes were considered along with their data types, further there would be a change to increase diagnosis accuracy. Thus, clustering the attributes along with their property can be done with co-clustering which is described in the next section.

4 Co-clustering Approach [11]

Co-clustering can be defined as simultaneous clustering of more properties of the dataset. Co-clustering produces a set of clusters from the original data and also a set of clusters according to their property. Co-clustering selects the features based on the subset of attributes and their properties, represented in a data matrix. Basically, there are two types of co-clustering approaches, one is Block co-clustering and the other one is Information Theoretic co-clustering. Among them, ITCC approach was proved to be optimal [11]. In the information theoretic co-clustering, the principle used was Bregman divergence. ITCC attempts to reduce the loss of information in the approximation of a data matrix X. The reduction in data loss can be maximized through a predefined distance called Bregman divergence function. For a given attribute data matrix X having a class of random variables, a matrix approximation scheme P and the co-clustering (R, C) will be defined. The main objective is to reduce the attribute loss or to minimize the Bregman Information by the approximation of $\widehat{X}$ for the defined co-clustering scheme (R, C). The Bregman information of X is defined by [13]:

$$BI_\phi(X) = E\left[log\frac{X}{E[X]}\right] \tag{1}$$

Here, the approximation of matrix scheme is defined through Bregman divergence d_ϕ and the expected value for an optimal co-clustering as follows

$$(R, C) = \text{arg min } E[d_\phi(X, \widehat{X})] \tag{2}$$

Here, Bregman divergence was directly related to the Euclidean distance and can be expressed as

$$d_\phi = (x_1 - x_2)^2 \tag{3}$$

Here, the Bregman divergence was evaluated by calculating the Euclidean distance between the attributes' magnitudes. The Euclidean distance will give the difference between the magnitudes of attribute and does not consider the divergence of attribute properties. Considering the divergence of attribute property, by adding a weight factor, would increase the accuracy even further. This proposed approach is given in next section.

5   Weighted ITCC Approach

The approach of conventional co-clustering using ITCC was observed to be effective. However, in this approach clustering is made based on the divergence factor of two observations (x_1, x_2) using Euclidean distance, where the magnitudes are compared to obtain a distance factor, and the relational property of data is not being explored. Cleveland data set, contains type attributes, in addition to the value attributes, that reflects the nature of its occurrence for all attributes categorized as discrete and continuous. However, the ITCC approach does not consider this nature of attribute variation for clustering. Hence, in this proposed method, a weighted ITCC coding for co-clustering of data type is presented. The approach of co-clustering using weight factor has been defined by allocating a weight value to the distinct class attributes in the sub cluster.

Let w_i be the allocated weight for each sub cluster C_i, where the sub clusters are derived using the method of medical and computer based selected feature set, clustered using the data type property with five distinct class of effect, where the class attributes is defined as (Table 2).

$$C_i \in (H_0, S_1, S_2, S_3, S_4) \tag{4}$$

The allocated weights are assigned based on the severity of effect for each class attribute. The ITCC based co-clustering has been successfully applied over

Table 2 Class Label attribute with its distinct values

Class type (C_i)	Definition	% of effect	Condition	Associated weight (w_i)
H_0	Class healthy	0	Normal	1
S_1	Sick-1 class	0–10	Initial stage	2
S_2	Sick-2 class	10–30	Abnormal-low	3
S_3	Sick-3 class	30–60	Abnormal-high	4
S_4	Sick-4 class	60–90	Critical stage	5

Cleveland heart disease dataset, where the dataset is clustered into sub clusters based on Bregman divergence Information criterion $I_\varnothing(X)$, satisfying the convergence problem,

$$\arg \min \left(E[d_\varnothing(X, \widetilde{X}]) \right) \qquad (5)$$

where $X \in C_i$ and $\widetilde{X}$ is the new co-cluster formed from the sub cluster C_i. Here, $d_\varnothing$ is defined as the divergence operator given by the Euclidean distance of the two observing elements (x_1, x_2) as,

$$d_\varnothing(x_1, x_2) = (x_1, x_2)^2 \qquad (6)$$

As observed, continuous data types are more effective in deriving a decision than the discrete ones, and so including type factor and integrating the co-clusters based on the class attribute leads to a finer co-clustering. Therefore, in this approach, weighted co-clustering computes an aggregate weight of each of the class label (w_i) as,

$$W = \sum_{i=1}^{n} w_i \qquad (7)$$

where w_i is the weight assigned to a class attribute for n sub classes under observation. While considering co-clustering of each class, the aggregated weight value is computed, and the co-clustering limit of $\frac{W_c}{N}$ is set, where w_c is the total weight value allocated to each class attribute,

$$W_c = \sum_{i=1}^{N} w_i \qquad (8)$$

where, N is the number of classes. The convergence problem is then defined by,

$$\arg \min \left(E[d_\varnothing(X, \widetilde{X})] \right) \Rightarrow \left(\min(W) > \frac{W_c}{N} \right) \qquad (9)$$

The solution to the convergence problem is defined here by two objectives, with minimum estimated divergence based on Euclidean distance, subjected to the condition having a minimum aggregated weight value lower than half of the total class weight value. A limit of $\frac{W_c}{N}$ is considered, to have a co-clustering objective of minimum number of similar elements entering into a class, without breaking the relational property among the observing attribute. The operational algorithm for the weighted co-clustering approach is

ALGORITHM: W-ITCC
Output   : Co_cluster classes (CC_i)
Method M_0:
1. for each type attribute,
2. Compute sub-clusters, C_i $\in$ (Medical, computer selected Features) Where i, is the class attributes =1 to 5 for considered 5 classes.
3. Cluster the selected attributes from class C_i based on data type passed,
4. C_{ico}=1; if data type attribute = 'continuous'
5. C_{ico}=0; if data type attribute = 'discrete'
6. Formulate updated cluster, $C_{iupdt} = C_i \cap C_{ico}$, where i=1to5
7.   Perform   Co-clustering   over C_{iupdt} using w_co_clustering:M_1.

Method M_1: W_co-clustering (C_{iupdt}, w_i)
1. Allocate weight (w_i) to each C_{iupdt} $C_{iupdt} \Leftrightarrow w_i$
2. Compute Bregman Divergence using Euclidean Distance,
$d_\emptyset(x_1,x_2)= (x_1- x_2)^2$
3. Compute Bregman Index, $I_\emptyset(X)= E(\log\left(\frac{X}{E(X)}\right))$ Where $I_\emptyset(X)$is Bregman Index, satisfying the criterion, $\arg\min(I_\emptyset(X))$
4. Compute aggregated weight, W, defined by, $W = \sum_{i=1}^{n} w_i$ where, i is the sub cluster, selected by Bregman divergence criterion,
5.   Compute the limiting bound value W_c defined by, $W_c = \sum_{i=1}^{N} w_i$ where N is total number of sub clusters.
6. Applying the weight convergence criterion,
$$\arg\min(E[d_\emptyset(X,\bar{X})] \Rightarrow (\min(W) > \frac{W_c}{N})$$
The sub-cluster are co-clustered.
end M_1
end M_0
end

Table 3 Accuracy measurements for the proposed and earlier approaches

Input	Accuracy (%)				
	MFS [4]	CFS [4]	MFS_CFS_Co	ITCC [11]	W-ITCC
Healthy(0)	69.75	67.25	74.21	76.11	85.14
Sick1(1)	68.24	71.52	73.41	77.17	82.16
Sick2(2)	67.25	70.42	72.15	78.52	84.68
Sick3(3)	69.45	73.21	74.54	79.54	82.24
Sick4(4)	71.25	73.54	76.32	79.12	80.64

6 Experimental Results

The performance of our proposed approach was verified through accuracy in Table 3. The accuracy, sensitivity, specificity, Recall, precision and F-measure of previous approaches as compared to W-ITTC were evaluated and projected in the Table 4.

It is observed that, the accuracy of the proposed W-ITCC is better compared with ITCC [11], MFS_CFS_Co, CFS [4] and MFS [4]. Table 4 gives the details about the performance of proposed and earlier approaches with respect to sensitivity, specificity, recall, precision, F-measure and Time. All the metrics were evaluated for MFS-CFS-Continuous, MFS-CFS-Continuous-ITCC and the proposed MFS-CFS-Continuous-Weighted-ITCC for all classes such as healthy, sick2, sick2, sick3 and sick4.

Figure 1 represents the Receiver Operating Characteristics (ROC) of the proposed approach for all classes. In Fig. 1, as the specificity increase, the sensitivity is also increasing and the ROC was independent of class. The next comparison has taken place through accuracy for proposed approach with earlier approaches. The obtained accuracy comparison plot was in the next bar graph in Fig. 2.

Figure 2 shows the plot of accuracy for the proposed approach for all classes. It also gives the comparative analysis of the proposed approach with earlier approaches for every class. From the above figure, it is observed that, for every class, the accuracy of the proposed approach is better when compared with earlier approaches.

Table 4 Performance table for MFC-CFS-Co, MFS-CFS-Co-ITCC and MFS-CFS-Co-W-ITCC

	Feature selection	Accuracy	Sensitivity	Specificity	Recall	Precision	F-measure	Time
Healthy	MFS-CFS-Co	51.254536	0.70	0.424255	0.61	0.514772	0.585056	0.2851
	ITCC [11]	73.454021	0.80	0.585915	0.89	0.686667	0.774621	0.1668
	W-ITCC	78.625464	0.84	0.288872	0.71	0.515706	0.596522	0.1451
Sick-1	MFS-CFS-Co	41.353608	0.72	0.285319	0.72	0.523784	0.587333	0.3473
	ITCC [11]	50.225258	0.58	0.242766	0.58	0.421769	0.496957	0.1380
	W-ITCC	70.081237	0.92	0.458809	0.91	0.645803	0.755802	0.1259
Sick-2	MFS-CFS-Co	51.524536	0.60	0.405255	0.60	0.512772	0.545056	0.2593
	ITCC [11]	61.134021	0.88	0.542915	0.87	0.654667	0.762621	0.1479
	W-ITCC	72.155464	0.72	0.287872	0.72	0.523706	0.552122	0.1625
Sick-3	MFS-CFS-Co	50.763608	0.72	0.247319	0.73	0.525687	0.585333	0.3587
	ITCC [11]	65.254021	0.90	0.541915	0.85	0.663667	0.774621	0.1473
	W-ITCC	74.041237	0.92	0.475809	0.88	0.678803	0.769802	0.1423
Sick-4	MFS-CFS-Co	52.743608	0.72	0.241319	0.72	0.545723	0.578333	0.3668
	ITCC [11]	67.564021	0.90	0.585915	0.87	0.642667	0.785621	0.1780
	W-ITCC	75.481237	0.92	0.456809	0.91	0.645803	0.774802	0.1528

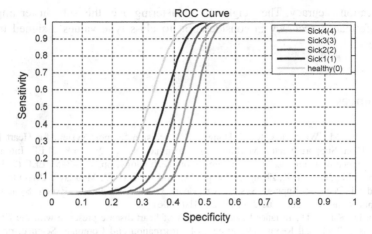

Fig. 1 Receiver operating characteristics

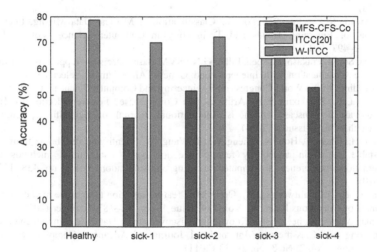

Fig. 2 Accuracy plot

7 Conclusion

A new approach of co-clustering logic based on the relational property of the data attributes and class values is presented. In the process of mining for heart disease analysis, the clustering operation based on the data type property for continuous and discrete is used. The sub-cluster formation results in optimal data clusters with medical and computer oriented feature selection, in consideration with its data type property. The data type property is observed to be a significant observation in appropriate clustering of dataset, and has shown a higher performance in classification accuracy. The weighted co-clustering over this sub-cluster improves the performance accuracy in consideration to class type values assigned to each class level.

References

1. Gudadhe, M., Wankhade, K., & Dongre, S.: Decision Support System for Heart Disease based on Support Vector Machine and Artificial Neural Network, IEEE International Conference on Computer and Communication Technology, pp. 741–745, IEEE (2010).
2. Yan, H.M., Jiang, Y.T., Zheng, J., Peng, C.L. and Li, Q.H.: A Multilayer Perceptron-Based Medical Decision Support System for Heart Disease Diagnosis, Expert Systems with Applications, Vol. 30, issue 2, pp. 272–281 (2006).
3. Aha, D., Kibler, D.: Instance-based prediction of heart-disease presence with the Cleveland database, Technical Report, Department of Information and Computer Science, pp. 88–07 (1995).
4. Nahar, Jesmin, Computational intelligence for heart disease diagnosis: A medical knowledge driven approach, Expert Systems with Applications, Vol. 40, issue 1, pp. 96–107, Elsevier (2013).
5. Rajeswari, K.: Feature Selection for Classification in Medical Data Mining, International Journal of Emerging Trends and Technology in Computer Science, Vol. 2, Issue 2, pp. 492–497 (2013).
6. Bhatia, Sumit, Prakash, Praveen, Pillai, G.N.: SVM Based Decision Support System for Heart Disease Classification with Integer-Coded Genetic Algorithm to Select Critical Features, Proceedings of the World Congress on Engineering and Computer Science, pp. 22–24 (2008).
7. Subanya, B., Rajalaxmi, R.R.: Artificial Bee Colony based Feature Selection for Effective Cardiovascular Disease Diagnosis, International Journal of Scientific & Engineering Research, Vol. 5, Issue 5 (2014).
8. Zhao, H., Chen, J., Hou, N., Zheng, C., & Wang, W.: Identifying metabolite biomarkers in unstable angina in-patients by feature selection based data mining methods, Second international conference on computer modeling and simulation, Vol. 1, pp. 438–442, IEEE (2010).
9. Fan, Y., & Chaovalitwongse, W.: Optimizing feature selection to improve medical diagnosis, Annals of Operations Research, Vol. 174, issue 1, pp. 169–183, Springer (2010).
10. Kumar, Varun, Rathee, Nisha: Knowledge Discovery from Database using an Integration of clustering and Classification, International Journal of Advanced Computer Science and Applications, vol. 2, No.3, pp. 29–33 (2011).
11. Ahmed, Mohiuddin, Mahmood, Abdun Naser, Maher, Michael J.: Heart Disease Diagnosis Using Co-Clustering, Scalable Information Systems, Lecture Notes of the Institute for Computer Sciences, Social Informatics and Telecommunications Engineering, Vol. 139, pp. 61–70, Springer (2014).
12. Govaert and Nadif: Clustering with block mixture models, Pattern Recognition, vol. 36, issue. 2, pp. 463–473, Elsevier (2003).
13. Banerjee, A., Dhillon, I., Ghosh, J., Merugu, S., and Modha, D. S.: A generalized maximum entropy approach to Bregman co-clustering and matrix approximation, Journal of Machine Learning Research, vol. 8, pp. 1919–1986 (2007).
14. UCI Cleveland heart disease data details. http://archive.ics.uci.edu/ml/machine-learningdatabases/heart-disease/heartdisease (2010).
15. Witten, I. H., & Frank, E., "Data mining: Practical machine learning tools and techniques", San Francisco: Morgan Kaufmann, Elsevier (2005).

Secure User Authentication with Graphical Passwords and PassText

Raj Mohammed Mohd, C. Shoba Bindu and D. Vasumathi

Abstract Most of the users access the Internet or web services by means of smart phones, which are capable of handling all kinds of computations. There is a necessity to provide mutual authentication among the clients (Mobile Hosts) and servers. In general, users authenticated to system or website by means of User ID and Password, a claim for user's identity. Passwords have their own vulnerabilities such as dictionary, brute-force, guessing, observation and spyware attacks etc. Users are the weakest link in any secure system because, they choose simple, short and easy to remember passwords. we proposes a strong password generation algorithm with help of PassText and Graphical Password concepts. PassText Password concept is used to generate a unique strong password with help of password images (graphical password). User need to remember only the password images instead of text. This paper analyzes the security issues with help of scyther tool.

Keywords Authentication · Mobile host · Web services · Passtext and graphical passwords · Scyther tool

R.M. Mohd (✉)
GITAM Hyderabad, Hyderabad, Telangana, India
e-mail: raaz.mohd@gmail.com

C.S. Bindu
JNTUCE Anantapuramu, Anantapuramu, Andhra Pradesh, India
e-mail: shobabindhu@gmail.com

D. Vasumathi
JNTUH CE, Hyderabad, Telangana, India
e-mail: vasukumar_devara@yahoo.com

© Springer Science+Business Media Singapore 2017 523
S.C. Satapathy et al. (eds.), *Proceedings of the First International Conference on Computational Intelligence and Informatics*, Advances in Intelligent Systems and Computing 507, DOI 10.1007/978-981-10-2471-9_51

1 Introduction

The current statistics of web users [1] shows that over 3.5 billion users are connected to internet in the world. It shows that 40 % world population has Internet connection due to evolution of mobile and web services offered in various sectors of life. Most of web based applications exchange the data or information over the net using password based authentication methods. These are also called as SFA (single factor authentication methods) and they have their own vulnerabilities like password guessing, dictionary, brute-force search, external observation, spyware or stolen-verifier attacks.

The security of a system is enhanced by adding additional factors like smart cards(TFA), devices, biometric tokens(iris, thumb print, palm print, finger print). These are also called as Token based authentication methods and provide more usability and security. We can also strengthen the security with help of generating a strong password with help of password generation algorithms. But, Existing PGA's are used to make a strong password and it should be recalled or memorize by the user. We can also use Mobile Device as token instead of smart card in remote user authentication methods to avoid stolen verifier attack as suggested by Misbah et al. scheme [2].

In order to achieve usability, we can recommend token based authentication methods, which are based on what the user has. These methods require token acquisition and recognition devices and less memory load on user. But, these are expensive in nature and there is a chance to have false positives when it is recognizing the user identity. We can also achieve more usability by means of Password Managers, which stores the passwords and help the user to login next time.

Theoretically, Graphical Passwords achieve more usability. Because, Users remember images rather than the text. These passwords achieve more security if it stored in the form of images (binary form) rather than their pixel values, rgb values, grey level values, image names etc.

PassText Password is proposed by Karen Renaud et al. [3] and generate a password by making some changes in the text file and store that entire text file as password of more length and hard to crack by means of automated tools. This method achieve usability by making few changes in a text file and also more security store the password in CLOBB format as suggested by Bindu et al. scheme [4].

In this paper, we propose a strong password generation algorithm (SPGA) takes text file and graphical passwords or a set of images (user should remember) and generate a strong password. This method is resistant to all password vulnerabilities such as dictionary, brute-force and guessing attacks, spyware attacks (internal observation with key loggers, mouse listeners and screen loggers), shoulder-surfing attacks (external observation with cameras and physical observation).

The rest of paper is organized as follows: Sect. 2 proposes user authentication protocol, Sect. 3 analyzes the security issues of the proposed protocol and finally concluded in Sect. 4.

2 Proposed Work

In this section, we propose a secure user authentication protocol, which describe a remote user authentication with Mobile Token, shoulder-surfing resistant graphical password and a SPGA algorithm which generates a strong password.

2.1 SPGA Protocol with Mobile Token

It consists of registration, login, and password change and recovery phases.
Registration:

Step 1: User 'Ui' register to AS with a registration request to AS by submits his mail ID to AS via Mail Server and generates and verify OTP.
Step 2: Mail Server sends user's profile data to AS. Then, AS sends mail to Ui with a registration link form.
Step 3: Ui chose a text file and set of images as password and generate strong password and submits it to AS.
Step 4: AS computes a, $h(PWi)$ Ni as follows:

$$a = h(\text{mail\_ID})$$
$$h(PWi) = h[h(I1)F\{Text\}||h(I2)F\{Text\}||h(I3)F\{Text\}]$$
$$Ni = h(PWi) \oplus h(x \oplus TIDi)$$

Where x is server's master secret key, TIDi is a token identifier.
$xF\{Y\}$: F is a function, which takes input Y as text file and x is hash index generated with hash of images and generate a password.

Step 5: AS personalizes an MT with the parameters Ni, $h()$, a, yi is server's secret key shared with each MT which is identified by TIDi and mapping of IDi.

Login Phase: User will login to the system to access the resources at server 'S' by the following steps.

Step 1: Ui login to Server 'S' to access the services. S redirects to AS.
Step 2: Ui submits ID to AS. AS retrieves TIDi and generate an image grid consisting of both passimage and non passimages along with text file chosen by user.

Step 3: Ui activates his MT and select pass images from Image grid and MT
         generate and compute strong password from SPGA algorithm as follows.

$$h(PWi) = h[h(I1)F\{Text\}\|h(I2)F\{Text\}\|h(I3)F\{Text\}]$$

Step 4: MT checks whether $h'(PWi) == h(PWi)$. If it holds, MT proceeds to
         compute $Bi = h'(PWi)$, $Bi = h(PWi) \oplus Ni$; $Ci = h(Bi \oplus r)$ and $E = Ey$
         (Ci, a) and sends E to AS, where E is encryption of message using yi,
Step 5· After receiving E, AS computes Dyi (Ci, a) using yi, where D is
         decryption. It retrieves TIDi using 'a' and also computes
         $Bi' = h'(x \oplus TIDi)$; $Ci = h(Bi' \oplus r)$
Step 6: AS checks if $Ci == Ci'$. If it holds then AS sends authentication suc-
         cessful message and forwarded to S and access the resources.

Password Change Phase: User can change his or her password after successful
login.

Step 1: Ui requests for change password to AS.
Step 2: User choose images and text file and submits to AS.
Step 3: AS sends Eyi $(h(PW))$ and request to change text password at MT to Ui.
Step 4: Ui activates MT and select new password 'p' like Step 3 of registration.
         MT computes $Nin = h(h(I) \oplus p)$ and replaces Ni with new Nin, $h(I)$, h
         (PWi).
Step 5: MT sends a confirmation message to AS.

Password Recovery Phase: If a user Ui forgets his password (MT is not lost), he
submits the reset password request to AS and described as follows.

Step 1: Ui submits his mail Id or ID.
Step 2: AS forwards the IDi to mail server and provide and verify OTP with Ui.
Step 3: Upon successful verification, mail server sends a positive ACK to AS.
         AS sends a mail to user with a password reset link.
Step 4: AS presents a new image grid for choose set of images as password and
         text file to upload.
Step 5: Ui chooses I1, I2, I3 and submits it to AS. AS sends Ey_i $(h(PWi))$ and a
         request to change password at MT to Ui.
Step 6: Ui activates MT& enters new password 'p'. MT computes
         $Ni = hn(PWi) \oplus p$ and replaces Ni with the new Nin, $h(PW)$, $h()$ sends a
         confirm message to AS.

If user Ui loses her mobile token or accidently deletes it, he has to get a new MT
from server by following the steps described in the registration process.

Fig. 1 Proposed Graphical
password method

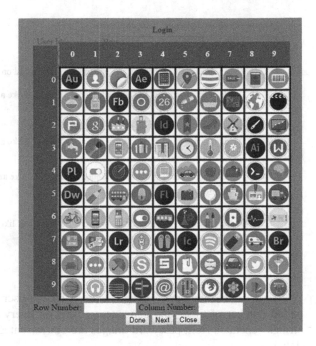

2.2 Graphical Passwords

In this section, we have used a cognition based graphical user authentication method which is resistant to observation attacks [5] proposed by us. This scheme display images on 10×10 grid and consists of both passimages and non-passimages. User selects the images by typing the position (row, column) of image, which changes randomly to avoid the observation attacks. This method facilitates more usability as compared to other user authentication methods by less login and registration times. The graphical password method is as shown in Fig. 1.

2.3 Strong Password Generation Algorithm

In this section, we describe the SPGA algorithm, which is used to generate a strong password with help of PassText Concept. Graphical password concepts achieve the usability of the user to memorize a graphical password easily. PassText concept enhance the security or strength of password with its length and modifications in a text file.

```
Input: Index of Hahsed Image , Text document
Output: Strong Password
Algorithm SPGA
  Begin
    For every image, calculate hash value and obtain index 0-99based on the algorithm requirements.
    Switch(index)as
      Case 0:  Made changes in a text file as mentioned with index 0 like add,replace,append a character in
              a word.
          Return String0
          Break;
      Case 1:  Made changes in a text file as mentioned with index 1 like add,replace,append even word
          Return String1
          Break;
      Case 2:  Made changes in a text file as mentioned with index 2 like add,replace,append odd word
          Return String2
          Break;
          ...................
      Case 99:  Made changes in a text file as mentioned with index3 like add,replace,append a sixth word.
          Return String-n
          Break;
  End
```

Based on the number of images chosen by the user, SPGA algorithm will be called number of times and generate a string for every image hash index and all strings generated from password images are concatenated, hashed and stored in the database.

3   Security Analysis of Proposed Method

This section analyzes security issues of the proposed scheme against various password vulnerabilities as well as protocol vulnerabilities.

3.1   Security Threats

The proposed scheme secure against protocol vulnerabilities such as replay, insider, stolen verifier attack, server spoofing attack (Phishing), fraudulent copying the Mobile Token, Denial service attack and storage of registration data.

- *Replay Attack*: In this scenario, the adversary intercepts the message transmitted between two principals and then sends it at a later time to gain access to the resources. In this, if the message Eyi (Ci, a), transmitted in step 10 of Login is intercepted and replayed by adversary, he cannot gain access to the resources because both the client and server checks for the freshness of nonce (r) every time whenever they receive a message. So, the attack will fail.

- *Insider Attack*: The insider attack is usually performed by an insider of the organization who has access to the sensitive data resources by revealing the user secret information to others. In the proposed scheme, the user's password is never stored on anywhere in a system. Moreover, none of the parameters required in protocol computation are stored in server in plain text form, instead, to avoid such attack, all the user credentials are stored in the database as message digest which is irreversible.
- *Stolen Verifier Attack*: In this scenario, an adversary can access to the database server and steals the password verifier and later uses it for offline guessing attack. But since, the scheme does not maintain a verifier table the attack cannot be performed.
- *Server Spoofing Attack*: It is a phishing attack, where an adversary act or pretends as legitimate server and obtain the user's sensitive information. An adversary perform this attack by creating a page which looks similar to a valid server page and then redirect the user's login request to fool the user. But, it is very difficult to create image grid along with user selected image during registration. Hence, the attack will fail.
- *Fraudulently copying the Mobile Token*: If an adversary gets access to the user's MT and wants to fraudulently copy the MT (.apk file) which stores the parameters Ni, $h(I)$, $h(Pwi)$, a, yi, he can neither retrieve the user's secret Pwi nor the server's master secret 'x' from the available parameters as the Pwi, x are stored as digest.
- *Denial of Service Attack*: Suppose, if an attacker who has control over the server, modifies any secret information of the user stored in the database server by replacing it with a newly created message digest, then the user will not be able to login even with his valid credentials. This is called as denial of service attack. In the proposed scheme, since there is no secret information stored on the server, the denial of service attack will not work.
- *Storage of Registration Data*: In the proposed scheme a user profile is maintained, which stores secret questions; answers to secret questions in message digest form so that even if the attacker gets access to the database he cannot figure out the answers of the secret questions the user has set.

The proposed scheme secure against password vulnerabilities such as brute-force search, dictionary, guessing and observation attacks (spyware, shoulder-surfing).

- *Secure against guessing attack*: It is more difficult for the attacker and the user to know what modifications have been done to the base document. So, it is secure against guessing attack. If adversary guess the all the images correctly then only they can login to the system and they didn't obtain the information about the password as well as text file chosen by the user.

- *Secure against brute force search attack*: Suppose if the user has chosen the text file of size 64 k with three modifications done the password space becomes $(216)^{3.}$ So the proposed scheme is resistant to brute force attack even with less number of images chosen or changes done in the file(back ground mode) makes a strong password.
- *Secure against dictionary attack*: There is no dictionary for what kind of changes done by the system (SPGA Algorithm) in a user chosen document. SPGA algorithm is not disclosed and even if it is disclosed it is difficult to obtain both text file and set of password images chosen by the user. So, it is difficult to carry out dictionary attack compared to classical passwords.
- *Secure against Shoulder-surfing attack.* This is also called as external or physical observation attack on graphical passwords. An adversary can observe login process of valid user. Compared to other graphical passwords it is less vulnerable to shoulder-surfing attack [6], where user select an image with help of row and column of image. If an adversary can able to observe both pictures and position of an image then only he can obtain the password images. Even though the changes made to the document by the user are viewed by the attackers, he will not be able to identify the index number given by the user as will not be displayed on the screen. So, the proposed scheme is secure against shoulder-surfing attack.
- *Secure against Spyware attack*: This is also called as internal observation attack done by either a key loggers [7] or key listening technique or by Mouse listeners. In the proposed system, User enters row and column of image that can be changed randomly on a grid. Adversary cannot able to obtain the password by means of keyloggers and mouse listeners. In Proposed system, we didn't type the changes in a text file rather we made changes with help of hash index of images and concatenate to form a strong password. The strong password again hashed and stored in database. Graphical password scheme reduces the chance of the password being revealed and hence it is secure against Spyware attack.

3.2 Formal Verification of Proposed Protocol Using Scyther Tool

This section discusses about the brief description about the scyther tool [8], which is used to verify the security protocols specified with .Spdl extension. The main objective of the scyther tool is to clearly distinguish the protocol descriptions from their behavior and its pseudo code is as shown below.

```
/*
 * SPGA-MT protocol
 */
// The protocol description
secret x : Function;
const equal : Function;
const hash : Function;
const XOR : Function;
//const r : Nonce;
const TID : Nonce;
hashfunction H1;

protocol SPGA-MT (C,S)
{
role C
{
const i,P,Images,TextFile;
fresh HA : Ticket;
var y : Nonce;
fresh r : Nonce;
send_1 (C,S,{C,r}pk(S));
recv_2 (S,C,{r,y,S,Images}pk(C));
send_3(C,S,{H1(XOR(XOR(H1(XOR(H1(i),P)),XOR(H1(XOR(H1(i),P)),H1(XOR(x
,TID)))),r)),y}pk(S),{H1(HA,y)}pk(S));
claim_i1(C,Secret,XOR(H1(H1(XOR(H1(i),P))),XOR(H1(XOR(H1(i),P)),H1(XO
R(x,TID)))));
claim_i2(C,Secret,H1(XOR(XOR(H1(XOR(H1(i),P)),XOR(H1(XOR(H1(i),P)),H1
(XOR(x,TID)))),r)));
claim_i3 (C,Secret,HA);
claim_i4 (C,Niagree);
claim_i5 (C,Nisynch);
claim_i6(C,Secret,Images);
}
role S
{
var HA : Ticket;
var r : Nonce;
fresh y : Nonce;
var Ci : Ticket;
const Images,TextFile;
recv_1 (C,S,{C,r}pk(S));
send_2 (S,C,{r,y,S,Images}pk(C));
recv_3(C,S,{Ci,y}pk(S),{H1(HA,y)}pk(S));
claim_r1 (S,Secret,H1(XOR(x,TID)));
claim_r2 (S,Secret,Ci);
claim_r3(S,Secret,H1(XOR(H1(XOR(x,TID)),r)));
claim_r4 (S,Secret,HA);
claim_r5 (S,Niagree);
claim_r6(S,Secret,Images);
}
}
```

The empirical analysis of the proposed method is as shown in the Fig. 2. The proposed method is resistant to all the password related threats and protocol threats. Scyther tool deals only with network protocol and not the password related threats.

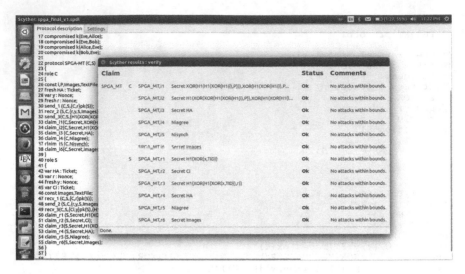

Fig. 2 Empirical analysis of proposed method

4 Conclusion

We propose a user authentication method using graphical passwords to achieve more usability and the security can be enhanced by means of PassText Password concept. We propose SPGA algorithm, which generate a strong password by making changes in a text file with help of set of images (graphical passwords) as passwords. The proposed method is resistant to both password vulnerabilities and protocol based threats and it also achieves more usability.

Acknowledgments We would like to thanks the people who involved in this research process directly or indirectly to analyze and implement the proposed scheme. We also acknowledge the AICTE for funding this project under Research Promotion Scheme with Usable Security Lab to complete this work at JNTU, Anantapuramu.

References

1. Statistics of Internet http://www.internetlivestats.com/internet-users/ Last accessed Janaury1, 2016.
2. Mohammed Misbahuddin et. al "A Unique-ID based Usable Multi-Factor Authentication Scheme for e-Services" Int'l Conf. Security and Management, SAM'15.
3. R. V. Yampolskiy,"Secure Network authentication with PassText", The IEEE International conference on Information and Technology ITNG'07, April 2007.
4. Shoba Bindu C " Secure Usable Authentication Using Strong Pass text Passwords" I. J. Computer Network and Information Security, 2015, 3, 57–64 Published Online February 2015 in MECS (http://www.mecs-press.org/).

5. Raj Mohammed, C. Shoba Bindu, P. Chandrasekhar Reddy and B. Satyanarayana, "A novel cognition based Graphical Authentication scheme which is resistant to shoulder-surfing attack", in Proceedings of 2nd International conference on information Processing, ICIP 2008.
6. C. Shoba Bindu, "Improved novel graphical password authentication scheme and usability study", i-manager's journal of Software Engineering, Vol. 3, No. 4, April–June 2009.
7. Nairit Adhikary, Rohit Shrivastava, Ashwani Kumar, Sunil Kumar Verma, Monark Bag, Vrijendra Singh, "Battering Keyloggers and Screen Recording Software by Fabricating Passwords", International Journal of Computer Network and Information Security, 2012, 5, 13–21, doi:10.5815/ijcnis.2012.05.02.
8. Cas Cremers, "The Scyther Tool Verification, Falsification, and Analysis of Security Protocols", Tool Paper, http://www.people.inf.ethz.ch/cremersc/downloads/papers/The Scyther Tool: Verification, Falsification, and Analysis of Security Protocols.pdf.

A Novel Hybrid Fuzzy Pattern Classifier Using Order-Weighted Averaging

Søren Atmakuri Davidsen and M. Padmavathamma

Abstract The order-weighted averaging operator is commonly used in decision-making processes, where its powerful yet simple nature allows to aggregate output from several data sources into a meaningful result. Key to the order-weighted averaging operator is assignment of weights. Several approaches have been suggested and their properties examined for decision-making, except in the area of heuristic search assignment. In this paper, weight assignment is experimentally examined for supervised classification tasks using a fuzzy pattern classifier with order-weighted averaging, using maximum entropy and two separate heuristic search approaches, namely genetic algorithm and pattern search. The experiment is conducted using a sample of 20 UCI data sets. Results are discussed and recommendations made for when and how to apply heuristic search for this type of classifier.

Keywords Classification · Fuzzy systems · Fuzzy classifier · Order weighted averaging · Heuristic search

1 Introduction

In fuzzy systems for decision-making, key concepts are assignment of fuzzy sets to give meaningful coverage of training examples, and, design of fuzzy rules to infer a decision. A key concept of fuzzy systems is the link to human linguistics, where the goal is to mimic as to make systems which we as humans understand the working of without the underlying mathematics. This principle works well in low-dimensional rule-bases, and simple decision-making aggregations such as a crisp Boolean conjunction works well.

S.A. Davidsen (✉) · M. Padmavathamma
Department of Computer Science, Sri Venkateswara University, Tirupati 517502, India
e-mail: sorend@acm.org

M. Padmavathamma
e-mail: prof.padma@yahoo.com

© Springer Science+Business Media Singapore 2017   535
S.C. Satapathy et al. (eds.), *Proceedings of the First International Conference on Computational Intelligence and Informatics*, Advances in Intelligent Systems and Computing 507, DOI 10.1007/978-981-10-2471-9_52

However, in higher dimensions where rules may not be meaningful and where the rules may be more complex, a crisp conjunction, disjunction or averaging may not be useful. In these cases fuzzy decision-making operators such as the order-weighted averaging operator, may make more sense. An operator as the order-weighted averaging operator allows to produce a decision-making aggregation from minimum (AND) to maximum (OR), and everything in between.

Learning an OWA operator from data was proposed by Filev and Yager [5] using a gradient descent method and a little later by Torra [12] using an active set method. However, both methods are focused on regression problems, and unfortunately are found to perform poorly on real-world problems, for example in the case of traffic speed estimation [1].

A recent survey by Liu has categorized the different methods of obtaining OWA weights [7]. They categorize OWA learning methods into the following:

- Optimization-based methods—optimize weights according to a specified level or *orness*, dispersion or other measure.
- Sample learning methods—methods where OWA weights are learned by sampling data with known outputs.
- Function-based methods—methods where OWA weights are generated by a function.
- Argument-dependent methods—methods where weights are construct OWA weights as functions of additional parameters, which makes the OWA weights unique to each input vector.
- Preference methods—methods where samples for learning OWA weights are ranked by experts.

Here the primary interest is sampling methods, where the OWA is learned from data in a classification context.

There are several fuzzy classifiers which have made use of OWA. The fuzzy pattern tree classifier [11] makes use of OWA as one of the possible binary aggregation operators, however only with two weight vectors $(0.8, 0.2)$ and $(0.2, 0.8)$. Also the maximum entropy OWA (MEOWA) has been used for aggregation in a fuzzy pattern classifier [9], but with assignment of *andness* instead of a learning approach.

In order to overcome the problems with useful weight assignment in classification, this paper proposes the use of heuristic search. A priori it is expected that learning weights will be useful in more complex problems. To the authors' best knowledge, this is the first time learned OWA operators have been evaluated in the context of classification problems.

The remainder of the paper is structured as follows: Sect. 3 introduces the order-weighted averaging operator and the fuzzy pattern classifier and propose a new classifier and how to learn its model. Section 4 analyze the classifier in an experimental evaluation on a sample of UCI data sets. Finally, Sect. 5 discuss the results and conclude the paper.

2 Classification and Fuzzy Pattern Classifier

Classification is a kind of supervised learning, where the task is predicting the class of an unseen example. It involves two steps:

- Learn a model from training examples;
- Use the learned model to predict class of new unseen examples.

For most learning algorithms, learning a model from training data involves searching an optimal model from a set of candidate models. Given a hypothesis space $\mathcal{H}$, training data $\mathcal{D} = \{(x^1, y^1), (x^2, y^2), \ldots, (x^n, y^n)\}$, and a loss function $\mathcal{L}$, search for a model h^* minimizing the loss, in other words which covers the training examples best:

$$h^* = \arg\min_{h \in \mathcal{H}} \sum_{i=1}^{n} \mathcal{L}(h(x^i), y^i) \tag{1}$$

Training examples are vectors $x^i = (x_1^i, x_2^i, \ldots, x_m^i)$, where each element is called a feature, and their corresponding class is one of k classes, $y^i \in \{1, \ldots, k\}$.

The fuzzy pattern classifier (FPC) is a classifier which uses concepts from fuzzy logic to construct class-wise fuzzy patterns, namely fuzzy sets to describe features, and fuzzy aggregation to determine combined features match of a pattern. FPC has been proposed several times in literature, this paper is consistent with the model of [6, 8].

FPC is defined by a set of fuzzy patterns and an aggregation operator (λ, F), where $F = \{F_1, \ldots, F_k\}$ one pattern for each class. Each fuzzy pattern F_i is a collection of fuzzy membership functions $\mu_{i,j}$, one for each input feature j, which describe the fuzzy set for the specific feature:

$$F = \begin{bmatrix} \mu_{1,1}, \mu_{1,2}, \ldots, \mu_{1,m} \\ \mu_{2,1}, \mu_{2,2}, \ldots, \mu_{2,m} \\ \ldots \\ \mu_{k,1}, \mu_{k,2}, \ldots, \mu_{k,m} \end{bmatrix} \tag{2}$$

In the matrix F each row describes the fuzzy pattern of a class. The aggregation operator λ is generally any fuzzy aggregation operator. In literature several aggregation operators have been suggested: in [8] the arithmetic was used as aggregation operator, [6] proposed to use the product, [2] used the geometric mean, and [9] suggested to use the power-mean with controllable *andness*.

Training the FPC means assigning values to λ and the membership functions $\mu_{i,j}$. Again, several heuristics have been proposed [6, 8, 9], as well as evolutionary methods [3]. Here the following heuristic is used:

$$\mu_{i,j} = \Pi(x; r, p, q), \tag{3}$$

where $r = \bar{x}_j$, $p = r - (\max x_j - \min x_j)$, and $q = r + (\max x_j - \min x_j)$. The Π membership function is defined as follows:

$$\Pi(x; r, p, q) = \begin{cases} 2^{t-1}\left((x-a)/(r-a)\right)^t & a < x \leq p \\ 1 - \left(2^{t-1}\left((r-x)/(r-a)\right)^t\right) & p < x \leq r \\ 1 - \left(2^{t-1}\left((x-r)/(b-r)\right)^t\right) & r < x \leq q \\ 2^{t-1}\left((b-x)/(b-r)\right)^t & q < x \leq b \\ 0 & \text{otherwise} \end{cases}$$

where $a = r - 2p$, $b = r + 2q$

and is a membership function with a Π shape, where r is the top-point and p, q are the midpoints ($\Pi(x) = 0.5$) on the down-slopes left and right side of the top-point. The fuzzifier parameter t is set to 2.

3 Order-Weighted Averaging for Classification

Often aggregation operators are used in some decision-making process, in order to combine multiple inputs into a single decision output. Classical decision can be made using Boolean AND (minimum) and OR (maximum). However, most problems require something in between, which is either *and-like* or *or-like*. This is the motivation behind Yager's order-weighted averaging (OWA) operator [14], which can have weights assigned for any mean between AND and OR aggregation.

Unique to the OWA operator is ordering of the input and assignment of weights, meaning a specific element in the input is not tied to a specific weight, but rather depends on the ordering operation. This provides a non-linear aggregation.

Definition 1 An OWA operator of dimension m aggregates as follows:

$$\lambda_{\text{OWA}}(x; v) = \sum_{i=1}^{m} \left(v_i x_{(i)}\right), \tag{4}$$

where

- $(\cdot)$ is a permutation of x such that $x_{(1)} \geq x_{(2)} \geq \cdots \geq x_{(m)}$
- $v = (v_1, v_2, \ldots, v_m) \in [0, 1]$, $\sum_{i=1}^{m} v_i = 1$

By careful assignment of weights, it is possible to use OWA for other well-known means, for example $(1, 0, \ldots, 0)$ gives minimum, $(0, \ldots, 0, 1)$ gives maximum, and $(\frac{1}{m}, \ldots, \frac{1}{m})$ the arithmetic mean.

To measure how *or-like* or *and-like* a specific OWA operator is, Yager introduced the *orness* of an OWA operator's weights. Andness is simply the negation of orness, so for orness $= 1$, andness $= 0$ and vice versa:

$$\text{orness}(v) = \frac{1}{m-1} \sum_{i=1}^{m} (m-i) \, v_i, \quad \text{andness}(v) = 1 - \text{orness}(v) \qquad (5)$$

Dispersion is a measure of how uniformly the OWA weights are distributed. Consider the weight vectors for the median $(0,0,1,0,0)$ and the arithmetic mean $(0.2, 0.2, 0.2, 0.2, 0.2)$. Both share an *orness* of 0.5, yet the arithmetic mean considers all input values while the median only a single. Here we use the normalized dispersion, as it makes more sense for comparing data sets with variable dimensions. The normalized dispersion is calculated as follows:

$$\text{ndisp}(v) = \frac{1}{\ln m} \left(-\sum_{i=1}^{m} v_i \ln v_i \right) \qquad (6)$$

Higher dispersion means more features are considered by the OWA operator. It has been shown that for a fixed *orness* and dimension m there is a unique solution for a weight vector with maximum entropy [4], this we denote the MEOWA operator.

In order to use search techniques without the OWA constraint $\sum_{i=1}^{m} v_i = 1$, the following weight mapping is used $w : \Omega \rightarrow v$ [5]:

$$v_i = \frac{1}{\sum_{i=1}^{m} e^{\Omega_i}} e^{\Omega_i}, \quad \Omega \in [0,1] \qquad (7)$$

3.1 Heuristic Search of OWA Weights

The search space for OWA weights is by definition infinite, so searching all possible combinations is not possible. However, several approximation methods exists, which mostly go under the name heuristic search. Heuristic search works by iteratively applying the heuristic to generate candidate solutions, and progressing with the best candidates using a *fitness* measure to evaluate each candidate. For searching OWA weights, this paper propose using two different heuristic search methods: Pattern and Genetic Algorithm. In the following the two are differentiated.

Pattern Search (PS) is an optimization method suitable for fast heuristic tuning of parameters [10]. It works by a bounded narrowing of each parameter in the search-space until a suitable solution is found. Genetic algorithm (GA) is another type of heuristic search, which uses principles from evolutionary theory, cross-over and mutation, to refine a solution. Their unique characteristics are as follows:

- PS: Iteratively seek to optimize a single parameter at a time. Here it means that the OWA operator will have a single weight v_i changed per iteration. PS has two control parameters: number of solution adjustments to perform (called evaluations), and number of solutions to produce.
- GA: Iteratively seek to optimize all parameters at once. For each iteration a new solution is generated by combining previous solutions, along with adding random changes within the search space. GA has several control parameters, most importantly the size of population (candidate solutions) and number of iterations.

For a complete description of both search techniques including their algorithms, please refer to [10, 13]. Here we denote both a function $S(f) \to \Omega$ where S is the search algorithm, f the fitness function, and Ω the final candidate solution which obtained the best fitness.

3.2 FPC with Learned OWA

To learn the OWA for a FPC, first we define a mapping from the candidate solution Ω to a candidate model h:

$$h = \mathrm{FPC}(\lambda, F), \quad \lambda = \lambda_{\mathrm{OWA}}(x; v), \quad v = w(\Omega), \tag{8}$$

where F are the membership functions as defined in Eq. 2 and v as defined by Eq. 7.

Once the candidate model is found, mean squared error is used as *fitness* (similarity) of the values predicted by the model and the optimal values of clear class boundaries. Mean squared error is defined as follows in this context:

$$\mathrm{fitness}(\Omega) = \frac{1}{nk} \sum_{i=1}^{n} \sum_{j=1}^{k} \left(h(x^i) - K_k(y^i) \right)^2, \tag{9}$$

where K is a mapping function such that:

$$K_k(y) = \begin{cases} 1 & \text{if } y = k \\ 0 & \text{otherwise} \end{cases}$$

As the fuzzy pattern classifier's model contains a single aggregation operator λ, it is clear that OWA can be used for this operator. The algorithm for training FPC with OWA is outlined in Algorithm 1.

Algorithm 1: Fuzzy pattern classifier with OWA training algorithm.

Data: $\mathcal{D}, S$, fitness—training data, search function, and fitness function.
Result: (λ, F)—FPC model.
for $i \leftarrow 1 \ldots k$ **do**
    $T = $ all examples in $\mathcal{D}$ belonging to class i;
    for $j \leftarrow 1 \ldots m$ **do**
        calculate r, p, q for j^{th} feature in T;
        $F_{i,j} = \Pi(x; r, p, q)$;
    end
end
$\Omega \leftarrow S(\text{fitness})$ using fitness function from Eq. 9;
$\lambda \leftarrow \text{OWA}(v_1, v_2, \ldots, v_m)$ using mapping Eq. 7;
return (λ, F)

For the FPC with MEOWA, weights are not searched, but only the *andness* parameter for the MEOWA operator. For simplicity we have used PS to do this search.

4 Experimental Evaluation

As described, three novel classifiers are proposed, FPC with MEOWA, PS searched weights, and GA searched weights, respectively FPC_{ME}, FPC_{PS}, and FPC_{GA}. The three are evaluated together with three classifiers using fixed *andness*, namely minimum, mean, and maximum, in an experimental setup. Each classifier was tested on the data sets using stratified tenfold cross validation ten times, meaning the mean results reported are from $(10 \times 10\,\text{CV})$. Accuracy is used as a general purpose metric for comparison, and Friedman with Nemenyi post hoc is used for statistical analysis.

For each classifier the following settings are used:

- FPC_{ME}: Number of iterations: 10 (to search for *andness*).
- FPC_{PS}: Number of iterations: 5 m, number of solutions: 10.
- FPC_{GA}: Number of generations: 10 m, number of chromosomes: 100.

The classifiers have been implemented in our Python open source fuzzy classifier library FyLearn, which is ready for immediate download and use.[1]

4.1 UCI Data Sets

Twenty data sets were sampled from the UCI machine learning repository. The basic properties of the data sets are found in Table 1. Data sets were sampled using the following criteria:

---

[1] Available from http://www.cs.svuni.in/~sorend/fylearn/.

Table 1 Data sets and their basic statistics: Number of Examples, number of Features, number of Classes and size of the Majority class

#	Data set	(E/F/C/M)			
1	Balance scale	625	4	3	0.46
2	Banknote authentication	1372	4	2	0.56
3	Blood transfusion service center	748	4	2	0.76
4	Bupa liver disorders	345	6	2	0.58
5	Climate model simulation crashes	540	18	2	0.91
6	Fertility diagnosis	100	9	2	0.88
7	Glass	214	10	6	0.36
8	Haberman	306	3	2	0.74
9	Indian liver patient	579	10	2	0.72
10	Ionosphere	351	34	2	0.64
11	Iris	150	4	3	0.33
12	Mammographic masses	830	5	2	0.51
13	Parkinsons	195	22	2	0.75
14	Pima indians diabetes	768	8	2	0.65
15	Satimage	6435	36	6	0.24
16	Telugu vowels	871	3	6	0.24
17	Vertebral column	310	6	2	0.68
18	Wheat seeds	210	7	3	0.33
19	Wine	178	12	3	0.40
20	Wisconsin breast cancer	683	9	2	0.65

- The data set contains only numerical values;
- The data set is suitable for classification.

For a few of the data sets, columns with missing values and example identifiers were dropped, so only numerical values and class remained. The used data sets are available for download.[2]

4.2 Accuracy

Mean accuracy is a measure of how many correct predictions were made by a classifier:

$$\text{accuracy}(h) = \frac{1}{n} \sum_{i=1}^{n} \begin{cases} 1 & \text{if } h(x^i) = y^i \\ 0 & \text{otherwise} \end{cases} \tag{10}$$

---

[2]Available from http://www.cs.svuni.in/~sorend/files/datasets.zip.

Table 2 Mean accuracy for each of the selected classifiers. **Bold** indicates highest scoring classifier

#	FPC_{min}	FPC_{mean}	FPC_{ME}	FPC_{GA}	FPC_{PS}	FPC_{max}
1	10.03(2.36)	53.30(4.71)	53.01(4.62)	53.38(4.59)	**53.49(4.49)**	43.95(8.82)
2	**85.13(0.79)**	72.81(2.80)	75.50(2.30)	75.40(2.42)	75.44(2.39)	54.84(1.10)
3	**68.62(8.55)**	66.61(7.90)	64.11(10.09)	65.15(8.94)	65.15(8.94)	60.25(3.96)
4	44.44(4.94)	**56.72(2.34)**	56.59(2.18)	56.39(2.21)	56.41(2.20)	50.89(3.33)
5	42.24(3.12)	**90.74(0.66)**	90.08(0.91)	90.08(0.97)	90.05(1.00)	61.58(2.47)
6	**88.00(0.38)**	87.44(1.46)	87.33(1.79)	87.44(1.46)	87.44(1.46)	83.28(9.07)
7	32.71(0.26)	41.18(1.94)	41.08(1.97)	41.37(1.98)	**41.42(2.00)**	30.36(5.87)
8	**71.78(2.58)**	68.25(6.81)	68.47(6.81)	68.26(5.97)	68.27(5.95)	65.36(3.67)
9	70.71(1.84)	70.56(1.93)	**70.83(1.57)**	70.68(1.57)	70.61(1.67)	55.48(3.93)
10	35.90(0.06)	**56.56(10.87)**	56.27(10.96)	56.34(10.79)	56.38(10.68)	52.77(4.31)
11	53.56(9.56)	**77.93(5.98)**	77.00(5.84)	77.70(5.82)	77.70(5.82)	68.44(4.94)
12	71.79(10.25)	80.50(2.70)	**80.58(2.62)**	80.50(2.68)	80.53(2.64)	65.51(9.23)
13	28.66(4.06)	75.39(2.67)	75.47(2.39)	**75.70(2.10)**	75.67(2.04)	67.93(6.67)
14	65.89(1.74)	70.96(1.86)	71.48(1.45)	71.74(1.66)	**71.80(1.69)**	55.51(2.45)
15	65.92(4.12)	68.87(3.11)	**71.54(2.65)**	70.12(2.91)	70.16(2.94)	39.33(2.36)
16	29.25(8.83)	49.29(3.45)	50.17(3.78)	50.56(3.88)	**50.61(3.94)**	39.53(4.64)
17	50.18(13.38)	64.91(10.13)	**64.98(10.24)**	64.71(9.44)	64.64(9.50)	61.18(7.23)
18	47.25(4.89)	**79.68(6.49)**	78.68(6.16)	79.47(6.14)	79.50(6.07)	61.32(4.81)
19	34.58(1.41)	69.92(7.71)	69.61(7.43)	69.83(7.15)	**69.98(7.05)**	50.49(6.16)
20	74.36(1.87)	**94.06(1.58)**	93.56(1.70)	93.96(1.60)	93.93(1.60)	83.47(3.40)

Here accuracy is reported after splitting into test and training sets using stratified tenfold splits. Table 2 show the accuracy obtained from selected classifiers.

On the reported accuracies, the Friedman test was used along with the Nemenyi post hoc statistic, shown in Table 3. The Friedman test reports average rankings of classifiers within 5 degrees of freedom, with $\chi^2 = 43.101$ and a p value = 0.00003525, well below $p < 0.05$ meaning there is statistical difference between the classifiers.

Using the Nemenyi post hoc statistic to find which classifiers are different (also Table 3), the results are not so encouraging. Only statistical differences are between FPC with means (arithmetic mean and OWA means) and the operators FPC_{min}, FPC_{max}, but, as these consider only a single feature in their decision this is as expected. The *andness* and normalized dispersion for the learned classifiers is seen in Fig. 2.

Looking at the individual data sets gives some further insight, the following observations have been made for the accuracy of the learned OWA classifiers:

- FPC_{ME} ranked last of the means-based classifiers. This is as expected, since each weight was not specified, but only the *andness*.

Table 3 Friedman with Nemenyi post hoc test. **Bold** indicates significant difference at $p < 0.05$

#	FPC_{min}	FPC_{mean}	FPC_{ME}	FPC_{GA}	FPC_{PS}	Mean rank
FPC_{min}	–	–	–	–	0-	4.60000
FPC_{mean}	**0.02182**	–	–	–	–	2.75000
FPC_{ME}	0.05919	0.99942	–	–	–	2.95000
FPC_{GA}	**0.03225**	1.00000	0.99994	–	–	2.82500
FPC_{PS}	**0.00323**	0.99407	0.94965	0.98459	-	2.42500
FPC_{max}	0.70448	**0.00007**	**0.00034**	**0.00013**	**0.00000**	5.45000

- FPC_{PS} performs best on the multi-class problems $\{1, 7, 16, 19\}$. The data set 14 is a case, where it is difficult to separate the classes due to high overlap and misinformation in the features (use of 0 as *not available* indicator), here FPC_{PS} performs best.
- FPC_{GA} was generally not a high scoring classifier, with only first rank on the data set 13, which is a data set with large number of features and only few training examples. From the accuracy results and the learned weights in Fig. 1, almost the same classifiers as FPC_{PS} were obtained. This is as expected, as the same fitness function was searched.

4.3 Learned OWA Operators

The last to see is the actual OWA weights learned for each data set. The mean weights have been graphically represented in Fig. 1. As one can notice, there is almost complete overlap of the GA and PS learned classifiers; both are equally good in searching appropriate weight values for the problem posed. The MEOWA-based classifier has, by its definition, a more linear curve than the searched OWA weights.

The mean *andness* of the learned OWA-based classifiers is also interesting, as it may indicate how different from the arithmetic mean operator the learned OWA weights are. Figure 2 show this graphically. As seen, the *andness* of the learned OWA weights are different for each data set, within the range $[0.31, 0.72]$.

The following observations have been made based on the mean learned *andness* level:

- Considering the data sets with the lowest learned *andness*, $\{2, 3, 9, 14, 15\}$, data sets 2 and 3 are both ranked number one in accuracy by the FPC_{min} classifier, which is as we would expect for all. However, these two data sets contain only 4 features each. When there are more features, the mean operators perform better, for data set 9 only little, while for data sets 14 and 15 the accuracy increase is within $[5.62, 5.91]$. In most cases the difference between the means-based classifiers is insignificant, except for data set 15 which is also the data set with the most features.

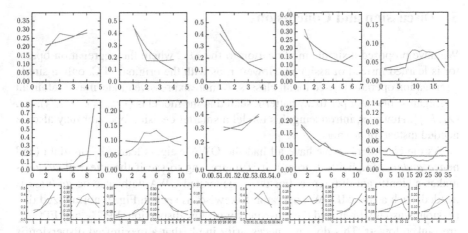

Fig. 1 Mean learned OWA operators for all data sets. x-axis is the weight positions, i, y-axis is the mean of the weights assigned for the given data set. Classifiers: ■ FPC_{MB}, ■ FPC_{GA}, ■ FPC_{PS} (Color figure online)

Fig. 2 Mean *andness* ■ and normalized dispersion ■ for each data set. Means are calculated using OWA weights from all trained classifiers FPC_{MB}, FPC_{PS}, and FPC_{GA} (Color figure online)

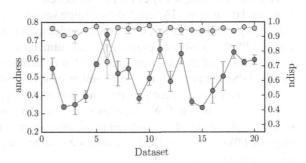

- Considering the data sets with the highest learned *andness*, $\{6, 11, 13, 18, 20\}$, data set 6 appears a curiosity. The learned OWA weights have low standard deviation, but the normalized dispersion is the lowest of all and with a high standard deviation. Looking at Fig. 1 it is observed that there is a large difference between the MEOWA and trained weights, which explains the curiosity. For three of the data sets the arithmetic mean-based OWA is ranked top, while the one by heuristically searched OWA, data set 13, all outrank the arithmetic mean as aggregation. Again this data set has the most features.
- Considering the data sets with the learned *andness* most near to 0.5 (arithmetic mean), $\{7, 8, 10, 12, 17\}$. As the difference between arithmetic mean and MEOWA is small in this group, results are expected to be very similar. Accuracy shows this, as all means-based classifiers have an accuracy ≤ 0.34 of each other.

5 Discussion and Conclusions

We have proposed a simple modification to the FPC where the aggregation operator is learned instead of assigned. However, as with the original FPC, only a single
aggregation operator is used for all classes. A more expressive model may contain an
aggregation operator per fuzzy pattern, such that the model is $\{(\lambda_1, F_1), (\lambda_2, F_2), \ldots,$
$(\lambda_k, F_k)\}$. However, introducing this model a smarter decision operator may also be
needed instead of $\arg\max$.

As the classifiers above have all had the OWA weights learned from data (and
hence *andness* implicitly learned), it is interesting to see if the effect of *andness*
makes a significant difference. For this a generated OWA with specific *andness* is
needed, such as the MEOWA operator. Likewise, as seen in Fig. 2, normalized dispersion is usually high, and the classifiers with low dispersion, FPC_{min} and FPC_{max}
are ranked lowest. This does not necessarily imply that a maximized dispersion is
best, and one focus area could be to produce aggregations with granular control of
dispersion covering the whole range $[0, 1]$.

The run-time complexity of the proposed heuristic search has not been discussed.
As noticed the PS and GA found almost same weights, so the one with least run-time
complexity could be chosen. Run-time complexity depends on the chosen configuration. For PS the number of evaluations and number of solutions $O(es)$ and the same
for GA, number of chromosomes and number of iterations $O(ci)$. Though practically,
we found PS to require a very small number of solutions, so it should be first choice.

Lastly, as the results are generally not too encouraging, it leads to a hypothesis
that constraints on the OWA weights could be a future direction. As an example
weights could searched and ordered in a monotonically increasing order before creating the OWA operator. Other examples could be judging type operators (such as
take minimum or maximum).

5.1 Conclusions

This paper has proposed the use of order-weighted averaging in a fuzzy pattern classifier. The order-weighted averaging has been introduced together with its main properties, and how to adapt it for classification. The fuzzy pattern classifier has been
introduced with an algorithm for adapting it to learned order-weighted averaging
weights. An experimental evaluation was included showing mixed results, but giving
the grounds for making recommendations on when to use the proposed classifier. To
summarize the findings, the proposed classifier is better suitable for problems with
larger number of features and pattern search the preferable search heuristic. Finally,
future research directions have been discussed.

References

1. C. Bachmann. "Multi-sensor data fusion for traffic speed and travel time estimation". MA thesis. University of Toronto, 2011.
2. S. F. Bocklisch. Process Analysis with Fuzzy Methods. Berlin Verlag Technik, 1987.
3. S. A. Davidsen, E. Sreedevi, and M. Padmavathamma. "Local and global genetic fuzzy pattern classifiers". In: Proc. Machine Learning and Data Mining MLDM. Hamburg, Germany: Springer Berlin Heidelberg, 2015.
4. D. Filev and R. R. Yager. "Analytic properties of maximum entropy OWA operators". In: Information Sciences 85.1–3 (1995), pp. 11–27.
5. D. Filev and R. R. Yager. "On the issue of obtaining OWA operator weights". In: Fuzzy Sets and Systems 94 (1998), pp. 157–169.
6. A. Ghosh, S. K. Meher, and B. U. Shankar. "A novel fuzzy classifier based on product aggregation operator". In: Pattern Recognition 41.3 (2008), pp. 961–971.
7. X. Liu. "A Review of the OWA Determination Methods: Classification and Some Extensions". In: Recent Developments in the Ordered Weighted Averaging Operators: Theory and Practice. Ed. by RonaldR. Yager, Janusz Kacprzyk, and Gleb Beliakov. Vol. 265. Studies in Fuzziness and Soft Computing. Springer Berlin Heidelberg, 2011, pp. 49–90.
8. S. K. Meher. "A new fuzzy supervised classification method based on aggregation operator". In: Proc. 3rd Int. IEEE Conf. on Signal-Image Technologies and Internet-Based Systems. 2008, pp. 876–882.
9. U. Mönks, H. L. Larsen, and V. Lohweg. "Aggregation operator based fuzzy pattern classifier design". In: Proc. Machine Learning in Real-Time Applications. 2009.
10. M. E. H. Pedersen. "Tuning and simplifying heuristical optimization". Ph.D. thesis. University of Southampton, UK, 2010.
11. R. Senge and E. Hüllemeier. "Top-down induction of fuzzy pattern trees". In: IEEE Transactions on Fuzzy Systems 19.2 (2011), pp. 241–252.
12. V. Torra. "On the learning of weights in some aggregation operators: the weighted mean and OWA operators". In: Mathware and Soft Computing 6 (1999), pp. 249–265.
13. D. Whitley. "A genetic algorithm tutorial". In: Statistics and Computing 4 (1994), pp. 65–85.
14. R. R. Yager. "On ordered weighted averaging aggregation operators in multicriteria decision making". In: IEEE J. on Systems, Man and Cybernetics 18 (1988), pp. 183–190.

Efficient Query Analysis and Performance Evaluation of the Nosql Data Store for BigData

Sangeeta Gupta and G. Narsimha

Abstract The voluminous amounts of data generated from the web applications and social networking and online auction sites are highly unstructured in nature. To store and analyze such data, traditional ways of using relational databases are not suitable. This yields path towards the acceptance of emerging nosql databases as an efficient means to deal with bigdata. This work presents an efficient nosql data store and proves its effectiveness by analyzing the results in terms of efficient querying by evaluating the performance estimation on read and write operations on simple and complex queries, also for storage and retrieval of increasing number of records. The results presented depict that the chosen nosql datastore—Cassansdra is efficient over the relational database—mysql and the other nosql databases—HBase and MongoDB, that leads to achieving cost saving benefits to any organization willing to use Nosql-Cassandra for managing Bigdata for heavy loads.

Keywords Mysql · Nosql · Bigdata · Hbase · Mongodb · Cassandra

1 Introduction

Enormous amounts of data flood across the internet and the storage capacities of the relational technologies have experienced inadequacy for the same. To store peta bytes of data, most of the organizations, particularly social networking sites and e-commerce sites are moving towards cloud to deploy their applications, but at increased security risks. This growing amounts of data which is too big and

S. Gupta (✉)
JNTUK, Kakinada, India
e-mail: ss4gupta13@gmail.com

G. Narsimha
CSE, JNTUHCEJ, Kondagattu, Karimnagar, India

© Springer Science+Business Media Singapore 2017   549
S.C. Satapathy et al. (eds.), *Proceedings of the First International Conference on Computational Intelligence and Informatics*, Advances in Intelligent Systems and Computing 507, DOI 10.1007/978-981-10-2471-9_53

complex to capture, store, process, and interpret is referred to as Bigdata. It is characterized by 4 Vs such as Volume, Velocity, Veracity and Variety [1]. The storage and analysis of such data can be made effective using the Nosql databases.

Cloud computing has evolved as a new computing paradigm, allowing end users to utilize the resources on a demand-driven basis, unlike grid and cluster computing which are the traditional approaches to access the resources. The foremost benefit of cloud is to pay only for the resources which users utilize. If there are an unexpected set of users bombarding for the resources, they would just have to pay for what they have been using. This usage is termed as elasticity of the cloud. Cloud provides a variety of service models such as Infrastructure as a service (IaaS), Platform as a Service (PaaS), Software as a Service (SaaS), Database as a Service (DaaS) and deployment models such as public, private, hybrid and community clouds. An application to be hosted on a scalable environment can use either of these models in a cost-efficient manner to reap their benefits. The other benefits provided by cloud can be utilized in terms of elasticity, scalability, efficiency, reusability [2].

Most of the modern world data are projected in the form of word documents, pdf files, audio and video formats and relational databases may not be suitable to serve such data. Also, using them for scalable applications impose heavy costs making them less attractive for deploying large-scale applications in cloud. An alternate approach is to use the emerging Nosql databases, which are not ACID compliant and which provide support to structured, unstructured, and semi-structured storage of massive data in terms of peta bytes. Nosql databases do not rely on a fixed schema, there are no join operations and they rely on CAP (Consistency, Availability, and Partitioning) features in contrast to the ACID properties supported by traditional databases. Eventual consistency is supported by a few Nosql types, where the updates may not propagate immediately across all the nodes in a cluster [3].

This paper presents a comparative analysis of various Nosql types such as Hbase, Mongodb, with Mysql and brings into light their limitations. It also presents a novel integrated Nosql, Cassandra which is advantageous and efficient over the mentioned Nosql species, by amalgamating their benefits and crossing out their shortcomings in terms of performance and scalability over the estimated read and write operations on the database for execution of simple and complex queries.

The paper is organized as follows: Sect. 2 presents background study of the work including the scenarios where the mentioned databases are being used; Sect. 3 presents a detailed comparative analysis with suggested solutions and integrated store development with performance estimation on read and write operations on simple and complex queries, The fourth section presents the results as obtained when inserting and retrieving records in multiples of hundred in both Mysql and Cassandra and shows the better performance of Cassandra over mysql. The fifth section concludes the work and throws light on the future enhancement.

2 Related Work

In this part of the section, several related works on mysql and nosql types are discussed and their limitations are observed.

Mysql has been used as a prominent relational database for storing data samples in a wide variety of applications. Naim et al. [4] have used mysql to store finger prints data for biometric, with the help of a virtual server. Tables created were person identification number, real end-points data and real branch-points data, which employ structured data storage. If the amount of information collected is drastically increased, this would require large number of tables to accommodate the growing data and also if the data storage is in form of text or image format rather than pixel data as in [4], then usage of mysql will become inappropriate.

Kulshreshta and Sachdeva [5] have compared the performance of mysql with DB4o database on sample hospital dataset and showed that object-oriented data-bases such as DB4o are always better as compared to relational databases such as mysql in terms of time taken to persist the data in the events of huge amounts of growing records. Though object-database deals well with respect to the huge data, they occupy large storage space.

The column-oriented data store HBase is a distributed database developed on top of Hadoop Distributed File System (HDFS), which adopts master–slave architecture with Name Node acting as a Master and Data Nodes acting as slaves. Vora [6] used the Nosql database HBase to perform random reads and writes on very large datasets in the form of image files and the results were proved to be better than using mysql on such data. Though the performance of HBase was shown to be better than mysql, the limitation of [5] presented the model to be appropriate to perform write-once read-many operations on the attributes, but not suitable to support multiple write operations, i.e., the files in HDFS were accessible efficiently in read mode but does not support multiple writes. Also, another limitation observed was, in order to use HBase, an in-depth understanding about Hadoop framework, MapReduce Programming model is required.

Zhao et al. [7] presented a comparison of Mongo DB, a document-based Nosql store with Mysql, highlighting the exceptional features of mongo DB like support for dynamic schemas, faster data integration, support for ad hoc queries, load balancing and automatic sharding and also depicted the support of mongo DB at relational calculus, achieving better performance than mysql. But the limitation of this approach is that there is no expertise in this area and no specialist tools are available to analyze data efficiently.

In this section, apart from identifying various areas of application of both Relational and Nosql databases, their limitations are brought into consideration and solutions are presented by using an integrated data store-Cassandra in the next section to overcome the mentioned limitations.

3   Cassandra—An Integrated Data Store

Cassandra is a novel integrated Nosql, used which aims at providing support to any kind of data (structured, semi-structured or unstructured) as emerging from the real-world social networking websites such as facebook, and e-commerce sites such as eBay concatenating the scalability aspects of BigData, leading to eventual consistency and providing an efficient way to solve complex queries by avoiding join operations. Cassandra is used to overcome the limitations possessed by the mentioned data stores and it integrates the benefits of Mysql, Hbase and MongoDB data stores. Hence, any modern-world application would be greatly benefitted in migrating their applications from Mysql to Cassandra. It uses peer-to-peer architecture, where all nodes are given equal priority in a cluster and the nodes are said to communicate with each other through gossip protocol. There is no single point of failure in Cassandra; hence there is no down time for running an application. The query language used to perform operation with the database is CQL (Cassandra Query Language).

3.1   Comparison of Nosql with Mysql

Relational Databases are confined to ACID properties in contrast to Nosql's CAP properties. Nosql databases support both strict and eventual consistency, in which changes need not propagate instantly across all the nodes in a cluster as compared to the relational databases, which provide support only to strict consistency. Nosql serves horizontal scalability aspect than the vertical scalability of relational model as in mysql [8].

   Table 1 depicts differences between relational and Nosql databases in terms of features like consistency, scalability, join operations and data formats.

Table 1 Comparison of relational with Nosql databases

	Supporting features			
Database	Consistency	Scalability	Data format	Joins
Relational (mysql)	Strict	Vertical	Structured	Complex tasks require joins
Nosql (Hbase, Mongodb, cassandra)	Strict/eventual	Horizontal	Structured, semi-structured, unstructured	No joins are to be performed

3.2 Querying Differences

Relational databases like mysql, oracle, etc., use SQL for storing, retrieving and manipulating data, whereas in nosql types, there is no single standard query language to meet varying users requirements. Querying data stored in nosql databases is specific to the data model. So, each nosql comes with its own query language like, Cassandra has CQL, HBase has HQL, etc.

To explain about the differences among mysql, hbase, Cassandra, we have considered sample tables titled journal and conference. The syntaxes used by various data stores vary as shown in Table 2 to perform insert, update and delete operations. We have also taken simple and complex queries to analyze these differences for data retrieval operation (select) as in Tables 3 and 4.

In the above example, the syntax for retrieving (reading) data from mysql and Cassandra are similar, while in mongoDB find() is used to retrieve the data, and in hbase, get is used for the same.

Table 2 Querying differences between mysql and nosql with insert, update and delete operations

Database	Insert operation	Update (write) operation	Delete operation
Mysql	Insert into journal values ('ieee',1234,'openaccess');	Update journal set jid = 1234 where jid = 1345;	Delete from journal where jname = 'mnuoo';
HBase	Put 'journal','row1','jid: a','ieee';	Same as insert	Disable 'journal';
MongoDB	Db.journal.insert ({jname:"ieee",jid:1234, accesstype:"openaccess"})	Db.journal.update ({}, {'$set':'jid':'jid'}});	Db.journal.remove ();
Cassandra	Insert into journal values ('ieee',1234,'openaccess');	Update journal set jid = 1234 where jid = 1345;	Update journal set jid = 1234 where jid = 1345

Table 3 Simple query: query to find the name of journal with id 1234

Database	Retrieval operation
Mysql	Select j.jname from journal j where j. jid = 1234;
HBase	Get 'journal','jname';
MongoDB	Db.journal.find({}, {"jname":1,"jid":0,"jtype":0});
Cassandra	Select jname from journal where jid = 1234;

Table 4 Complex query (joins/nested): query to find the journal names whose ids match with that of the ids in conference table

Database	Retrieval operation
Mysql	Select j.jname from journal j where jid in(select c.jid from conference c). Here jid in conference table is a foreign key
HBase	Get command can't be used to run the same query as in mysql, but if integrated with mapreduce code, the query will be executed
MongoDB	As in HBase, MongoDB also requires mapreduce command integration
Cassandra	Super column families and data denormalization can be done to execute complex queries in an efficient way

Cassandra Super column families and data denormalization can be done to execute complex queries in an efficient way.

To perform complex joins or nested queries, Mysql requires foreign keys to be created performing joins across multiple tables. But, this method may lead to increase in execution time in order to retrieve data from multiple tables, thereby degrading the overall performance. In HBase, complex joins are supported by integrating hbase code with mapreduce code using nested loops, which is again a time-consuming process. Mongo DB also uses mapreduce command to process such data.

In Cassandra, performing complex joins or nested queries requires denormalization of data into partitions, leading to efficient querying from a single replica node, rather than gathering the data from across the entire cluster. Thus, it provides an efficient mechanism to retrieve data in a simpler way, and also the speed of query execution is much better than in mysql, mongodb and hbase.

4 Result Evaluation for Unstructured Data Using Nosql and Mysql Data Store

This section presents the results of evaluating increasing number of records for both read and write operations using mysql and Cassandra. It also shows the better performance of Cassandra over mysql for write operations. However, the other nosql types have not been used due to time constraints.

4.1 Workload Generator

The workload is the key to performance benchmarking and stability analysis. One application is needed to generate continuous data for batch processing or high streaming real and live data. Web crawler is the chosen application, which generates data from various e-commerce sites, which is highly unstructured. The application will also generate the 'read' and 'write' requests to Nosql-Cassandra and Mysql databases, suitable for benchmarking.

4.2 Workload Executor

The Workload executor is run in two phases:

1. Load Phase ('Write' Phase)
2. Retrieve Phase ('Read' Phase)

The load phase workload working set is created from 100 records to 1 million records. These records are loaded to Cassandra and Mysql through JDBC connectivity. The client threads create multiple threads to load data in parallel in both Cassandra and Mysql databases. Increasing the number of threads can increase the throughput of the database.

The Retrieval phase works on data loaded in databases during the load phase. This phase generates some queries which read data from clusters. These queries can retrieve the small data set as well as large data sets with simple 'SELECT' to complex joint queries.

4.3 Statistics/Metrics Collection

The statistics are collected through logs by writing the application to database and dashboard. Timestamps are placed at regular intervals to monitor the performance and the results are recorded for load and retrieval phases with a varying number of records.

4.4 Performance Benchmarking

Benchmarking is referred to as the process of evaluating a system against some reference to determine the relative performance of the system. The basic primitive job of the database system remains generic, yet database systems exhibit different flavors and requirements based on the environment in which they operate. Also, database systems contribute hugely to the proper and efficient functioning of organizational and business information needs. Hence, selecting the right database with the right features is often a very critical decision.

4.4.1 Load Process

As part of Benchmarking, Bulk load was done ahead of each workload. Each database was allowed to perform non-durable 'writes' for this stage only to inject data as fast as possible.

Table 5 Write performance

Records (no. of inserts)	Cassandra WT (ms)	Records (no. of inserts)	Mysql WT (ms)
100	1	100	5
200	2	200	9
500	4	500	19
1,000	8	1,000	43
10,000	60	10,000	400
100,000	456	100,000	3,000

Fig. 1 Write performance for throughput versus response time

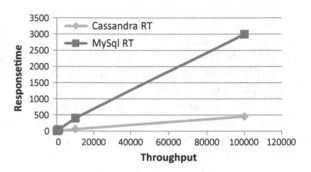

For low loads (100 records), 'write' requests showed a steady performance but with increasing load (increasing the number of records from 100 to 1,000, 10,000 and so on) with hundreds of requests on the same node, the performance scaled up and reached a maximum level at a certain peak point. Even though the throughput is distributed for low loads, it was observed to be stable for high loads.

The performance for 'writes' is similar in Mysql as in Cassandra, but more time is taken by Mysql showing that Cassandra performs much better 'writes' over Mysql.

The results were recorded as shown in Table 5 and graphically plotted as in Fig. 1.

4.4.2 Retrieval Process

During the retrieval phase, the time taken to retrieve the records increased drastically in Cassandra, while it was a gradual increase in Mysql with an increasing number of records (100, 1,000, 10,000, 100,000) on the same hardware configuration. Hence Mysql shows better results during the retrieval phase over Cassandra. The results for retrieval of records from both Cassandra and Mysql databases are as shown in Table 6 and Fig. 2.

Table 6 Read performance

Records (no. of retrievals)	Cassandra RT (ms)	Records (no. of retrievals)	Mysql RT (ms)
100	2	100	1
200	3	200	2
500	5	500	2
1,000	8	1,000	5
10,000	10	10,000	6
100,000	12	100,000	8

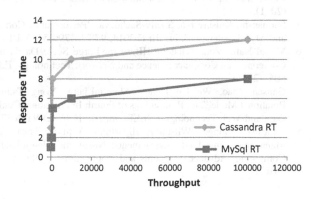

Fig. 2 Read performance for throughput versus response time

5 Conclusion and Future Work

Most of the organizations rely on structures databases like Mysql, which do not harness the requirements of scalability and availability of real-world data. The available set of Nosql databases support various aspects to meet the upcoming trends in growing data like support for eventual consistency, scalability, availability, and fault-tolerance. In this paper, Nosql databases Hbase and Mongodb are discussed and apart from mentioning their advantages as compared to Mysql, their limitations are also presented and solutions are suggested to overcome the limitations, heading towards the adoption of Integrated Nosql database-Cassandra. Modern world requirements in form of Bigdata can be efficiently analyzed and interpreted using this integrated nosql database with respect to query analyzation. Future work can be taken up to conduct more experiments on Cassandra which has performed far better compared to MySQL on write operations, but showed poor performance on read operations. Hence, showing improvement in the read operations with Cassandra making necessary modifications at the selection of appropriate algorithm, compression techniques can be taken up as the future work.

References

1. Venkat N Gudivada, Dhana Rao, Vijay V Raghavan,"Nosql systems for Big Data Management " in Proceedings of 10th World Congress on Services, IEEE, doi 10.1109/SERVICES.2014.42, pp:190–197(2014).
2. Thomas Sandholm, Dongman Lee, "Notes on Cloud Computing Principles", in: Sandholm and Lee Journal of Cloud Computing: Advances, Systems and applications, 3:21, Springer 2014.
3. Divyakant Agarwal, Sudipto Das, Amr EI Abbadi, "Bigdata and Cloud Computing: Current State and Future opportunities", in: EDBT 2011/ACM, March 22–24, Uppsala, Sweden (2011).
4. Nani Fadzlina Naim, Ahmad Ihsan Mohd Yassin, Wan Mohd Ameerul Wan Zamri, Suzi Seroja Sarnin, "Mysql Database for storage of fingerprint data", in: Proceedings of 13th International Conference on Modelling and Simulation, IEEE, doi:10.1109/UKSIM.2011.62, pp:293–298, (2011).
5. Sudhanshu Kulshreshta, Shelly Sachdeva, "Performance Comparison for Data Storage-DB4o and Mysql Databases", in: IEEE 2014, 978-1-4799-5173-4/14, (2014).
6. Mehul Nalin Vora, "Hadoop-HBase for Large Scale Data", in: Proceedings of International Conference on Computer Science and Network Technology, IEEE, December 24–26, pp: 601–605, (2011).
7. Gansen Zhao, Weichai Huang, ShunlinLiang, Yong Tang "Modelling MongoDB with Relational Model", in: Proceedings of Fourth International Conference on Emerging Intelligent Data and Web Technologies, IEEE, doi:10.1109/EIDWT.2013.25, pp:115–121 (2013).
8. Katarina Grolinger, Wilson A Higashino, Abhivav Tiwari and Miriam AM Capretz"Data Management in Cloud Environments: Nosql and Newsql data stores",in Journal of Cloud Computing: Advances, Systems and Applications, Volume 2, doi:10.1186/2192-113X-2-22, (2013).

Performance Evaluation of QoS-Aware Routing Protocols in Wireless Sensor Networks

Jayashree Agarkhed, Patil Yogita Dattatraya and Siddarama R. Patil

Abstract Wireless Sensor Networks (WSNs) consists of densely populated large number of spatially distributed configurable sensors, to meet the requirements of industrial, military, precision agriculture and health monitoring applications with ease of implementation and maintenance cost. Transmission of data requires both energy and quality of service (QoS) aware routing to ensure efficient use of the sensors and effective access of gathered information. In turn routing technique must provide reliable transmission of data without compromising QoS requirements of applications. We have addressed different routing protocol categories with range of QoS metrics to be achieved, to improve performance of WSNs applications.

Keywords Qos challenges · Qos-aware routing · Qos metric · Routing techniques · WSN issues

1 Introduction

WSN is an emerging technology that supports variety of applications to provide security, surveillance and environmental monitoring, health monitoring, etc. Sensor nodes have twofold capabilities. Primarily they are capable to sense variety of environmental conditions such as pressure, sound level, temperature, humidity variations, vibrations, etc. [1]. Secondly they also have limited computing capability.

J. Agarkhed (✉) · P.Y. Dattatraya · S.R. Patil
P.D.A College of Engineering, Gulbarga 585102, Karnataka, India
e-mail: jayashreeptl@yahoo.com

P.Y. Dattatraya
e-mail: agyogita@gmail.com

S.R. Patil
e-mail: pdapatil@gmail.com

© Springer Science+Business Media Singapore 2017   559
S.C. Satapathy et al. (eds.), *Proceedings of the First International Conference on Computational Intelligence and Informatics*, Advances in Intelligent Systems and Computing 507, DOI 10.1007/978-981-10-2471-9_54

WSNs have following unique characteristics which should be considered in the design of protocols to meet demanding requirements of applications.

• Availability of limited resources with capability constraints.
• Availability of battery with limited amount power and its difficulty of recharging.
• Interoperability problems due to heterogeneous nature of WSN.
• Huge initial deployment investment demand long lifetime of WSN.
• Handling of node failure, leads to difficultly of handling of topology change.

The stringent requirements of real-time applications include end-to-end delay, bandwidth availability and reduced packet losses. Dealing with real-time data demands both energy efficiency and Quality of Service (QoS) assurance to ensure efficient usage of WSNs resources and accuracy of the collected information. Communication protocols for WSNs must therefore be specially designed to efficiently operate under these constraints. In order to satisfy the energy constraints and the QoS requirements for the WSNs, clustering has been a widely used approach for organization of sensors into clusters [2]. QoS is concerned with communication quality that is characterized by noticeable communication delay, jitter introduced while transmission, insufficient bandwidth, and loss rate. Traditional metrics, however, cannot fully characterize the QoS in WSNs because of the diverse characteristics of WSN applications. Providing QoS support in wireless sensor networks is an emerging area of research. Due to various resource constraints like processing power, memory, bandwidth and power sources in sensor networks, QoS support in WSNs is a challenging task.

The rest of the paper has been organized as follows: Sect. 2 details about the previous related work. The Sect. 3 specifies challenges for QoS-aware routing design. Section 4 presents performance comparison. Section 5 concludes the paper.

2 Related Work

A good survey of routing techniques are well discussed in [3]. In this section along with classical routing techniques, protocols tailored for applications that needs certain QoS requirements to satisfy, are discussed.

A. Classical Routing Techniques

From a large body of related work in routing techniques in WSN, proposed protocols can be classified as shown in Fig. 1.

Figure 1 shows classification of routing protocols based on,

• **Network structure**: Classified as flat-based routing where all nodes play same role, in hierarchical routing, different nodes play different role as cluster head, non-cluster node, relay node, etc., and in location-based routing, data transmission require location information.

Fig. 1 Classical routing
technique

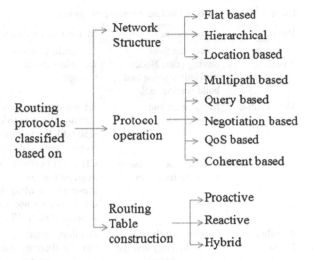

- **Protocol operation**: Routing protocols are classified as multipath-based routing, query-based routing, negotiation-based, QoS-based routing protocols, etc., depending on application requirements.
- **Routing table construction**: Protocols classified as proactive if routing path information has to be maintained prior to application data transmission requirement. This leads to maintenance of routing table at each node. Reactive protocols compute route on application demand while hybrid protocols combine features of both proactive and reactive.

Further classification can be, location based, data centric based, hierarchical based, mobility based, multipath based, QoS-based protocols [3]. These protocols have been differentiated by their unique characteristic and key features as shown in Table 1. Cluster-based and multipath-based routing protocols achieve maximum QoS requirements.

B. **QoS-Based Routing Techniques**

QoS-based routing protocols can be used according to application requirements satisfying different QoS parameters. QoS has multiple meanings and perspectives from different research and technical communities, point of view [4]. QoS defined as measurements of application reliability to achieve energy efficiency. An alternative definition of QoS is to support spatial resolution [5] and satisfying QoS parameters specific to the application, such as sensor node measurement, deployment type, Coverage method and number of sensor nodes active at given instant. QoS also refers to meeting application needs while efficiently using all network resources like bandwidth and power usage.

Table 1 Classification of different routing protocols

Protocol	Characteristics	Method to find route	Key features
Location based	Source discover routing path. Node performs sensing and build routing table	Use node location or incoming signal strength	GPS device used to find location
Data centric	Negotiation and resource adoption	Intermediate sensors forward the data to the sink which aggregates its own data and route	Controls the redundancy. Efficient utilization of bandwidth
Hierarchical	Provide efficient use of the scare resources	CH (CH) perform data aggregation and transmission while the other nodes sense and forward data to CH	Avoid collision, resource-based scheduling, uniform energy dissipation
Mobility based	Ensures that the sink is used efficiently and the responsibilities are distributed at an optimum level	Position, speed, moving direction and transmission range, is used to indicate a sensor node's location	Energy efficient, reliable, provide energy conservation
Multipath based [8, 9]	Node forwards its data in different routes through the network links	Perform path discovery considering traffic distribution and path maintenance	Support load balancing, congestion avoidance, reduce route inquiries and node failure and has lower routing overhead
QoS based	Provide service with minimum delay, high reliability and maximum fault tolerance	Use shortest path to increase nodes life. Use multipath algorithm or multiple sink to reduce end-to-end delay	Energy efficient, reliable and provide increased throughput

C. QoS Requirements

In [6] the author discusses about QoS requirements in WSNs and presents a survey of some of the QoS-aware routing techniques in WSNs. The middleware approach has been used for QoS support deployment of WSNs. In [7] the author discussed specific characteristics and constraints in wireless sensor networks listing QoS factors required for developing security applications for such resource constraint networks. The author has evaluated secure wireless sensor network satisfying QoS using Hawk nodes.

In [8] the author has focused on the performance analysis on QoS parameter and their analysis on the basis of their delay, throughput and data transfer rate, etc. In [9] the author presented different definitions of QoS in traditional networks with various parameters and new techniques. The author also provided introductory review of WSN and its unique characteristics such as severe resource constraints and reviews of QoS implementations in protocol layer stack of WSN.

In [10] the author presents a low interference energy-efficient multipath routing protocol (LIEMRO) to improve the QoS requirements of event-driven applications.

In [11] the author presents a comprehensive taxonomy on the existing multipath-based routing protocols designed for wireless sensor networks and discuss each protocol category. The author has discussed about operations of different protocols in detail with emphasis on their disadvantages and advantages. These routing techniques demand fulfilling QoS requirements considering various constraints in WSN as discussed in next section.

In [12] the author presented algorithm, InRout to select best route for achieving QoS requirements of network and applications considering limitations of sensor nodes. The basic idea used was to combine information at each node and multiple routes to satisfy QoS-based routing with low memory utilization and lower overhead. To select best route Q-learning has been used, that take into account network conditions and application settings. Cross-layer approach has been applied to make network and MAC layers to work together for reserving end-to-end bandwidth with low end-to-end delay.

3   Design Considerations for QoS-Aware Routing

Findings of extensive literature survey, motivates to design efficient routing protocols considering enormous issues in WSNs.

A. Issues in WSN

The routing protocols must consider various issues that need to be addressed in WSN are, achieving high bandwidth for information transmission, computing tasks with low energy consumption, satisfying user's demand for QoS provisioning, supporting node mobility, congestion control along with congestion detection and congestion mitigation, providing reliability measures, end-to-end packet recovery schemes, dealing with cache ACK/NACK, scalability support, synchronization, data caching, data aggregation to control traffic, reducing computational overhead, providing data security and availability along with support of authorized access and user authentication, confidentiality [12–14]. QoS-aware routing protocols must deal with various challenges due to unique characteristics of WSN as discussed in next section.

B. Challenges in QoS Provisioning

Characteristics of WSN lead to limitations of QoS.

- **Resource limitation**: Due to limited amount of bandwidth and memory, computing power and data transmission power, sensor nodes demand QoS methods to be energy efficient, less complex and light in computation.
- **Data redundancy**: Data redundancy improves reliability but leads to more energy consumption for information sensing, data transmission and processing

Table 2 QoS metrics for efficient routing protocol design

Performance metric	Definition	Formula	Abbreviation used	Desired value
Packet delivery ratio	P_{dr} is ratio of number of data packet delivered from source node to the destination node	$P_{dr} = \frac{1}{c}\sum_{f=0}^{c}\frac{R_f}{N_f}$ or $P_{dr} = \frac{\text{total packet received}}{\text{total packet transmitted}}$	P_{dr}: is the packet delivery ratio C: total no. of connections f: unique flow id serving as index R_f: count of packet received from f N_f: count of packet transmitted to f	More
Load fairness	Load fairness denoted as L_f is count of messages moved forward from sensors while routing	$Lf = \left[\sum_{i=1}^{n}(Mi)\right]^2 \div \left[n\sum_{i=1}^{n}(Mi)^2\right]$	Mi: no. of messages forwarded by node i n: no. of nodes involved in routing	High
Energy per message	Specified as energy required for transmitting n-bit message and energy required to receive n-bit message	$E_{tr}(n,d) = E_{ele} \times n + E_{amp} \times n \times d^2$ $E_{rec}(n) = E_{ele} \times n$	Etr: energy required to transmit n bits Erec: energy required to receive n bits Eele: dissipated energy by radio to run transmitter or receiver circuitry Eamp: required energy by transmit amplifier d: distance	Less
Expected transmission time	E_{tt} is time required to transmit the packet. It includes throughput for calculation	$E_{tt} = E_{tx} \times \frac{S}{B}$ $E_{tx} = \frac{1}{d_f \times d_r}$	E_{tx}: transmission count expected S: packet size B: bandwidth of link d_f: forward delivery ratio	Less

(continued)

Table 2 (continued)

Performance metric	Definition	Formula	Abbreviation used	Desired value
			d_r: reverse delivery ratio	
End-to-end delay	Is the time difference between time when packet was sent from source and time when packet arrive at destination	$D = \frac{1}{N}\sum_{i=1}^{s}(ri - si)$	D: end-to-end delay; N: no. of successfully received packets; r_i: packet i received time; s_i: packet i sent time; i: unique packet id	Low
Throughput	T is average number of packets successfully delivered to destination per unit time	$T = \frac{p}{t}$ Measured in bits/seconds	p: number of packets; t: time in seconds	High
Residual energy	Energy for sensing, communication and processing. Calculated at the end of simulation	$E_{rem} = E_0 - \sum_{i=1} E_i$	E_0: initial energy; E_i: energy of sensor node at ith transmission time	High
Packet dropping rate	Defined as number of data packets that are dropped during transmission from source node to sink	$P_{drop} = P_{tx} - P_{rx} \times E_{ele} \times n$	P_{drop}: Packet dropping rate; P_{tx}: total packet sent; P_{rx}: total packet received	Zero or Less

of data. Data redundancy can be decreased by aggregating data. This requires complex QoS-aware routing techniques to be designed.

- **Frequent topology change**: Topology change occurs due to node or link failure or due to node or sink mobility. Proactive protocol for networks with frequent topology change is not desirable. To support QoS, new technique must incur complex reactive method for routing.
- **Efficient energy utilization**: Sensor nodes can be scheduled to be in active or in sleep mode to improve battery life of each node. More the nodes in active mode, creates hole. To avoid holes, nodes are scheduled to sleep. This in turn causes increased end-to-end delay and minimized throughput resulting trade-off between some of the QoS parameters.

To overcome these certain issues, routing techniques are evaluated for better performance considering QoS metric as shown in Table 2. The QoS metric facilitates to design best routing protocols, sustaining node with maximum residual energy, use of less energy per message transmission, negligible packet drop and negligible end-to-end delay satisfying load fairness.

4 Performance Comparison

Routing protocol must satisfy maximum number of performance parameters for its low-cost implementation and maintenance. To analyze the performance of routing protocols, they have been compared using different QoS performance metrics as shown in Tables 3 and 4. Routing protocols in Table 4 are based on concepts of basic routing protocols mentioned in Table 3. The comparison provide guideline to use routing technique with specific network structure, protocol operation and route construction method to achieve better performance.

Table 3 Basic routing protocols

Protocol	Routing protocol type	Key features	End-to-end delay	Energy consumption	Overhead
AODV [25]	QoS based	Reactive	0.03299	Less, since it keeps track of only next hop for a routing instead of entire route	Least
DSDV [25]	QoS based	Proactive	0.01044	More since routing table updated periodically or immediately when topology change	Less
DSR [25]	QoS based	Reactive	0.01291	More since it maintains the whole path to destination in routing table, unlike AODV that keeps only next hop node information	More

Table 4 Routing protocol with different QoS requirements

Protocol	Routing protocol type	Key features	End-to-end delay	Energy consumption	Overhead
MR2 [15]	Multipath	Interference awareness, reactive	More	Less	More
HCMR [16]	Multipath	Reactive	Less	More	Least
SPIN [19]	Multipath	Reactive	More	Less	Least
3R [20]	Multipath	Reactive	Least	More	Less
LEACH [17]	Hierarchical	Proactive	More	Medium	Least
CBEEQR [18]	Hierarchical	Reactive	Less	Less	More
CBRMN [21]	Mobility based	Reactive	Less	Less	Less
MBC [22]	Mobility based	Reactive	Least	Least	Least
STAR [23]	QoS based	Proactive	Less	Less	Less
RIP [24]	QoS based	Proactive	Least	Least	Least

5 Conclusion

WSNs need to provide different levels of Quality of Services (QoS) based on different demands of various types of applications. Routing protocols are designed to satisfy different QoS parameters to provide better performance and to increase lifetime of network considering WSNs challenges and issues. The comparison highlights design of efficient routing protocols satisfying specific QoS metric for specific application using specific routing technique.

References

1. Patil Yogita Dattatraya, Jayashree Agarkhed, "A Review on Various Issues and Applications in Wireless Sensor Networks," *International Journal of Science and Research*, 2015, ISSN (Online): 2319–7064
2. Dali Wei et al., "Energy Efficient Clustering Algorithms for Wireless Sensor Networks," proceedings in 13th International workshop on computer aided modeling, analysis and design of communication links, ICC 2008.
3. K. Akkaya, M. Younis, " A survey on routing protocols for wireless sensor networks," *Elsevier Journal of Ad Hoc Networks*, vol.3 issue 3, 2005, pp. 325–349.
4. Chen and P.K. Varshney, "QoS Support in Wireless Sensor Networks: A Survey," Proc. Int'l Conf. Wireless Networks, CSREA Press, 2004, pp. 227–233.

5. Iyer, R. and L. Kleinrock, "QoS Control for Sensor Networks," IEEE International Comm. Conference (ICC 2003), Anchorage, AK, May 11–15.
6. Bhuyan, Bhaskar, et al. "Quality of Service (QoS) provisions in wireless sensor networks and related challenges." *Wireless Sensor Network* 2.11 (2010): 861.
7. Pazynyuk, Tamara, et al. "Qos as means of providing wsns security."*Networking, 2008. ICN 2008. Seventh International Conference on.* IEEE, 2008.
8. Ranjan, Nikhil, and Garima Krishna. "Wireless Sensor Network: Quality of Services Parameters and Analysis." *Proceedings in Conference on Advances in Communication and Control Systems-2013.* Atlantis Press, 2013.
9. Korkalainen, Marko, et al. "Survey of wireless sensor networks simulation tools for demanding applications." *Networking and Services, 2009. ICNS'09. Fifth International Conference on.* IEEE, 2009.
10. Radi, Marjan, et al. "Interference-aware multipath routing protocol for QoS improvement in event-driven wireless sensor networks." *Tsinghua Science & Technology* 16.5 (2011): 475–490.
11. Radi, Marjan, et al. "Multipath routing in wireless sensor networks: survey and research challenges." *Sensors* 12.1, 2012, 650–685.
12. Villaverde, Berta Carballido, Susan Rea, and Dirk Pesch. "InRout–A QoS aware route selection algorithm for industrial wireless sensor networks." *Ad Hoc Networks* 10.3, 2012, 458–478.
13. Preeti Sharma, "A Review of Attacks on Wireless Sensor Networks," Information Systems and Communication, Vol. 3, pp. 251–255, 2012S9
14. Gaurav Sharma, Suman Bala, Anil K. Verma, "Security Frameworks for Wireless Sensor Networks-Review," Proceedings in 2nd International Conference on Communication, Computing &., vol 6, 2012.
15. MoufidaMaimour, "Maximally Radio-Disjoint Multipath Routing for Wireless Multimedia Sensor Networks," *Proceedings in 4th ACM workshop on Wireless multimedia networking and performance modeling,* 2008, pp. 26–31.
16. Jayashree Agarkhed, G. S. Biradar, V. D. Mytri, "Energy Efficient Interference Aware Multipath Routing Protocol in WMSN," *Proceedings in India Conference (INDICON),* IEEE 2012, pp. 1–4.
17. W. Heinzelman, A.Chandrakasan and H.Balakrishnan, "An application-specific protocol architecture for wireless microsensor networks," IEEE *Transaction on Wireless Communications,* 2002, vol. 1, no. 4, pp. 660–670.
18. Jayashree Agarkhed, G. S. Biradar, V. D. Mytri, "Cluster Based Energy Efficient QoS Routing in Multi-Sink Wireless Multimedia Sensor Networks," *Electronics and applications (ICIEA),* IEEE 2012, pp. 731 – 736.
19. Kulik et al., "Negotiation-based protocols for disseminating information in wireless sensor networks," *Wireless Networks,* 8(2/3), 2002, pp. 169–185.
20. M. Krogmann et. al., "Reliable, real-time routing in Wireless Sensor and actuator Networks", ISRN Commun. Netw, 2011, pp. 8, http://dx.doi.org/10.5402/2011/943504 (Article ID 943504).
21. Samer et al., "Cluster based routing protocol for mobile nodes in wireless sensor network," *3rd International Conference on Quality of service in Heterogeneous Wired/Wireless Networks, Wareloo, Ontario, Canada,* Aug. 2006, pp. 233–241
22. S. Deng, J. Li L. Shen, "Mobility-based Clustering Protocol for Wireless Sensor Networks with mobile nodes," *IET Wireless Sensor Systems,* Jan. 2011, ISSN 2043-6386.
23. Sahabul Alam, Debashis De, Anindita Ray, "Analysis of Energy Consumption for IARP, RIP and STAR Routing Protocols in Wireless Sensor Networks," *Proceedings in 2nd International conference on Advances in Computing and Communication Engineering (ICACCE),* 2015, IEEE, pp. 11–16.

24. Meena Kaushal et al., " Performance Analysis of DSR,DYM,OLSR and RIP Protocols of Manet using CBR and VBR Transmission Traffic Mode," *International Journal of Current Engineering and Technology*, vol.3, issue 4, Oct. 2013, pp. 1541–1548.
25. Rajeshkumar, V., & Sivakumar, P. (2013). Comparative Study of AODV, DSDV and DSR Routing Protocols in MANET Using Network Simulator-2. *International Journal of Advanced Research in Computer and Communication Engineering*, 2, 12.

Agent-Based QoS Routing for Intrusion Detection of Sinkhole Attack in Clustered Wireless Sensor Networks

Gauri Kalnoor, Jayashree Agarkhed and Siddarama R. Patil

Abstract Secured transmission of data is the most challenging and critical issue in Wireless Sensor networks (WSN). QoS parameters such as energy efficiency, power consumption, end-to-end delay and so on plays an important role for system performance. Clustering approach is an effective way that can enhance system and network performance. In Clustered Wireless Sensor Network (CWSN), any type of attacks that are malicious and harmful to the network can occur and security of CWSN is affected. One of the major attack that may occur in CWSN is sinkhole attack that needs to be detected. An agent-based protocol is used to detect or prevent such type of attack. This can help improve the performance of the network that includes QoS parameters. An intrusion detection system (IDS) is configured using agent-based protocol to detect if an intrusion occurs.

Keywords Energy efficiency · Intruder attacks · Intrusion detection system · Qos · WSN · Multipath routing

1 Introduction

WSN [1] consists of spatially distributed, tiny and low-powered sensor nodes that are autonomous in nature. These sensors monitors environmental conditions mainly temperature, sound and pressure. The most important applications [2, 3] of WSN needs the ability to operate in an unattended harsh environments where monitoring the entire network cannot be easily scheduled. Military applications such as battlefield surveillance are motivated by the development of WSN. Nowadays such

G. Kalnoor (✉) · J. Agarkhed · S.R. Patil
P.D.A College of Engineering, Gulbarga 585102, Karnataka, India
e-mail: kalnoor.gauri@gmail.com

J. Agarkhed
e-mail: jayashreeptl@yahoo.com

S.R. Patil
e-mail: pdapatil@gmail.com

© Springer Science+Business Media Singapore 2017
S.C. Satapathy et al. (eds.), *Proceedings of the First International Conference on Computational Intelligence and Informatics*, Advances in Intelligent Systems and Computing 507, DOI 10.1007/978-981-10-2471-9_55

571

networks are used in industrial applications, machine health monitoring and so on. In WSN, clustering approach is used to improve performance of the network. Clustered WSN is characterized by aggregation of data by electing cluster head in each clustered network and reduces the cost of the network. Two most important methods in WSN for clustering are: selection of cluster head (CH) periodically and assigning one or multiple clusters for each node. Sensor nodes [4] are grouped together into clusters to achieve scalability and high energy efficiency. High energy consumption occurs when the data is to be sent to higher distance within the network with lot of time spent on transmitting the data. Intrusion is an activity which affects the QoS of the network that compromises with the integrity and energy efficiency of WSN. It provides security to the network by identifying threats by following the intruders [5]. The nodes monitor the neighbor nodes and collaborative sharing of data leads to successive detection of sinkhole attack.

2   Related Work

Exclusive literature survey is done that explains the requirements of QoS in WSN intrusion detection.

2.1   Data Mining Techniques for Intrusion Detection in CWSN

The authors in [6] proposed the hybrid IDS which is lightweight and distributed. In WSN, this type of IDS is considered to be the second line of defense. The central agent in IDS is discussed that performs detection of an intruder accurately using data mining techniques [7] such as Classification and Regression Tree (CART), C4.5.

2.2   Integrated IDS for CWSN

The authors in [8] showed that the CH and the sink are mostly attacked by the enemies, and this makes security necessary. In CWSN, due to different probabilities of attacks, individual IDS are designed for sink, sensor nodes and CHs. The Intelligent Hybrid IDS (IHIDS) is proposed in the CH, and it decreases the energy consumption [9] and lifetime of CWSN is prolonged. The main aim of CH is to detect attacks and avoid wasting of resources.

2.3　Decentralized IDS in CWSN

In [10], the authors discussed different types of attacks, as WSN are deployed in unprotected and open environment. Few measures for prevention are used to protect against such types of attacks. But for few attacks [11], the preventive measures cannot be designed, for ex, wormhole attack.

2.4　Rule-Based IDS for Selection of Attribute in CWSN

The authors in [12] have proposed a new rule-based algorithm for attribute selection for detecting intruders and different types of attacks in WSN. The algorithm used to achieve this goal is SVM-based enhanced multiclass algorithm. The proposed method achieves high detection accuracy and false alarm rate is reduced. The authors in [13] have presented a mechanism called Statistical En-route Filtering (SEF) which helps to detect false reports and drops them.

3　Detection of an Intruder Against Sinkhole Attack

In this section, we discuss the sinkhole attack and explain how an intruder can be detected in WSN whenever a sinkhole attack occurs.

3.1　Sinkhole Attack

In a Sinkhole attack [14, 15], the main aim of an adversary is to create a network traffic at particular area, through the node that is compromised, and then a sinkhole is created at the center with an adversary. Figure 1 shows the sinkhole attack where a node is compromised and attracts the other nodes in the surrounding region.

3.2　IDS Against Sinkhole Attack

The IDS is designed such that the sinkhole attack is detected based on the secure routing protocols based on clustering. By reducing the cost of communication and energy consumption is optimized in this type of architecture. This also minimizes the number of nodes that leads to increase in network life. To detect a sinkhole attack, the signature-based IDS model is designed. This technique is used because of the mobility of the sink throughout the WSN. In CWSN, an IDS agent is

Fig. 1 Sinkhole attack

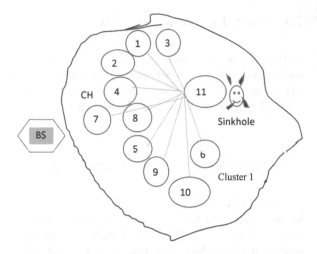

considered to be the node at a leader level. In each cluster, an agent is elected in IDS such that the attack can be detected with high accuracy. Here, the system becomes active only whenever important or abnormal event occurs in the clustered network. This helps to avoid wastage of energy. The two phases of the system are discussed below.

3.2.1 Initialization Phase

In this phase, the initialization of the node takes place in two steps.

Step 1: The construction of rule for detection uses signatures or patterns by the local or global agent in the system. Once the data are collected, the detection rate is calculated in two different levels: at sink level, the detection rate of the cluster is initialized to zero and once the data packets transmitted are incremented, the detection frequency is initialized. At the agent level, the IDS agent is designed when the local detection rate is initialized in each cluster, incrementing the data transmitted to the sink.

Step 2: Mobility of a sink: The maximum detection rate is calculated by the sink and then coordinates with the new position of the sink. The sink advertises the new position to all the IDS agents.

3.2.2 Detection Phase

In this phase, the detection of intruder takes place once the advertised message is received to the node outside the area of sink movement. In this case, each agent receives an advertisement from the sink with a particular time period. If an advertisement is received outside this time period, a false advertisement is considered and a detection alarm will be sent to the IDS agent in CWSN. The advertisement is verified as follows using Eq. 1.

$$\text{Verification Agent Rate } (VAR) = \frac{\text{Send} - \text{time}}{\text{Time} - \text{Period}} * 100 \qquad (1)$$

$$\text{If } VAR = \text{mobility} - \text{check}$$

$$\text{True Advertisement}$$

$$\text{else}$$

$$\text{False Advertisement}$$

Here, mobility-check represents the number of moves made by the sink within the time period. Send-time is the an advertisement is sent.

4 Agent-Based QoS Routing Protocol

In this section, the routing protocol with agent based on QoS metrics is discussed improving the performance of the network. In the agent-based routing model, the agent ID, type of agent, source and destination node ID, agent hop distance, start time, end time and current node ID are the data structure used [16]. In the network design, the QoS metrics such as delay, bandwidth, loss of packet, energy consumed are defined. The model is proposed with forward agent and reverse agent in the updation of a routing table takes place. The behavior of the IDS agent is shown in Fig. 2.

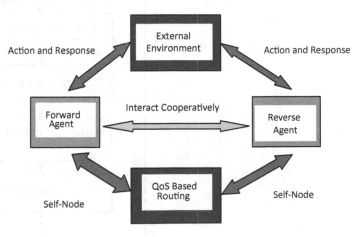

Fig. 2 IDS agent abstraction

In CWSN, CH acts as the gateway between different clusters and the each sensor node acts as the router. This model uses First in First out (FIFO) scheduling algorithm. Based on this algorithm, the sinkhole attack is detected in the clustered-based WSN. The routing table is stored in every router and discovers the path of packet transmission. In CWSN [17], clustering is performed by the sensor nodes, the election procedure is held to elect a CH for each cluster and then the intrusion is detected to provide security for the network. The clustering formation is explained in Fig. 3.

The clustering is formed using key generation algorithm in which encryption and decryption of data are considered to prevent the intruder attack, especially the sinkhole attack. Initially, the network with many number of sensor nodes and other nodes are classified in number of small or large clusters. The key generation technique can be proposed in the cluster formation model of WSN. For every cluster formed in the network, the CH is elected by using election method so that all the nodes in the cluster get the chance of being elected as a CH. Initially, the network with many number of sensor nodes and other nodes are classified in number of small or large clusters. The key generation technique can be proposed in the cluster formation model of WSN. For every cluster formed in the network, the CH is elected by using election method so that all the nodes in the cluster get the chance of being elected as a CH. The mobility of the sink [18] is determined to

Fig. 3 Clustering in WSN

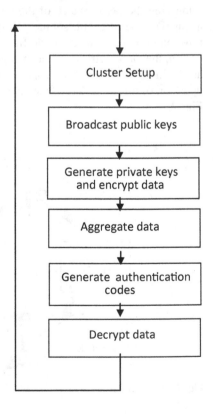

detect if a sinkhole attack has occurred throughout the network. The forward and reverse IDS agent acts as the router between the communicating nodes. Thus, the detection rate and accuracy increases as the network is secured by two lines of defense. The QoS parameters such as energy conserved, power conserved are less compared to other types of model. Finally, an efficient path is found to reach the destination without compromising the QoS of the network.

In the proposed algorithm, each path is recorded in the routing table from source to a particular node and also whenever new nodes are added to vertex list V. We have considered the algorithm such that, the loops are avoided and efficiency of the network is obtained.

Figure 4 shows the design of agent-based QoS routing protocol for detecting an intruder when sinkhole attack occurs.

The algorithm applied for agent-based routing is discussed below:

Agent Based Routing Algorithm:
Step 1: *Start*
Step 2: *Clustering set up and elect CH.*
Step 3: *V: Set of nodes in the sensor network.*
   s: Source node, D: Destination node.
   S: Set of nodes in the attacked area
   R: Set of root nodes, R ∈ ∅
Step 4: *for each V ∈ S,*
    Check the correct Route
     if current root = best root
      IDS agent forwards the packet
    else
      detect if sinkhole attack is occured
Step 5: *Calculate the cost of the path discovered with*
   Detection rate and FAR.
Step 6: *Drop packets, if an intruder is detected by IDS agent.*
Step 7: *Calculate the number of packets dropped in the network.*
Step 7: *For each optimised path , update the link cost*
   in the routing table along with S and D.
Step 8: *End.*

In the algorithm, every path from source to a particular node is recorded and new elements are added to vertex list V. We have considered the algorithm such that, the loops are avoided and efficiency of the network is obtained.

Fig. 4 Flowchart for
agent-based routing protocol

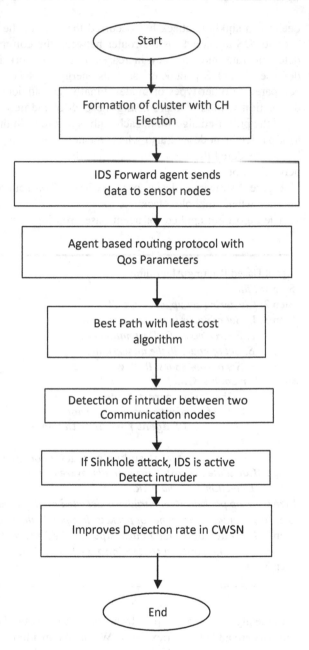

5 Performance Results and Discussions

The performance of CWSN is analyzed and the results obtained are discussed.

5.1 Performance Analysis

In this section, we discuss the sensor network performance once the proposed model is used for detecting intruder.

The performance of the network is observed based on the QoS parameters for IDS model proposed. Most of the parameters are considered for improving the QoS of CWSN.

Based on the applications of WSN, the requirements of QoS is discussed as below:

1. **Density of the network**: It is the analysis of behavior of proposed IDS model when large set of sensor nodes are added and the networks grow to a large extent. The density D is calculated using Eq. 2.

$$D(R) = \frac{(N\pi R^2)}{A},$$ (2)

where

   N   is the number of nodes in CWSN
   R   is the range of transmission
   A   is the region in which N sensor nodes are scattered.

2. **Load of the Network (L)**: It is the total number of events generated by the nodes over the network per unit of time. This impacts on consumption of energy and bandwidth. At time t, the load on network L is determined using Eq. 3.

$$L(t) = N_{event} + N_{attack},$$ (3)

where

   N_{event}:   The number of new events generated in the network,
   N_{attack}:   The number of sinkhole attacks generated

3. **Energy Consumption (E)**: Because of congestion in the network and overhead of communication, the energy consumption becomes an important QoS parameter. E is defined as the total sum of energy consumed to transmit data (E_T), energy consumed to receive data (E_R), energy for carrier sensing (Ec) and energy in sleep mode (Es). The energy consumption of the network is carried out using Eq. 4.

$$E = E_T + E_R + Ec + Es \tag{4}$$

4. **Detection Rate of Sinkhole attack**: It represents the rate of detection when sinkhole attack is generated. Sinkhole detection rate (SD_R) is calculated using Eq. 5.

$$SD_R = \frac{N_D}{N_G} \times 100 \tag{5}$$

N_D: Number of sinkhole attacks detected
N_G: Number of sinkhole attacks generated and injected in the sensor network

5. **Efficiency (EF):** It determines the time needed by the IDS agent using agent-based routing protocol to detect the first occurrence of sinkhole attack. It is calculated by Eq. 6.

$$EF = DT - OT, \tag{6}$$

where

DT   is the time taken to detect first sinkhole attack
OT   is the time for performing the sinkhole attack

6. **Average Efficiency** (Av_{EF}) is the time needed for IDS agents to detect all generated sinkhole attacks in the sensor network. It is calculated using Eq. 7.

$$Av_{EF} = \frac{\sum_{i=1}^{n} EF_n}{n}, \tag{7}$$

where n is the number of sinkhole attacks occurred in the network.

5.2   Results

- The number of sensor nodes is increased with IDS, the energy consumption is analyzed. The energy consumption increases as the density increases but with acceptable rate. Once the density increases, the number of neighboring nodes for each sensor nodes also increases with increase in cost of communication. But, IDS designed in the network does not add on to the energy consumption as in our model, IDS is active only whenever an event occurs.
- By increasing the number of detection events, sinkhole attacks are not generated. Sensor nodes are selected randomly when a new event occurs. The energy consumption increases with increase in traffic load. When load of traffic is high, many nodes are involved in the process of data dissemination and some IDS become active to check the sink's advertisements for detecting sinkhole attacks eventually.

- If the number of sinkhole attacks and number of data generated events are fixed, the rate of detection is high but increases the density of the network.
- In the proposed IDS, the efficiency is analyzed. Here, minimum time is required to detect the first sinkhole attack, and with increase in sinkhole attacks, the network is not impacted with efficiency.

The performance results are discussed in Table 1.

Table 1 Performance results

QoS Parameters	Performance Results			Sinkhole attack Detection rate
Network Density	As network density increases, the energy consumed also increases.			96.09%
	Network Density(y-axis)	Energy(x-axis)		
	0,79	10 mw/h		
	1,57	20 mw/h		
Traffic or network load without IDS	As the network r-load increases, the energy consumption also increases.			60.15%
	Network Load (y-axis)	Energy (x-axis)		
	20	48 mw/h		
	25	55 mw/h		
Traffic or network load with IDS	With increase in network load, energy consumed increases but very less variation when IDS is used.			94.05%
	Network Load (y-axis)	Energy (x-axis)		
	20	46 mw/h		
	25	48 mw/h		
Efficiency	With increase in density, efficiency is increase with IDs, but if IDS is not applied, the efficiency decreases.			98%
	Density (y-axis)	Efficiency (x-axis)		
	2,36	0,7		
	1,57	0,7		

6 Conclusion

The performance of sensor network is increased by taking into consideration the QoS parameters when a number of sinkhole attacks are detected. The agent-based routing protocol is designed with forward and reverse IDS agents running in the network. The clustered WSN requires high security and the intruders need to be detected along the route of data transmission. The false advertisements from the sink are detected and hence the detection rate is increased by improving the performance of the network.

References

1. Ahmed, A. S., & Rajeswari, A. (2012, April). Intrusion detection in heterogeneous Wireless Sensor Networks with an energy efficient localization algorithm. In *Recent Trends In Information Technology (ICRTIT), 2012 International Conference on* (pp. 389–394). IEEE.
2. Rassam, M. A., Maarof, M. A., & Zainal, A. (2012). A survey of intrusion detection schemes in wireless sensor networks. *American Journal of Applied Sciences, 9*(10), 1636.
3. Alrajeh, N. A., Khan, S., & Shams, B. (2013). Intrusion detection systems in wireless sensor networks: a review. *International Journal of Distributed Sensor Networks, 2013*.
4. Shen, W., Han, G., Shu, L., Rodrigues, J. J., & Chilamkurti, N. (2011). A new energy prediction approach for intrusion detection in cluster-based wireless sensor networks. In *Green Communications and Networking* (pp. 1–12). Springer Berlin Heidelberg.
5. Goyal, D., & Tripathy, M. R. (2012, January). Routing protocols in wireless sensor networks: a survey. In *Advanced Computing & Communication Technologies (ACCT), 2012 Second International Conference on* (pp. 474–480). IEEE.
6. He, D., Chen, C., Chan, S. C., Bu, J., & Yang, L. T. (2013). Security analysis and improvement of a secure and distributed reprogramming protocol for wireless sensor networks. *Industrial Electronics, IEEE Transactions on, 60*(11), 5348–5354.
7. Coppolino, L., D'Antonio, S., Garofalo, A., & Romano, L. (2013, October). Applying data mining techniques to intrusion detection in wireless sensor networks. In *P2P, Parallel, Grid, Cloud and Internet Computing (3PGCIC), 2013 Eighth International Conference on* (pp. 247–254). IEEE.
8. Yan, K. Q., Wang, S. C., & Liu, C. W. (2009, March). A hybrid intrusion detection system of cluster-based wireless sensor networks. In *Proceedings of the International MultiConference of Engineers and Computer Scientists* (Vol. 1, pp. 18–20).
9. Wang, S. S., Yan, K. Q., Wang, S. C., & Liu, C. W. (2011). An integrated intrusion detection system for cluster-based wireless sensor networks. *Expert Systems with Applications, 38*(12), 15234–15243.
10. Boubiche, D. E., & Bilami, A. (2012). Cross layer intrusion detection system for wireless sensor network. *International Journal of Network Security & Its Applications, 4*(2), 35.
11. Mohamed Mubarak, T., Sattar, S. A., Rao, G. A., & Sajitha, M. (2011, March). Intrusion detection: An energy efficient approach in heterogeneous WSN. In *Emerging Trends in Electrical and Computer Technology (ICETECT), 2011 International Conference on* (pp. 1092–1096). IEEE.
12. Anand, K., Ganapathy, S., Kulothungan, K., Yogesh, P., & Kannan, A. (2012). A rule based approach for attribute selection and intrusion detection in wireless sensor networks. *Procedia Engineering, 38*, 1658–1664.

13. Shamshirband, S., Amini, A., Anuar, N. B., Kiah, M. L. M., Teh, Y. W., & Furnell, S. (2014). D-FICCA: A density-based fuzzy imperialist competitive clustering algorithm for intrusion detection in wireless sensor networks. *Measurement, 55*, 212–226.
14. Sheela, D., Naveen, K. C., & Mahadevan, G. (2011, June). A non-cryptographic method of sink hole attack detection in wireless sensor networks. In *Recent Trends in Information Technology (ICRTIT), 2011 International Conference on* (pp. 527–532). IEEE.
15. Sharmila, S., & Umamaheswari, G. (2011, July). Detection of sinkhole attack in wireless sensor networks using message digest algorithms. In *Process Automation, Control and Computing (PACC), 2011 International Conference on* (pp. 1–6). IEEE.
16. Chen, S., & Nahrstedt, K. (2013). Distributed quality-of-service routing in ad hoc networks. *Selected Areas in Communications, IEEE Journal on, 17*(8), 1488–1505.
17. Sivakumar, R., Sinha, P., & Bharghavan, V. (2014). CEDAR: a core-extraction distributed ad hoc routing algorithm. *Selected Areas in Communications, IEEE Journal on, 17*(8), 1454–1465.
18. Lee, W. C., Hluchyi, M. G., & Humblet, P. A. (2014). Routing subject to quality of service constraints in integrated communication networks. *Network, IEEE, 9*(4), 46–55.

An Efficient Probability of Detection Model for Wireless Sensor Networks

Seetaiah Kilaru, S. Lakshmanachari, P. Krishna Kishore, B. Surendra and T. Vishnuvardhan

Abstract In the wireless sensor networks, probability of detection is a challenging task. There are various factors like coverage area, node density and efficiency of the sensors affecting the performance. In this paper, we proposed efficient intruder detection algorithms in a wireless sensor network. We developed a mathematical expression for the sensing range, node density and other required parameters and it is useful for any designer to predict the detection probability. We used homogeneous network environment with uniform distribution of sensor nodes to find the detection probability. We evaluated the proposed detection probability methods in both single-sensing and multi-sensing scenarios and observed that the results are satisfactory.

Keywords Wireless sensor · Intruder detection · Single sensing · Node density · Network topology · Degree of coverage

S. Kilaru (✉)
BVRIT Hyderabad College of Engineering for Women, Hyderabad, India
e-mail: dr.seetaiah@gmail.com

S. Lakshmanachari · P.K. Kishore · B. Surendra · T. Vishnuvardhan
Institute of Aeronautical Engineering, Hyderabad, India
e-mail: lakshmansiddi@gmail.com

P.K. Kishore
e-mail: Krishna.boinapalli@gmail.com

B. Surendra
e-mail: suri1253@gmail.com

T. Vishnuvardhan
e-mail: vishnuvardhanreddytalla@gmail.com

© Springer Science+Business Media Singapore 2017
S.C. Satapathy et al. (eds.), *Proceedings of the First International Conference on Computational Intelligence and Informatics*, Advances in Intelligent Systems and Computing 507, DOI 10.1007/978-981-10-2471-9_56

1 Introduction

Technical advancements in wireless communication system design and advancements in microprocessor technology created sensor nodes. The communication between these nodes made possible to design wireless sensor network. Now, the current research is focusing on intelligent wireless sensor networks (WSN). The function of these intelligent sensor networks includes monitoring, controlling and surveillance. In this method, each sensor senses the information based on its physical design and forwards the collected data to the assigned user terminal. This paper focused on the methodology to find the intruder in the defined area using surveillance sensor nodes. For this application, traditional characterization of WSN is not suitable and hence there is a need to configure in customized application. We considered homogeneous network to place the sensors by considering various parameters like sensing range, density of the sensors and field of interest. Any unauthorized person can identified using the single-sensor node or multi-sensor nodes. Several works on this issue were carried previously, but there exists a problem with respect to coverage. In [1–3] the authors proposed an efficient way to use WSN to track and monitor for surveillance applications. Some works were focused on finding the expression to quantify the coverage enhancing the deployment quality [4, 5] and they related coverage analysis with quality of service. To mitigate the problems in coverage problems, several algorithms were proposed to reduce the energy consumption in WSN [6, 7].

2 System Model and Topology

2.1 Basic Information

If we consider any sensor, the sensing range determines the range of the sensor which can effectively collects the data. For the node N_i which is centered at ξ_i, the sensor range is defined as

$$R_{\text{SENS}} = \left\{ \xi_j \in \mathfrak{R}^2 : \left| \xi_i - \xi_j \right| \leq R_{\text{SENS}} \right\}$$

Every sensor has transmission range based on its physical features and we defined the possible transmission range as

$$R_{\text{TRANS}} = \left\{ \xi_j \in \mathfrak{R}^2 : \left| \xi_i - \xi_j \right| \leq R_{\text{TRANS}} \right\}$$

In the above equation the distance between two sensors were represented as $\xi_i - \xi_j$ and is called as Euclidian distance [8, 9].

Let us consider two nodes N_i and N_j and are located at ξ_i and ξ_j, respectively. The distance between these two nodes is represented as d_{ij}. The two nodes can collaborate with each other and is represented as S_{N_i} and S_{N_j} with respect to node 1 and node 2, respectively. The two nodes are collaborative with each other only when

$$d_{ij} = \left| \xi_i - \xi_j \right| \leq 2R_{TRANS}$$

The set can be represented as $S_{col}(N_i)$ and is represented as

$$S_{col}(N_i) = \bigcup_{\left\{ N_j: \left| \xi_i - \xi_j \right| \leq 2R_{TRANS} \right\}} S_{N_j}$$

To maintain a continuous stream of information from the sensors, we should keep sensors with intersecting area. We also called it as overlapped (ol) area and it is equivalent to intersection between S_{N_i} and S_{N_j}

$$S_{ol} = S_{N_i} \cap S_{N_j}$$

Figure 1 represents $S_{col}(N_i)$ and S_{ol}. Each node has many neighbor nodes and we can represent these nodes in mathematical expression as

$$N_i = \left\{ N_j \in G: \left| \xi_i - \xi_j \right| \leq 2R_{TRANS} \right\}$$

2.2 Network Topology

In this paper, we assumed that there are N nodes distributed over a region A. The region has edge length L and we assumed uniform distribution of users. The two neighbor nodes are connected with each other if the distance between them is less than the Euclidean distance. The region is to be covered only when each point in that region should be covered by at least one sensor. For simplified analysis, in this paper we assumed that all existing nodes in the design have equal sensing range and

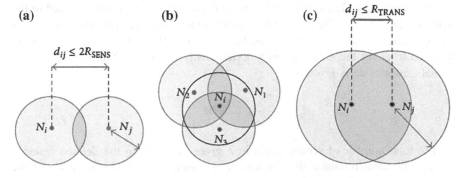

Fig. 1 **a** Individual sensor nodes **b** Overlapped scenario of sensor nodes **c** Communicating sensor nodes

transmission range. If we draw a graph with all nodes in their reachable region, it is called as covering-reachability graph and is represented as G. G is a function of set of nodes, their sensing ranges and their transmission ranges.

2.3 Degrees and Coverage

For perfect monitoring of any situation, each location in defined region should be covered and sensed by at least one sensor. The coverage model depends on the application and it will vary from scenario to scenario. Sometimes, it is mandatory to observe the same location with many sensors to achieve the efficiency in the design process. The location covered by k nodes to collect information is called as k-coverage region, where k represents number of nodes. Let an area A with region $\Re^2$ with degree of coverage k, where every point of the defined region to be covered by at least k nodes as described in [10, 11].

3 Detection Model in WSNs

The main aim of this paper is to identify the unauthorized person entry into defined field of interest. The best deployment scenario should be identified prior to the real deployment. When we choose random deployment, then the quality of the network is poor [12]. High detection probability is possible with the efficient selection of nodes placement and their range characteristics.

3.1 Sensing Model Probability

Initially consider a random network topology where nodes are distributed randomly in defined location. The quality of the node deployment was directly related to the detection probability. The sensing probability mainly depends on two factors; they are application and type of device used as a sensor [3]. Assume that, N nodes are required to cover the area S_{area} with node density λ. The probability of the individual sensor node is given as

$$p = \frac{\pi R_{\text{SENS}}^2}{S_e}$$

The total number of sensor nodes in a given area within the defined distance R_{SENS} will form binomial distribution. If the number of nodes is less and the probability of detection is high, then the binomial distribution is converted as Poisson distribution. Then the mean value of the new distribution is given as

$$Np = \frac{N\pi R_{\text{SENS}}^2}{S_{\text{area}}}$$

If the intruder is entered into the area, then the sensing model probability with k sensor nodes follows the Poisson process and is given as

$$p(n=k) = \frac{(s\lambda)^k}{k!} e^{-S\lambda}$$

In the above expression, S represents the area in which intruder entered with the projection l. We consider a random projection as shown in Fig. 2.

The area covered S is given as

$$S = 2R_{\text{SENS}} l + \pi R_{\text{SENS}}^2$$

Hence, the probability further can be expressed as

$$p(n=k) = \frac{\left((2R_{\text{SENS}} l + \pi R_{\text{SENS}}^2)\lambda \right)^k}{k!} e^{-\left((2R_{\text{SENS}} l + \pi R_{\text{SENS}}^2)\lambda \right)}$$

The intruder travel distance is calculated as

$$l = \int_a^b \sqrt{(f'(x) + (g'(x))^2} \, dx$$

As we already mentioned, the node density is also one of the critical parameters to determine the quality of the deployment. Node density will directly affect the

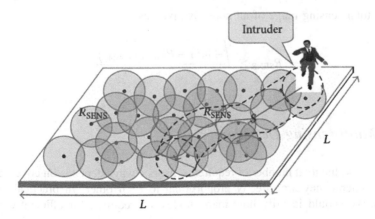

Fig. 2 Sample intruder trajectory

quality of the system. If node density is increases, the probability of connectivity is also increases. To improve the node density accuracy, the partial node density and is represented as

$$\lambda = \sum_{i=1}^{m} \lambda_i = \sum_{i=1}^{m} \frac{n_i}{S_i},$$

where λ is called as the resultant node density. For the same coverage degree, it is advised to maintain same node density throughout the area.

3.2 Single-sensing Detection

Let $\bar{p}$ is a probability of an event that no sensor can detect information and is represented as $\overline{P} = e^{-\lambda x}$. Then, the probability of an event that at least one sensor can detect information is equal to the complement of $\overline{P}$. The sensing model probability can be represented as

$$P(n \geq 1) = 1 - \overline{P} = 1 - e^{-\lambda x}$$

If any intruder enters into the location, the time to detect him should not cross the threshold distance and is given as

$$P(n \geq 1), \ 0 \leq l \leq l_{THR} = 1 - e^{-\left(2R_{SENS}l_{THR} + \pi R_{SENS}^2\right)x}$$

If the intruder is detected immediately, then we can sense him using the expression

$$P_{sensing}(nbrs \geq 1, \ l_{THR} = 0) \geq P_{thresholdsensing}$$

The total sensing range of all nodes is given as

$$R_{SENS} \geq \sqrt{\frac{-\ln\left(1 - P_{thresholdsensing}\right)}{\lambda \pi}}$$

3.3 Multi-sensing Detection

In a network, the total number of required sensors always depends on coverage area and in general they are directly proportional to each other. In practical circumstances, we should identify how many nodes are required for efficient coverage before deployment. If the degree of coverage is k, then the probability of detection when the intruder can be identified within the threshold distance is given as

$$P(n \geq k,\ 0 \leq l \leq l_{THR}) = 1 - \sum_{i=0}^{k-1} \frac{(s\lambda)^i}{i!} e^{-s\lambda},$$

where s is the surface covered by intruder.

4 Simulation Results

We are evaluating the proposed model with the help of MATLAB. Here, we assumed sensor nodes were placed in a homogeneous environment. We evaluated the scenario with the help of both single-sensing and multi-sensing detection scenarios. Here, there are N nodes to cover the defined location and they are distributed uniformly. We considered two coverage areas 100×100 m^2 and 200×200 m^2. Figure 3 shows the intrusion detection probability.

From Fig. 3, it is observed that, if N increases, the sensing range and detection probability increases. We can also note that, if we know the sensing range in advance, we can find the number of required sensor nodes to cover the region uniformly.

Figure 4 shows the intrusion detection probability as a function of distance with respect to the node availability rate. If the intrusion distance increases, then the probability of detection is also increases. Figure 5 shows the multi-sensing scenario with respect to intruder distance. We can observe that the probability of detection increases with the increase in distance. We can also observe that the probability of detection of single sensing is higher than the multi-sensing. This is due to the reason that multi-sensing will take strict consideration to identify the intruder as explained above.

Fig. 3 Intrusion detection probability that at least one node can detect the intruder

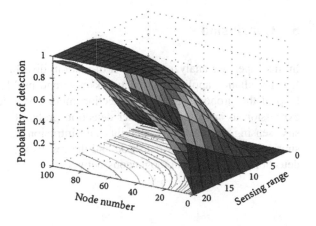

Fig. 4 Probability of
detection versus intrusion
probability

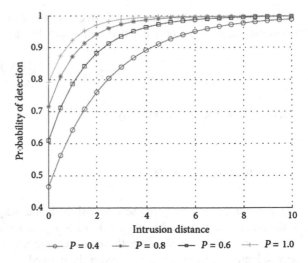

Fig. 5 Comparison of
sensing detection probabilities

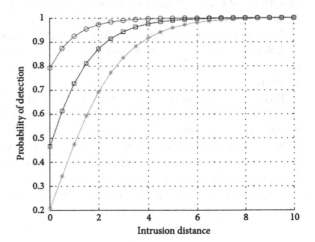

5   Conclusion

In this paper, we analyzed the effect of random distribution of sensors in a network
for surveillance applications. Network coverage problem was addressed with the
help of the proposed intrusion detection model. We developed the mathematical
expression to analyze the intruder detection probability by considering node density, sensing range, node availability and intrusion distance. We finally compared
single- and multi-sensing detection and the results were really helpful in designing
detection model for a homogeneous network.

References

1. Kilaru, Seetaiah. "Ability of OFDMA in Handling Interference of Femto Cells Under Random Access Process." Journal of Engineering Science and Technology Review 7.2 (2014): 133–136.
2. Kilaru, S.; Harikishore, K.; Sravani, T.; Anvesh, C.L.; Balaji, T., "Review and analysis of promising technologies with respect to Fifth generation networks," in Networks & Soft Computing (ICNSC), 2014 First International Conference on, vol., no., pp. 248–251, 19–20 Aug. 2014, doi:10.1109/CNSC.2014.6906653.
3. Kilaru, Dr Seetaiah, B. Rohith Kumar, S. Jaipal Reddy, Ch. Revanth, M. Sai Greeshma. "A Critical Review on the Implementation of MIMO with CSIT and CSIR." Wireless Communication [Online], 8.1 (2016): 53–57. Web. 25 Feb. 2016.
4. Seetaiah kilaru. " Public Safety Communication using Relay node in LTE- Advanced Technology ", International Journal of Engineering Research & Technology (IJERT), ISSN: 2278-0181, Vol 2(11), 2013, http://www.ijert.org, pp 1767–1772.
5. Kilaru, Seetaiah, and Aditya Gali. "Improving Quality of Service of Femto Cell Using Optimum Location Identification." International Journal of Computer Network and Information Security(IJCNIS) 7.10 (2015): 35–43. doi:10.5815/ijcnis.2015.10.04.
6. Seetaiah Kilaru, Y V Narayana and Gandhiraja R "Energy efficiency in Cognitive Radio Network: Green Technology towards next generation networks" Springer Microelectronics, Electromagnetics and Telecommunications, pp 305–313, 2016. doi:10.1007/978-81-322-2728-1_27.
7. Kilaru, Seetaiah, and Y. Ashwini Prasad. "K Sai kiran." NV Sarath Chandra published "Design and Analysis of Heterogeneous networks" International Journal of Applied Engineering Research (IJAER), ISSN: 0973-4562.
8. Kilaru, S., Prasad, Y. A., Kiran, K. S., & Chandra, N. S. (2014). Design and Analysis of Heterogeneous Networks. International Journal of Applied Engineering Research, 9(17), 4201–4208.
9. Kilaru, S., Balaji Tungala, A. L., & Hemanth, V. (2014). Effective Analysis of Positioning System for Group of Micro Air Vehicle Swarms Based on Audio Frequency. Journal of Engineering Science and Technology Review, 7(4), 169–172.
10. Seetaiah Kilaru, "Mathematical Analysis of Interference Issues and Study of Path Loss, Fading and Shadowing Effects on Cognitive Radio Networks" CiiT international journal of wireless communications. Vol 7, No 8 (2015) PP: 251–255.
11. Kilaru, Seetaiah, & Adithya Gali. "Review and Analysis of Macrocell Coverage Expansion Technologies." Wireless Communication [Online], 7.10 (2015): 334–337. Web. 11 Jan. 2016.
12. KILARU, Seetaiah, PULI, Bhargava, PULI, Shwetha. "Possibilities of implementation of synchronous Ethernet in popular Ethernet version using timing and interference constraints" Leonardo Electronic Journal of Practices and Technologies, vol 15, No 27, pp 185–197, 2015.

Incremental Updating of Mined Association Rules for Reflecting Record Insertions

N. Satyavathi, B. Rama and A. Nagaraju

Abstract Many algorithms came into existence for mining association rules. Since the databases in the real world are subjected to frequent changes, the algorithms need to be rerun to generate association rules that can reflect record insertions. It causes overhead the algorithm needs to scan entire database every time and repeat the process. Incremental updating of mined association rules is challenging. Recently, Deng and Lv proposed an algorithm named FIN (Frequent Itemsets using Nodesets) for fast mining of frequent itemsets. They proposed a data structure named Nodesets which consume less memory. In this paper, we proposed an algorithm named FIN_INCRE based on FIN which updates mined association rules without reinventing the wheel again. When new records are inserted, only the nodes in the data structure are updated adaptively using the concept of pre-large itemsets that effectively avoid re-scanning original data set. We built a prototype application to demonstrate the proof of concept. The empirical results reveal that the proposed algorithm improves the performance significantly.

Keywords FIN · FIN_INCRE · Incremental data set · Itemset

N. Satyavathi (✉)
Department of Computer Science and Engineering,
Vaagdevi College of Engineering, Warangal, Telangana, India
e-mail: satya15_n@yahoo.co.in

B. Rama
Department of Computer Science,
Kakatiya University, Warangal, Telangana, India
e-mail: rama.abbidi@gmail.com

A. Nagaraju
Department of Computer Science and Engineering,
Rajasthan Central University, Ajmer, Rajasthan, India
e-mail: kits.nagaraju@gmail.com

© Springer Science+Business Media Singapore 2017
S.C. Satapathy et al. (eds.), *Proceedings of the First International Conference
on Computational Intelligence and Informatics*, Advances in Intelligent Systems
and Computing 507, DOI 10.1007/978-981-10-2471-9_57

595

1 Introduction

Frequent itemset mining is one of the widely used algorithms in data mining domain. It has its utility in the decision-making, as it can generate patterns that give rise to actionable knowledge. Many frequent itemset mining algorithms came into existence. Recently two data structures were used for improving mining mechanism of frequent itemsets. They are known as Node-list and N-list. They make use of pre-order and post-order to represent itemset. The algorithms based on these data structures consumed more memory. Deng and Lv [1] proposed a new data structure known as Nodeset which is proved to be more efficient for extracting frequent itemsets. Algorithm 1 in [1] is used to construct POC (Pre-order Coding) tree for discovering frequent itemsets. The sample transaction database and the constructed POC tree are shown in Table 1 and Fig. 1, respectively.

The POC tree for the data present in Table 1 is as shown in Fig. 1. The tree is constructed for further processing while discovering frequent itemsets.

Deng and Lv [1] proposed an algorithm known as FIN for fast mining of association rules. The algorithm makes use of either pre-order or post-order, so as to improve performance. In this paper, we proposed a new algorithm known as FIN_INCRE which follows incremental approach to update association rules. This is our main contribution in this paper. The remainder of the paper is structured as follows. Section 2 reviews the literature on frequent itemsets, association rule

Table 1 Sample transaction database

ID	Items	Ordered frequent items
1	a, c, g, f	c, f, a
2	e, a, c, b	b, c, e, a
3	e, c, b, i	b, c, e
4	b, f, h	b, f
5	b, f, e, c, d	b, c, e, f

Fig. 1 POC tree

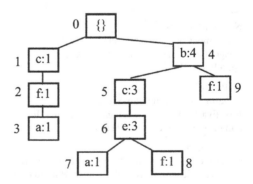

mining, and updating association rules, dynamically. Section 3 presents the proposed framework and the underlying algorithm. Section 4 presents the experimental results, while Sect. 5 concludes the paper besides providing directions for future work.

2 Related Works

There are many existing algorithms that are used to mine frequent itemsets. They are explored in [2]. Such algorithms can be broadly classified into A-priori-like [3] and FP-growth like [4] methods. A-priori kind of methods has a common feature known as priori probability. A-priori kind of methods can reduce size of candidates, thus achieving high performance. At the same time, such methods are highly expensive, as they scan data set every time, the algorithm is executed [5]. The FP-Growth method, on the other hand, can generate frequent itemsets without generating candidates, thus proving efficiency. Its performance is also due to the data structure known as FP-tree. However, FP-trees are complex and not efficient with sparse data sets.

Later on, N-list and Node-list are the two data structures explored in [6, 7] and the trees associated with them include Pre–Post and PPV, respectively. These trees are highly efficient than a priori, while relatively better than FP-Growth. Nevertheless, the N-list and Node-list consume more memory. To overcome this drawback Deng and Lv [1] proposed Nodeset structure which uses either pre- or post-order code. The proposed algorithm for frequent itemset mining in [1] is called FIN. FIN can generate frequent itemsets faster. However, FIN has no provision for updating association rules without scanning entire database.

To overcome the above issues, in this paper, we proposed a novel algorithm named FIN_INCRE based on FIN. Our algorithm can work with incremental data sets, so that it can scan only the modified records. This will result in the reduction of memory usage and execution time significantly.

3 Proposed Methodology

The proposed methodology is based on the FIN algorithm proposed by Deng and Lv [1]. The proposed methodology is shown in Fig. 2. The framework takes data set as input and performs association rule mining in incremental fashion. For the first time, the underlying algorithm of the framework FIN_INCRE acts on full data set. Once association rules are mined, the subsequent attempts are made incrementally to update association rules that have been generated. The nodesets data structure proposed in [1] was based on POC tree which is known as Pre-Order Coding tree. We used the POC tree construction algorithm proposed in [1] for generating POC tree.

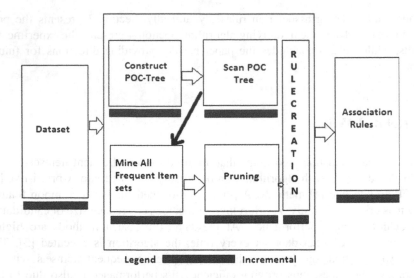

Fig. 2 Overview of the proposed framework

As shown in Fig. 2, it is evident that there are different modules in the proposed framework. The POC tree construction module constructs POC tree with frequent 1-itemsets. The scanning POC tree modules generate frequent 2-itemsets and their related nodesets. Mine, all frequent itemsets module generates all frequent itemsets, while pruning removes the frequent itemsets that do not satisfy the given support threshold. Then, the rules are generated. When the data set is subjected to new records, then the whole methodology becomes incremental in nature. The underlying algorithm is FIN_INCRE. The data sets used for experiments are presented in Table 2. All the modules in the proposed framework are incremental aware. They participate in incremental approach while updating generated association rules.

These data sets are collected from UCI machine learning repository. We modified FIN algorithm to update generated association rules based on the changes made to data set. The algorithm consumes less memory and takes very less time for execution. Our algorithm FIN_INCRE is as shown below. This algorithm is used once, FIN obtains frequent itemsets into the underlying data structure.

Table 2 Summary of data sets used

Database	Avg. length	#Items	#Trans
Mushroom	23	119	8124
Connect13	43	130	67557
T25I10D100 K [11]	25	990	99822

Algorithm 1. FIN_INCRE

Algorithm: FIN_INCRE
Inputs: Incremental Dataset *D'*, support *sup*, POC-tree *tree*
Output: Updated association rules

01 Initialize vector *FIS, MFIS*
02 Obtain frequent item sets associated with *tree* to *FIS*
03 Generate frequent item sets *FIS'* from *D'*
04 Scan the tree with pre-order traversal
05 For each *fis* in *FIS*
06   For each *fis'* in *FIS'*
07   IF *fis* and *fis'* are compatible THEN
08   merge to update item sets into *MFIS*
09   ENF IF
10   End For
11 End For
12 For each *mfis* in *MFIS*
13   For each *r* in *R*
14   IF support >= *sup* THEN
15   IF *mfis* is related to *r* THEN
16   update *R*
17   END IF
18   END IF
19   End For
20 return *R*

As can be seen, the proposed algorithm does not scan the entire database. Instead, it considers only changes in the data set and modified the tree which holds the frequent itemsets. At the same time, there will be pruning taking place to ensure the resultant patterns satisfy the given support.

4   Experimental Results

Experiments are made with our prototype application. The difference between FIN and the proposed FIN_INCRE is found using three data sets collected from UCI [8]. The two algorithms have shown different performance. The FIN_INCRE outperforms the FIN algorithm, as it does the association rule mining incrementally.

As shown in Fig. 3, the horizontal axis represents minimum support, while the vertical axis represents execution time in seconds. The minimum support considered includes 5, 10, 15, 20, and 25. There is huge performance difference between the FIN and FIN_INCRE algorithms. This difference is due to the fact that FIN_INCRE does the incremental generation of association rules. In other words, the FIN_INCRE updates association rules instead of generating them from the scratch.

As shown in Fig. 4, the horizontal axis represents minimum support, while the vertical axis represents execution time in seconds. The minimum support considered includes 40, 45, 50, 55, and 60. There is huge performance difference between the FIN and FIN_INCRE algorithms. This difference is due to the fact that FIN_INCRE does the incremental generation of association rules. In other words, the FIN_INCRE updates association rules instead of generating them from the scratch.

As shown in Fig. 5, the horizontal axis represents minimum support, while the vertical axis represents execution time in seconds. The minimum support considered includes 1, 2, 3, 4, and 5. There is huge performance difference between the FIN and FIN_INCRE algorithms. This difference is due to the fact that FIN_INCRE does the incremental generation of association rules. In other words, the FIN_INCRE updates association rules instead of generating them from the scratch.

Fig. 3 Shows the performance comparison with mushroom data set

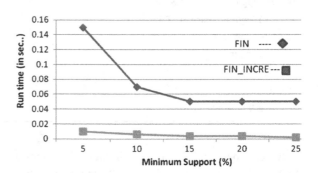

Fig. 4 Shows the performance comparison with connect data set

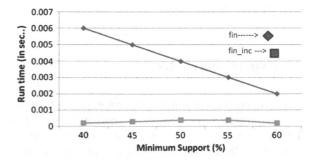

Fig. 5 Shows the performance comparison with T25I10D100 K data set

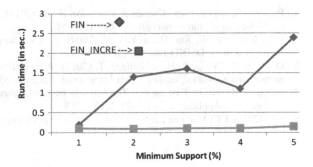

5 Conclusion and Future Work

Association rule mining is one of the widely used data mining activity. It is achieved in two phases. In the first phase, frequent itemsets are generated from data set. In the second phase, association rules are generated. Two statistical measures, such as support and confidence, are used to improve the quality of association rules. The problem with existing algorithms is that they scan entire data set to generate frequent item sets. Every time, they are executed, they produce frequent itemsets by scanning entire data set. However, data sets in the real world change frequently. After generating association rules, the new insertions to the data set are to be considered for updating generated association rules. This is a challenging task. In this paper, we proposed an algorithm named FIN_INCRE which is based on the FIN algorithm. Our algorithm is incremental in nature. When new records are inserted, it updates the nodes in the tree while performing required pruning to update rules. This improves the performance significantly in terms of time and the memory usage. Our prototype application demonstrates the proof of concept. Empirical results revealed that that the proposed algorithm is able to update rules based on the incremental data set rather than entire data set besides outperforming FIN which takes entire database with less memory overhead and execution time. Our future work is to investigate further on dynamically updating association rules in the context of big data.

References

1. Deng, Z and Lv, S. (2014). Fast mining frequent itemsets using Nodesets. Elsevier, Expert Systems with Applications, 41, p 4505–4512.
2. N.Satyavathi, B.Rama, and A. Nagaraju, IJIRSET," Dynamically updating association rules: Present state-of-the-art of index support for frequent itemset mining" Volume 4, issue July 2015 ISSN (online): 2319–8753, ISSN (print): 2347–6710.
3. Agarwal, R., & Srikant, R. (1998). Fast algorithm for mining association rules. In VLDB'94 (pp. 487–499).

4. Bay Vo, Tuong Le, Frans Coenen & Tzung-Pei Hong. (2013). Mining frequent itemsets using the N-list and subsume concepts. 52(10), 1868–8071.
5. Han, J., Cheng, H., Xin, D., & Yan, X. (2007). Frequent itemset mining: current status and future directions. DMKD Journal, 15(1), 55–86.
6. Deng, Z. H., Wang, Z. H., & Jiang, J. J. (2012). A new algorithm for fast mining Frequent itemsets using N-lists. Science China Information Sciences, 55(9), 2008.
7. Deng, Z. H. (2014). Fast mining Top-Rank-K frequent patterns by using node-lists. Expert Systems with Applications, 41(4–2), 1763–1768.
8. UCI (2016). UCI Machine Learning Repository. <https://archive.ics.uci.edu/ml/datasets.html>.

Pattern Anonymization: Hybridizing Data Restructure with Feature Set Partitioning for Privacy Preserving in Supervised Learning

MD. Riyazuddin and V.V.S.S.S. Balaram

Abstract Majority of the data available for knowledge discovery and information retrieval are prone to identity disclosure. The major act to disclose the identity is through exploring the pattern of attributes involved in data formation. The existing benchmarking models are anonymizing the data either by generalizing, deleting the sensitive attributes, or adding noise to the data. Either of these approaches is not guaranteed in optimality and accuracy in results that obtained from the mining models applied on that data set. The deviation in results often causes falsified decision-making, which is unconditionally not acceptable in certain domains like health mining. To fill the gap, here, we proposed a novel hybridization of feature set partitioning and data restructuring to achieve the pattern anonymization. The model is particularly aimed to restructure the data for supervised learning. To the best of our knowledge, pattern anonymization is first of its kind that attempted to anonymize the patterns rather individual attributes. The experiment results also indicate the scope of robustness and scalability of the supervised learning on restructured data.

Keywords Privacy preserving · Anonymization · Identity disclosure · Data mining · Pattern anonymization · Data restructuring · Feature set partitioning · K-anonymity · L-diversity

MD.Riyazuddin (✉)
Muffakham Jah College of Engineering & Technology,
Banjara Hills, Hyderabad, Telangana, India
e-mail: riyaz.mdr1@gmail.com

V.V.S.S.S.Balaram
Sreenidhi Institute of Science & Technology,
Ghatkeser, Hyderabad, Telangana, India

© Springer Science+Business Media Singapore 2017   603
S.C. Satapathy et al. (eds.), *Proceedings of the First International Conference on Computational Intelligence and Informatics*, Advances in Intelligent Systems and Computing 507, DOI 10.1007/978-981-10-2471-9_58

1 Introduction

The exploration of eligible data patterns [1] is a key process of information retrieval and knowledge discovery. Data mining is one of the significant concepts, which is generally used for information retrieval and knowledge discovery by identifying the eligible patterns from the given data [2–4]. The considerable consequence of these data-mining algorithms and models is identity and personal information discloser.

Above 90 % of the participants of a survey [5] aware that data related to any individual can be shared or sell by a company only under the permission given that individual, henceforth the sharing of personal information by an individual to an eligible organization is hassle free. The data-mining applications often apply on these personal and sensitive data that leads to violate individual privacy. Henceforth, often, the implementations of these mining models are restricted [6–8].

The objective of this study is related to securing the privacy of the data from leakages possible in data mining. The organizations should protect the privacy of the customer's personal information, which could be revealed unofficially due to patterns discovered by data-mining activities [9].

It has been learnt that removal of sensitive identities related to personal information is not adequate solution for privacy protection [10]. The data publishing for statistical analysis are another domain, which is sensitive to this privacy leakage. The considerable study was done and landed with solutions like restricting queries and perturbation the data to protect privacy in data publishing [11]. However, none of these methods are compatible to preserve privacy in data mining.

The predictive mining strategies, such as classification, cannot be possible on query restricted and perturbed data set. Since the predictive mining models require to explore the possible associations between attributes. Henceforth anonymizing the data is considered to be the best approach to prevent privacy leakages in predictive mining models. Many of such solutions [12–16] can be found in the literature of past decade. The most successful privacy protection strategy was k-anonymity [17, 18] that anonymizes the record of each individual, such that it cannot be distinguished from minimum K other individuals. This is done mostly by generalizing the sensitive attributes of the records or eliminating these sensitive attributes from the given data set or including noise, such that no individual personal information is leaked.

The similar objective has been considered in this research article. The aim is to explore the constraints of the existing models and defining a data anonymization approach toward privacy preserving data mining.

The rest of the paper is organized as follows. Sect. 2 contains the exploration of the strengths and constraints of the existing benchmarking models and that followed by Sects. 3 and 4, which contains detailed projection of the proposed model. Sect. 5 presents the experimental setup and performance analysis that followed by the Sect. 6, which concludes the proposal.

2 Related Work

K-anonymity is optimized in [19], which is done by performing a search to identify the attributes possible to allow the privacy leakage. A multi-objective method for hiding sensitive association rules is devised in [20]. This model is an evolutionary strategy, since it is using GA to identify the sensitive rules. This model is optimal to preserve privacy and deliver extremely significant rules. The main constraints of these models [19, 20] are the need of prior knowledge of the sensitive attributes of the data set and computational complexity of search is nonlinear.

The concept of feature set partitioning introduced in [21] is aimed to magnify the scalability of the supervised learning. The feature set partitioning decomposes the actual feature set into multiple subsets and further builds each subset level classifier. Further ensembles all these classifiers to recognize the class of the target record [22]. Generalizing the feature during the feature selection is the main objective of feature set partitioning strategy. The classifier will be constructed using the representative features. The empirical study of the model evinced that the optimality is proportionate to the formation of minimal number of subsets from maximal number of features, which is also a significant constraint for data sets with sparse and divergent feature set. The privacy leakage is another biggest constraint of this supervised learning by feature set partitioning.

To overcome the constraint of privacy leakage observed in classification strategy called feature set partitioning [21], Matatov et al. [23] extended the model devised in [21]. The model proposed in [23] is achieving optimal k-anonymity for each feature subset to prevent identity leakage of the data set. This model is also an evolutionary strategy, as it is using genetic algorithm to identify the optimal feature subsets. The empirical study evincing that the model is optimal to achieve K-anonymization for defined feature subsets. The considerable limits of the model are nonlinear computational complexity in optimal feature subset discovery and requirement of prior domain knowledge to identify the feature's sensitivity toward privacy leakage, which is an essential factor to define cost function of the genetic algorithm. The anonymization is relevant and specific to the ensemble classification by partitioned feature set proposed in [21].

The existing benchmarking models that aimed to privacy preserving data mining are anonymizing data by generalizing, obliterating the quasi fields of the records in given data set, or restructuring the data set by adding noise. The process of generalizing or obliterating the field values may achieve optimal anonymity but causes severe violation in mining results. On other method of adding noise also causes considerable deviation in mining results, which is due to the noise included. Another specific hurdle in any of the existing models is the compulsion of the prior knowledge about the quasi attributes. Hence, any of the benchmarking models minimizes their negative impact on mining results, if and only if the anonymization done in the context of the data set under the close monitoring of domain experts and specific to particular mining algorithm. In most of the cases, prior knowledge of the data and close monitoring of the domain experts is big constraint. Hence, it is obvious to conclude the need of an optimal strategy for privacy preserving data mining that works without prior knowledge of the data given and close monitoring of the domain experts. On the other dimension, the anonymization to preserve the privacy should not violate the originality of the mining results.

In order to this, we proposed a hybrid approach that combines the feature set partitioning and anonymizing through data set restructuring by including trivial records. The proposed model is anonymizing the patterns observed in given data set, since the feature patterns are the primary factors those leads to identity leakage. In addition, the other unique feature of the proposal is that it compatible to any of the supervised learning-based mining algorithm. To the best of our knowledge, model defined here is first of its kind that retains the structure of the actual records while adding the trivial records to achieve pattern disclosure.

3   Pattern Anonymization by Feature Set Partitioning and Data Set Restructuring

3.1   Notations Used in Model Exploration

- Data Set: A set of records and each record contain values for fixed number and order of attributes associating to a class label.
- Class Label: An attribute representing the state of the record
- Attribute Set: A set of attributes represents few or all attributes those labels the different fields of the records.
- Value Set: The set of values in a record representing an attribute set is known as Value set
- Trivial Record: A record that contains trivial values for one or more attributes.
- Trivial Value Set: A set of values in a trivial record represents an attribute set, such that one or more values are trivial.

3.2   Feature Set Partitioning

Let consider all the attributes of the given data set as a set A. Furthermore, all possible subsets from A of different sizes will created as a set S.

Similarly, create a set R that contains the subsets formed from each record r of the given data set D, which is done based on the subsets defined from A. A subset rs_i contains the values from r of the attributes found in subset $\{s_i \, \exists \, s_i \in S\}$.

Finding coverage of each subset $\{r_j s_i \, \exists \, r_j \in D \wedge s_i \in S\}$, which represents the number of records contains $r_j s_i$

Prune the subsets from R under bide rule [citation required] is as follows:

Let a set $r_j s_i$ with coverage k and a set $r_p s_q$ with coverage k', if $\left(r_j s_i \subseteq r_p s_q\right)$ and $(k \equiv k')$, then $r_j s_i$ can be discarded.

3.3 Data Set Restructuring

Let K be the optimal number of records representing each feature set partition to achieve K-anonymity.

If any of the value set $r_j s_i$ from R, which is representing values for attribute set s_i in record, r_j is with coverage k less than K, then $K - k$, the number of trivial records will be formed, such that each trivial record with trivial value set $tr_j s_i$ for attribute set s_i replaced by respective value set $r_j s_i$. This is done by updating existing trivial records if any otherwise creates new trivial records.

Furthermore, each trivial record will be associated to a class label randomly from all possible class labels, which is done to achieve maximum possible diversity in class label representation.

4 Process Model of the Feature Set Partitioning and Data Restructuring

For set of size n, possible number of subsets are $2^n - 1$, which excludes the empty set [citation required]. Henceforth, finding $2^{n-1} - 1$ attribute sets $S = \{s_1, s_2, s_3, \ldots . s_{|S|}\}$ from the set of attribute labels A of size n, which excludes class label attribute.

Order S by the size and order of attributes

Let R be an empty set

For each record $\{r_i \exists r_i \in D \forall i = 1, 2, ..., |D|\}$ that represents the values for all attributes of the set A Begin

Prepare $2^{n-1} - 1$ value sets $r_i S = \{r_i s_1, r_i s_2, ..., r_i s_{|S|}\}$ from values of record r that excludes the class label value.

Order $r_i S$ in the order of S

$R \leftarrow r_i S$ //move $r_i S$ to R

End

4.1 Further Find All Possible Subsets from All Entries of the R as Follows

Let $\overline{R}$ be the empty set

For each $\{r_i S \exists r_i S \in R \forall i = 1,2,3.... | D |\}$ Begin

$$\overline{R} \leftarrow \{\overline{R} \cup r_i S\}$$

End

Let C be the set of all possible unique class labels observed for records in dataset D

4.2 Further Find the Coverage of the Each Entry of $\overline{R}$ as Follows

$k\overline{R} \leftarrow \phi$ // is an empty map

$l\overline{R} \leftarrow \phi$ // is an empty map

For each $\left\{rs_i \exists rs_i \in \overline{R} \forall i = 1,2,...,\left|\overline{R}\right|\right\}$ Begin

For each $\{c \exists c \in C\}$ Begin

// C is a set of class label $\{c_i \exists c_i \in C \forall i = 1,2,...,|C|\}$

$l = 0$

$k = 0$

For each $\{r_j \exists r_j \in D \forall j = 1,2,...,|D|\}$ Begin

If $\left(rs_i \subseteq r_j\right)$ then Begin

$k = k+1$

If$\left(r_j(cl) \equiv c\right)$ then begin

// cl is the class label of the record r_j

$l = l+1$

End

End

End

$k\overline{R}\{rs_i\} \leftarrow k$

$l\overline{R}\{rs_i(c)\} \leftarrow l$

End

End

Sort $\overline{R}$ in ascending order of the subset size and order of attributes

4.3 Further Prune $l\overline{R}$ and $k\overline{R}$ as Follows

For each $\left\{rs_i \exists rs_i \in \overline{R} \forall i = 1, 2, ..., \left|\overline{R}\right| - 1\right\}$ Begin

   For each $\left\{rs_j \exists rs_j \in \overline{R} \forall j = i+1, i+2, ..., \left|\overline{R}\right|\right\}$ Begin

      If $\left(\left(k\overline{R}\{rs_i\} \equiv k\overline{R}\{rs_j\}\right) \& \& \left(rs_i \subseteq rs_j\right)\right)$ then Begin

         // delete key rs_i and respective value from $k\overline{R}$

         $k\overline{R} \leftarrow k\overline{R} \setminus \{rs_i\}$

      End

      For each $\left\{c_m \exists c_m \in C \forall m = 1, 2, ..., |C|\right\}$ Begin

         If $\left(\left(l\overline{R}\{rs_i(c_m)\} \equiv l\overline{R}\{rs_j(c_m)\}\right) \& \& \left(rs_i \subseteq rs_j\right)\right)$ then Begin

            // delete key $rs_i(c_m)$ and respective value

            $l\overline{R} \setminus rs_i(c_m)$

         End

      End

   End

End

Let $\{keys\}_k$ be the set of all keys exists in $k\overline{R}$

Let $\{keys\}_l$ be the set of keys exists in $l\overline{R}$

//Anonymity

Let K be the expected number of records representing a given feature partition fp for optimal anonymity

  //Diversity

The number of records representing a given feature partition fp associable to class labels $\{C\} - c$ for optimal anonymity.

4.4   Restructuring Data for K-Anonymity of the Feature Sets Is as Follows

// Sort $\{keys\}_k$ in ascending order of the size of all subsets exists as keys

Let $\overline{D} \leftarrow \phi$ //The new empty dataset.

//A record $\left\{ tr \exists tr \in \overline{D} \right\}$ that contains trivial value for at least one attribute $\{a \exists a \in A\}$ is called trivial record $\widehat{tr}$.

// the number of attributes contains trivial values in tr indicates $\left\| \overset{..}{tr} \right\|$

For each $\{rs_i \exists rs_i \in \{key\}_k \ \forall i = 1, 2, ..., | \{keys\}_k |\}$ Begin

If $\left(k\overline{R}\{rs_i\} < K \right)$ then Begin

$\qquad p = K - k\overline{R}\{rs_i\}$

If $\left(\overline{D} \equiv \phi \right)$ then $\overline{D} \leftarrow \left\{ tr \exists \left\| \widehat{tr} \right\| \equiv |A| \right\}$

// tr is a record with trivial values for all attributes of A

$\qquad idx = 0$

$$\text{For each trivial record} \begin{pmatrix} \left(tr \exists tr \in \overline{D} \right) \&\& \\ \left(trs_i \notin \{keys\}_k \right) \&\& \\ \left(\left\| \widehat{tr} \right\| > | trs_i | \right) \&\& \\ \left(idx < p \right) \end{pmatrix} \text{Begin}$$

// trs_i is the values set for subset $\{s_i \exists s_i \in S\}$ in tr ,

// $\left\{ trs_i \notin \{keys\}_k \right\}$ indicates that all values in trs_i are trivial and

// $\left(\left\| \widehat{tr} \right\| > | trs_i | \right)$ indicates that number of attributes in tr with trivial values must be greater than the size of the trs_i

$\quad \{trs_i \exists trs_i \in tr\} \leftarrow rs_i$ // replacing trivial values of the set trs_i by the values of the set rs_i

$\qquad idx = idx + 1$

$\qquad$ End

If $\left(idx < p \right)$ then for each $\{q \forall q = idx, idx + 1, idx + 2, ..., p\}$

$\overline{D} \leftarrow \left\{ tr \exists \left\| \widehat{tr} \right\| \equiv |A| \right\}$ // tr is a record with trivial values for all attributes of A

$\{trs_i \exists trs_i \in tr\} \leftarrow rs_i$ // replacing trivial value set trs_i of record tr by the value set rs_i

$\qquad$ End

$\qquad$ End

$\qquad$ End

4.5 Add Label to All Trivial Records in $\overline{D}$ as Follows

For each $\left(tr \ni tr \in \overline{D} \right)$ Begin

    // assigning a label selected randomly from class label set, this is done to achieve maximal diversity of associability between attribute values and class labels

        $tr(cl) \leftarrow rand(\{C\})$

    End

    $\overline{D} \leftarrow D$ // adding all the records from D to restructured dataset $\overline{D}$

5 Experimental Setup and Results Analysis

The experiments were conducted to assess the compatibility of the restructured data set toward supervised learning. The impact of K-anonymity with maximal possible diversity is a proven strategy toward privacy preserving [24]. Hence, the experiments conducted here were not aimed to explore the optimality of the K-anonymity and maximal diversity.

The accuracy, robustness, and scalability of the results obtained from restructured data set are assessed through statistical metrics [25] called precision, sensitivity, and accuracy, respectively, which are estimated using the count of truly classified and count of falsely classified.

Since the assessment metrics called computational and resource complexity also included in the performance analysis, a computer with i5 processor, 4 GB RAM, and Nvidia 4 GB graphics card is used. The implementation was done in CUDA [26]. Statistical metrics analysis was done using explorative language R [27]. The input and obtained results were explored in Table 1.

Table 1 Particulars of the input data set and results obtained

No of features	14
No of records in original data set	303
No of records in restructured data set	512
No of groups formed from original records	16
No of groups formed from restructured data set (original and trivial records)	23
No of groups after pruning the trivial records	16
Truly classified no of records	301
Falsely classified records	2
Precision	0.99339934
Sensitivity	1
Accuracy	0.994

5.1   The Data Set

The objective of the proposed model (Pattern Anonymization Approach) is to perform the optimal supervised learning on restructured data set that protects from pattern disclosure. To assess the scalability and supervised learning accuracy, we adopt the heart disease data set [28]. We initially classified the data set by classification tool J48 [29] and obtained prior knowledge of the possible groups of records.

5.2   Performance Analysis

The classes predicted by the proposed pattern anonymization approach (PAA) were assessed, by comparing the classification of the original records under restricted data set, which include trivial records also. The Metric values indicate that classification of original records after restructuring the data set is significant (precision is 0.99339934 that indicates the truly classified records ratio). The sensitivity of true and trivial record classification is also considerably high (sensitivity is 1 that indicates no trivial records included into actual groups of the original records). The overall classification optimality is observed as best, since almost the 100 % of the records grouped into relevant labels under the given restructured data set and experimental setup (accuracy is 0.994).

Fig. 1 Feature set partitioning and data set restructuring completion time observed for divergent count of input records

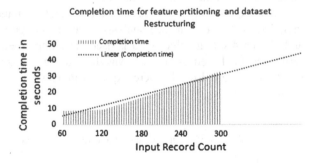

Fig. 2 Memory used for feature set partitioning and data set restructuring of divergent count of input records

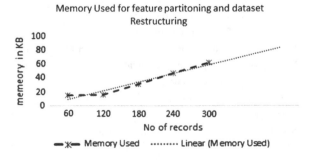

The computational complexity and resource cost are also assessed, which is done under divergent count of records given as input. The time complexity observed to be linear for given divergent count of records as input (see Fig. 1). The memory usage of data set restructuring with feature set partitioning also being noticed as linear for given divergent count of input records (see Fig. 2).

6 Conclusion

Privacy preserving supervised learning is prime objective of the model proposed here in this paper. In this context, many of existing models succeeded to prevent privacy disclosure under certain constraints, such as prior knowledge of the data to identify the sensitive attributes, compromising at optimality of the mining results due to sensitive attribute eradicating, and generalizing or nonlinear complexity observed in the process of sensitive feature identification. The proposed model is hybridizing the data restructuring with feature set portioning to achieve privacy preserving for any of existing supervised learning-based mining model. The best part of this model is that no prior knowledge of the data is required to anonymize and computational complexity is observed that was linear. The important factor to adopt this model is that it cannot violate the mining results, which is a biggest constraint of the existing models. The motivation gained from this model drives our future research to minimize the computational complexity to much minimal that compared with the present model. In the other direction of future research, the similar anonymization strategy can be devised for rule mining and unsupervised learning strategies.

References

1. U. Fayyad, G. Piatetsky-Shapiro, P. Smyth, From data mining to knowledge discovery: an overview, in: Advances in Knowledge Discovery and Data Mining, AAAI Press, Menlo Park, CA, 1996, pp. 1–31.
2. H. Chen, Intelligence and security informatics: information systems perspective, Decision Support Systems 41 (3) (2006) 555–559.
3. D. Martens, L. Bruynseels, B. Baesens, M. Willekens, J. Vanthienen, Predicting going concern opinion with data mining, Decision Support Systems 45 (4) (2008) 765–777.
4. T.S. Raghu, H. Chen, Cyberinfrastructure forhomeland security: advances in information sharing, data mining, and collaboration systems, Decision Support Systems 43 (4) (2007) 1321–1323.
5. S. Greengard, Privacy: entitlement or illusion? Personnel Journal 75 (5) (1996) 74–88.
6. M. Kantarcioglu, J. Jin, C. Clifton, When do data mining results violate privacy? in: Proc of the 10th ACM SIGKDD International Conference on Knowledge Discovery and Data Mining, ACM, New York, NY, 2004, pp. 599–604.

7. C. Clifton, M. Kantarcioglu, J. Vaidya, Defining privacy for data mining, in: H. Kargupta et al. (Eds.), Proc. of the National Science Foundation Workshop on Next Generation Data Mining, Baltimore, Maryland, 2002, pp. 126–133.

8. M. Feingold, M. Jeffords, M. Leahy, Data Mining Moratorium Act of 2003, US Senate Bill (proposed), 2003.

9. L. Cao, C. Zhang, Domain-driven, actionable knowledge discovery, Intelligent Systems 22 (4) (2007) 78–88.

10. Zhu, Dan, Xiao-Bai Li, and Shuning Wu. "Identity disclosure protection: A data reconstruction approach for privacy-preserving data mining." Decision Support Systems 48.1 (2009): 133–140.

11. N.R. Adam, J.C. Wortmann, Security-control methods for statistical databases: a comparative study, ACM Computing Surveys 21 (4) (1989) 515–556.

12. A. Amiri, Dare to share: protecting sensitive knowledge with data sanitization, Decision Support Systems 43 (1) (2007) 181–191.

13. D.S. Chowdhury, G.T. Duncan, R. Krishnan, S.F. Roehrig, S. Mukherjee, Disclosure detection in multivariate categorical databases: auditing confidentiality protection through two new matrix operators, Management Science 45 (12) (1999) 1710–1723.

14. R. Garfinkel, R. Gopal, P. Goes, Privacy protection of binary confidential data against deterministic, stochastic, and insider threat, Management Science 48 (6) (2002) 749–764.

15. S. Menon, S. Sarkar, Minimizing information loss and preserving privacy, Management Science 53 (1) (2007) 102–116.

16. S. Menon, S. Sarkar, S. Mukherjee, Maximizing accuracy of shared databases when concealing sensitive patterns, Information Systems Research 16 (3) (2005) 256–270.

17. P. Samarati, Protecting respondents' identities in microdata release, IEEE Transactions on Knowledge and Data Engineering 13 (6) (2001) 1010–1027.

18. L. Sweeney, k-Anonymity: a model for protecting privacy, International Journal on Uncertainty, Fuzziness and Knowledge-based Systems 10 (5) (2002) 557–570.

19. Bayardo R. J., Agrawal R.: Data Privacy through Optimal k-Anonymization. Proceedings of the ICDE Conference, pp. 217–228, 2005.

20. Dehkordi, M. N., Badie, K., & Zadeh, A. K. (2009). A novel method for privacy preserving in association rule mining based on genetic algorithms. Journal of software, 4(6), 555–562.

21. L. Rokach, O. Maimon, Theory and application of feature decomposition, in: Proc. of the First IEEE International Conference on Data Mining, IEEE Computer Society, Washington, DC, 2001, pp. 473–480.

22. E. Menahem, L. Rokach, Y. Elovici, Troika – an improved stacking schema for classification tasks, Information Sciences 179 (24) (2009) 4097–4122.

23. Matatov, N., Rokach, L., & Maimon, O. (2010). Privacy-preserving data mining: A feature set partitioning approach. Information Sciences, 180(14), 2696–2720.

24. Chen, Rui, et al. "Privacy-preserving trajectory data publishing by local suppression." Information Sciences 231 (2013): 83–97.

25. Powers, D. M. (2006). Evaluation: from precision, recall and F-measure to ROC, informedness, markedness and correlation. *23rd International conference on machine learning*. Pitsburg.

26. Nvidia. (2008). C. U. D. A. Programming guide.

27. Ihaka, R. &. (1996). R: a language for data analysis and graphics. Journal of computational and graphical statistics, 299–314.

28. https://archive.ics.uci.edu/ml/machine-learning.

29. Patil, Tina R., and S.S. Sherekar. "Performance analysis of Naive Bayes and J48 classification algorithm for data classification." International Journal of Computer Science and Applications 6.2 (2013): 256–261.

Effective Visiting Schedule Generation in a Tourist Recommender System Using Hadoop

T. Ragunathan, Sudheer Kumar Battula, Jorika Vedika and M. NagaRatna

Abstract E-commerce has changed the way which the users select and purchase items. Most of the e-commerce applications deployed in the Web today use recommender systems to recommend items to the online users based on their earlier purchases. The tourist recommender systems discussed in the literature, so far, cover, regarding the best routes from one city to another city by including the tourist spots and beautiful scenery-based sites and the destination tourist spots by accepting images or description of the tourist spots as input. In this paper, we have proposed the architecture for a tourist recommender system and then a novel scheduling algorithm for preparing the visit schedules in a city for the tourists based on user requirements and we have implemented the same using Hadoop framework.

Keywords MapReduce · HDFS · Tourist recommender system

1 Introduction

E-commerce has changed the way which the users select and purchase items. Most of the e-commerce applications [1] deployed in the Web today use recommender systems to recommend items to the online users based on their earlier purchases. Collaborative filtering techniques are discussed in the literature [2] to recommend

T. Ragunathan (✉) · S.K. Battula · J. Vedika
Department of Computer Science and Engineering, ACE Engineering College,
Hyderabad, India
e-mail: deanresearch@aceec.ac.in

S.K. Battula
e-mail: sudheer.itdict@gmail.com

J. Vedika
e-mail: v.jorika@gmail.com

M. NagaRatna
Department of Computer Science and Engineering, JNTUCEH, Hyderabad, India
e-mail: mratnajntu@gmail.com

© Springer Science+Business Media Singapore 2017   615
S.C. Satapathy et al. (eds.), *Proceedings of the First International Conference on Computational Intelligence and Informatics*, Advances in Intelligent Systems and Computing 507, DOI 10.1007/978-981-10-2471-9_59

items to the users based on their purchase behaviour. Content-based filtering techniques [3] are also discussed in the literature which uses user profiles to recommend items. Both collaborative- and content-based filtering techniques are combined to form hybrid techniques and these technique improve the quality of recommendations by combining the best features of both the recommendation techniques.

The tourist recommender systems discussed in the literature, so far, cover, regarding the best routes from one city to another city by including the tourist spots and beautiful scenery-based sites and the destination tourist spots by accepting images or description of the tourist spots as input. To the best of our knowledge, these systems were not implemented using any distributed computing technique, and also, we did not find any recommender system for recommending places of visit in a city, visit schedules, and the mode of transport for optimizing the expenditure of the tourists.

In this paper, we have proposed the architecture for a tourist recommender system and then a novel distributed algorithm for preparing the visit schedules in a city for the tourists based on user requirements by following parallel and distributed computing techniques. The proposed system, namely tourist recommender system (TRS), recommends efficient tour schedules for the tourists, so that more places can be visited in a city by the tourists within the stipulated number of days stay planned by them. We have developed a distance matrix based on the place of stay in the city and then generated efficient tour schedules to the users by considering different modes of transport.

We have implemented our algorithm using Hadoop framework (distributed computing framework) in a ten-node Hadoop cluster. We generated tour schedules for the users based on their constraints given as input to the system. TRS is a novel system and is very useful for the tourists who are planning to visit various places of cities by optimizing number of days of their stay in those cities. This system also recommends effective schedules, so that more places will be covered in a day based on the user requirements.

This paper is divided into five sections: Sect. 2 covers the related work. Section 3 discusses the proposed system architecture and algorithms. Section 4 discusses the prototype system that we developed at our research center and results. Section 5 covers the conclusion and future works.

2 Related Work

In the literature, many techniques are discussed for recommender systems to improve the quality of recommendations.

Next, we discuss the techniques which are discussed in the literature.

A content-based filtering technique was discussed in [4] which maintains the profiles of the users for recommending items. In these techniques, for new users, items are recommended based on their interests which can be found through their profiles. For the old users, items are recommended based on their previous purchase pattern.

Collaborative filtering techniques are discussed in [5, 6]. The idea of collaborative filtering is to recommend items based on the similarity between users and their purchased items. In user-based collaborative filtering techniques, similar users are identified based on their common interests, and then, items purchased by a user are recommended to similar users. In item-based collaborative filtering techniques [7–9], based on the items purchased by the users, similar users are found and then items are recommended. In hybrid filtering technique discussed in [10], both content-based and collaborative filtering techniques are combined to recommend items to the users.

Context-aware [11], semantic-based [12], and peer-to-peer [13] approaches are also proposed in the literature for effective recommendation of items to the users. In [14], the authors propose an intelligent route recommender system. In this paper, the route which is having more sceneries is recommended to the users. The main criteria used to select a route is the visibility of scenic sights between one tourist spot and another. In this system, the authors did not consider the cost effectiveness for proposing the routes.

In the paper [15], tourist sites are recommended to the users based on the text or images provided by them. The text or images about the tourist site given by the users are compared with already available text and images of tourist sites in the system and similar tourist sites are recommended to the users. The system proposed in [16] recommends the tourist places to the users based on the log data maintained in the system and the feedback given by the users of the system, and also, the authors propose to collect information, regarding tour packages from various tour agencies and recommend the best available tour package to the users based on their requirement. This system did not consider the tourist sites which are not included by the agencies and did not discuss, regarding multi-mode transport facility, which are used by the users to visit various places in a cost effective manner.

3 Proposed System Architecture and Algorithms

In this section, first, we discuss regarding the architecture of the proposed system. Next, we discuss the algorithm that we propose for generating schedules.

3.1 The Architecture of the Proposed System

The main goal of TRS is to recommend effective visiting schedules to the tourists by obtaining their requirements. We have considered the hadoop distributed file system (HDFS) and map reduce programming paradigm supported in Hadoop [17] for proposing the architecture. The architecture of the proposed system is shown in Fig. 1. This system consists of five important components which are as follows.

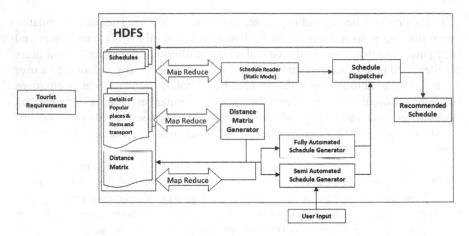

Fig. 1 Architecture of tourist recommender system

1. Schedule Reader
2. Schedule Dispatcher
3. Distance Matrix Generator
4. Fully Automatic Schedule Generator
5. Semi-automatic Schedule Generator

Schedule Reader: Manual as well as system generated visiting schedules are stored in the HDFS which can be read by the Schedule Reader (SR) and then delivered to the Schedule Dispatcher (SD) if the user has opted for "static" mode of scheduling.
Schedule Dispatcher: The SD dispatches the generated schedules to the users. This component stores the generated schedule to the HDFS if that schedule is not already there.
Distance Matrix Generator: A distance matrix is maintained for each city and is a two-dimensional array which stores the distance from the source given by the user to all famous places in city. This matrix can be generated manually and the system can also generate this automatically using the distance matrix generator (DMG) component. Note that by passing source place and destination place to the Google Maps API which return the distance between the source and destination, this detail is stored in the form of matrix (distance matrix) to the HDFS. The DMG passes this distance information to the fully automatic scheduler or semi-automatic scheduler based on the option given by the user.
Fully Automatic Schedule Generator: Once the source and destination places are given as input by the users, this scheduler generates the schedules automatically by reading the distance matrix values and items available in various places of the city. The generated schedule is then passed to the SD.

Semi-automatic Schedule Generator: Once the source and destination places are given as input by the users, this scheduler generates the schedules by reading the distance matrix values and items available in various places of the city and by considering the feedback given by the users. The generated schedule is then passed to the SD.

3.2 Proposed Algorithm

We have developed the algorithm by considering two modes, namely (i) Static (ii) Dynamic. In the static mode, the schedules already available in the HDFS and the schedules which satisfy the user requirements will be delivered to the users. In the dynamic mode, the schedules can be generated without the user intervention (fully automatic schedule) or by obtaining the feedback from the users (semi-automatic schedule).

Assumptions and Abbreviations

  (i) Google API is used to find distance between two places and for displaying the maps.
 (ii) Src-Source Place of tourist.
(iii) Atime-Arrival time of tourist on the first day.
 (iv) *Fplaces*—It is an array to store famous places of the given city.
  (v) *length*—It gives the length of the array *Fplaces* and this denotes the number of places to be visited by the users
 (vi) Distancematrix—It is a two-dimensional array to store distance of all famous places of the city

 (vii) Ctime-Current time
(viii) Etime-Estimated Time.
 (ix) Dtime—The maximum time up to the user prefers to visit the places in a day
  (x) n—number of days of visit in the city.
 (xi) *Schedule*—it is the list which consists of places of visit and the estimated visiting time for the corresponding places

1: **procedure** SCHEDULER(*Src, Atime, Dtime, n*)
2:     Read famous places of the city (*Src*) from HDFS and store the same in the array *Fplaces*
3:     $r \leftarrow 0$
4:     $s \leftarrow 0$
5:     *Fplaces*(0) $\leftarrow$ *Src*
6:     *mode* $\leftarrow$ *selectmode*
7:     /*select mode is the input given by the user*/
8:     **if** *mode* = *static* **then**
9:         return Matched schedule from HDFS

```
10:     else
11:         if mode = semiautomatic then
12:             Fplaces(0) ← UserSelectedPlaces
13:                 /* user has to manually select the places of visit which is given as
        User Selected Places*/
14:             end if
15:         while n! = 0 do
16:             while r <=Fplaces.length do
17.                 while s <=Fplaces.length do
18:                     Distancematrix(r)(s)=distance(Fplaces(r), Fplaces(s))
19:                         /* distance function interacts with Google Maps to give the
        distance between two places*/
20:                     s ← s + 1
21:                 end while
22:                 r ← r + 1
23:             end while
24:             while Ctime <= Dtime do
25:                 places ← least distance from the source using distance matrix.
26:                 Etime ← calculate estimated time from
27:                 the source to destination
28:                 if Schedule! = places not in Schedule then
29:                     Schedule ← add places and Etime
30:                     Ctime ← Ctime + Etime
31:                 end if
32:             end while
33:             n ← n − 1
34:         end while
35:     end if
36:     Display the contents of the schedule to the user.
37: end procedure
```

The scheduler takes source, destination, arrival time, maximum time up to which scheduling can be done in a day, and number of days of visit in the city. After this, the user has to select "Static" or "Dynamic" mode for preparing schedules. If the user has selected "static mode", then already existing schedules stored in the HDFS which meet the criteria of the user will be selected and displayed. If user has selected "dynamic mode" and "semi-automatic" option, then based on the visiting places available and distance matrix values, schedules are selected and shown to the user for selection. Note that, only user selected places are included in the schedule. If the user selected "dynamic mode" and "fully automatic" option, then based on the visiting places available and distance matrix values, schedules are automatically generated and shown to the user for selection. Note that, in dynamic mode (both for semi- or fully automatic options), the following method is used for generating the schedule. Initially, current time (Ctime) will be the starting time and it is checked with destination time (Dtime). If Ctime is less than Dtime, a visiting place (p) which is very near to the source is found and this place will be added into the Schedule provided that it is not already there. Next, p becomes the source place. After this, estimated visiting time to visit p is added to the Ctime, and then, next nearest place (np) to p will be found by going through the details in the distance matrix. Next, np will be added into the Schedule provided that it is not already there. This step is repeated until Ctime becomes greater than or equal to Dtime. Note that, these steps are repeated for all n days for preparing the complete schedule.

4 Prototype System and Results

Prototype Environment We developed the prototype of the TRS in a Hadoop Cluster which consists of 10 Datanodes and 1 Namenode and Secondary NameNode. Each node has got one Intel(R) Core(TM) i3-2120 CPU @ 3.30 GHz with 2 GB RAM, and ubuntu 14.0.4 64-bit operating System with Hadoop 1.0.4. We used Java 8 to develop this application. For designing front-end Web pages, we used HTML, CSS, and JS. For the server-side programming, we used Java Servlets.

In this system the input data given by the user is submitted to TRS. Figure 2 shows the input form used by the user for entering input details. In static mode, for the given

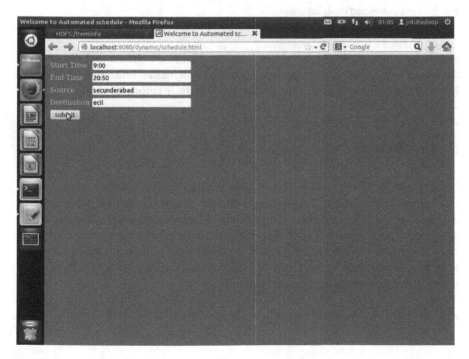

Fig. 2 User input form

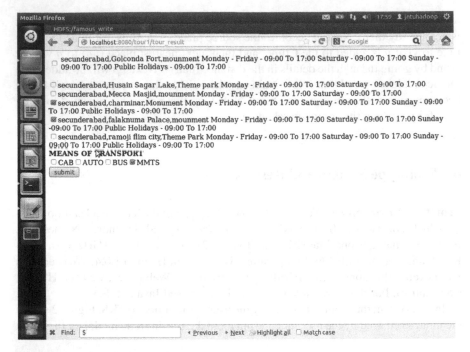

Fig. 3 Semi-automated mode schedule

user input, already existing schedules from the HDFS are chosen and displayed to the user. Figure 3 shows the schedule generated in dynamic semi-automatic mode. In Fig. 4 shows the schedule generated in dynamic fully automatic mode.

5 Conclusion

In this paper, we have proposed the architecture for a tourist schedule recommender system and then a novel distributed algorithm for preparing the visit schedules in a city for the tourists based on user requirements. We have implemented this algorithm using map reduce paradigm of Hadoop framework for generating visiting schedules. Note that the map reduce programs that we developed in the prototype system perform parallel and distributed processing of information, regarding the cities stored in the HDFS for generating visiting schedules.

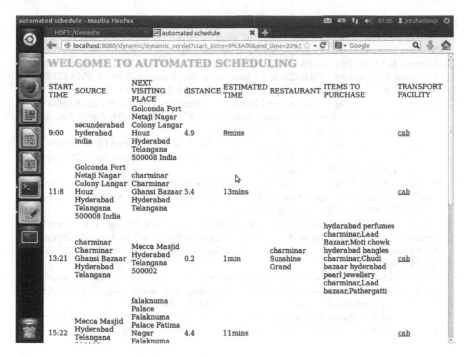

Fig. 4 Fully automated schedule generation

References

1. J. B. Schafer, J. A. Konstan, and J. Riedl, E-Commerce recommendation applications, Data Mining and Knowledge Discovery, vol. 5, no. 1, pp. 115–153, Janaury 2001.
2. Ying-Wei Chen; Xin Xia; Yong-Ge Shi, "A collaborative filtering recommendation algorithm based on contents' genome," in Information Science and Control Engineering 2012 (ICISCE 2012), IET International Conference on, vol., no., pp.1–4, 7–9 Dec. 2012
3. Iwahama, K.; Hijikata, Y.; Nishida, S., "Content-based filtering system for music data," in Applications and the Internet Workshops, 2004. SAINT 2004 Workshops. 2004 International Symposium on, vol., no., pp.480–487, 26–30 Jan. 2004
4. M. Balabanovic, Y. Shoham, "Content-based, collaborative recommendation", Communications of the ACM 40 (3) (1997) 66–72.
5. Adomavicius, G.; Tuzhilin, A., "Toward the next generation of recommender systems: a survey of the stateof-the-art and possible extensions," Knowledge and Data Engineering, IEEE Transactions on, vol.17, no.6, pp.734,749, June 2005
6. D. M. Pennock, E. Horvitz, S. Lawrence, and C. L. Giles, Collaborative filtering by personality diagnosis: a hybrid memory- and model-based approach, in Proceedings of the 16th Conference on Uncertainty in Artificial Intelligence (UAI '00), pp. 473–480, 2000.
7. M. Sarwar, G. Karypis, J. A. Konstan, and J. Reidl, Item-based collaborative filtering recommendation algorithms, in ACM www 01, pp. 285–295, ACM, 2001.
8. M. Deshpande and G. Karypis, Item-based top-N recommendation algorithms, ACM Transactions on Information Systems, vol. 22, no. 1, pp. 143–177, 2004.
9. G. Linden, B. Smith, and J. York, Amazon.com recommendations: item-to-item collaborative filtering, IEEE Internet Computing, vol. 7, no. 1, pp. 76–80, 2003.

10. "Hybrid Collaborative Filtering and Content-Based Filtering for Improved Recommender System" by Kyung-Yong Jung, Dong-Hyun Park and Jung-Hyun Lee in Lecture Notes in Computer Science, 2004, Volume 3036/2004, Springer, pp. 295–302
11. Kumara, B.T.G.S.; Incheon Paik; Ohashi, H.; Yaguchi, Y.; Wuhui Chen, "Context-Aware Filtering and Visualization of Web Service Clusters," in Web Services (ICWS), 2014 IEEE International Conference on, vol., no., pp.89–96, June 27 2014–July 2 2014
12. Lecue, F., "Combining Collaborative Filtering and Semantic Content-Based Approaches to Recommend Web Services," in Semantic Computing (ICSC), 2010 IEEE Fourth International Conference on, vol., no., pp.200–205, 22–24 Sept. 2010
13. SongJie Gong; HongWu Ye; Ping Su, "A Peer-to-Peer Based Distributed Collaborative Filtering Architecture," in Artificial Intelligence, 2009. JCAI '09. International Joint Conference on, vol., no., pp.305–307, 25–26 April 2009
14. Kawai, Y.; Zhang, J.; Kawasaki, H., "Tour recommendation system based on web information and GIS," in Multimedia and Expo, 2009. ICME 2009. IEEE International Conference on, vol., no., pp.990–993, June 28 2009–July 3 2009
15. Liangliang Cao; Jiebo Luo; Gallagher, A.; Xin Jin; Jiawei Han; Huang, T.S., "Aworldwide tourism recommendation system based on geotaggedweb photos," in Acoustics Speech and Signal Processing (ICASSP), 2010 IEEE International Conference on, vol., no., pp.2274–2277, 14–19 March 2010
16. Xinyu Li; Zhongchun Mi; Zhenmei Zhang; Jiani Wu, "A location-aware recommender system for Tourism mobile commerce," in Information Science and Engineering (ICISE), 2010 2nd International Conference on, vol., no., pp.1709–1711, 4–6 Dec. 2010
17. M. Bhandarkar, "MapReduce programming with apache Hadoop," Parallel and Distributed Processing (IPDPS), 2010 IEEE International Symposium on, Atlanta, GA, 2010,

A Study and Mapping of Radiation Levels from Mobile Towers in and Around Bangalore

N. Raghu, N. Krishna Murthy, K. Nagendra and V.N. Trupti

Abstract High-frequency radiations from mobile towers are known to have adverse effect on life of human beings, live stock, and birds. The rapid development of mobile communication has led to the installation of mobile towers in heavily habitat areas, and the cluster of towers being put up by different network providers has led to increased levels of radiation. At certain areas, this level may have reached dangerous levels as to cause long-term effect. The paper aims to map the levels of radiation in selected areas of possible vulnerability and to locate high-risk areas along with the measured radiation levels.

Keywords Electromagnetic radiation · Electromagnetic field (EMF) · Radio frequency (RF) · Very high frequency (VHF) · Extremely low frequency (ELF)

1 Introduction

In recent days, there have been discussions about air pollution, noise pollution, water pollution, and soil pollution, but a new form of pollution, namely electric pollution is causing concern, but it is not recognized as much as other pollutions. Electric pollution is the presence of higher levels of electromagnetic radiation (EM radiation)

N. Raghu (✉) · N.K. Murthy · K. Nagendra · V.N. Trupti
Electrical and Electronics Engineering Department, School of Engineering
and Technology, Jain University, 562112 Ram Nagar, Karnataka, India
e-mail: n.raghu@jainuniversity.ac.in

N.K. Murthy
e-mail: krishnamurthy.access@gmail.com

K. Nagendra
e-mail: kbnagendra@gmail.com

V.N. Trupti
e-mail: trupti.vrn@gmail.com

© Springer Science+Business Media Singapore 2017
S.C. Satapathy et al. (eds.), *Proceedings of the First International Conference
on Computational Intelligence and Informatics*, Advances in Intelligent Systems
and Computing 507, DOI 10.1007/978-981-10-2471-9_60

and electromagnetic fields (EMF) at certain locations which have hazardous short-term and long-term effects on human beings and other living creatures [1].

Even sun rays which are a form of electromagnetic (EM) radiation can be harmful as noticed in cases, where people resorting to sun bathing for long duration have developed skin problems and in extreme cases leading to cancer. Another area where EM radiation was known to cause deleterious effect was X-rays. X-rays has such useful application in the field of medicine, and industry was found to be dangerous after repeated exposures, as was observed by early radiologist, now, X-rays are employed with due care and precaution to avoid its ill effects [2].

In the above context, microwave radiation from mobile towers has drawn the attention of and concern of the users. This problem is to be addressed urgently, and the detailed study of the levels of radiation at potential risky areas, the health problems associate with people, and living creatures in such areas is to be conducted. Such studies have been done in advanced countries and many recommendations have been put forth to limit the radiation to certain level, so that its ill effects are tolerable [3–5]. In our country, no such recommendation exists, and because of this, there is need to do extensive survey of risky areas and their associated radiation levels together with health problems noticed, so that a consensus may be arrived to recommend to limit the radiation to levels tolerable under the existing conditions as a first step towards safety precautions against radiation hazard. Furthermore, continuous studies are required to review the levels suggested and eventually lead to legislation for control of radiation levels in the larger interest of human safety.

The radiation effects cause from very high frequency (VHF) with low power and extremely low frequency (ELF) with high power [6]. This paper deals with very high frequency with low-power measurement of radiation levels and its effects.

2 Present Scenario of Mobile Towers

The mobile towers transmit microwave radiation in the range of 869–894 MHz (CDMA), 935–960 MHz (GSM 900), and power density is in vicinity is about 4.7 W/m^2 and in 1805–1880 MHz (GSM 1800) range is 9.2 W/m^2, and nowadays, 3G network has entered in few cities and 4G is expected in near future [5]. Presently, in India, about seven lakhs mobile towers have been installed. In Bangalore, thousands of mobile towers are placed, and also in most of these towers, four-to-five transmitters are installed due to which the radiation level has increased. A number of places can easily be seen to have been exposed to high levels of radiation and are high-risk areas [7, 8].

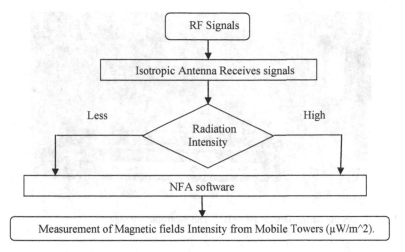

Fig. 1 Measurement of radiation level

It is reported that people are staying close to mobile towers and facing the transmitters and are still constructing high-rise buildings facing the towers and in close proximity to them [1]. Thus, one can see that there are already large number potentially risky areas, where buildings exist and new buildings are being constructed continuously.

3 Methodology

The step-by-step procedure followed to measure electromagnetic fields intensity (μW/m^2) is as shown in the flow chart below (Fig. 1). The available radio frequency signals in free space were collected using isotropic antenna. The radiation level present in the radio signals is loaded to the HF59B^{TM} device which is analyzed using NFA soft.

4 Measurement Setup

To measure the radiation levels of these sites, a measurement setup consisting of HF59B^{TM} device and isotropic antenna is used (Fig. 2). This device receives the maximum radiation signals available in free space which are emitted from mobile

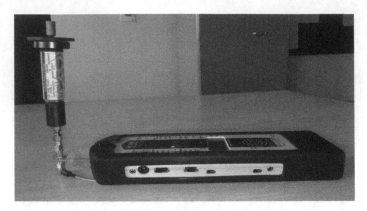

Fig. 2 Radiation measuring device HF59B™ and isotropic antenna

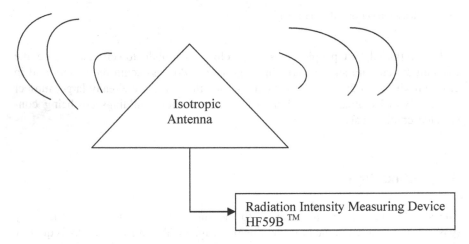

Fig. 3 Experimental setup to measure radiation levels

towers from different directions at several places. The measurement of radiation setup is as shown below (Fig. 3). This survey was conducted at noon between 12:00 pm and 4 pm.

5 Effects on Human Beings

The radiations emitted from mobile towers have a direct impact on human life. These radiations are classified as low-level and high-level radiation. The exposure to these radiations has certain biological effects which are listed below.

1. Low-level radiation exposure effects are every small and they are not detected.
2. In high-level radiation, when there is less exposure, it causes skin damage, nausea, vomiting, fatigue, sleep disturbances [9], headache [1], hearing problem, dizziness [2], bone marrow damage, damage in white blood cell [10], and damage to cells lining the small intestine.
3. Normal exposure of radiation (0.002 µW/m^2) which causes sleep disorders, abnormal blood pressure, weakness, limb pain, joint pain [10], digestive problem [10], and also even damage to the DNA of human body has been reported in some cases [2, 11, 12].
4. In high-level radiation with prolonged exposure, it has resulted in the following hazards.

   (a) Among children, it has resulted in leukemia [1, 2], growth retardation, small head, brain size, and mental retardation.
   (b) Among adults, breast cancer, bladder colon, liver cancer, lung cancer, stomach cancer, and ovarian cancer have occurred [2, 4, 6].

5. In many cases, loss of hair has been noticed among people exposed to medium level of radiation.

6 Radiation Levels Measured and Analysis of the Data Obtained

The radiation norm given by ICNIRP (International Commission for Non-Ionizing Radiation Protection) guidelines in India at 1998 for safe power density is given by f/200, where frequency (f) is in megahertz, has been considered [5].

The reason for selecting is that particular region is mainly because its densely populated area and it also includes major public places, such as hospitals, schools, factories, temples, etc. The route for measurement in Bangalore was started using a ground position system (GPS) instrument for navigation from Minerva Circle to konanakunte cross for the first set of readings, as shown in route map (Fig. 4), and measured less than 2 kHz radiation levels are showed in (Fig. 5), more than 2 kHz radiation levels are showed in (Fig. 6). The second set of readings was measured from konanakunte cross to Kagglipura, as shown in route map (Fig. 7) and measured less than 2 kHz radiation levels are showed in (Fig. 8), more than 2 kHz

Fig. 4 GPS route from Minerva circle to konanakunte cross

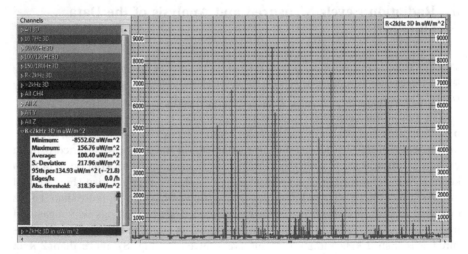

Fig. 5 Radiation levels from Minerva circle to konanakunte cross (<2 kHz)

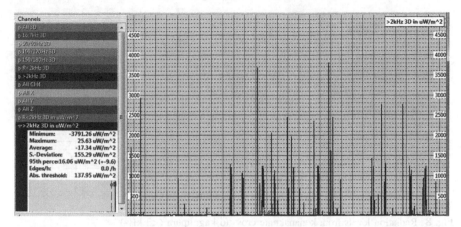

Fig. 6 Radiation levels from Minerva circle to konanakunte cross (>2 kHz)

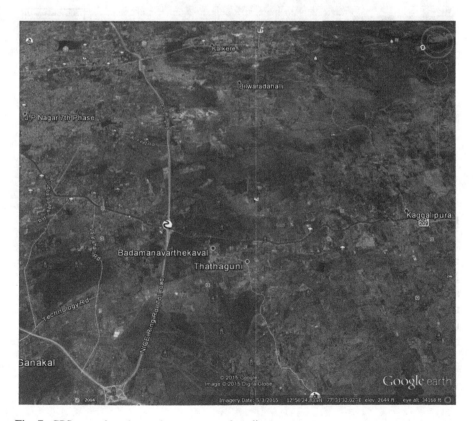

Fig. 7 GPS route from konanakunte cross to kagglipura

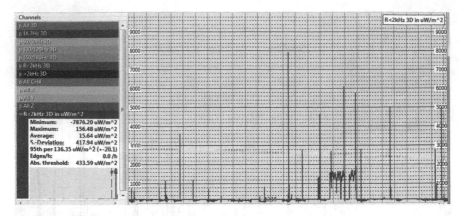

Fig. 8 Radiation levels from konanakunte cross to kagglipura (<2 kHz)

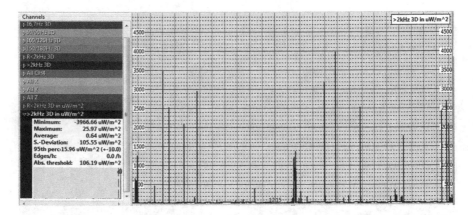

Fig. 9 Radiation levels from konanakunte cross to kagglipura (>2 kHz)

radiation levels are showed in (Fig. 9). The third set of readings was measured from Minerva circle to City Railway station as shown in route map (Fig. 10) and measured less than 2 kHz radiation levels are showed in (Fig. 11), more than 2 kHz radiation levels are showed in (Fig. 12). Table 1 show the radio frequency radiation (RFR) levels in various places in the range less than 2 kHz and greater than 2 kHz around the Bangalore region.

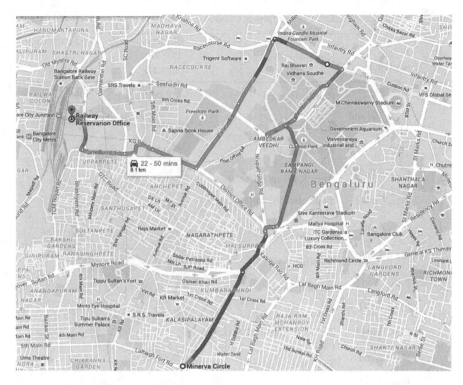

Fig. 10 GPS route from Minerva circle to city railway station

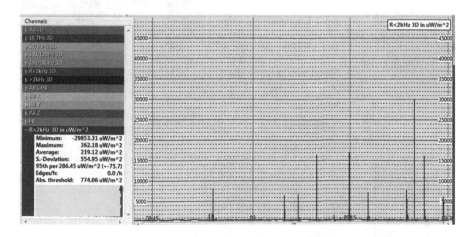

Fig. 11 Radiation levels from Minerva circle to city railway station (<2 kHz)

Table 1 Estimation of radiated power (power density) from mobile towers at various locations

Sl. no.	Figure reference	Name of the place	Electromagnetic field intensity ($\mu W/m^2$)	
			<2 KHz	>2 KHz
1.	Figs 10 and 11	Railway station	30000	2900
2.	Figs 10 and 11	Raj bahavan	17000	3100
3.	Figs. 10 and 11	K R circle	16000	3900
4.	Figs. 4 and 5	Ande Ka Fanda, JP nagar	8600	2200
5.	Figs 7 and 8	Yellamma dasappa institute of technology	7900	1450
6.	Figs. 4 and 5	Minerva circle	7800	2900
7.	Figs. 4 and 5	Sarakhi signal	7400	3800
8.	Figs 7 and 8	Art of living	6100	200
9.	Figs. 4 and 5	Konanakunte cross	6200	1350
10.	Figs 7 and 8	Shell petroleum kanakapura road	5800	2500
11.	Figs. 7 and 8	Kagglipura bustand	5000	400
12.	Figs. 4 and 5	Bansankari	4600	1200
13.	Figs. 4 and 5	J P nagar link road	4000	100
14.	Figs. 7 and 8	KSIT Engineering college	2800	200
15.	Figs. 4 and 5	Jaraganahalli govt. school	2400	1250
16.	Figs. 4 and 5	Metro station, jayanagar	400	100
17.	Figs. 4 and 5	Canara bank (South end road) jayanagar	300	1300
18.	Figs. 4 and 5	Southend circle	250	900
19.	Figs. 4 and 5	Nanda takies	200	100

Very high radiation level [4].

High radiation level [4].

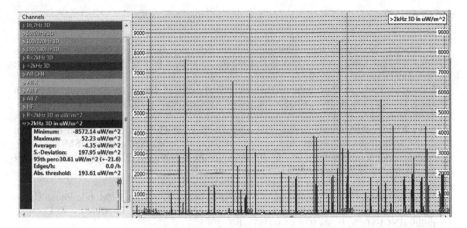

Fig. 12 Radiation levels from Minerva circle to city railway station (>2 kHz)

7 Conclusions

Serious health effects wave been noticed at radiation levels as low as 100 μw/m². However, it is abnormally low level to be fixed, because the mobile operators lose connectivity. Therefore, many countries have adopted practical levels to be fixed immediately to 100,000 μw/m². This should be gradually reduced to 10,000 μw/m², within a couple of years. As seen from the measurements, many areas level greater than 10,000 μw/m² and may approach to 100,000 μw/m². Thus, the situation is sounding alarm and authorities have to wake up to limit the radiation level to 10,000 μw/m² in the interest of human safety and also to avoid the installation of mobile towers on school buildings, hospitals, and residential houses in close vicinity and facing such premises.

8 Future Scope of Investigation

More intensive survey of radiation together with a survey of associated health problems in such areas in cooperation with medical institutions is urgently needed to know the of current state of hazardous level.

Acknowledgments This work is carried out at project laboratory department of Electrical and Electronics Engineering and the authors would like to thank all the faculty members of Department of Electrical & Electronics Engineering and management of School of Engineering & Technology, Jain University.

References

1. B. Blake Levitt, Henry Lai, "Biological effects from exposure to electromagnetic radiation emitted by cell tower base stations and other antenna arrays" in NRC Research Press, Rev. 18: 369–395 (2010).
2. Lalrinthara Pachuau, Zaithanzauva Pachuau, "Study of Cell Tower Radiation and its Health Hazards on human body", in IOSR Journal of Applied Physics (IOSR-JAP), e-ISSN: 2278–4861. Volume 6, 2014.
3. Neha Kumar, Prof. Girish Kumar, "Biological Effects of Cell Tower Radiation on Human Body" in International Symposium on Microwave and Optical Technology (ISMOT), 2009.
4. Mobile Telecommunications and health research programme (MTHR) Report 2007, http://www.mthr.org.uk/documents/MTHR_report_2007.pdf.
5. Prof. Girish Kumar, "Report on Cell Tower Radiation", 2010.
6. R. Seetharaman, G.S. Uthayakumar, N. Gurusamy and N. Kumaravel, "Mobile Phone Usage and Cancer" in 10th International Conference on Electromagnetic Interference & Compatibility (INCEMIC), ISBN 978-81-903575-1-7, 2008.
7. Mohit kaushal, Tanvir Singh and Amit kumar, "Effects of mobile towers radiations and case studies from different countries pertaining the issue" in International Journal of applied Engineering research, ISSN 0973-4562 Vol. No. 11, 2012.
8. V S Tanwar," Living Dangerously in Indian Cities: an RF Radiation Pollution Perspective", in 9th International Conference on Electromagnetic Interference and Compatibility (INCEMIC)", ISBN 978-1-4244-5203-3, 2006.
9. Dimitris J. Panagopoulos, "Electromagnetic Interaction between Environmental Fields and Living Systems Determines Health and Well-Being", in Electromagnetic Fields: Principles, Biophysical Effects, ISBN: 978-1-62417-063-8, 2013.
10. Firstenberg, "Radio wave packet", 2001.
11. Levitt B, Lai H, "Biological effects from exposure to electromagnetic radiation emitted by cell tower base stations and other antenna arrays", Environ. Rev. 18: 369–395, 2010.
12. Dimitris J. Panagopoulos, "Analyzing the Health Impacts of Modern Telecommunications Microwaves", in Advances in Medicine and Biology, ISBN: 978-1-61122-790-1, 2011.

Low-Cost Smart Watering System in Multi-soil and Multi-crop Environment Using GPS and GPRS

Sudheer Kumar Nagothu and G. Anitha

Abstract Nowadays, it is a common practice to cultivate various crops in a field. Normal watering system will not work well when there is multi-crop in the field, because each crop requires different levels of watering. The same problem exists when it is multi-soil land. When moisture sensor is used in this context, many numbers of sensors are placed at different crop locations, and their data are analyzed to water the plants. To collect the data from various sensors, and processes them, and to turn on/off, the sprinklers or some other watering system based upon the data is a complex process. Each moisture sensor requires some power source for its operation (normally battery is used). Here, an idea is proposed to use a robot with moisture sensor and GPS. The robot will move around the field, which will test the moisture for every 10 m, it has moved, based up on the moisture data, the sprinklers system in that region will be on/off. Using weather data provide by meteorological department, watering to the crop can be adjusted.

Keywords GPS · GPRS · Low-cost watering system · Weather data

1 Introduction

Water is a precious resource, which should be carefully used. Farmers are watering the plants, without understanding the amount of water required for the plants. When various crops are planted in the field, each crop requires various levels of moisture, for example, paddy may require heavy moisture compare with maize, etc. The level of watering also will vary based upon the stage of crop. To solve all these problems, an idea is proposed for multi-crop and multi-soil land irrigation. Figure 1 shows the

S.K. Nagothu (✉) · G. Anitha
Division of Avionics, Department of Aerospace Engineering, Anna University Chennai MIT Campus, Chennai 600044, India
e-mail: sudheernagothu@gmail.com

© Springer Science+Business Media Singapore 2017
S.C. Satapathy et al. (eds.), *Proceedings of the First International Conference on Computational Intelligence and Informatics*, Advances in Intelligent Systems and Computing 507, DOI 10.1007/978-981-10-2471-9_61

637

Fig. 1 Block diagram of robot

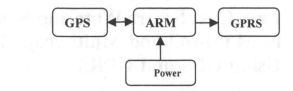

Fig. 2 Hardware kit of navigation for robot

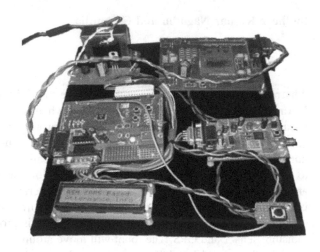

block diagram of the equipment in the robot, and Fig. 2 represents the hard kit of navigation required for the robot.

The subsystems GPS and GPRS are integrated along with moisture sensor using ARM microcontroller. The moisture readings are sent to the server along with the position, where the measurement is made, which is used to regulate water flow to the crops.

Normally, in these types of applications, many moisture sensors are used. It is not only a costly affair to maintain those sensors using batteries, but also it will be very problematic to find the location of sensor once we forgot the location where it is buried. Special scanners need to be used to detect them, which consumes a lot of time and manpower. Integrated GPS and GPRS system are used in many applications [1–3]. Here one such type of application in agriculture is discussed.

2 System Working

Initially, the position of the various crops in agriculture field is noted in terms of latitude and longitude. Various sprinklers are placed in all different crops. The robot will move around the field, which uses sense and avoid technology to avoid if there is any block in front of it, but it will move the entire field using GPS available in it.

For every 20-m distance, it has traveled, the moisture sensor will be made to penetrate into earth, and the moisture content is sent to the server along with position where the content is measured. This position is checked against the crop position database, i.e., to check what crop is there in that location, and whether moisture level in that soil position is optimum or not, and if watering is required for that crop, respective sprinkler system can be on.

The moisture sensor robot readings can be compared with crop position and sprinkler system position using haver-sine formula. Here, a square type of coverage architecture is followed to avoid the ambiguity. So, at any time, the robot will be in anyone square block. The hexagonal type of division may be more useful to cover the entire area, but normally crops are planted in square format only. Equal-size square boxes are considered here to reduce complexity.

The sprinkler is located at the center of the square block, which can cover up to 10-m radius. So, normally for every 20-m distance, a sprinkler system is available.

$$a = \sin^2\left(\frac{\phi_2 - \phi_1}{2}\right) + \cos(\phi_2) * \cos(\phi_1) \sin^2\left(\frac{\lambda_2 - \lambda_1}{2}\right)$$

$$d = 2 * R_E * a \tan 2\left(\frac{\sqrt{a}}{\sqrt{(1-a)}}\right)$$

where ϕ_1 and λ_1 are latitude and longitude of robot position, ϕ_2 and λ_2 are latitude and longitude of crop position, and R_E is radius of earth (around 6371000 m), and all angles are in radians. Distance in meters is given by d.

The working model of robot is given in Fig. 3. The agriculture field is shown in Fig. 4, which is subdivided into various parts a, b c d, e, f, etc., as shown in Fig. 5.

Fig. 3 Robot in agriculture field

Fig. 4 Agriculture field

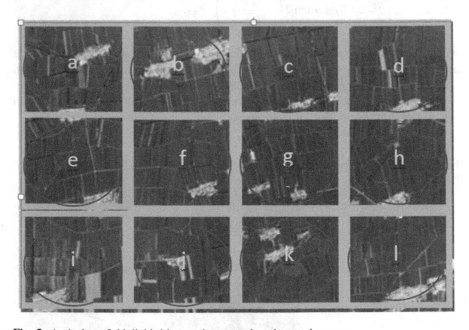

Fig. 5 Agriculture field divided into various parts based upon the crop

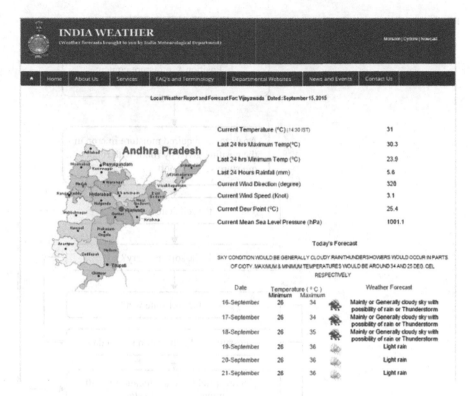

Fig. 6 Weather data from Indian Meteorological Department

The position of the robot is compared with the center position of the crop (Sprinkler position), which results in a distance less than or equal to 10 m with respect to the sprinkler. The moisture content from the robot is analyzed, along with weather data. The required weather data can be gathered from the Website http://www. indiaweather.gov.in/, for Indian users, which is shown in Fig. 6. These data can be stored in the database. The weather data for the next 5 days can be predicted, and watering can be adjusted based upon moisture sensor data and weather prediction. When the soil moisture is very dry, watering can be done immediately irrespective of the weather. But when soil is dry, weather data can be taken into consideration to water the plants. If there is any prediction that rain will occur in immediate 2 to 3 days, watering can be delayed. The flow of the working model is shown in flow chart (Fig. 7).

As shown in the flow chart, when soil state is dry, weather data predicted for the immediate days are checked, and if rain is predict in next 2 to 3 days, watering can be stopped. When the soil state is very dry, immediate watering will occur using sprinkler.

Fig. 7 Flow chart

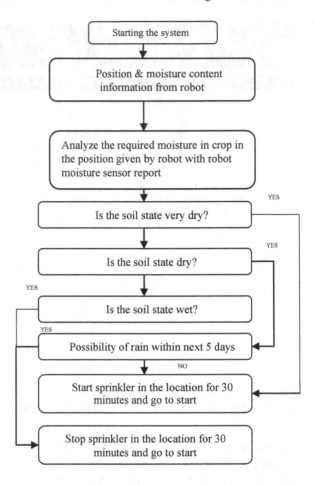

3 Conclusion

The multi-crop and multi-soil watering system has been proposed in a low cost by using a single moisture sensor. The problem of forgetting the buried moisture sensor position can be solved with this technique, since we are using a single sensor which will move around the field, whose position is known for each and every instant. By using GPS, a well-structured system is proposed, which will ON/OFF the sprinkler in the respective location by analyzing the moisture content and crop on that location along with weather data.

References

1. Sudheer Kumar Nagothu, Om prakash kumar, G Anitha, "Autonomous monitoring and attendance system using inertial navigation system and GPRS in predefined location", 2014 3rd International Conference on Eco-friendly Computing and Communication Systems (ICECCS), Year: 2014, Pages: 261–265, doi:10.1109/Eco-friendly.2014.60.
2. Sudheer Kumar Nagothu, G.Anitha, Annapantula Sudhakar, "Navigation aid for pepole (Joggers and runners) in unfamiliar urban environment using Inertial Navigation", 2014 Sixth International Conference on Advanced Computing (ICoAC), Year: 2014, Pages: 216–219, doi:10.1109/ICoAC.2014.7229713.
3. Sudheer Kumar Nagothu, Om prakash kumar, G Anitha, "GPS Aided Autonomous Monitoring and Attendance System" Fourth International Conference on Recent Trends in Computer Science & Engineering (ICRTCSE 2016), Procedia Computer Science Volume 87, 2016, Pages 99–104 doi:10.1016/j.procs.2016.05.133.

An Image Encryption Technique Using Scan Based Approach and Image as Key

Shrija Somaraj and Mohammad AliHussain

Abstract In this paper, a new algorithm for image encryption using the scan method is proposed. Using SCAN language, it is possible to generate a wide range of scanning paths based on the spatial accessing methodology. The proposed algorithm is implemented using different gray and color images, and the experimental results and security analysis indicate the advantages of the proposed algorithm. The original image can be reproduced using this algorithm without any loss of information. The algorithm is simple and fast as compared with other recent approaches. It is secure enough to be used in a wide range of applications, as it has passed all the security requirements. This paper presents an overview of the encryption and decryption process using the proposed algorithm. The implementation of the algorithm is done in MATLAB and tested on various gray and color images.

Keywords Image encryption · Image decryption · Security analysis · Scan patterns

1 Introduction

Security of information has become a major requirement in today's scenario as data and information can be accessed through varied sources, such as network, Web, cloud, and many more. Many ways of securing information by encryption have already been proposed like Data Encryption Standard (DES), RSA Algorithm,

S. Somaraj (✉)
Research and Development Centre,
Bharathiar University Coimbatore, Chennai, India
e-mail: shrijamadhu@yahoo.co.in

M. AliHussain
Department of Computer Science and Engineering,
Andhra Loyola Institute of Engineering and Technology,
Vijayawada, AP, India
e-mail: alihussain.phd@gmail.com

© Springer Science+Business Media Singapore 2017   645
S.C. Satapathy et al. (eds.), *Proceedings of the First International Conference on Computational Intelligence and Informatics*, Advances in Intelligent Systems and Computing 507, DOI 10.1007/978-981-10-2471-9_62

Blowfish algorithm, Chaos-Based Methods, Hash-Based Methods, Scrambling-Based Methods, and many other symmetric and asymmetric methods [1–5].

In 1986, Bourbakis proposed a combined compression, encryption, and hiding system which were based on the SCAN language [6]. "SCAN" refers to the different ways of scanning a 2D image. The SCAN language can generate (nxn)! scanning paths for an image of nxn size based on a 2-D spatial accessing method. The SCAN algorithm can determine an optimal scanning path which minimizes the number of bits used for encoding the scanning path and also the bit sequence. This method can compress an image by specifying a suitable scanning path for the image in an encoded form. After compression, the encrypted image is generated by rearrangement of bits of the compressed image. A set of scanning paths are used for the rearrangement of the bits which are kept secret. The set of scanning paths being used forms the key for encryption. The level of security achieved is high and it is highly impossible to find key using currently available computational technologies. The SCAN methodology is suitable for compression, encryption, and hiding information in multimedia-based applications [7–17].

The SCAN language has some partition patterns and scanning patterns. Both have some transformations also.

1.1 Partition Patterns

There are three basic partition patterns that include (Fig. 1)

- B-type partition patterns
- Z-type partition patterns
- X-type partition patterns

Each basic partition pattern has eight different transformations. These depend on the initial point and the final point which B-type partition pattern with its eight transformations can be defined from B0 to B7 (Fig. 2).

Z-type partition pattern with its eight transformations can be defined from Z0 to Z7 (Fig. 3).

X-type partition pattern with its eight transformations can be defined from X0 to X7 (Fig. 4).

Fig. 1 Basic partition patterns B type, Z type, and the X type

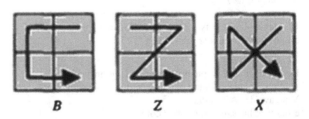

B   Z   X

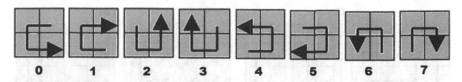

Fig. 2 Transformations of B-type partition pattern B(0–7)

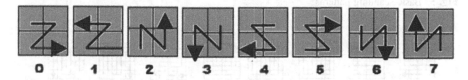

Fig. 3 Transformations of Z-type partition pattern Z(0–7)

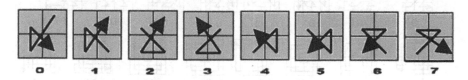

Fig. 4 Transformations of X-type partition pattern X(0–7)

1.2 Basic Scanning Pattern

There are four basic scanning patterns

- Spiral—(S) Pattern
- Continuous Orthogonal—(O) Pattern
- Continuous Diagonal—(D) Pattern
- Continuous Raster—(C) Pattern

Figure 5 shows the four basic scanning patterns (a) represents the Spiral—S pattern, (b) represents the Continuous Orthogonal—O pattern, (c) represents the Continuous Diagonal—D pattern, and (d) represents the Continuous Raster—C pattern. The scanning patterns can be rotated through angles of $0°$, $90°$, $180°$, and $270°$. If these are represented as S0, S2, S4, and S6, then the reverse patterns can be represented as S1, S3, S5, and S7 which is shown in Fig. 6. Eight transformations of scanning pattern C are shown in Fig. 6a, eight transformations of scanning pattern D are shown in Fig. 6b, eight transformations of scanning pattern S are shown in Fig. 6c, and eight transformations of scanning pattern O are shown in Fig. 6d.

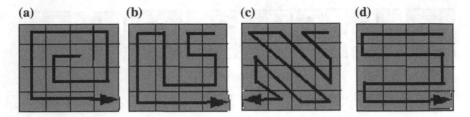

Fig. 5 Basic scanning patterns

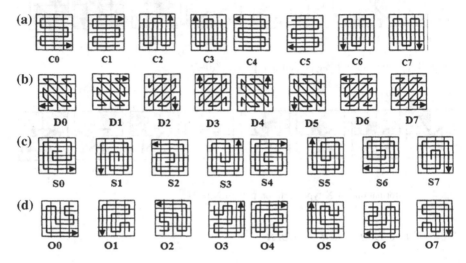

Fig. 6 Showing the eight transformations: **a** scanning pattern C (C0–C7), **b** scanning pattern D (D0–D7), **c** scanning pattern S (S0–S7), and **d** scanning pattern O (O0–O7)

2 Methodology

2.1 Algorithm for Encryption

Step 1. Take original image and key image of the same size.

Step 2. Select a scanning pattern from the four basic patterns Spiral (S), Continuous Diagonal (D), Continuous Orthogonal (O), and Continuous Raster (C) for original image and store the generated image.

Step 3. Select a scanning pattern from the four basic patterns Spiral (S), Continuous Diagonal (D), Continuous Orthogonal (O), and Continuous Raster (C) for key image and store the generated image.

Step 4. Perform XOR (bitxor() in MATLAB) on the images generated in Step 2 and Step 3 to get the resultant image.

2.2 Algorithm for Decryption

Step 1. Take Encrypted Image.
Step 2. Select the key image (secret key) used in encryption process.
Step 3. Select a scanning pattern for key image from the four basic patterns Spiral
(S), Continuous Diagonal (D), Continuous Orthogonal (O), and Continuous
Raster (C), and store the generated image.
Step 4. Perform XOR (bitxor() in MATLAB) on the images in Step 1 and Step 3 to
get an intermediate image.
Step 5. Select the same scanning pattern from the four basic patterns Spiral (S),
Continuous Diagonal (D), Continuous Orthogonal (O), and Continuous
Raster(C) for image in Step 4, as done for original image in the encryption
process.

Let the original image be I, key image be K, S and S1 are two different scanning
patterns, and E is the encrypted image; then, the process of encryption and
decryption can be represented as follows:
Encryption Process

$$I + S \rightarrow I1 \tag{1}$$

$$K + S1 \rightarrow K1 \tag{2}$$

$$E \leftarrow I1 \oplus K1 \tag{3}$$

Decryption Process

$$K + S1 \rightarrow K1 \tag{4}$$

$$D \leftarrow E \oplus K1 \tag{5}$$

$$D + S \rightarrow I \tag{6}$$

In Eq. (1) scanning pattern S is applied on original image I which produces the
image I1. Equation (2) represents the scanning pattern S1 applied on key image K
generating the K1 image. Next, in Eq. (3), bitwise XOR operation is applied on the
scanned I1 and K1 images. Equations (4)–(6) represent the decryption process,
where first scanning pattern applied on key image is the same as encryption; next,
Eq. (5) shows the application of bitwise XOR on the encrypted image E and the
scanned key K1, Finally, Eq. (6) shows the application of the same scanning pattern
used in encryption process on the image D generated in the previous step (5) which
results in getting back the original image.

Fig. 7 Gray image encryption using scan method: **a** key image, **b** original images, **c** encrypted images, and **d** decrypted original images

3 Experimental Results

The proposed algorithm is implemented in MATLAB 7.0 and the images are taken from USC-SIPI database [18]. The algorithm is suitable for encrypting both gray and color images. It can also work on different file formats like tiff, bmp, jpeg, pgm, png, etc. In Fig. 7, encryption and decryption of gray images using the proposed method are shown, where (a) shows the key image, (b) represents different original plain images, (c) shows corresponding encrypted images, and (d) shows the decrypted images. In Fig. 8, encryption and decryption of color images using the proposed method are shown, where (a) shows the key image, (b) represents

(a)

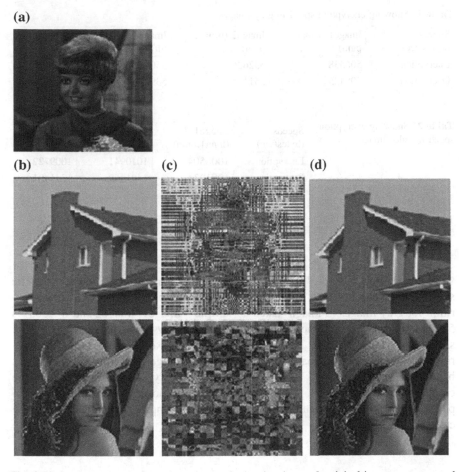

(b)   **(c)**   **(d)**

Fig. 8 Color image encryption using scan method: **a** key image, **b** original images, **c** encrypted images, and **d** decrypted original images

different original plain images, (c) shows corresponding encrypted images, and (d) shows the decrypted images.

4 Performance Analysis

The execution speed of an algorithm is also an important aspect for analyzing the performance of the algorithm. Both encryption and decryption speeds can be measured based on the size (bytes) encrypted or decrypted and the time (sec) taken for encryption and decryption. The proposed scheme was implemented using MATLAB 7 and the execution time was measured on an Intel Core 2 Duo processor

Table 1 Showing encryption speed in gray images

Speeds (bytes/sec)	Image1 (papav. pgm)	Image2 (tulips. pgm)	Image3 (cameraman. tiff)	Avg
Encryption	506538	552626	550167	536443
Decryption	508425	551185	539924	533178

Table 2 Showing encryption speed in color images

Speeds (bytes/sec)	Image1 (lena1.bmp)	Image2 (4.1.05.tiff)	Avg
Encryption	1008504	1010941	1009722
Decryption	1281334	1401469	1341401

Table 3 Comparison of encryption speeds

Speeds (bytes/sec)	AES	MASK	SCAN
Encryption	8100.87	48907.46	550167

with processing speed of 3.0 GHz and 3 GB RAM, running Windows 8.1. The simulation results in Table 1 show that the average execution speed of the algorithm for gray images (3 images taken as sample—Image1, Image2 and Image3) is 536443 bytes/second for encryption and 533178 bytes/second for decryption. The simulation results in Table 2 show that the average execution speed of the algorithm for color images (2 images taken as sample—Image1 and Image2) is 1009722 bytes/s for encryption and 1341401 bytes/s for decryption. Table 3 shows the comparative study of AES, MASK, and SCAN (proposed) algorithms. The proposed algorithm shows a better encryption speed as compared with the others in the comparison.

5 Conclusion

From the experimental results, we conclude that the proposed method of image encryption using SCAN approach gives very good results for both color and gray images. To keep it simple, only single scan pattern is used at a time, but if required a hybrid approach, a combination of two scan patterns and partitions can also be employed. With the increase in complexity, the encrypted image appears more distorted as compared with the single-level encryption. It has high processing speed. The proposed method of encryption uses simple XOR operation and integer arithmetic which can be easily implemented in the hardware. The simulation results show that the proposed algorithms have great performance in terms of security, speed, and sensitivity.

References

1. Pareek, N. K., Patidar, V., Sud, K. K.: Image encryption using chaotic logistic map. J. Image and Vision Computing, 926–934 (2006).
2. Zhou, Y., Cao, W., Philip Chen, C. L.: Image encryption using binary bitplane. J. Signal Processing 100,197–207 (2014).
3. Zhu, Z. L., Zhang, W., Wong, K. W., Yu, H.: A chaos-based symmetric image encryption scheme using a bit-level permutation. J. Information Sciences 181, 1171–1186, Elsevier (2010).
4. Zhang, C. Y., Zhang, W. X., Weng, S. W.: Comparison of Two Kinds of Image Scrambling Methods Based on LSB Steganalysis. J. of Information Hiding and Multimedia Signal Processing, vol. 6, no. 4 (2015).
5. Yuan, H., Jiang, L.: Image Scrambling based on Spiral Filling of Bits. J. International Journal of Signal Processing, Image Processing and Pattern Recognition vol. 8, no. 3, pp. 225–234 (2015).
6. Bourbakis, N.: A Language for Sequential Access of Two Dimensional Array Elements. IEEE Workshop on LFA, pp 52–58, Singapore(1986).
7. Rad, R. M., Attar, A., Atani, R. E.: A New Fast and Simple Image Encryption Algorithm Using Scan Patterns and XOR. J. International Journal of Signal Processing, Image Processing and Pattern Recognition,vol. 6, no. 5, pp. 275–290 (2013).
8. Li,C.,Lo,K.T.: Cryptanalysis of an Image Encryption Scheme Using Cellular Automata Substitution and SCAN. In: PCM 2010, Springer-Verlag Berlin Heidelberg, LNCS 6297, pp. 601–610 (2010).
9. Panduranga, H. T., Naveen Kumar, S. K.: Hybrid approach for Image Encryption Using SCAN Patterns and Carrier Images.J. International Journal on Computer Science and Engineering, vol. 02, no. 02, pp. 297–300 (2010).
10. Bourbakis, N., Alexopoulos, C.: A Fractal Based Image Processing Language – Formal Modeling. J. Pattern Recognition Journal, vol 32, no 2, pp. 317–338 (1999).
11. Alexopoulos, C., Bourbakis, N., Ioannou, N.: Image Encryption Method Using a Class of Fractals. J. Journal of Electronic Imaging, pp 251–259 (1995).
12. Maniccam, S. S., Bourbakis, N.: "Image and Video encryption using SCAN Patterns. Pattern Recognition, vol.37, pp. 725–757 (2004).
13. Chen, C. S., Chen, R. J.: Image Encryption and Decryption using SCAN Methodology. In: Proc. PDCAT, IEEE (2006).
14. Kachriset, C.:A reconfigurable logic based processor for the scan image and video encryption algorithm. In: IJPP, vol.31, no.6, pp. 489–506 (2003).
15. Bourbakis, N.: Image Data Compression Encryption Using G-SCAN Patterns. IEEE Confon SMC, pp. 1117–1120 (1997).
16. Maniccam, S. S., Bourbakis, N.: Lossless image compression and encryption using SCAN. J. of Pattern Recognition, vol. 34, no. 6, pp. 1229–1245 (2001).
17. Maniccam, S. S., Bourbakis, N.: SCAN Based Lossless Image Compression and Encryption. In: The IEEE International Conference on Information Intelligence and Systems, ICIIS, Washington, DC, (1999).
18. USC-SIPI Image Database, http://sipi.usc.edu/database/.

IPMOntoShare: Merging of Integrated Pest Management Ontologies

Archana Chougule and Vijay Kumar Jha

Abstract Integrated pest management (IPM) is a combination of different techniques to increase crop production in eco-friendly manner. Minimizing use of pesticides with IPM will reduce risk of human diseases and will also reduce environmental risks. Various computerized systems are used for IPM, where agricultural experts provide their pest management knowledge as input for decision-making. Integrated pest management knowledge if represented as ontology, it can be shared by heterogeneous agricultural computerized systems. This paper presents a tool to develop IPM ontology using upper IPM ontology and domain specific crop IPM ontology. Tool is named IPMOntoDeveloper. IPM ontologies developed by distinct agricultural experts can be integrated into one to enrich knowledge base of IPM practices for specific crop. This paper presents a system named IPMOntoShare to merge IPM ontologies developed by various agricultural experts. It combines several approaches of ontology matching, including name matching and structure matching.

Keywords Integrated pest management · Knowledge sharing · Ontology development · Ontology merging

1 Introduction

Integrated pest management is the need of time. It is observed that excessive use of pests has adverse effect on human health and also on environment. A number of diseases are caused because of toxic pesticides used to control pests on crops. As pesticides are sprayed evenly on complete crop field, it can affect soil, running

A. Chougule (✉) · V.K. Jha
Birla Institute of Technology, Mesra, Ranchi, India
e-mail: chouguleab@gmail.com

V.K. Jha
e-mail: vkjha@bitmesra.ac.in

© Springer Science+Business Media Singapore 2017   655
S.C. Satapathy et al. (eds.), *Proceedings of the First International Conference on Computational Intelligence and Informatics*, Advances in Intelligent Systems and Computing 507, DOI 10.1007/978-981-10-2471-9_63

water and air at the crop field and across the crop field. It also affects other species which are not targeted by the pesticide.

To reduce these risks, integrate pest management (IPM) technique is used. The use of computerized systems to implement IPM techniques is obvious. As these systems are of varying kind, there must be a way to represent knowledge which can be easily shared among heterogeneous systems. One of the best approaches to represent domain knowledge is through ontologies. Formal representation of concepts, relationships, assumptions, and constraints in specific domain can be presented as ontology. The ontology can be used as a classification tool in specific domain as it defines structure and hierarchy of concepts in the domain. Hence, an easy to use tool for agricultural experts named IPMOntoDeveloper is proposed here. It uses ontology of core IPM techniques as upper ontology for the development of crop specific IPM ontology. While building ontology, viewpoints of developers can be different. To support sharing of IPM ontologies developed by various agricultural experts, ontology merging is required. Creating a new ontology from two source ontologies is called as ontology merging. This paper introduces IPMOntoShare, a system for mapping and merging of two IPM ontologies. Comparison of similarity of concepts at each level of ontology is discussed in detail.

This paper is organized as follows. Section 1 details ontology merging approaches proposed by various authors. Section 2 describes working of IPMOntoDeveloper, a tool for IPM ontology development. Merging of IPM ontologies using IPMOntoShare is detailed in Sect. 3. In Sect. 4, we demonstrate the use of IPMOntoShare for integrating IPM ontologies of rice pests. The conclusion is given in Sect. 5, and references are at the end of the paper.

2   Literature Survey

A number of approaches for ontology merging are proposed by various authors. This section discusses some of those approaches. Li et al. [1] provide approach to ontology merging using concept lattice technique which comes under formal concept analysis. They define matrix for ontology matching, which mentions similarity between instances, definition, and structure of ontologies, respectively. They propose assigning of weights to all these parameters for measuring similarity. They take threshold value from user for deciding similarity. Based on threshold value and calculated similarity value, it is decided whether two ontologies are similar. The next step is to build concept lattice using concept lattice construction algorithm. Using generated concept lattice, global ontology is generated. The relationships in ontologies are constructed at last step.

Cuevas Rasgado et al. [2] present special notations for ontology merging called as ontology merging (OM) notation. They also present ontology merging algorithm where merging takes place automatically without user intervention. They provide list of labels to be used in ontology which identify the description of the concepts and relations among them. In OM algorithm, similarity of concept from one ontology is measured with the concept from other. Similar concepts are added once to a new third ontology, and all differing concepts from both ontologies are added as new concept in third ontology. They also consider relation similarity. For concepts which are synonyms of each other, they use COM algorithm of OM. They also mention about removing nested relations. An ontology merging method based on WordNet is proposed by Kong et al. [3]. Equality of concepts is measured using WordNet. They also measured values of sets of concepts using Jaccord coefficient and most-specific parent method. Based on these measured values, they reconstructed the hierarchy. Target driven merging of ontologies is proposed by Raunich et al. [4]. The approach is based on equivalence matching between a source taxonomy and target taxonomy for merging. They use integrated concept graph to adapt and extend properties of merged taxonomy.

Ontology merging using machine learning techniques is discussed by Richardson et al. [5]. The use of hierarchical clustering algorithm for ontology learning and the use of Bayesian theorem, cosine, and KL divergence functions are discussed in this paper. The use of description logic and description graph for merging domain ontologies is proposed by Gupta et al. [6]. Combination of lexical, semantic, and rule-based methods is described in [7]. For semantic matching, OpenCyc and WordNet are used for finding synonyms. Simple hearst patterns and propositional formulae are used for rule based matching. The concepts are merged only if the similarity value is above a user-defined threshold. Similar approach is mentioned in [8]. Detail algorithms for ontology merging, attribute merging, relation merging, and superclass merging are explained in the paper. Prompt plug-in available with protégé can be used for ontology merging. Semi-automatic approach of merging is use by Prompt. Detail algorithm used by Prompt is described by de Araujo et al. [9]. Clustering technique for merging multiple ontologies is proposed by de Araújo et al. [10]. They generate similarity matrix by matching classes and properties. Consistency checking of ontology mappings using ontology evaluation knowledge within the semantic knowledge of merging system is proposed by Fahad and Abdul Qadir [11]. Fully automated ontology matching using upper ontologies is proposed by Mascardi et al. [12]. Three algorithms namely uo_match, structural_uo_match, and mixed_match are implemented for ontology matching. They used SUMO-OWL as upper ontology for running experiments. The implementation is done using the Alignment API. Computing similarity between ontologies using change weights semantic graph is put forward by Yang et al. [13]. They combined name-based and structure-based approach. They calculated name similarity matrix and used it as initial values for edges in change weights semantic graphs. Kremen et al. [14] introduced OWL 2 ontology merging tool named OWLDiff. It is open source and can be used as plug in with Protégé and NeOn toolkit. To reduce error in merged ontology, compatibility of mappings should be checked. Using Galois connection for deciding compatible and incompatible mappings is described by Abbas et al. [15]. Analysis and preservation of disjoint knowledge before actually merging two ontologies is given by Fahad et al. [16].

3 IPMOntoDeveloper

There are core techniques in IPM as biological, chemical, cultural, mechanical, and physical. For developing IPM ontology for specific crop, one or more of these techniques should be mentioned as IPM type for specific crop pest.

IPM ontology can be developed from text descriptions of ontologies as mentioned in [18]. IPMOntoDeveloper is a tool developed using java language as mentioned in [18]. For developing IPM ontology for any crop pest, the upper IPM ontology mentioned in Fig. 1 is used, and ontology shown in Fig. 2 is used as basic domain ontology for any crop pest. Agricultural experts can derive various ontologies from domain ontology of crop specific IPM. These ontologies are then passed as input to IPMOntoShare for generating merged IPM ontology.

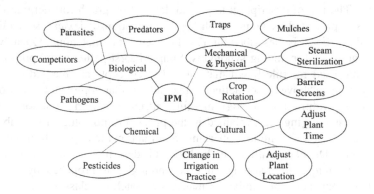

Fig. 1 Upper IPM ontology

Fig. 2 Domain ontology for crop-specific IPM

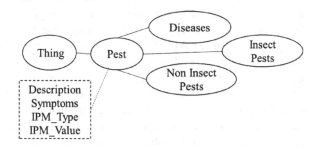

4 IPM Ontology Merging with IPMOntoShare

This section describes stages of merging two candidate IPM ontologies developed by IPMOntoDeveloper. Here, it is assumed that the two IPM ontologies developed by agricultural experts are for the same crop. In ontology merging process, upper IPM ontology shown in Fig. 1 is considered as the basis of merging. The type of IPM for each crop pest is mentioned by agricultural expert [18].

As pest is the basic concept in ontologies generated by IPMOntoDeveloper, ontology matching is performed at level of pest individuals. A top-down approach for ontology merging is adopted. Therefore, at stage 1, similarity matching is done by comparing at class level, i.e., comparing basic types of crop pests. At stage 2, comparison is done at individual level, and at last stage 3, data properties and their values are compared. Details of each stage are as follows:

Stage 1: For comparison at class level, IPMOntoShare first does semantic comparison of pest types. Before checking for name similarity, each word in ontology is converted to its base form using stemming algorithm. Porter's Stemmer is for this purpose. If pest types have similar names, then it is represented as one node in merged IPM ontology. If names are different, then it is checked whether the two terms are synonyms with the help of AGROVOC and WordNet dictionary. AGROVOC [19] is an agricultural vocabulary provided by Food and Agricultural Organization (FAO). It is available in multiple languages. Pest-type similarity

Fig. 3 System flow diagram for proposed approach

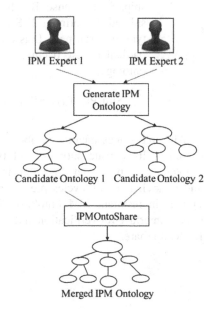

IPM Expert 1   IPM Expert 2

Generate IPM Ontology

Candidate Ontology 1   Candidate Ontology 2

IPMOntoShare

Merged IPM Ontology

computation is also done at structural level using is-a and is-part-of relationships mentioned in candidate IPM ontologies. (Fig. 3).

Stage 2: At this stage of similarity computation, pest individuals of each pest type are compared using the same technique mentioned above. If pest individuals are matching, then a single pest individual is added as leaf node under the same pest type in IPM merged ontology else they are added as separate pest individual to common parent pest type. In formula 1, pest individual is represented as PI and P_Merge is crop pest type in merged ontology.

$$\text{If } Sim(PI_1, PI_2) == = 1 \text{ then } Add(PI1, \text{P\_Merge})$$
$$\text{Else } Add(PI1, \text{P\_Merge}) \text{ AND } Add(PI2, \text{P\_Merge}) \tag{1}$$

Stage 3: Last stage is of matching crop pest properties and their values. Properties of crop pests like IPM type, symptoms, growth stage are stored at leaf nodes of candidate ontologies. For each candidate ontology generated by IPMOntoDeveloper, IPM type is compared first. The IPM type is one of the IPM techniques mentioned in upper IPM ontology. If type of IPM is matching, then only IPMOntoShare proceeds to the next stage of matching, i.e., matching of symptoms. Pests have tendency to occur at specific development stages of crop. Matching development stage of crop in which that pest occurs is also done. If all property values are matching, then it is concluded that those pest individuals are similar as given if formula 2. To compare similarity at structural level, relationships like isCausedBy, hasSymptom, and isAppliedRemedy are considered. In formula 2, S represents pest symptoms, IT represents IPM type for crop pest, D is crop development stages of crop in which that particular pest occurs, and PI is pest individual in candidate pest ontology.

$$Sim(PI_1, PI_2) = Sim(S_1, S_2) + Sim(IT_1, IT_2) + Sim(D_1, D_2) \tag{2}$$

All three stages mentioned above are followed iteratively to get final merged ontology. It is made sure that all types and individuals of pests in candidate ontologies are preserved. Last step in IPM ontology merging process is checking and removing of any cycles present and removing of any repetition of pest individuals in resultant IPM ontology. This task is done manually by agricultural expert. Interface for verification and editing of merged IPM ontology is provided by IPMOntoShare.

5 Implementation

IPMOntoDeveloper and IPMOntoShare are developed using java language. User friendly interfaces are provided to agricultural experts for easy development and merging of IPM ontologies.

To analyze results of IPMOntoShare, two agricultural experts were asked to develop IPM ontology for pests on rice using IPMOntoDeveloper merge then using IPMOntoShare. A part of developed and merged ontologies are shown in Figs. 4 and 5 to demonstrate stages of ontology merging by IPMOntoShare.

In Fig. 4a, pest-type stem-borer has Dark_headed as pest individual in ontology by expert-1 and the same is not present in ontology by expert-2, as shown in Fig. 4b. Hence, it is added as separate pest individual in merged ontology, as shown in Fig. 4c.

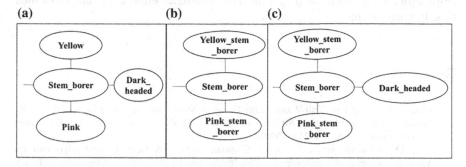

Fig. 4 Merging at pest individual level. **a** Rice IPM ontology by expert-1. **b** Rice IPM ontology by expert-2. **c** Merged rice IPM ontology generated by IPMOntoShare

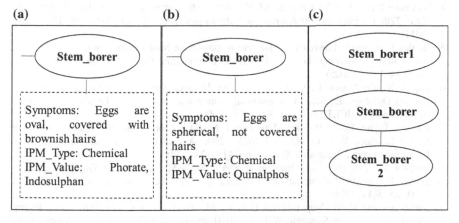

Fig. 5 Merging at pest properties level. **a** Rice IPM ontology by expert-1. **b** Rice IPM ontology by expert-2. **c** Merged rice IPM ontology by IPMOntoShare

In Fig. 5, Symptom and IPM_Value are different for pest individual stem-borer. Hence, it is added as Stem-borer1 and Stem-borer2 in merged ontology, as shown in Fig. 5c.

6 Conclusion

As types of tools and techniques used for IPM are varying, representing IPM knowledge as ontology makes it possible to share IPM knowledge. This paper described in detail how such ontology can be constructed. To integrate such various ontologies in one combined ontology for each crop, ontology merging process is discussed. This paper demonstrates how ontology merging techniques can be utilized. Performance of IPMOntoShare system illustrated more clearly with example of developing and merging IPM ontology for rice crop. The discussed approach helps agricultural experts to play with IPM ontologies effortlessly and saves their time to great extent.

References

1. Shixiang Li, Hong Fan, Yuli Wang, Liang Hong: A Model and Approach for Heterogeneous Ontology Automatic Merging. In: International Joint Conference on Computational Sciences and Optimization. pp. 214–217. (2009)
2. Alma Delia Cuevas Rasgado, Adolfo Guzman Arenas: A language and Algorithm for Automatic Merging of Ontologies. In: 15th International Conference on Computing, CIC'06. pp. 180–185. (2006)
3. Hyunjang Kong, Myunggwon Hwang, Pankoo Kim: A New Methodology for Merging the Heterogeneous Domain Ontologies based on the WordNet. In: Proceedings of the International Conference on Next Generation Web Services Practices (NWeSP'05). (2005)
4. Salvatore Raunich, Erhard Rahm: ATOM: Automatic Target-driven Ontology Merging. In: IEEE 27th International Conference on Data Engineering (ICDE), 2011. pp. 1276–1279. (2011)
5. Bartley Richardson, Lawrence J. Mazlack: Approximate Metrics For Autonomous Semantic Web Ontology Merging. In: The 2005 IEEE International Conference on Fuzzy Systems. pp. 1014–1019. (2005)
6. Rajesh Kumar Gupta, B. D. Chaudhary: An Instance Based Methodology for Merging Domain Ontology. In: Second International Conference on Emerging Trends in Engineering and Technology, ICETET-09. pp. 848–855. (2009)
7. Julia M. Taylor, Daniel Poliakov, Lawrence J. Mazlack: Domain-Specific Ontology Merging for the Semantic Web. In: NAFIPS 2005 - 2005 Annual Meeting of the North American Fuzzy Information Processing Society. pp. 418-423. (2005)
8. Guanyu Li, Zhenghai Luo, Jianshuang Shao: Multi-Mapping Based Ontology Merging System Design. In: 2nd International Conference on Advanced Computer Control (ICACC), 2010. pp. 5–11. (2010)
9. Sanjay Kumar Malik, Nupur Prakash, S. A. M. Rizvi: Ontology Merging using Prompt plug-in of Protégé in Semantic Web. In: 2010 International Conference on Computational Intelligence and Communication Networks; pp. 476–481. IEEEXplore. (2010)

10. Fabiana Freire de Araújo, Fernanda Lígia R. Lopes, Bernadette Farias Lóscio: MeMO: A Clustering-based Approach for Merging Multiple Ontologies. In: IEEE 2010 Workshops on Database and Expert Systems Applications. pp. 177–180. (2010)
11. Muhammad Fahad, Muhammad Abdul Qadir: Similarity Computation by Ontology Merging System: DKP-OM. In: 2nd International Conference on Computer, Control and Communication, 2009. IC4 2009. pp. 1–6. (2009)
12. Viviana Mascardi, Angela Locoro, Paolo Rosso: Automatic Ontology Matching via Upper Ontologies: A Systematic Evaluation. In: IEEE Transactions on Knowledge and Data Engineering, Vol. 22, No. 5. pp. 609–623. (2010)
13. Feng Yang, Lei Liu: A flexible Approach for Ontology Matching. In: International Conference on Computational Intelligence and Software Engineering, 2009. CiSE 2009; pp. 1–4. (2009)
14. Petr Kremen, Marek Smid, Zdenek Kouba: OWLDiff: A Practical Tool for Comparison and Merge of OWL ontologies. In: 22nd International workshop on Database and Expert Systems Applications, IEEEXplore. pp. 229–233. (2011)
15. Muhammad Anu Abbas, Giuseppe Berio: Ontology Merging: Compatible and Incompatible Ontology Mapping. In: Ninth International Conference on Semantics, Knowledge and Grids. IEEEXplore. pp. 129–134; (2014)
16. Muhammad Fahad, Nejib Moalla, Abdelaziz Bouras: Disjoint-Knowledge Analysis and Preservation in Ontology Merging Process. In: Fifth International Conference on Software Engineering Advances. IEEEXplore. pp. 422–428.; (2010)
17. Archana Chougule, Vijay Kumar Jha, Debajyoti Mukhopadhyay: AgroKanti: Location-Aware Decision Support System for Forecasting of Pests and Diseases in Grapes. In: Information Systems Design and Intelligent Applications. Volume 433 of the series Advances in Intelligent Systems and Computing. pp. 677–685. Springer Link. (2016)
18. AGROVOC Thesaurus: http://aims.fao.org/agrovoc#.VF29AvmUc2U

Graph Based Word Sense Disambiguation

Neeraja Koppula, B. Padmaja Rani and Koppula Srinivas Rao

Abstract Word-sense disambiguation is an open challenge in natural language processing. It is the process of identifying the actual meaning of the word based on the senses of the surrounding words of the context in which it is used. Knowledge-based approaches are becoming most popular than other approaches for word-sense disambiguation. Knowledge-based approaches do not require large volumes of training data instead uses Lexical knowledge bases to construct undirected graphs. In this paper, traditional Page Rank algorithms and random walk approaches are compared extensively.

Keywords Word-sense disambiguation · Knowledge-based approach · Page Rank · Random walks approach · Graph-based approach

1 Introduction

Human language has many ambiguous words, which leads to misunderstanding. In NLP, the main focus of intensive research is word-sense disambiguation, which is an open challenge task. Word-sense disambiguation automatically chooses the correct sense to the ambiguous word. The sense of the word depends on the surrounding context. WSD is useful in several tasks, such as machine translation and information retrieval. The WSD systems are supervised approaches, unsupervised approaches, and knowledge-based approaches.

N. Koppula (✉) · B. Padmaja Rani
Department of CSE, JNTUCEH, Hyderabad, India
e-mail: kneeraja123@gmail.com

B. Padmaja Rani
e-mail: padmaja_jntuh@jntuh.ac.in

K.S. Rao
Department of CSE, CMRCET, Hyderabad, India
e-mail: ksreenu2k@gmail.com

© Springer Science+Business Media Singapore 2017

665

S.C. Satapathy et al. (eds.), *Proceedings of the First International Conference on Computational Intelligence and Informatics*, Advances in Intelligent Systems and Computing 507, DOI 10.1007/978-981-10-2471-9_64

Supervised approaches [1] use the corpus, which needs a large hand tagged data with senses, which is expensive, and the accuracy is more than the other WSD systems. However, it suffers from knowledge acquisition, which is a bottle neck. The accuracy is about 70 %, whereas in unsupervised approach, the accuracy is about 50 %.

Knowledge-based WSD approach also requires a corpus with some annotated data, which is also known as lexical knowledge base (LkB), this LKB is also used by the unsupervised approaches to find the correct sense of the word, and hence, this approach does not require any training data. In this article, we are focusing on knowledge-based methods.

There are two main methods: graph-based approach and similarity-based approach. In graph-based approach, it is represented as graph, each node is represented as sense, and the edge between them is represented as a relation between the senses, by applying depth first search assigning correct sense to the ambiguous word. In similarity-based approach, the senses which have a similarity sense and the sense having maximum similarity are assigned as the correct sense of the word.

2   Related Work

Knowledge-based WSD assigns a sense to an ambiguous word depending on the surrounding context. One word can have many senses, and the intended sense is chosen by calculating the relatedness among sense using semantic similarity metric. In this, senses are compared in a pairwise fashion, and the computation grows exponentially with the no of senses, i.e., for a sequence of n words where each word has k senses, we need to consider Kn sense sequences, which is a major drawback.

Recently, graph-based method for knowledge-based WSD [2] has gained much importance in NLP. These methods will find and exploit the structural properties of the graph underlying a particular LKB using some graph-based techniques. All these methods use some versions of WordNet as a LKB.

Graph-based WSD systems [3] are performed over a graph all nodes are represented as senses and edges between them are represented as relations among pair of senses. The relations (semantic, co-occurrence, etc.) may have weights attached to them. By applying ranking algorithm to the graph and then assigning the sense with highest rank to the corresponding word. Developing smaller subgraphs online for each target word and to extract subgraph nodes and relations particularly relevant for the set of senses from a given input context and analyzed most relevant nodes are chosen as the intended senses of the words.

A complete weighted graph is formed with the synsets of the words in the input context by applying a text rank algorithm [4]. The weight of the links joining two synsets is calculated by executing Lesk's algorithm [5]. Once the complete graph is built a random walk algorithm, Page Rank is executed over the graph, and the words are assigned to the most relevant synset.

3 WordNet

WordNet is an LKB, which can be represented as a graph; an LKB is a set of concepts and relations between them, plus a dictionary, which contains words and their senses. A list of words each of them linked to atleast one concept of the LKB. Given a WordNet, we can build an undirected graph where nodes represent the LKB concepts and relation between concepts is undirected edge.

Monosemous words will be related to one sense and polysemous words are related to two senses or several.

4 Algorithms

The random walk algorithm PageRank [6] is a method used for ranking a node in the graph, and this ranking is done according to their structural properties.

The main idea about this algorithm is whenever an edge from v_i to v_j exists in a graph, a vote from vertex i to vertex j is produced, and the rank of node j increases. The strength of the vote from i to j also depends on the rank of node i. The final rank of node i represents the probability of a random walk over the graph ending on node i in the graph.

Let G be a graph with vertices $(v_1, v_2, ..., v_n)$, let d_i be the out degree on node i, and let M be the probability matrix

$$M_{ij} = 1/d_i \quad \text{if an edge } i \text{ to } j \text{ exists}$$
$$= 0 \quad \text{otherwise}$$

The Page Rank vector over a graph G is given as follows:

$$P = cMP + (1 - c)v. \tag{1}$$

In the above equation, v is a stochastic vector and c in the damping factor varies between 0 and 1, the first term in the equation is the voting scheme, and the second term is smoothing factor, which makes any graph to be irreducible and thus converges stationary distribution.

The Page Rank is actually calculated by applying iterative algorithm that computes Eq. 1 successively until it converges to some threshold value. We need to slightly modify the Eq. 1 [7] to discard the effect of dangling nodes, i.e., a node without out links.

5 Random Walks

Page Rank is calculated by applying an algorithm on Eq. 1 repeatedly until it converges to given threshold value is achieved or the number of iterations is executed.

Random walks WSD can be applied on two methods.

5.1 Static Page Rank

This method is context independent; it returns the most predominant sense from the relative senses. Context is used as a baseline, if we apply traditional Page Rank algorithm to the WordNet. All the concepts are ranked according to their PageRank Value. We get a context independent word senses. Given a target word, it checks the relative senses and output the highest rank.

5.2 Personalized Page Rank

This method is purely based on the content words, and we are using LKB. Given an input text depending on the content words, a relative sense is assigned to a target word. Given a sentence in the target word, all the remaining content words are related to the concepts in LKB, and after the disambiguation process, all the concepts in LKB receive a score. For the target word, choose the appropriate and associated concept with maximum score.

The main drawback in this method is if a target word has two senses related by semantic relations, those senses will reinforce each other.

6 Evaluation Measures

WSD literature has used several measures of evaluation

1. Precision
2. Recall
3. F1 Measure

Precision is the percentage of correctly disambiguated instances divide by the no of instances disambiguated.

Recall is the percentage of correctly disambiguated instances divided by the total no of instances to be disambiguated.

Finally, F1 measure combines both the measures precision and recall, which is the harmonic mean of the two measures. The data sets used here are monosemous and polysemous words only.

The comparison graph with WSD and without WSD is shown clearly in Fig. 1.

6.1 Comparisons

To analyze the performance factors, the parameters which are used to measure the accuracy in the Page Rank algorithm are damping factor and no of iterations to converge.

Efficiency of using full graph versus subgraphs. Using different versions of WordNet. Table 1 compares the accuracy of the best graph-based methods: Mihalcea et al., Agirre and Soroa [8], and the method of Sinha and Mihalcea.

Fig. 1 Comparison graph with WSD and without WSD

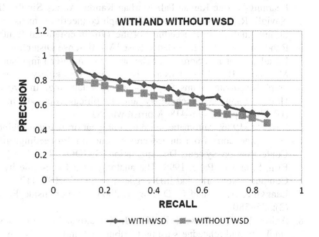

Table 1 Comparison with related work

	Senseval 2 accuracy	Senseval 3 accuracy
Mih05	54.2	52.2
Agi09	59.5	57.4
Sinha07	56.4	52.4
WT-A	58.67	52.67
FS	63.7	61.3

7 Conclusions

In this paper, we present a method for knowledge-based word-sense disambiguation about random walks algorithm to choose correct sense of the ambiguous word using LKB, and we are using full graph of WordNet efficiently. We have performed an extensive analysis between our Page Rank algorithms, i.e., graph-based methods. These algorithms can be easily ported to other languages with good results with requirement of WordNet in that language.

Acknowledgments I express my deep gratitude to all the authors in the reference section, who greatly helped to improve the article in this shape.

References

1. A Hybrid Approach To Word Sense Disambiguation Combining Supervised & Unsupervised Learning- Aloke Ranjan Pal, Anirban Kandu, Abhay Singh –IJAIA-July 2013 pages 89–100.
2. Navigli, R. and M. Lapata. 2007. Graph connectivity measures for unsupervised word sense disambiguation. In Proceedings of the 17th International Joint Conference.
3. Random Walks For Knowledge Based Word Sense Disambiguation – E Agirre, Oier Lopez De Lacalle, Aitor and Soroa Association For Computer Linguistics 2014. Pages 57–84.
4. Mihalcea, R. 2005. Unsupervised Large-vocabulary word sense disambiguation with graph-based algorithms for sequence data labeling. In Proceedings of the Conference on Human Language Technology and Empirical Methods in Natural Language Processing (HLT'05), pages 411–418, Morristown, NJ.
5. Lesk, M. 1986. Automatic sense disambiguation using machine readable dictionaries: How to tell a pine cone from an ice cream cone. In Proceedings of the 5th Annual International Conference on Systems Documentation (SIGDOC'86).
6. Brin, S. and L. Page. 1998. The anatomy of a large-scale hyper textual Web search engine. Computer Networks and ISDN Systems, 30(1–7):107–117.
7. Langville, A. N. and C. D. Meyer. 2003. Deeper inside PageRank. Internet Mathematics,1 (3):335–580.
8. Agirre, E., E. Alfonseca, K. Hall, J. Kravalova, M. Pasca, and A. Soroa. 2009. A study on similarity and relatedness using distributional and WordNet-based approaches. In Proceedings of the North American Chapter of the Association for Computational Linguistics - Human Language Technologies conference (NAACL/HLT'09), pages 19–27, Boulder, CO.

Enhanced Scaffold Design Pattern for Seculde Multi-tenant SaaS Application

Nagarajan Balasubramanian and Suguna Jayapal

Abstract Internet has sophisticated the life of a populist. The cloud is an Internet-based technology where data are location independent. It provides "on demand" service to the customer on pay-per-use model. The multi-tenancy is an essential characteristic in cloud, which has motivated many researchers to contribute their work. Scaffold is a platform, where programmers can specify and integrate application and database. The pattern is a reusable component used to develop applications. NoSQL document data model has an advantage of storing the values in a tiered storage. In this contribution, a pattern language for the scaffold is proposed, which will be useful for the researchers to develop malleable multi-tenant SaaS application.

Keywords Design pattern · Scaffold · NoSQL data model · Sharding · XML document

1 Introduction

Cloud can be expanded as common location independent online utility that is available on demand. It has sophisticated the work-life balance easy and eloquent.

Multi-tenancy is an architectural pattern in which a single instance of software is run on the service provider's infrastructure and multiple tenant access the same instance. A tenant is the organizational entity which rents SaaS solution. It entails some important characteristic, such as customization, scalability and availability. A multi-tenant application is a malleable, if it holds above conditions [1].

N. Balasubramanian (✉)
Research and Development Cell, Bharathiar University, Coimbatore 641046, India
e-mail: cloud.nagarajan@gmail.com

S. Jayapal
Department of Computer Science, Vellalar College for Women, Erode, India
e-mail: sugunajravi@yahoo.co.in

© Springer Science+Business Media Singapore 2017   671
S.C. Satapathy et al. (eds.), *Proceedings of the First International Conference on Computational Intelligence and Informatics*, Advances in Intelligent Systems and Computing 507, DOI 10.1007/978-981-10-2471-9_65

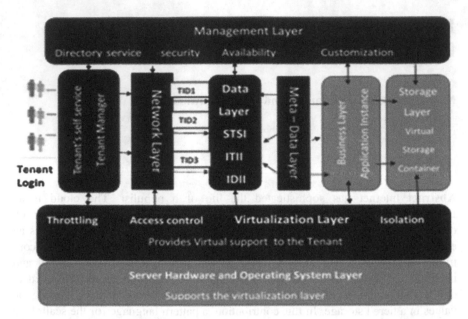

Fig. 1 A virtual adaptive scaffold for multi-tenant SaaS data model [2]

Scaffold is a platform, in which the programmer can specify how application and database can be interoperable. A virtual scaffold for multitenant SaaS data model is shown in Fig. 1. The motivation behind the scaffold is to address the concern of cloud skeptics, an efficient architectural platform, which will perform the create, read, update, and delete (CRUD) operations during an application development [2].

A pattern language is a reusable and implementation component specific to a virtual scaffold. Patterns provide an independent solution, for challenge related to application and database interoperability [3]. The metadata layer will hold the master copy of data [2]. The design pattern provides a guidance to store and retrieve data to and from a metadata layer, by proper authentication and configuration [4].

The business layer will execute the snippet code for accessing the data. Thus, patterns will be useful component, which helps to easily implement and develop an application by providing efficient interoperability between data layer, metadata layer and business layer [2].

The rest of this paper is organized as follows. In the next section, a related study on literature is carried out. Section 3 is eloquent elaboration of state-of-the-art studies on pattern formulation, and NoSQL data model has been done. Section 4 discusses on a complete implementation of scaffold design pattern with a sketch, snippet code, and annotation. Finally, a summary of conclusion with short description of future work is pinpointed.

2 Related Work

The literature survey is carried out systematically on contributions related to design pattern and data base. The scaffold is a platform where the application and database can be interoperable.

Bezemer et al. [1] has presented a position paper elucidating the importance of multi-tenancy in industry. The authors have suggested a tenant-specific multi-tenant pattern with some snippet code.

Strauch et al. [3, 5] have eloquent the ways and means by which application data layer can be moved over the cloud. They identified the challenges during implementation of cloud deployment model. An application layer was proposed in [3] and implemented in [5]. They instilled confidentiality pattern for a cloud-based application development. The authors claimed that their solution is reusable for any cloud deployment. This contribution is useful and helps to understand how data can be moved from data layer to application layer.

Fehling et al. [6] had provided an architectural pattern language. They provided pattern identification and format for a pattern language. The lists of catalog provided by the authors are useful for new pattern development.

Mietzner et al. [7] combined different multi-tenant patterns and evaluated a set of patterns that can be used to design, develop and deploy process aware service-oriented SaaS applications. The customization provided in [7] is also useful and related to this contribution.

Adewojo et al. [8] argued that tenant-isolated component only will represent a compromised implementation between the shared component and dedicated component They provide UML diagrams and snippet coding in [8]. This contribution helps us to better understand practical implementation of multi-tenant pattern language.

Li et al. [9] provided a performance comparison on key-value store implementation of SQL and NoSQL. Geroge et al. [10] has provided an elaborate comparison between several NoSQL. While Leavitt et al. [11] has entailed that organizations nowadays are compelled to work with Big Data, and hence, they are looking for alternative solution to handle the data through NoSQL data models. Kanwar et al. [12] has argued that for the distributed environment with huge volume of data, NoSQL was the solution. While Aghl et al. [13] have presented a comparison of SQL and mongoDB.

The above literatures [10–13] create spur for researcher to tradeoff their ideas from relational SQL to non-relational NoSQL.

Pallavi et al. [11] presented a storage model and its XML support to implement multi-tenancy and have suggested an architectural overview for conceptual multi-tenant SaaS.

To conclude with, a combined literature is conducted starting with multi-tenant architecture then defining design pattern and finally with the NoSQL data model. It helps to understand the concepts and creates motivation and spur to identify design pattern for multi-tenant component gateway pattern.

3 State-of-the-Art Studies

Section 3.1 is going to study on Patterns formulation, and Sect. 3.2 discusses about the NoSQL data model. This state-of-art study gives an eloquent understanding on multi-tenant component gateway pattern.

3.1 Pattern Formulation

Formulation of a significant pattern is the first step toward a solution. It includes a set of sections. The context section sets stage where the pattern takes place. The problem section explains what the actual problem is. The force section describes why the problem is difficult to solve. The solution section explains the solution in detail. The consequence section demonstrates what happen when solution is applied [6]. The next step is identification of relevant and irrelevant detail for a pattern [6]. This helps a data layer of a scaffold to hold critical data safe in the meta-data layer.

This can be carried out by pruning the pseudonymous and anonymous data [5], since they are more vulnerable in nature. The critical data can be handled by the data layer either by pseudonymizer or anonymization technique.

Pseudonymizer is a technique to provide a masked version of the data to the public while keeping the relation to the non-masked data in private. This enables processing of non-masked data in the private environment when required [5]. Anonymization is a technique to provide a reduced version of the critical data to the public while ensuring that is impossible to relate the reduced version to the critical data [9]. Thus, by identification of the critical data, an abstract description for a pattern can be realized [5]. The next step is to decide the design pattern associated for format which will include relation to other pattern, annotations, and snippet code [6].

3.2 NoSQL Data Model

The reason for pattern definition is to support a scaffold, where database and application can be inter-operable [2]. The data model provides back end support for any scaffold to develop multi-tenant applications.

The data can be stored in data layer in any one form—independent data base and independent instance (IDII), independent tables and shared database (ITSI), and finally, shared instance and shared tables [2].

The traditional database is becoming obsolete among the organization because of enormous volume of data [11]. These Big Data compel the organization to concentrate alternative way of storing their data in "silo" storages. The NoSQL helps to store the data in tiers with sufficient authentication [11].

The structured data in a relational database is predefined by name, layout, and design. They have certain rules to access the data in the database. SQL is a convenient language for structured data which satisfies Atomicity Consistency Isolation and Durability (ACID), but struggling with growth of huge data [14].

NoSQL data model works with principle, Basically Available Soft state Eventual consistency (BASE). By this, distributed storage of data is possible in NoSQL data model and the immediate data can be available in the cache [12]. In addition, any NoSQL should also satisfy Consistency, Availability, and Partition tolerance (in short CAP) rule. Consistency[C] equivalent to having a single up-to-date copy of the data, high availability [A] of the data (for updates), and tolerance to network Partitions [P] [14]. The data model for the NoSQL includes key-value, column-oriented and document and graph stores [9].

Key-value stores are based on key-value pairs, which resemble an associative map or a dictionary. The values are based on the key-value stored and follows two of the CAP rule, namely, consistency and availability. They are useful for providing use cases and resemble more like a structured database [9].

Column-oriented database stores are derived from Google Big table, in which the data are stored in a column-oriented way. Prior definition of column is not essential. This increases the flexibility in storing any data type. This data store provides flexible and powerful indexing [9].

Graph database originated from graph theory and use graphs on their data model. It can efficiently store the relationships between different data nodes. These databases are specialized in handling highly interconnected data, therefore very efficient in traversing relationships between different entities [5]. In document databases, the database stores and retrieves documents, which can be XML, JSON or BSON. These documents are self-describing, hierarchical tree data structures which can consist of maps, collections, and scalar values [9].

Document databases follow a master and slave node functioning. Mongo DB is an open source document database, which satisfy the CAP rules [13]. They can be used to write snippet code to establish inter-operability in the scaffold.

4 Malleable Scaffold Design Pattern

The motivation behind scaffold design pattern is to provide a relativity and implementation between various patterns used in, multi-tenant component gateway pattern.

Multi-tenancy is an architecture in which single instance of software application serves multiple tenants. To claim a multi-tenant application is malleable, it should satisfy certain characteristics such isolation, availability, scalability, and customizability [1].

A design pattern should be sketched to isolate the tenant, use share table, and share instance among tenant logged in, provide provision for scalability and customization for the tenant. The metadata layer should be properly maintained to provide data to the tenant. Figure 2 shows a multi-tenant common gateway pattern.

The tenant can login into as shared component,[1] tenant-isolated component (see Footnote 1) and dedicated component (see Footnote 1) [8]. The Authorization service pattern (see Footnote 1) help to identify the tenant. Gatekeeper Pattern[2] protects applications and services using a dedicated host instances and prune pseudonymous and anonymous persons using this pattern. Valet key pattern (see Footnote 1) will restrict direct access to a client for a specific resource or service. The identification of the tenant is necessary because to prune the pseudonymous and anonymous data. Federated identity pattern (see Footnote 1) provides the external identity for a tenant by assigning Tenant ID (TID).

The tenant into the scaffold with valid TID is considered as isolated tenant. The data model for these tenants is independent database and independent instance, independent tables and shared database instance or shared tables and shared database instance. Any malleable multi-tenant works good with shared table and shared instance, since multi-tenancy is sharing single instance by multiple tenant [2]. Tenant login as an isolated component can share the instance, customize them according to their needs and store in a distributed environment. The NoSQL can provide cache memory to store the data [12].

Compute and Query Responsibility Segregation (CQRS) Pattern (see Footnote 2) can be applied in a scaffold to perform CRUD operation [15]. This pattern segregates the operation that read data (Queries) from the operation that update data (command) by using separate interfaces. Thus, the pattern states there should be complete separation between "command" methods that perform actions and "query" method returns data. This implies that the data models used collaborated with other pattern to provide results in a scaffold [4] (Fig. 3).

A sample code is shown below (Fig. 4).

Event sourcing pattern (see Footnote 2) use an append-only store to record the full series of event that describe history of actions taken on data. This pattern can be applied through the mongoDB. **Sharding pattern (see Footnote 2)** can be used to divide a data store into a set of horizontal partition. In mongoDB, a driver is provided through which the tenant can access the data. Config in the network layer of the scaffold will segregate the command and query run by the tenant [13]. The data layer will shard, the Shared table and shared instance through the query router. Thus, sharding can be used by the tenant with shared component where the data is stored across number of machines as the XML documents [16]. An isolated tenant can create a collection and insert the data as follows (Fig. 5, Tables 1 and 2).

The isolated tenant logged in can customize their data by updating the collection using the following snippet code (Fig. 6).

The corresponding XML document for the updating is as below (Table 3).

---

[1]http://cloudcomputingpatterns.org.

[2]http://womdpwsazure.com.

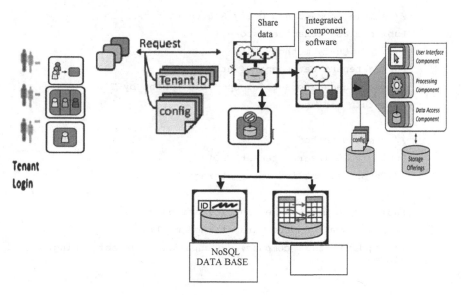

Fig. 2 Multi-tenant common gateway pattern

Fig. 3 Workflow in CQRS
design pattern

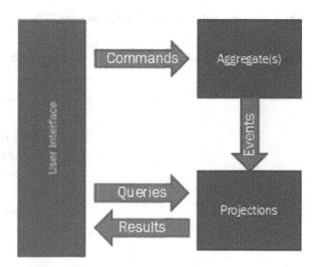

Thus, scaffold can implement the customization of tenant data, for the isolated tenant by executing application logic in the business layer. The application in the scaffold uses Activex Data Object (ADO).NET to provide data access to .NET users. The driver provided in the shard will help to interoperate with any language. Thus, the researcher can use these design patterns as easy-to-use benchmark into the scaffold, and develop malleable multi-tenant application.

```
// simple Command and Query Segregation example
Public class Tenantdatastore {
      // Query method
Public tenant Gettenant1 (int TID) {
// query data storage for specific tenant by TID
// return tenant
}
// command Method
Public void Insert (tenant TID)
// Insert tenant into data storage
}
Public void updatename (int id, stirng name) {
// find tenant in the data storage by TID
// Update the company, occupation, dateofjoining,
performance)

   }
}
```

Fig. 4 A snippet code for tenant identification using CQS

Fig. 5 Database creation and insertion for Tenant1

```
db.create.collection("Tenant1");
   db.Tenant1.insert([ {name:"john",
   ... age:34,
   ... occupation:"mongodbadmin",
   ... company:"it4u",
   ... country:"India"},
   ... {  name:"jack",
   ... age:35,
   ... occupation:"oracleadmin",
   ... country:"UK",
   ... company:"cloud4u"}])
```

Table 1 STSI table for Tenant1

Tenant ID	Name	Age	Occupation	Company	Country
1	John	34	mongodbadmin	It4u	India
1	Jack	35	oracleadmin	Cloud4u	UK

Table 2 XML document for Tenant1 documents

TID	CUSTOMER INFORMATION

1	`<TENANT1 ID = '1'>`
	`<NAME > JOHN </NAME>`
	`<AGE > 34 </AGE>`
	`<OCCUPATION > MONGODBADMIN </OCCUPATION>`
	`<COMPANY > IT4U </COMPANY>`
	`<COUNTRY > INDIA </COUNTRY>`
	`</TENANT1>`

1	`<TENANT1 ID = '1'>`
	`<NAME > JACK </NAME>`
	`<AGE > 35 </AGE>`
	`<OCCUPATION > ORACLEADMIN </OCCUPATION>`
	`<COUNTRY > UK </COUNTRY>`
	`<COMPANY > CLOUD4U < COMPANY>`
	`</TENANT1>`

Fig. 6 Snippet code for updation of a documents in collection

```
db. Tenant1.update({name="John"},
{
$set:{dateofjoining:"23feb2012",
Performance="good"
}
}
)
```

Table 3 XML document for Tenant1 information after updation

TID	Tenant1 INFORMATION

1	`<TENANT1 ID = '1'>`
	`<NAME > JOHN </NAME>`
	`<AGE > 34 </AGE>`
	`<OCCUPATION > MONGODBADMIN </OCCUPATION>`
	`<COMPANY > IT4U </COMPANY>`
	`<COUNTRY > INDIA </COUNTRY>`
	`<DATAOFJOINING > 23FEB2012 </DATEOFJOINING>`
	`<PERFORMANCE > GOOD </PERFORMANCE>`

5 Conclusion

In this contribution, a multi-tenant common gateway pattern is elaborately discussed. This pattern gives spur and motivation for the researcher, to design a malleable multi-tenant scaffold which will provide CRUD on the data available from the STSI table and eventually store the data in a distributed virtual container. Furthermore, NoSQL supports the tiered storage. The future work include includes, experimentation of this pattern and thus provide easy-to-use malleable multi-tenant applications which will be useful for all walks of life.

References

1. Cor-Paul Bezemer, Andy Zaidman and Aad't Hart, "Enabling Multi-Tenancy: An Industrial Experience Report", proceedings of the 26th IEEE international Conference on software maintence (ICSM), 2010, IEEE.
2. Nagarajan Balasubramanian and Dr. J. Suguna, "A Virtual Scaffold for Storage Multi-Tenant SaaS Data Models", international journal of Applied Engineering Research, volume 10, Number 20 (2015) pages 40775–40780.
3. Steve Strauch, Uwe Breitenbuecher, and Tobias Unger, "Cloud Data Pattern for Confidentiality", Proceedings of the 2nd International Conference on Cloud Computing and Service Science, CLOSER 2012, Portugal, SciTePress, 2012.
4. Alex Homer, John Sharp and Trent Swanson "Cloud Design Pattern", Microsoft press 2014.
5. Steve Strauch, Vasilios Andrikopoulos and Frank Leymann, "Using Patterns to Move the Application Data Layer to the Cloud", Proceedings of the 5th International conference on Pervasive Patterns and Applications, PATTERNS 2013, Spain.
6. Christoph Fehiling, Frank Leymann and Walter Schupeck. "An Architectural Pattern Language of Cloud-base Applications", proceeding of the 18th conference on pattern language of programs.
7. Ralph Mietzner, Tbias Unger, Robert Titze and Frank Leymann, "Combining Different Multi-Tenancy Pattern in Service – Oriented Applications" proceedings of the 13th IEEE Enterprise Distributed Object Conference, Published by IEEE, 2009, Pages 131–140.
8. A.A. Adewojo, J.M. Bass and I.K. Allison. "Enhanced cloud patterns: A case study of multi-tenancy patterns", International conference on information society (I- society 2015).
9. Yishan Li and Sathiamoorthy Manoharan, "A Performance Comparison of SQL and NoSQL databases", published by IEEE, 2013.
10. Bogdan George Tudorica and Cristian Bucur, "A Comparison between several NoSQL databases with comments and notes".
11. Neal Leavitt, "Will NoSQL Databases Live Up to their promise?" Published by the IEEE computer society, 2010. http://www.leavcom.com/pdf/NoSQL.pdf.
12. Renu Kanwar, Prakriti Trivedi and Kuldeep Singh "NoSQL, A Solution for distributed database management system", International journal of Computer Application, volume 67, No. 2 – April 2013.
13. Rajat Aghl, Sumeet Mehta and Navdeep Bphra, "A comprehensive comparison of SQL and mongoDB databases", International journal of scientific and research publication, volume 5, issue 2, February 2013.
14. Eric Brewer, "CAP Twelve years Later: how the "Rules" Have changed" published by IEEE Computer Society, February 2012, Pages 23–29.
15. Gerg Yourng, Command and Query Responsibility Segregation, http://cqrsinfo.com.
16. Pallavi GB and Dr. P. Jayarekha, "Multi-tenancy in SaaS – A Comprehensive survey", International journal of scientific and engineering research, volume 7, issue 7, July 2014.

Modelling of Three-Phase Asynchronous Motor Using Vector Control Theory

P. Karthigeyan, K. Vinod Kumar, T. Aravind Babu,
Y. Praveen Kumar and M. Senthil Raja

Abstract This paper presents the modelling of three-phase asynchronous motor using vector control. The purpose of using this control theory with respect to other methodologies as it holds the advantage of accessing all the internal variables required for machine operation. The control theory employed here is vector control. This theory can also be used without estimators. Park, Clarke, and Krause and its Inverse transformations have been used in this modelling. This modelling is done in MATLAB/SIMULINK platform by initializing the variables and parameters of three-phase induction motor, and the output of speed and torque is verified.

Keywords Squirrel cage induction motor · Direct quadrature · Park transformation · Clarke transformation · Krause transformation · Torque · Speed

1 Introduction

Induction motors is asynchronous motor which is operated by electromagnetic induction. The most commonly used induction motor is cage rotor. In three-phase induction motor, emf is dependent upon slip and rotor current. Vector control theory adopted here is the d-q control theory which involves abc–dqo and dqo–abc transformation. To determine the currents and voltages, three-phase voltage is

P. Karthigeyan (✉)
I.I.T Madras, Chennai 600036, India
e-mail: Karthigeyan.iitm@gmail.com

K.V. Kumar · T.A. Babu
SRIT, Ananthapuramu 515701, Andhra Pradesh, India

Y.P. Kumar
CBIT, Kadapa 516360, Andhra Pradesh, India

M.S. Raja
Vi Institute of Technology, Chennai 603108, India

© Springer Science+Business Media Singapore 2017   681
S.C. Satapathy et al. (eds.), *Proceedings of the First International Conference on Computational Intelligence and Informatics*, Advances in Intelligent Systems and Computing 507, DOI 10.1007/978-981-10-2471-9_66

converted to two-phase voltage and from two-phase voltage to three-phase voltage, and finally fed back to the source to determine the output.

2 Modeling of Squirrel Cage Induction Motor

Figure 1 given below is schematic diagram of three-phase squirrel cage asynchronous motor. The speed and torque of motor is estimated using the flux linkages equations, current equations. The various transformations are also used wherever necessary. The explanation for each block is given below. The inputs are three-phase voltages, fundamental frequency. The outputs are torque, and rotor speed. In d-q control theory, the three-phase variables are converted to two phase rotating frame which is used for flux linkage calculations. In addition, transformation of three-phase voltages to two-phase and the two-phase currents back to three-phase is done to determine the outputs. The individual blocks in simulation are explained below.

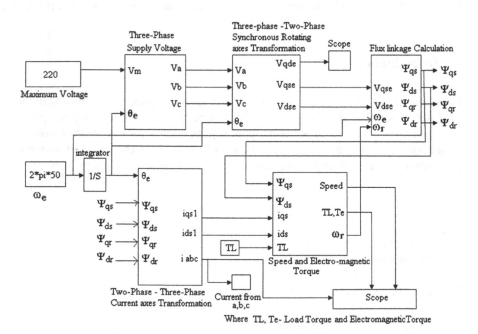

Fig. 1 Schematic diagram of squirrel cage induction motor

2.1 Current Equations

$$iqs = \frac{1}{Xls}\left(\Psi qs - \Psi mq\right) \tag{7}$$

$$ids = \frac{1}{Xls}\left(\Psi ds - \Psi md\right) \tag{8}$$

$$iqr = \frac{1}{Xlr}\left(\Psi qr - \Psi mq\right) \tag{9}$$

$$idr = \frac{1}{Xlr}\left(\Psi dr - \Psi md\right) \tag{10}$$

$$Te = \frac{3}{2}\left(\frac{P}{2}\right)\frac{1}{\omega b}\left(\Psi ds\, iqs - \Psi qs\, ids\right) \tag{11}$$

$$Te - TL = J\left(\frac{2}{P}\right)\frac{d\omega r}{dt} \tag{12}$$

2.2 Flux Linkage Equations

$$\frac{d\Psi qs}{dt} = \omega b\left[vqs - \frac{\omega e}{\omega b}\Psi ds + \frac{Rs}{Xls}\left(\frac{X^*ml}{Xlr}\Psi qr + \left(\frac{X^*ml}{Xls} - 1\right)\Psi qs\right)\right] \tag{13}$$

$$\frac{d\Psi ds}{dt} = \omega b\left[vds - \frac{\omega e}{\omega b}\Psi qs + \frac{Rs}{Xls}\left(\frac{X^*ml}{Xlr}\Psi dr + \left(\frac{X^*ml}{Xls} - 1\right)\Psi ds\right)\right] \tag{14}$$

$$\frac{d\Psi qr}{dt} = \omega b\left[-\frac{(\omega e - \omega r)}{\omega b}\Psi dr + \frac{Rs}{Xlr}\left(\frac{X^*ml}{Xls}\Psi qs + \left(\frac{X^*ml}{Xlr} - 1\right)\Psi qr\right)\right] \tag{15}$$

$$\frac{d\Psi dr}{dt} = \omega b\left[-\frac{(\omega e - \omega r)}{\omega b}\Psi qr + \frac{Rs}{Xlr}\left(\frac{X^*ml}{Xls}\Psi ds + \left(\frac{X^*ml}{Xlr} - 1\right)\Psi dr\right)\right] \tag{16}$$

2.3 Speed and Electromagnetic Torque

The speed and torque is calculated is using the currents, flux linkages, and the load torque as inputs. The following equations are modeled shown in is to calculate those parameters as outputs.

$$\frac{d\omega r}{dt} = \left(\frac{P}{2J}\right)(Te - TL) \tag{17}$$

2.4 Three-Phase Supply Block

This block is made up of embedded block where the program is embedded in it. The transformation is given by

$$\begin{bmatrix} Van \\ Vbn \\ Vcn \end{bmatrix} = \begin{bmatrix} +\frac{2}{3} & +\frac{2}{3} & -\frac{1}{3} \\ +\frac{1}{3} & -\frac{1}{3} & +\frac{2}{3} \\ -\frac{1}{3} & -\frac{1}{3} & +\frac{2}{3} \end{bmatrix} \begin{bmatrix} Vab \\ Vbo \\ Vco \end{bmatrix} \tag{18}$$

Three-phase voltages are obtained from unit vectors by giving Vm and θe as inputs.

2.5 Three-Phase to Two-Phase Synchronous Rotating Axes Transformation Block

KRAUSE TRANSFORMATION

$$\begin{bmatrix} V^s qs \\ V^s ds \end{bmatrix} = \begin{bmatrix} 1 & 0 & 0 \\ 0 & -\frac{1}{\sqrt{3}} & \frac{1}{\sqrt{3}} \end{bmatrix} \tag{20}$$

$$Vqs = V_{qs}^S \cos\theta_e - V_{ds}^S \sin\theta e$$
$$Vds = V_{qs}^S \sin\theta_e + V_{ds}^S \cos\theta \tag{21}$$

2.6 Three-Phase to Two-Phase Synchronous Current Axes Transformation Block

Now, the rotating axes is transformed into stationary axes using the below equations. These equations are embedded as program in the embedded block. Using Clarke's transformation, the two-phase is transformed into three-phase currents (Fig. 2).

$$i^s qs = Vqs \cos\theta_e - Vds \sin\theta e$$
$$i^s ds = Vqs \sin\theta_e + Vds \cos\theta e \tag{22}$$

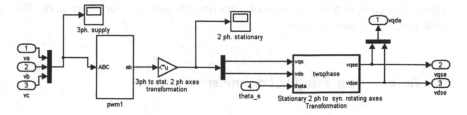

Fig. 2 Three-phase to two-phase synchronous rotating axes transformation block

$$
\begin{bmatrix} i_a \\ i_b \end{bmatrix} = \begin{bmatrix} 1 & 0 \\ -\frac{1}{2} & -\sqrt{\frac{3}{2}} \\ -\frac{1}{2} & \sqrt{\frac{3}{2}} \end{bmatrix}
$$

3 Simulation Results

3.1 Procedure to Run the Model

The three-phase asynchronous machine is simulated with the parameters of three-phase 220 V and frequency of 50 Hz. Before simulating, the parameters of the machine need to be initialized using M-file. This M-file should be debugged and

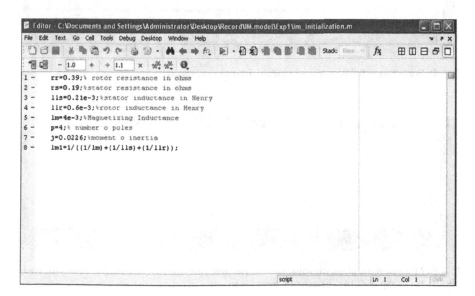

Fig. 3 Induction machine model initialization file

added to the path, so that execution of Simulink file does not show any error. The M-file program is shown below:

3.2 Initialization File for Parameters

(See Figs. 3, 4 and 5).

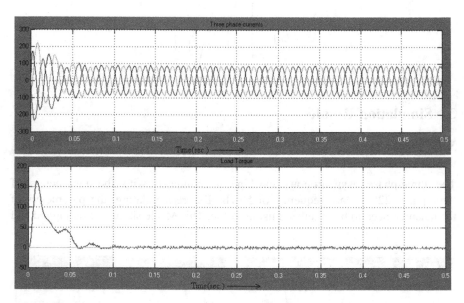

Fig. 4 Wave forms for three-phase currents, load torque, speed

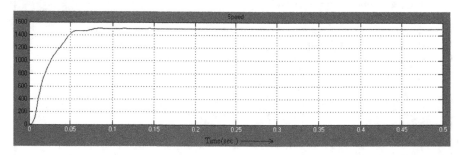

Fig. 5 Machine three-phase currents, load torque, and rotor speed. The machine comes to steady state at small slip of 0.14 s because of inertia load

4 Conclusion

Thus, the modelling of three phase asynchronous machine has been implemented using vector control (direct-quadrature control theory). At a rated slip of 0.14 s, the current, speed, and torque are calculated and verified using this vector control theory.

An Unsupervised Classification of Printed and Handwritten Telugu Words in Pre-printed Documents Using Text Discrimination Coefficient

N. Shobha Rani and T. Vasudev

Abstract Classification of handwritten and printed text in pre-printed documents enhances the performance of optical character recognition technologies. The objective of work presented lies in devising an approach to perform automatic classification of printed and handwritten text at word level, which is inherently found in pre-printed documents. The proposed work consists of three stages to perform the classification of printed and handwritten words in Telugu pre-printed documents. The stage one encompasses the feature computation from the segmented words, stage two determines text discrimination coefficient, and finally, the classification of printed and handwritten text using a decision model is accomplished in stage three. The statistical and geometrical moment features are computed with respect to the text block under consideration, and furthermore, these features are employed for determination of text discrimination coefficient. The results of experimentation are proved to be promising and robust with an accuracy of around 98.2 %.

Keywords Printed word · Handwritten word · Classification · Pre-printed documents · Geometric features · Discrimination coefficient

N.S. Rani (✉)
Maharaja Research Foundation, Maharaja Institute of Technology,
University of Mysore, Mysuru, Karnataka, India
e-mail: n.shoba1985@gmail.com

N.S. Rani · T. Vasudev
Department of Computer Science, Amrita Vishwa Vidyapeetham,
Amrita University, Mysuru Campus, Mysuru, Karnataka, India
e-mail: vasu@mitmysore.in

© Springer Science+Business Media Singapore 2017   689
S.C. Satapathy et al. (eds.), *Proceedings of the First International Conference on Computational Intelligence and Informatics*, Advances in Intelligent Systems and Computing 507, DOI 10.1007/978-981-10-2471-9_67

1 Introduction

Machine intelligent classification of textual components refers to the process of interpreting whether a text block is printed or handwritten. In monolithic documents, the aspect of classification of text blocks does not arise, where as it is an accentuated aspect essentially in pre-printed type documents [1], since pre-printed documents are composed of both printed and handwritten text and it is quite significant to classify the text blocks for an optical character recognition (OCR) [2]. The classification of these blocks also simplifies the classification and recognition stages of OCR due to the systematic organization of character database. The various categories of pre-printed documents possess both printed and handwritten texts which are in use at various organizations. Some of them are application forms for jobs/admissions, bank cheques, receipts, compliance forms, encumbrance forms, birth/death certificates and various other requisition forms used in Government/ Private organizations, etc. All these categories of documents require separation of printed and handwritten text blocks before subjecting to recognition by OCR. Some specific application systems include the extraction of handwritten text blocks from pre-printed forms [3] also requires the knowledge of differential interpretation between printed and handwritten text blocks. In this work, we mainly focus on the classification of text blocks generally found in Telugu forms/certificate documents used in various Government/Private organizations of Andhra Pradesh.

A very few attempts are tracked in the literature related to the word segmentation and classification of printed and handwritten words. The summary of the observations drawn from the literature reported is discussed subsequently.

Mark et al. [4] had proposed an approach based on pictographic recognition technology that employs the features like mathematical graph, topology and other geometrical features to discriminate the individual characters within the strings of English and Arabic handwritten script. The work has focused on extraction of the Arabic word segments from the document rather than discrimination of printed and handwritten portions. Ranjeet et al. [5] had employed the statistical and structural features of the word to discriminate the handwritten and printed Hindi words with an accuracy of 94.1 %. The algorithm considers width of segmented blocks to determine the word is handwritten or not and it may fail if handwritten word is consisting of uniform spaces between characters. Shirdhonkar et al. [6] proposed an approach for discrimination of printed and handwritten English words using the state features of patches in image segments. The algorithm employs neural network and SVM classifier with an accuracy of about 96 % for printed words and more than 80 % in the case of handwritten words. Rajesh et al. [7] had devised an approach based on the structural features of image and obtained an accuracy of about 95 % for separation of machine printed text and handwritten Hindi text. Lincoln et al. [8] had proposed an algorithm using data mining-based classification techniques and attained more than 80 % of accuracy. Hangarge et al. [9] had proposed a methodology for classification of printed and handwritten text in a few south Indian scripts using statistical features and have achieved an accuracy of around 99 %.

Samir et al. [10] had proposed an approach for classification of printed and handwritten words using grayscale feature vector and fed to a decision tree classifier for classification. The feature thresholds are identified based on feature ranking and obtained an accuracy of 96.4 %.

To the best of our knowledge, we have revised various works and observed that most of the classification is performed using supervised learning models with a cluster of features representing for both printed and handwritten text. In this work, we focus on developing an unsupervised dynamic text discrimination coefficient that classify the printed and handwritten text at word level specifically from Telugu pre-printed document forms.

2 Proposed Methodology

The proposed method for classification of printed and handwritten Telugu words is carried out in three stages. Stage 1 performs the feature computation by employing the geometric moments [11] and statistical features [12] from the words segmented. The density of black pixels and features computed are employed to determine the dynamic text discrimination coefficient in stage 2. Finally, the classification of printed and handwritten words through a decision model is accomplished in stage 3. Figure 1 details the three stages of proposed methodology.

2.1 Feature Computation

In this stage, the process of word segmentation initiates by acquiring a document image as an input. Furthermore, the horizontal projection profile features [13] of the document image are employed for segmentation of lines. The lines segmented are subject to computation of vertical projection profile features. Further words are segmented using the maximum zero valley features of each text line. The segmented words are subject to the features computation. The proposed system computes the center of mass features, the second-order geometrical moment M_{20}, the second-order complex moment C_{11}, statistical features, and density of black pixels.

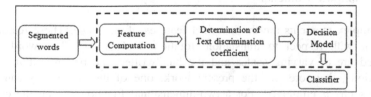

Fig. 1 Block diagram showing the stages for classification

2.1.1 Geometrical Moments

Geometric moment features are used extensively in various invariant and variant recognition of image patterns. These moments represent some fundamental geometric properties of image $f(x, y)$ like central moments, mass with area and center of mass. In general, two-dimensional geometrical moments of order $(p+q)$ are defined by [14]:

$$M_{pq} = \sum_{a1}^{a2} \sum_{b1}^{b2} x^p y^q f(x, y), \tag{2.1}$$

where $p, q = 0, 1, 2 \ldots \infty$ and $x^p y^q$ represent a basis function.

In the proposed method, the mass with area, center of mass properties and one of the second-order geometrical moments are employed, since the mass with area is a varying factor with respect to printed and handwritten words. The printed words are of smaller mass with area when compared to handwritten words. Thus, mass with area is a variant property from printed word to a handwritten word. The mass with area of a binary image represents a zeroth-order moment given as:

$$M_{00} = \sum_{a1}^{a2} \sum_{b1}^{b2} f(x, y). \tag{2.2}$$

The center of mass is the distribution of foreground intensities is focused at one point and can be variant with respect to printed and handwritten word. It also describes the unique position of an image with respect to a reference point. The center of mass with respect to x and y coordinates is given as:

$$\bar{x} = \frac{M_{10}}{M_{00}}, M_{10} = \sum_{a1}^{a2} \sum_{b1}^{b2} x.f(x, y); \quad \bar{y} = \frac{M_{01}}{M_{00}}, M_{01} = \sum_{a1}^{a2} \sum_{b1}^{b2} y.f(x, y), \tag{2.3}$$

where M_{10} and M_{01} are the first-order moments in X-direction and Y-direction, respectively. Furthermore, the second-order moments M_{20} in X-direction describes the orientations of an image with respect to the principle axis. The second-order moments in X-direction is given by:

$$M_{20} = \sum_{i=1}^{m} (x_i - \bar{x})^2. \tag{2.4}$$

2.1.2 Complex Moments

The complex moments are the invariant features of an image. These features are invariant with respect to translation, rotation, and scaling. The word images extracted are of varying scales and the handwritten can be present in varying orientations; therefore, for the present work, one of the complex second-order moments C_{11} is employed. For a two-dimensional function $f(x, y)$, the complex moments [14] of order (p, q) is given by:

$$C_{pq} = \sum_{a1}^{a2} \sum_{b1}^{b2} (x+jy)^p (x-jy)^q f(x,y), \tag{2.5}$$

where p and q are non-negative integers and $j = \sqrt{-1}$.

2.1.3 Statistical Features

The statistical quantities like mean, standard deviation, kurtosis and variance describe the aggregate quantities with respect to the intensity distribution of the pixels in the image. The mean intensity level of a two-dimensional image function $f(x,y)$ with intensity level r_i is given by:

$$Mean(f) = \frac{1}{mn} \sum_{x=1}^{m} \sum_{y=1}^{n} r_i p(r_i), \tag{2.6}$$

where $p(r_i)$ represents number of instances of intensity level r_i with respect to a pixel $f(x,y)$ in an image. The standard deviation of a two-dimensional image function $f(x,y)$ with mean, $Mean(f)$ is given by:

$$Std(f) = \sqrt{\frac{\sum_{i=1}^{m} \sum_{j=1}^{n} (f(x_i,y_j) - Mean(f))^2}{N-1}}. \tag{2.7}$$

The variance and kurtosis of a two-dimensional image function $f(x,y)$ are given by the following:

$$Var(f) = [Std(f)]^2 \tag{2.8}$$

$$Kur(f) = \frac{\sum_{i=1}^{m} \sum_{j=1}^{n} (f(x_i,y_j) - Mean(f))^4}{[Std(f)]^4 (m*n)} - 3. \tag{2.9}$$

The density of black pixels in a binary image is the count of pixels with their intensity level as black and computed using:

$$D(f) = \sum_{k=1}^{n} [p_k]_0, \tag{2.10}$$

where $[p_k]_0$ is the pixel with intensity level as '0'.

The proposed methodology employs the center of mass $\bar{x}$ with respect to 'x' coordinate position, second-order geometric moments M_{20} and second-order complex moments C_{11} for text classification. The moment features computed with respect to printed and handwritten text samples of varying lengths and types. The text samples include both handwritten and printed text.

2.2   Determination of Text Discrimination Coefficient

In conventional methods of classification, the standard classifiers like KNN, SVM, HMM and other unsupervised models are used for classification of printed and handwritten text. In the proposed method, a text discrimination coefficient is used to determine the dynamic threshold using the features computed in stage 1. The text discrimination coefficient is determined by employing the statistical and moments features. This coefficient is a dynamic threshold that discriminates the printed word from handwritten word.

The statistical features $D(f)$, $Kur(f)$, $Var(f)$, and $Std(f)$ of a two-dimensional image function $f(x, y)$ are adapted in devising the text discrimination coefficient. The product of density of black pixels, 'n' times the kurtosis and standard deviation is summed up with the variance features of the word to determine the text discrimination coefficient.

Let τ denotes the text discrimination coefficient and is defined as:

$$\tau = [D(f)*n.Kur(f)*Std(f)] + Var(f) \qquad (2.11)$$

subject to $4 \leq n \leq 10$.

The 'n' value can vary from 4 to 10, is the called as the text discrimination range for τ. The defined text discrimination coefficient is compared with the sum of the second-order geometric moments M_{20} and the second-order complex moment C_{11} is denoted as:

$$MC = M_{20} + C_{11}. \qquad (2.12)$$

After performing repeated analysis with different types of printed and handwritten text samples, we have deduced that, the text discrimination coefficient τ should be greater than the sum of moments MC for printed text and τ should be less than the sum of moments MC for handwritten text. The empirical relation for the text classification is defined in:

$$\tau > MC \quad => \quad \text{``Printed word''} \qquad (2.13)$$

$$\tau \leq MC \quad => \quad \text{``Handwritten word''}. \qquad (2.14)$$

The hypothesis testing between the text discrimination coefficient τ and the sum of moments MC is performed using Z-test [15] to determine whether one variable influences the other or not. The Z-test for population means is performed separately on printed and handwritten text samples with respect to the text discrimination coefficient τ and sum of moments MC. In the case of printed text, the number of text samples considered is $n = 500$ for which $\sigma = 37104.6$, $\bar{x} = 59074.39$ and $\mu = 24426.3$, where $n, \sigma, \bar{x}, \mu_0 \in \tau$ and $\mu \in MC$. The null hypothesis for testing is H_0, which represents the average of text discrimination coefficient τ, i.e., $\mu = \mu_0$ and

Fig. 2 *Right* tailed test

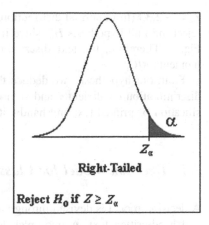

For printed test samples of size $n = 500$, the $Z_{cal} = 20.88$ and level of significance, $\alpha = 5\%$, $Z_\alpha = 1.65$ (from normal distribution table) and at $\alpha = 1\%$, $Z_\alpha = 2.33$ (from normal distribution table). In both the cases, we observe that, $Z_{cal} > Z_\alpha$. Thus, we reject the null hypothesis H_0, since it falls in the rejection region as mentioned in Fig. 2. Therefore, the text discrimination coefficient τ is greater than the sum of moments MC. Similarly, for handwritten text samples of size, $n = 650$ for which $\bar{x} = 408061.4$ and $\mu = 18867182$ and $n, \sigma, \bar{x}, \mu_0 \in \tau$ and $\mu \in MC$. The null hypothesis H_0 is the same as printed text, whereas the alternate hypothesis is defined as H_1, which represents average of text discrimination coefficient τ, $\mu_0 < \mu$. In this case, the $Z_{cal} = -1284.27$ and at level of significance., $\alpha = 5\%$, $Z_\alpha = -1.65$ and $\alpha = 1\%$,

The alternate hypothesis is H_1, which represents average of text discrimination coefficient τ, i.e., $\mu_0 > \mu$. The Z-test statistics for testing is given by:

$$Z_{cal} = \frac{\bar{x} - \mu}{\sigma/\sqrt{n}}. \tag{2.15}$$

Fig. 3 *Left* tailed test

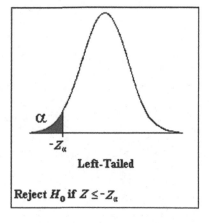

$Z_\alpha = -2.33$ (from normal distribution table). In both the cases, $Z_{cal} < Z_\alpha$. Thus, we reject the null hypothesis H_0, since it falls in the rejection region as mentioned in Fig. 3. Therefore, the text discrimination coefficient τ is less than the sum of moments MC.

From our hypothesis, we deduce that, the dynamic threshold computed by text discrimination coefficient τ and sum of moments MC are reliable enough in classification of printed text and handwritten text.

2.3 Decision Model for Classification of Text

A decision model is devised through a decision tree for the classification of printed and handwritten text. A two-level decision is performed by employing the text discrimination coefficient τ and the features computed. From the center of mass features with respect to x-coordinate and the standard deviation of the text, it is inferred that, the $Std(f)$ is always greater than center of mass with respect to x-coordinate $\overline{X}$ for printed text and vice versa. The text discrimination coefficient is further employed at another level classification. Figure 4 represents the proposed decision model for classification.

Let D_1 represents the relation at level one and D_2 represents the relation at level 2 which is given by (2.16) and (2.17). The PW and HW represents the printed text and handwritten text, respectively.

$$D_1 = MC > \tau \tag{2.16}$$

$$D_2 = Std(f) \geq \overline{X} \tag{2.17}$$

Fig. 4 Decision model for classification

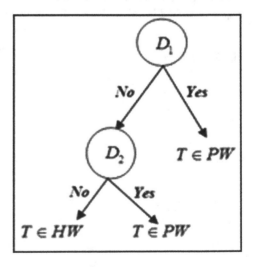

The classification of printed and handwritten words using decision model is elucidated in the subsequent section as experimental analysis.

3 Experimental Analysis

In the proposed methodology, the experimentation is carried out with 650 handwritten text samples and 500 printed text samples. The data sets are generated by segmentation of words in Telugu pre-printed application documents. The documents are pertaining to the regions of Anantapur district of state Andhra Pradesh. All the datasets are of resolution 300 dpi and above. The dimensions of the data sets are not scaled prior to feature computation. In the proposed system, precision and recall are the two basic measures that are employed for evaluating the accuracy of the system. The precision is the ratio of true positives t_p to the sum of true positives and false negatives $t_p + f_p$, whereas the false positive f_p are the incorrect text samples that are intuitively labeled by classifier as correct one. The recall R is the ratio of true positives t_p to the sum of the true positives and false negatives $t_p + f_n$ and is given by the following.

$$\text{The precision } P = \frac{t_p}{t_p + f_p} \tag{3.1}$$

$$\text{The recall } R = \frac{t_p}{t_p + f_n} \tag{3.2}$$

The false negative f_n is the correct samples that are intuitively labeled by classifier as incorrect one. The Table 1 shows the experimental statistics of the proposed methodology.

The computational outcomes of text discrimination coefficient τ and sum of moments MC with respect to text samples of printed and handwritten words of varying lengths are as shown in Figs. 5 and 6.

From Figs. 6 and 7, it is evident that, the text discrimination coefficient τ and sum of moments MC seem to be inversely related in the cases for both printed as well as handwritten text. The analysis is performed even on the text with numeric and alpha-numeric characters which includes some English alphabets. The rate of text discrimination is very promising for the text samples in Telugu script where as adequate and consistent in case of numerals and English characters. Figures 7 and 8

Table 1 Experimental statistics of proposed system

Type of text	No. of text samples	t_p	Precision (P) (%)	Recall (R) (%)
Printed	500	489	97.1	95
Handwritten	650	641	98.2	96.9
Total	1150	1130	98.2	90.18

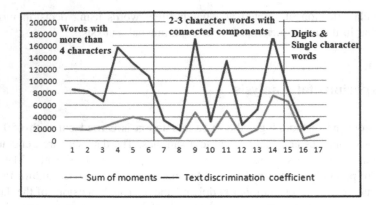

Fig. 5 τ versus *MC*-printed text

Fig. 6 τ versus *MC*-handwritten text

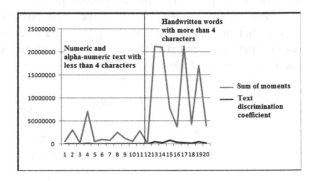

Fig. 7 $\overline{X}$-Printed versus handwritten text

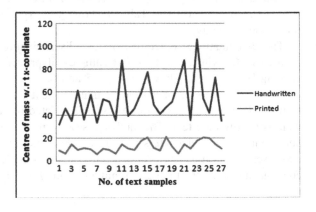

represent the center of mass $\overline{X}$ and standard deviation features with respect to printed and handwritten text samples, respectively. The method is also tested with text blocks of English and Kannada for which the results are equally good in classification.

Fig. 8 $Std(f)$-Printed versus hand written text

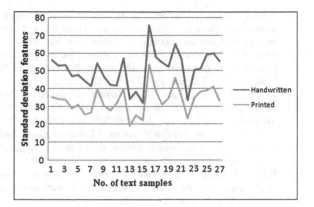

The $\overline{X}$ and $Std(f)$ features intuitively represent that, $Std(f) > \overline{X}$ in the case of printed text and vice versa. Thus, in the proposed system, these features are employed at the second level of decision model for discrimination of text.

4 Conclusion

The proposed system mainly focused on developing a dynamic threshold determination technique for discriminating printed and handwritten Telugu text. The sum of moment's features is correlated with the devised text discrimination coefficient to classify the text. The decision model is devised in two levels employing the text discrimination coefficient at one level and the statistical features at another level. The proposed algorithm is robust with respect to the text samples of Telugu script and consistent with the numerals and English alphabets. The method can also be extended toward the other South Indian scripts like Kannada and needs to be revised for the printed text images with very low resolution and more amount of distortions. Thus, the text discrimination coefficient devised is considered as a variant factor for printed and handwritten text classification.

References

1. Shobha Rani N., Vasudev T.: A Generic Line Elimination Methodology using Circular Masks for Printed and Handwritten Document Images, Emerging research in computing, information, communication and applications, Elsevier Science and technology, vol. 3, (2014).
2. Sadagopan Srinivasan., Li Zhao., Lin Sun., Zhen Fang, Peng Li.., Tao Wang., Ravishankar Iyer., Ramesh Illikkal., Dong Liu.,:Performance Characterization and Acceleration of Optical Character Recognition on Handheld Platforms, IEEE International Symposium on Workload Characterization (IISWC), (2010).

3. Suman V Patgar., Vasudev T.,: An unsupervised intelligent system to detect fabrication in photocopy document using geometric moments and gray level co-occurrence matrix, International journal of computer applications, Vol. 74(12), 29–34, (2013).
4. Mark A Walch., Donald T Gantz.,: Pictographic recognition technology applied to distinctive characteristics of handwritten Arabic text, Proceedings of Symposium on Document Image Understanding Technology, 173–186, (2005).
5. Ranjeet Srivastava., Ravi Kumar Tewari., Shashi Kant.,: Separation of machine printed and handwritten text for Hindi documents, International Research Journal of Engineering and Technology (IRJET), Vol. 2(2), pp. 704–708, (2015).
6. M.S. Shirdhonkar., Manish B Kokare.,: Discrimination between printed and handwritten text in documents, International journal of Computer Applications, Recent Trends in Image Processing and Pattern Recognition", 131–134, (2010)
7. Rajesh Pathak., Ravi Kumar Tewari.,: Distinction between machine printed text and handwritten Text in a document, International Journal of Scientific Engineering and Research (IJSER), Vol. 3(7), pp. 13–17, (2015).
8. Lincoln Faria da Silva., Aura Conci., Angel Sanchez.,: Automatic discrimination between printed and handwritten text in documents, XXII Brazilian Symposium on Computer Graphics and Image Processing (SIBGRAPI), (2009).
9. Mallikarjun Hangarge., K.C. Santosh., Srikanth Doddamani., Rajmohan Pardeshi.,: Statistical Texture Features based Handwritten and Printed Text Classification in South Indian Documents, International Conference on Emerging Trends in Electrical, Communications and Information Technologies, Elsevier, vol. 1(32), 215–221, (2012).
10. Samir Malakara., Rahul Kumar Dasa., Ram Sarkarb., Subhadip Basub., Mita Nasipuri.,: Handwritten and printed word identification using gray-scale feature vector and decision tree classifier, International Conference on Computational Intelligence: Modeling Techniques and Applications(CIMTA), Procedia Technology 10, 831–839, (2013).
11. Simon Xinmeng Lia.,: Image analysis by moments, Thesis, Department of electrical and computer engineering, University of Manitoba, Winnipeg, Canada, (1993).
12. Yan Qiu Chen., Mark X. Nixon., David W. Thomas.,: Statistical geometrical features for texture classification, Pattern recognition, Elsevier Science, vol. 28(4), 537–552, (1995).
13. Mohammed Javed., P. Nagabhushan., B.B. Chaudhuri.,: Extraction of projection profile, run-histogram and entropy features straight from run-length compressed text documents, IAPR Asian conference on pattern recognition, IEEE proceedings, 813–817, (2013).
14. Li. S., Moon chuan Lee., Chi Man Pun.,: Complex zernike moments features for shape based image retrieval, IEEE Transactions on Systems, Man and Cybernetics, Part A: Systems and Humans, Vol. 39(1), 227–237, (2008).
15. Franz Faul., Edgar Erdfelder., Axel Buchner., Albert-Georg Lang.,: statistical power analyses using G*Power 3.1: Tests for correlation and regression analyses, Behavior Research Methods, Springer, Vol. 41(4), 1149–1160, (2009).

Neural Nets for Thermodynamic Properties

R.V.S. Krishnadutt and J. Krishnaiah

Abstract Thermal power plants are controlled and monitored in real time using state-of-the-art distributed control systems (DCS). Both at real-time control and online monitoring of these plants, thermodynamic properties of water/steam along with their partial derivatives are required in computation and optimization of process. Particularly during data reconciliation and process optimization, the equipment models appear as non-linear constraints which require Jacobians/Hessians of thermodynamic properties. Thermodynamic properties of steam/water are complex functions (Gibbs and Helmholtz functions), and they are repeatedly called during non-linear optimization. The Jacobians/Hessians of these properties are numerically approximated involving further repeated calls to the functions. Legacy approaches involve approximating these functions using higher-order polynomial algorithms or look-up tables. These methods have the limitation of high computational time or need large memory. However, using artificial neural networks (ANNs), these limitations can be overcome, and in this paper, it is demonstrated that ANNs with back-propagation neural networks (BPN) are an effective means to improve the computational performances. The computational time is reduced by a factor of nearly four times.

Keywords Artificial neural networks · Thermodynamic properties · Jacobians · Hessians

R.V.S. Krishnadutt (✉) · J. Krishnaiah
Corporate R&D, BHEL, Vikasnagar, Hyderabad 500093, India
e-mail: krishnadutt@bhelrnd.co.in

J. Krishnaiah
e-mail: krishnaiah@bhelrnd.co.in

© Springer Science+Business Media Singapore 2017   701
S.C. Satapathy et al. (eds.), *Proceedings of the First International Conference on Computational Intelligence and Informatics*, Advances in Intelligent Systems and Computing 507, DOI 10.1007/978-981-10-2471-9_68

1   Introduction

Thermal power plants of large capacities (200–1000 MW) are instrumented and controlled using state-of-the-art distributed control systems (DCS). The working medium in these plants is water/steam in turbines and boiler sections and flue gas, coal and air in combustion part of boiler. Nearly 2000 analog sensors and around 10,000 digital signals are used in monitoring and controlling the process/equipment. Monitoring involves online computation of energy conversion efficiencies of equipments, like boiler, turbines, electrical generators and motors, and effectiveness of different heat exchangers. Monitoring also involves signal validation as the raw values from sensors are either outliers or non-numbers. Extensive research is focused on data validation and reconciliation. Shanker Narasimhan [1] has covered various classical approaches of data reconciliation in industrial processes. Optimization of plant for best operating plant is also a part of modern DCS. Both data reconciliation and performance optimization involve solving non-linear optimization problem subjected non-linear constraint equations [1]. Constraint equations represent equipment models in terms of continuity, momentum and energy balances. These involve repeated usage of thermodynamic properties like enthalpy, entropy, specific volume and their partial derivatives of first (Jacobian) and second order (Hessian). Thermodynamic properties of water and steam are represented using complex Gibbs functions [3], and they are subdivided into regions described in Table 1. For each region in Table 1, separate function approximations are used. Using the conventional steam table representation, to get a property value along with Jacobian and the corresponding Hessian involves repeated calls, nearly eleven times, to the function. This is computationally expensive and not amenable for use in real-time controllers having limited memory and high response time. Azimian et al. [2] has reported using artificial neural nets (ANNs) for evaluation of

Table 1 Different ANNs for different regions of water and steam

Sl. no	Net name	Description of the ANN model trained/testing	Training data		Test data	
			P (kg/cm2) (Range/Step)	T (deg. C) (Range/Step)	P (kg/cm2) (Range/Step)	T(deg. C) (Range/Step)
1	SubCNet	Subcooled region	(0.1–2.95)/ 0.05 (41–170)/1	(40–132)/1 (250–350)/1	(0.125–2.725)/ 0.05 (41.5-169.5)/1	(40.5-129.5)/1 (250.5 - 349.5)/1
2	SatFNet	Saturated fluid region	(0.1–3)/0.05 (3–170)/1	(Ts) 45.41-132.84 (Ts) 132.84-350.67		
3	SatVNet	Saturated vapour region	(0.1–3)/0.05 (3–170)/1	(Ts) 45.41–132.84 (Ts) 132.84–350.67		
4	SupHNet	Superheated region	(0.1–2.95)/ 0.05 (3–170)/1	(46–200)/1 (250– 545)/1	(0.125–2.725)/ 0.05 (3.5– 169.5)/1	(50.5–199.5)/1 (250.5–544.5)/ 1

thermodynamic properties of steam/water. Hataitep et al. [4], on the other hand, used neurofuzzy net (NF) approach to arrive at decision boundary for different regions and proposed different NFs for different regions. However, the ambiguity near interface of boundaries is not resolved due to fuzzy set. Chouai et al. [5] has proposed ANNs with back-propagation for obtaining thermodynamic properties of refrigerant fluids. Lilja et al. [6] has used different nets for different phases of steam/water for different properties and achieved a maximum mean squared error of 0.0045 on training set. However, the above literature does not address both Jacobian and Hessians of these properties. Ferrari et al. [7], however, has given a general formulation of smooth function approximation using neural networks wherein Jacobian is used as convergence criterion and does not directly give function value and its higher-order derivatives simultaneously. In this paper, ANN with BPN is used to obtain directly the thermodynamic properties along with their Jacobians and Hessians, and different neural network models are proposed for each region.

2 Feature Vector and Architecture of ANN for Thermodynamic Properties of Steam and Water

Typical thermodynamic properties are saturation pressure/temperature, enthalpy, entropy, specific volume, density and specific heat. Regions of interest cover compressed water, saturated water, saturated steam and superheated steam. Occasionally, choked flow calculations are also needed. All these properties are formulated in terms of Gibbs and Helmholtz free energy, and different formulations are proposed for different regions [1]. Jacobians of a property are obtained through numerical derivatives, Eq. 1, either using left side, right side or central difference. However, in the way the numerical derivative is obtained, there can be ambiguity, i.e., the property and its derivatives may not fall in the same region. This can result in inconsistency and non-convergence of non-linear optimization solution.

Table 2 shows the results of enthalpy calculation near saturated water and saturated steam conditions and its derivatives with respect to pressure and temperature. The inconsistency is avoided by preserving the state of the substance. This resulted in maintaining the magnitude of derivatives irrespective of the method of derivative calculation, viz. backward, central and forward derivatives.

The input feature vector taking care of all regions is shown in Eq. (1).

$$I = [p, t, x]^T, \tag{1}$$

where p, t and x are pressure, temperature and quality of steam, respectively.

The corresponding output vector is:

Table 2 Inconsistency in computing derivatives

S. no	Region	Preserving state	Enthalpy [kCal/kg] (P, T)	Partial derivatives with pressure [kCal/cm^2]			Partial derivatives with temperature [kCal/kg-°C]		
				Backward	Central	Forward	Backward	Central	Forward
1	Saturated liquid	No	45.43446 (0.1, 45.41953)	–5716637	–2858318	0.020449	0.998671	2858315	5716630
		Yes	45.43446 (0.1, 45.41953)	194.9019	194.8182	194.7345	0.998775	0.998776	0.998778
2	Saturated vapour	No	617.0974 (0.1, 45.41953)	–7.56082	–2858318	–5716629	57.6630	2858315	0.463318
		Yes	617.0974 (0.1, 45.41953)	82.84811	82.81154	82.77497	0.42455	0.424551	0.424552

$$O = \left[h, \frac{\partial h}{\partial p}, \frac{\partial h}{\partial t}, \frac{\partial^2 h}{\partial p^2}, \frac{\partial^2 h}{\partial p \partial t}, \frac{\partial^2 h}{\partial t^2} \right]^T \qquad (2)$$

The non-linear mapping from input vector I to O is achieved through ANNs, Eq. 3 which is dependent on structure and activation functions of ANN.

$$O = f(I) \qquad (3)$$

Mapping function f is obtained through training using the pair $[I^k, O^k]$ for $k = 1...M$ sets. This assumes that the f preserves both function and derivative continuities. For a typical single layer net, the mapping function f is shown in Eq. (4) read in conjunction with Fig. 1, from which it can be inferred that the non-linearity arises out of the non-linear activation function. The approximation of functional relationship existing among the input feature vector $I = [I^k]$ and output vector $O = [O^k]$, $k = 1, 2, ...$ M, samples is approximated by Eq. (3). And these activation functions preserve the derivative continuity in the original functional relationship existing between I and O. In normal ANN with gradient descent algorithm used for function approximation alone, the derivative continuity is not inherently satisfied. However, the training set involves both Jacobian and Hessian; hence, the net shall learn not only the functional values but their derivatives (Fig. 2).

Separate nets are designed for different properties and regions and each of these nets also have same architecture as shown in Fig. 2. Selection of a particular net is described in Fig. 3. For a given feature vector I, in the first phase,

$$O = o^T \sigma [WI + b] + B, \qquad (4)$$

where $\sigma(x) = \frac{1}{(1+e^{-x})}$

A neural net model approximates the saturation line which is used for selection of appropriate nets like SubCNet, and SupHNet, each representing subcooled

Fig. 1 Architecture of a single layer in an ANN

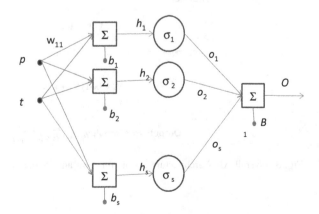

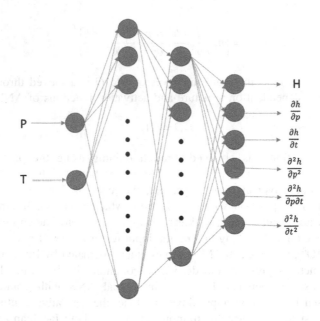

Fig. 2 Two-layer net for prediction of function and its derivatives

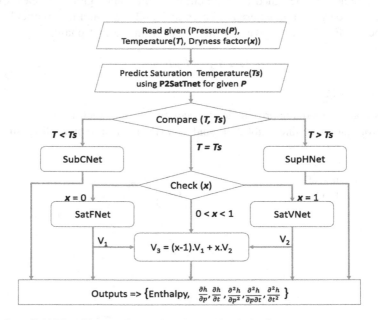

Fig. 3 Overall ANN architecture for the function and its derivatives

region, and superheated regions, respectively, shown in Fig. 3. Saturation and wet regions are handled based on dryness fraction x, and they lead to SatVNet and SatFNet. Each of these regions is modeled using feed-forward nets with sigmoidal activation function.

3 Neural Net Implementation of Thermodynamic Properties

The above architecture considers two hidden layers, first being designed with 16 neurons while second with 10 neurons. The design follows the empirical approach of

$$N_{h1} = 2 * (N_i + N_o) \& N_{h2} = 2 * N_o ,\tag{5}$$

where N_{h1}, N_{h2}, N_i and N_o are number of neurons in the first hidden layer, neurons in the second hidden layer, number of inputs and number of outputs, respectively. This is obtained from a study of various combinations. Each of these nets is trained using data generated using REFPROP [8]. Table 1 details different ANN models and ranges of training and test data sets.

Figures 4 and 5 show the Hinton diagrams of one of the ANN nets for the function value (here, enthalpy) and its derivative, respectively, while Figs. 6 and 7 show similar diagrams for superheated region. Hinton diagrams are plots of connection weights showing positive weights green in color and negative weight red in color. In Fig. 4, the net is trained with input vector and the associated function value as output, whereas Fig. 5 shows when the net is trained with input vector and

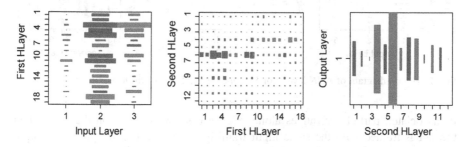

Fig. 4 Hinton diagram of ANN layers for subcooled enthalpy model

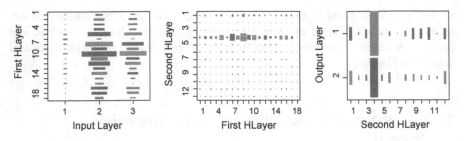

Fig. 5 Hinton diagram of ANN layers for Jacobians model of subcooled Enthalpy

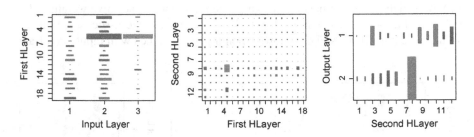

Fig. 6 Hinton diagram of ANN layers for superheated enthalpy model

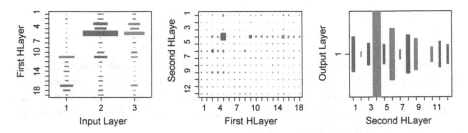

Fig. 7 Hinton diagram of ANN layers for Jacobians model of superheated enthalpy

its associated partial derivatives alone. Figure 6 shows the Hinton diagram when
the net is trained with the same input training set but with corresponding output
function and its associated partial derivatives simultaneously. Figures 8 and 9 show
the prediction accuracy both during training and testing phases, respectively.
Figure 10 shows the surface plot of the enthalpy and its two derivatives as functions
of pressure and temperature.

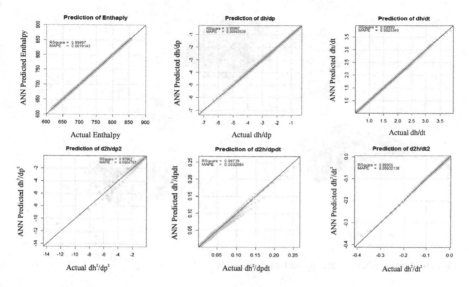

Fig. 8 ANN prediction of enthalpy and its derivatives with the training data

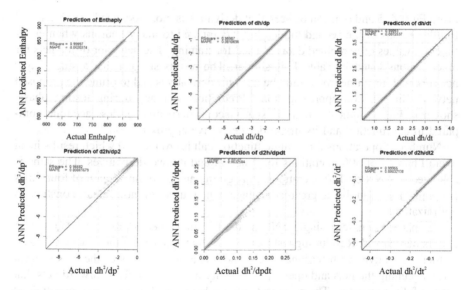

Fig. 9 ANN prediction of enthalpy and its derivatives with test data

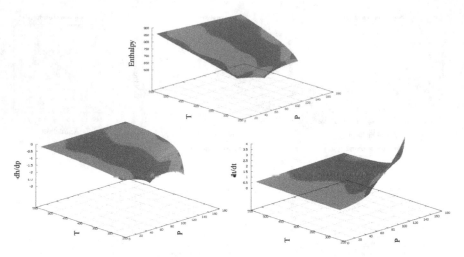

Fig. 10 ANN predicted enthalpy and its derivatives

4 Results and Conclusions

From Figs. 4, 5 and 6, it can be seen that derivative is more localized in the first and fourth neurons of the second hidden layer compared to many neurons when either function values or combined data are used for training. The behavior of the nets for other regions listed in Table 1, is similar. The results suggest that a single-layer network with fewer neurons could be used for derivatives and to prune the proposed network for function approximations. Prediction accuracy during testing phase, shown in Fig. 9, indicates that Hessian (second derivative) is less accurate compared to the function and its first-order derivative approximations.

Number of operations for a single function call is around 100 which results in an overall burden of 500 operations for function and its two derivatives. The same is achieved in ANN by 120 operations, thus giving an approximate gain of four times per call. This can be a great advantage when used in non-linear constrained optimization.

The paper discusses a single ANN model for simultaneous prediction of function and its derivatives. Results suggest that the ANNs approximate the function as well as their derivatives with reasonable accuracy. Hinton diagrams of the nets suggest further pruning the nets and open up possibility of designing feed-forward nets that are not fully connected. The proposed nets are being used in non-linear constrained optimization with substantial reduction in computational burden to reach convergence.

Acknowledgments The authors thank Management of BHEL for facilitating the investigations reported in the paper and according necessary approvals for publication. The authors also express gratitude to Ms. Hiral, Research Engineer, BHEL (R&D) for generating the data using REFPROP.

References

1. Shankar Narasimhan, Data Reconciliation and Gross Error Detection—An Intelligent use of process data, Gulf Publishing company, Houstin, Texas, (2000)
2. Azimian, AR; Arriagada, Jaime LU and Assadi, Mohsen Generation of steam tables using artificial neural networks, LU (2004) In Heat Transfer Engineering 25(2). p. 41–512.
3. C.A. Meyer, R.B.McClintock, G.J. Silvestri, R.C. Spencer, Jr, ASME Steam Tables, Sixth Edition, ASME, New York, 1993
4. Hataitep Wonggsuwarn, Modelling of Thermodynamic Properties based on Neurofuzzy System for Steam Power Plant, Proc. Of the World Congress on Engineering and Computer Science, Vol. II, WCECS, Oct.20–22, San Fransisco, USA.
5. Chouai, A., S. Laugier, and D. Richon. Modeling of thermodynamic properties using neural networks: Application to refrigerants. Fluid Phase Equilibria 199, no. 1, 53–62 (2002):
6. Lilja, Reijo, and Jari J. Hamalainen. Modeling of thermodynamic properties of substances by neural networks. In Neural Networks, 1999. IJCNN'99. International Joint Conference on, vol. 6, pp. 3927–3932. IEEE, 1999.
7. Silvia Ferrari, Robert F.Stengel, Smooth Functions Approximation Using Neural Networks, IEEE Trans. On Neural Networks, Vol.16, No.1, Jan.2005.
8. Lemmon, E. W., M. L. Huber, and M. O. McLinden. REFPROP: Reference fluid thermodynamic and transport properties. NIST standard reference database 23, no. 8.0 (2007).

Parallelization of Synthetic Aperture Radar (SAR) Image Formation Algorithm

P.V.R.R. Bhogendra Rao and S.S. Shashank

Abstract Synthetic aperture radar (SAR)-based platforms have to process increasingly large number of complex floating-point operations and have to meet hard real-time deadlines. However, real-time use of SAR is severely restricted by computation time taken for image formation. One of the classical methods of reducing this computation time to make it suitable for real-time application is multi-processing. A successful attempt has been made by the authors to develop and test a parallel algorithm for synthetic aperture radar image formation, and the results are presented in this paper.

Keywords Synthetic aperture radar · SAR image formation · Digital signal processing · Parallel computing · Distributed computing

1 Introduction

Synthetic aperture radar is increasingly used in environmental monitoring, disaster management, military and defense, remote sensing etc., owing to its special benefits like all weather, day and night imaging capabilities over optical imaging [1]. SAR images became more and more valuable for earth and remote observation, since they deliver more and different information than regular optical images. However, the process of generating an image from the sensed raw data, is highly computationally complex process and consumes good amount of time before it can be used. Also, interpreting and processing such images is difficult and requires advanced algorithms and computation architectures. As further increase in the clock frequency in von Neumann architecture is no longer feasible, one way to increase the

P.V.R.R. Bhogendra Rao (✉)
Defence Research and Development Laboratory, Hyderabad, India
e-mail: bhogendra.rao@gmail.com

S.S. Shashank
BrahMos Aerospace Private Limited, Hyderabad, India
e-mail: shashankss3s@gmail.com

© Springer Science+Business Media Singapore 2017
S.C. Satapathy et al. (eds.), *Proceedings of the First International Conference on Computational Intelligence and Informatics*, Advances in Intelligent Systems and Computing 507, DOI 10.1007/978-981-10-2471-9_69

processing power is to switch to alternatives like parallel computing machines, hardware accelerators such as AltiVec extensions with VSIPL-compliant library and GPGPU-based systems.

1.1 Basic SAR Geometry

Synthetic aperture radar produces a two-dimensional (2-D) image. One dimension of the image is called range (or cross-track) and is a measure of the "Line-of-sight" distance from the radar to the target [2, 3]. The other dimension is called azimuth (or along-track) and is perpendicular to range.

The basic SAR processing geometry is shown in Fig. 1. Sensor transmits pulses and receives the echoes in a direction perpendicular to sensor motion. The received signal is digitized and stored as complex samples for each pulse, which forms the SAR raw data. These data are 2-D matrices which carry both amplitude and phase information of ground returns. One dimension of the array represents distance in the slant range direction between the sensor and the target, and is commonly referred to as the range direction. The other dimension represents the along-track or azimuth direction [1].

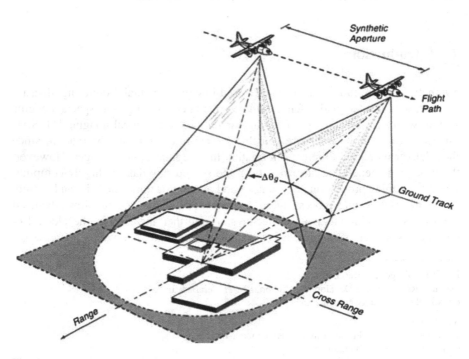

Fig. 1 Basic SAR geometry (*Image source* [4])

The processing of SAR data requires two-dimensional convolutions and FFTs. As the range and azimuth timescales are different, the range and azimuth directions can be treated separately during processing. In each dimension, the data are convolved with a matched filter. A series of signal and image processing steps is carried out to extract information from the signal.

This paper is organized into six further sections. Section 2 summarizes the work so far carried out by researchers. Section 2 describes the synthetic aperture radar wavefront image formation algorithm [4] as well as theoretical considerations regarding SAR echo model. Necessary prerequisites for frequency domain operation are stated as well. Section 3 describes the issues in the design of parallel algorithm and the factors that influence the runtime performance of the algorithm. Section 4 describes the experimental setup for parallel computing of SAR imaging. Section 5 describes simulation results which are used to validate the theoretical and practical considerations. Finally, Sect. 6 concludes with a concise discussion about the whole paper.

2 SAR Wavefront Algorithm

Figure 2 shows the block diagram of wavefront reconstruction algorithm [4]. The input to the algorithm is 2D complex matrix from pre-processor module, where the radar echo is range compressed using de-ramped technique. In range compression step, the received signal is correlated with a matched filter, also called range compression filter, which is, at most, a delayed version of transmitted pulse.

One-dimensional FFT [5, 6] in azimuth direction converts the data into 2D-frequency domain. In azimuth (slow time) compression step, data across all pulse returns from a given range are correlated with a matched filter.

The data are multiplied with a 2D complex reference phase function computed for a selected range, usually the scene center. The data are mapped from slant range to down range using sinc interpolation along range frequency axis. A two-dimensional IFFT converts the RCM-corrected data into fully focused SAR image.

The process of getting high-resolution images with SAR wavefront reconstruction image formation algorithm is highly complex involving large number of computations like FFTs [5, 6], convolution or match filtering in two (range and azimuth) directions, matrix multiplications and interpolation [7] for 32-bit floating-point calculations and requires high computation time.

One way is to use GPGPU [8–13] platforms using CUDA processor. This way, it works well for small scene and sufficient work is progressed on it; however, in case of real-time implementation for large scene area to achieve high throughput, it is better to go for parallel computing technique. Parallel processing technique gives the way to achieve high image precision in real time.

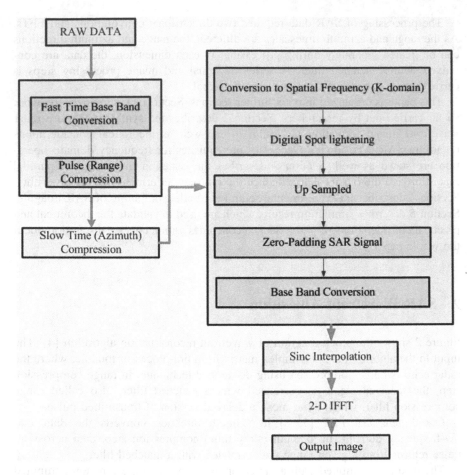

Fig. 2 Wavefront reconstruction algorithm

3 Parallel Implementation

The parallel algorithm was implemented using VSIPL standard specification [14]. Data-independent vector view computations and matrix view computations were completely parallelized by spawning a thread on the other core which resulted in speed up of almost 2x. When computations were analyzed for each function call, it was observed that FFTs consumed more CPU time compared to other operations.

3.1 Parallel Implementation of FFTs

Following are the assumptions and optimization made to speed up the computation performance for FFTs.

- Keep the size of the vector or matrix nearest to power of two, which will substantially speed up the computation performance by a large amount.
- If matrix FFTs are performed, divide the matrix into two equivalent subviews (submatrices) by dividing them in row-wise or column-wise based upon the direction of FFT (azimuth side or range side). This type of implementation can be easily done without moving data to new variables using VSIPL subviews which are very easy to use and implement. The divided submatrices are applied FFT operation in parallel and the resultant computed FFT matrix is obtained.
- If the algorithm uses FFT shifts, the following method can be applied to achieve the same computation performance as stated above. Consider the quadrants of a matrix as ~A, ~B, ~C, ~D on which FFT needs to be applied.

  - FFT on row-side, shown in Fig. 3.

    Divide the complex matrix into two subviews row-wise.
    A, B quadrants remain in submatrix1 and C, D remain in submatrix2.
    Swap A, B quadrants and apply FFT for the swapped submatrix1.
    Swap D, C quadrants and apply FFT for the swapped submatrix2.
    Reswap the submatrix1 quadrants.
    Reswap the submatrix2 quadrants.
    Since submatrix1 and submatrix2 computations are completely independent of each other, swapping and FFT operations can be applied in parallel.

  - FFT on column-side, shown Fig. 4.

    Divide the complex matrix into two subviews column-wise.
    Now A, C quadrants remain in submatrix1 and B, D remain in submatrix2.
    Swap A, C quadrants and apply FFT for the swapped submatrix1.
    Swap B, D quadrants and apply FFT for the swapped submatrix2.
    Reswap the submatrix1 quadrants.
    Reswap the submatrix2 quadrants.

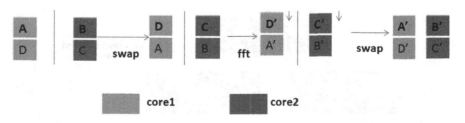

Fig. 3 Row FFT division

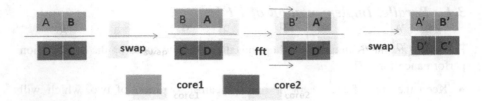

Fig. 4 Column FFT division

Since submatrix1 and submatrix2 computations are completely independent of each other, swapping and FFT operations can be applied in parallel.

- In SAR algorithms, a set of FFTs and IFFTs are applied. If FFT's or IFFT's remain in the same direction, then their complete computation can be spawned as separate threads or it can even be offloaded to other nodes. Further division of matrix into subviews on the other nodes will speed up the performance tremendously.

3.2 Parallel Implementation of Wavefront Algorithm

Design and Implementation SAR wavefront algorithm computations were divided equally on three nodes. Out of three nodes, one of them is a master node which communicates with slave nodes (other two nodes) and is responsible for synchronization and data transfers. Figure 5 shows master–slave relationship between the processor nodes.

A shared memory using PCIE bus was established for data communication. Any node can communicate with any other node using PCIE bus.

Each node, in turn, runs two threads one on each core. Data-independent computations were done in parallel. For synchronization between the nodes, hardware semaphores were used. Whenever parallelism is not possible to achieve due to dependency on results of computation on other node, the results must to be brought back to the master node.

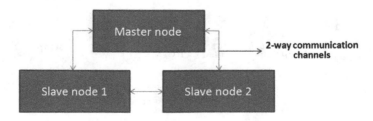

Fig. 5 Master–slave architecture

After the dependency is solved, the data must be again distributed to slave nodes. Figure 6 shows parallel SAR wavefront algorithm implementation.

Performance Assessment Use of AltiVec SIMD unit for signal processing computations uses existing library such as VSIPL and enhances the performance of large data set based computations of algorithms tremendously.

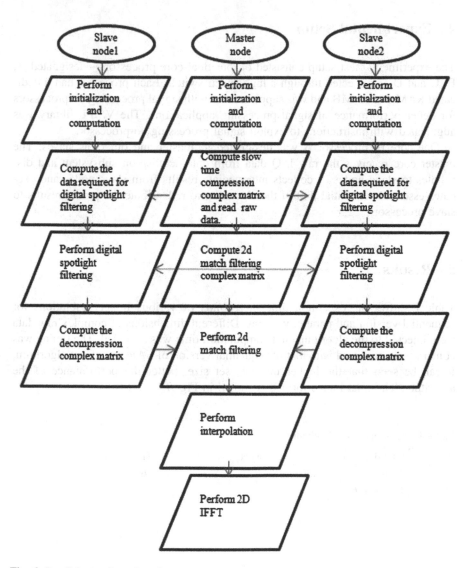

Fig. 6 Parallel wavefront data flow

VSIPL adds an abstraction layer, which makes it portable across diverse platforms and processor architectures. Porting on to other platforms is merely recompiling the code with specific VSIPL-compliant library. Use of POSIX threads for parallel computation speeds up the performance.

4 Experimental Setup

The experimental test setup consisted of four dual-core processors—designated A, B, C and D—connected through a high-speed switch. Each processor had a dedicated RAM of 512 MB and was supplemented with signal processing co-processors for better performance of signal processing applications. The VSIPL library was augmented with instructions to exploit signal processing co-processor.

One core of processor A was designated as master and others as slaves. The master core receives the raw I, Q data from data acquisition subsystem and distributes to slave nodes. It collects intermediate results from slave nodes and executes essential sequential part of the algorithm to redistribute the partial results to slave processors.

5 Results

Table 1 shows the execution time in seconds of various image resolutions for sequential version and parallel versions. Different dimensions of complex raw data were injected to the algorithm, and computation time was recorded. The timing was comaratively analyzed with that of sequential version of SAR wavefront algorithm. It can be seen that the higher the data set size, better the performance of the algorithm. The final image formed is shown in Fig. 7.

Table 1 Performance of algorithm

S. no	Raw data size	Time for sequential version (s)	Time for parallel version (s)
1.	4096 × 2048	150	46
2.	186 × 176	5	4

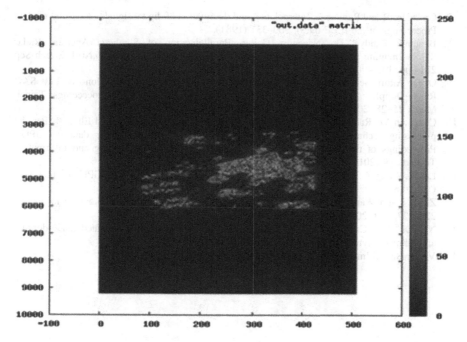

Fig. 7 Image formed

6 Conclusion

In this paper, we developed a method for parallel computing technique for SAR imaging algorithms. The computational complexity of the proposed method for wavefront reconstruction algorithm is significantly lower than the traditional one, without any performance loss in terms of imaging quality. This algorithm was implemented in MATLAB, and in C using signal processing libraries.

References

1. Carrara W.G., Goodman R.S. and Majewski R.M., Spotlight Synthetic Aperture Radar Signal Processing Algorithms, Artech House, 1995.
2. Curlander J.C. and McDonough R.N., Synthetic Aperture Radar Systems and Signal Processing, J. Wiley & Sons, USA. 1991.
3. Cumming I.G. and Wong F.H., Digital Processing of Synthetic Aperture Radar Data: Algorithms and Implementation, Artech House Publishers, 1st ed. 2005.
4. Soumekh. M, Synthetic Aperture Radar Signal Processing With MATLAB, Wiley-Blackwell, 1999.
5. MATLAB Documentation: fft-Discrete Fourier Transform. Version 7.9.0.529 (R2009b) & (R2015a), The Mathworks, Inc.
6. Stolt R.H., Migration by fourier transform, Geophysics, vol. 43, no. 1, p. 23, 1978.

7. Crochiere R and Rabiner L, Interpolation and decimation of digital signals - a tutorial review, Proceedings of the IEEE 69(3), 300–331 (1981).
8. Bhaumik Pandya, Dr. Nagendra Gajjar, Parallelization of Synthetic Aperture Radar (SAR) Imaging Algorithms on GPU, proceedings of IJCSC Volume-5,N0.1 March-Sep 2014 pp. 143–146 ISSN-0973-7391.
9. Ozgur Altun, Selcuk Paker, and Mesut Kartal. Realization of Interpolation-free Fast SAR Range-Doppler Algorithm Using Parallel Processing on GPU, PIERS Proceedings, Taipei, March 25–28, 2013.
10. Chapman W, Ranka S, Sahni S, Schmalz M, Majumder U, Moore L and Elton B. Parallel processing techniques for the processing of synthetic aperture radar data on GPUs, Proceedings of the IEEE International Symposium on Signal Processing and Information Technology (2011).
11. Liu B., Wang K., Liu X. and Yu W., An efficient SAR processor based on GPU via cuda, 1–5, Oct. 2009.
12. Zhao S. and Wang R., A GPU based range-doppler algorithm for SAR imaging in opencl, 224–227, Oct. 2011.
13. Ning X., Yeh C., Zhou B., Gao W. and Yang J. Multiple-GPU accelerated range-doppler algorithm for synthetic aperture radar imaging, 698–701, May 2011.
14. Vector Signal Image Processing Forum, http://www.vsipl.org.

Author Index

© Springer Science+Business Media Singapore 2017
S.C. Satapathy et al. (eds.), *Proceedings of the First International Conference on Computational Intelligence and Informatics*, Advances in Intelligent Systems and Computing 507, DOI 10.1007/978-981-10-2471-9

Printed in the United States
By Bookmasters